# The 3-D Global Spatial Data Model

## Data Model

### Principles and Applications

### Second Edition

# The 3-D Global Spatial Data Model

## Principles and Applications

### Second Edition

Earl F. Burkholder

**CRC Press**
Taylor & Francis Group
Boca Raton London New York

CRC Press is an imprint of the
Taylor & Francis Group, an **informa** business

CRC Press
Taylor & Francis Group
6000 Broken Sound Parkway NW, Suite 300
Boca Raton, FL 33487-2742

First issued in paperback 2019

ISBN-13: 978-1-4987-2216-2 (hbk)
ISBN-13: 978-0-367-87299-1 (pbk)

### Library of Congress Cataloging-in-Publication Data

Names: Burkholder, Earl F.
Title: The 3-D global spatial data model : principles and applications / Earl
F. Burkholder.
Other titles: Three dimensional global spatial data model
Description: Second edition. | Boca Raton : CRC Press, 2017. | Includes
bibliographical references and index.
Identifiers: LCCN 2016058605 | ISBN 9781498722162 (hardcopy : alk. paper)
Subjects: LCSH: Spatial data infrastructures--Mathematics. | Spatial systems.
| Three-dimensional imaging.
Classification: LCC QA402 .B866 2017 | DDC 910.285--dc23
LC record available at https://lccn.loc.gov/2016058605

**Visit the Taylor & Francis Web site at**
**http://www.taylorandfrancis.com**

**and the CRC Press Web site at**
**http://www.crcpress.com**

*This book is dedicated to Dr. Kurt W. Bauer, Executive Director of the Southeastern Wisconsin Regional Planning Commission (SEWRPC) from 1962 to 1997. I had the opportunity to work for Dr. Bauer on several seminal projects that are referenced in this book. Dr. Bauer provided unequaled insight into the role of spatial data and the surveying profession in the development of civil infrastructure and insisted that our professional duties are always to those we serve. Dr. Bauer became a valued mentor in addition to his being a demanding client.*

# Contents

Preface to the Second Edition .................................................................................. xix
Preface to the First Edition .................................................................................... xxi
Acknowledgments................................................................................................... xxiii
Author ....................................................................................................................... xxv
List of Abbreviations............................................................................................... xxvii
Introduction.............................................................................................................. xxxi

**Chapter 1**  The Global Spatial Data Model (GSDM) Defined............................. 1

     Introduction .................................................................................................. 1
     The GSDM ................................................................................................... 2
        Functional Model Component......................................................... 3
        Computational Designations ........................................................... 5
        Algorithm for Functional Model .................................................... 10
        Stochastic Model Component.......................................................... 14
        The GSDM Covariance Matrices ................................................... 14
        The GSDM 3-D Inverse................................................................... 16
     BURKORD™: Software and Database............................................ 18
     Summary ....................................................................................................... 18
     References .................................................................................................... 19

**Chapter 2**  Featuring the 3-D Global Spatial Data Model................................... 21

     Introduction .................................................................................................. 21
     The GSDM Facilitates Existing Initiatives................................................ 22
        U.S. National Academy of Public Administration Reports .......... 22
        National Oceanic and Atmospheric Administration ...................... 23
        Coalition of Geospatial Organizations ........................................... 24
     Other Applications....................................................................................... 25
        Dynamic Environments .................................................................. 25
        Static Environments........................................................................ 25
     Information Provided by the GSDM............................................................ 26
     Summary ....................................................................................................... 26
     References .................................................................................................... 27

**Chapter 3**  Spatial Data and the Science of Measurement .................................. 29

     Introduction .................................................................................................. 29
     Spatial Data Defined.................................................................................... 29
     Coordinate Systems Give Meaning to Spatial Data .................................. 30
        Spatial Data Types .......................................................................... 32

Spatial Data Visualization Is Well Defined ........................................34
Direct and Indirect Measurements Contain Uncertainty...................34
    Fundamental Physical Constants Are Held Exact .....................34
    Measurements/Observations Contain Errors ............................35
Measurements Used to Create Spatial Data .....................................35
    Taping ......................................................................................35
    Leveling ...................................................................................35
    EDMI.......................................................................................35
    Angles......................................................................................36
    GPS and GNSS........................................................................36
    Remote Sensing .......................................................................38
    Photogrammetry ......................................................................38
    LiDAR ......................................................................................38
    Logistics ..................................................................................38
    Errorless Spatial Data .............................................................39
Sources of Primary Spatial Data .......................................................41
    Observations and Measurements ............................................41
    Errorless Quantities ................................................................42
Derived Spatial Data Are Computed from Primary Spatial Data.......42
Establishing and Preserving the Value of Spatial Data .....................43
Summary .............................................................................................44
References ...........................................................................................44

**Chapter 4**   Summary of Mathematical Concepts ...............................................47

Introduction .......................................................................................47
Conventions ........................................................................................48
    Numbers ..................................................................................48
    Fractions ..................................................................................48
    Decimal....................................................................................48
    Radian......................................................................................49
    Sexagesimal.............................................................................50
    Binary ......................................................................................50
    Unit Conversions ....................................................................51
    Coordinate Systems .................................................................51
    Significant Digits ....................................................................52
        Addition and Subtraction...................................................52
        Multiplication and Division ..............................................53
        Avoid Mistakes by Working with Coordinate Differences .......54
Logic....................................................................................................54
Arithmetic............................................................................................55
Algebra ................................................................................................55
    Axioms of Equality (for Real Numbers $A$, $B$, and $C$)....................56
    Axioms of Addition (for Real Numbers $A$, $B$, and $C$) ...................56
    Axioms of Multiplication (for Real Numbers $A$, $B$, and $C$)...........56
    Boolean Algebra ......................................................................56

Geometry .................................................................................................... 56
    Point ................................................................................................... 57
    Distance ............................................................................................. 57
    Dimension .......................................................................................... 57
    Line .................................................................................................... 57
    Plane .................................................................................................. 57
    Angle .................................................................................................. 58
    Circle ................................................................................................. 58
    Ellipse ................................................................................................ 58
    Triangle .............................................................................................. 58
    Quadrilateral ...................................................................................... 58
    Rectangle ........................................................................................... 59
    Square ................................................................................................ 59
    Trapezoid ........................................................................................... 59
    Parallelogram ..................................................................................... 59
    Polygon .............................................................................................. 59
    Pythagorean Theorem ....................................................................... 59
Solid Geometry ........................................................................................ 60
    Sphere ................................................................................................ 60
    Ellipsoid ............................................................................................. 60
    Cube ................................................................................................... 60
    Polyhedron ......................................................................................... 60
    Tetrahedron ........................................................................................ 60
    Pyramid .............................................................................................. 60
    Equation of a Plane in Space ............................................................ 60
    Equation of a Sphere in Space .......................................................... 61
    Equation of an Ellipsoid Centered on the Origin ........................... 61
    Conic Sections ................................................................................... 61
    Vectors ............................................................................................... 62
Trigonometry ............................................................................................ 62
    Trigonometric Identities ................................................................... 63
    Law of Sines ...................................................................................... 64
    Law of Cosines .................................................................................. 65
Spherical Trigonometry ........................................................................... 65
Calculus .................................................................................................... 68
    Example .............................................................................................. 68
    Differential Calculus Equations ........................................................ 70
    Integral Calculus Equations .............................................................. 70
Probability and Statistics ......................................................................... 71
    Introduction ....................................................................................... 71
    Standard Deviation ............................................................................ 72
    Measurement ..................................................................................... 73
    Errors ................................................................................................. 73
        Blunders ........................................................................................ 74
        Systematic Errors ......................................................................... 74
        Random Errors .............................................................................. 74

Error Sources ............................................................................... 74
  Personal.................................................................................. 74
  Environmental........................................................................ 75
  Instrumental .......................................................................... 75
Accuracy and Precision ............................................................. 75
Computing Standard Deviations.................................................. 76
Standard Deviation of the Mean ................................................. 76
Confidence Intervals ................................................................. 77
Hypothesis Testing ................................................................... 78
Matrix Algebra......................................................................... 79
Models........................................................................................ 80
  Functional .............................................................................. 80
  Stochastic............................................................................... 80
Error Propagation ....................................................................... 81
Error Ellipses.............................................................................. 87
Least Squares.............................................................................. 88
  Linearization .......................................................................... 89
  Procedure for Nonlinear Solution.............................................. 90
Applications to the GSDM ............................................................ 90
References ................................................................................. 91


Chapter 5    Geometrical Models for Spatial Data Computations ........................ 93

Introduction ............................................................................... 93
Conventions............................................................................... 94
Two-Dimensional Cartesian Models .............................................. 97
  Math/Science Reference System ............................................... 98
  Engineering/Surveying Reference System ................................. 98
Coordinate Geometry .................................................................. 99
  Forward.................................................................................. 99
  Inverse................................................................................. 100
  Intersections......................................................................... 100
    Line-Line (One Solution or No Solution If Lines Are Parallel)....102
    Line-Circle (May Have Two Solutions, One Solution, or
      No Solution)..................................................................... 102
    Circle-Circle (May Have Two Solutions, One Solution, or
      No Solution)..................................................................... 103
  Perpendicular Offset .............................................................. 104
  Area by Coordinates .............................................................. 104
Circular Curves......................................................................... 106
  Definitions ........................................................................... 106
  Degree of Curve.................................................................... 106
  Elements and Equations ......................................................... 107
  Stationing............................................................................. 109
  Metric Considerations ........................................................... 110

Area Formed by Curves ..................................................................... 111
Area of Unit Circle .......................................................................... 112
Spiral Curves .......................................................................................... 113
Spiral Geometry .............................................................................. 113
Intersecting a Line with a Spiral ................................................... 116
Computing Area Adjacent to a Spiral .................................................... 117
Radial Surveying ................................................................................... 118
Vertical Curves ...................................................................................... 121
Three-Dimensional Models for Spatial Data ......................................... 124
Volume of a Rectangular Solid ....................................................... 124
Volume of a Sphere ......................................................................... 124
Volume of Cone ............................................................................... 125
Prismoidal Formula ......................................................................... 126
Traditional 3-D Spatial Data Models .............................................. 128
The 3-D GSDM ...................................................................................... 128
References .............................................................................................. 129

**Chapter 6**  Overview of Geodesy ........................................................................ 131

Introduction: Science and Art ............................................................... 131
Fields of Geodesy .................................................................................. 131
Goals of Geodesy .................................................................................. 132
Historical Perspective ............................................................................ 137
Religion, Science, and Geodesy ..................................................... 138
Degree Measurement ....................................................................... 139
Eratosthenes .................................................................................... 139
Poseidonius ...................................................................................... 140
Caliph Abdullah al Mamun ............................................................ 140
Gerardus Mercator ........................................................................... 140
Willebrord Snellius .......................................................................... 141
Jean Picard ....................................................................................... 141
Isaac Newton ................................................................................... 141
Jean-Dominique and Jacques Cassini ............................................. 142
French Academy of Science ............................................................ 142
Meter ................................................................................................ 143
Developments during the Nineteenth and Twentieth Centuries ....... 143
Forecast for the Twenty-First Century .................................................. 145
References .............................................................................................. 146

**Chapter 7**  Geometrical Geodesy ........................................................................ 147

Introduction ........................................................................................... 147
The Two-Dimensional Ellipse ............................................................... 149
The Three-Dimensional Ellipsoid ......................................................... 154
Ellipsoid Radii of Curvature ........................................................... 154
Normal Section Radius of Curvature ............................................... 155
Geometrical Mean Radius ............................................................... 155

Rotational Ellipsoid .......................................................................... 155
  Equation of Ellipsoid ................................................................... 155
  Geocentric and Geodetic Coordinates ....................................... 156
BK1 Transformation ........................................................................ 157
BK2 Transformation ........................................................................ 158
  Iteration ......................................................................................... 158
  Noniterative (Vincenty) Method ................................................. 159
  Example of BK1 Transformation ................................................ 160
  Example of BK2 Transformation—Iteration ............................. 161
  Example of BK2 Transformation—Vincenty's Method
  (Same Point) ................................................................................. 162
  Meridian Arc Length .................................................................... 163
  Length of a Parallel ...................................................................... 166
  Surface Area of Sphere ................................................................ 166
  Ellipsoid Surface Area ................................................................. 167
The Geodetic Line ........................................................................... 169
  Description ..................................................................................... 169
  Clairaut's Constant ....................................................................... 170
  Geodetic Azimuths ....................................................................... 172
  Target Height Correction ............................................................. 174
  Geodesic Correction ..................................................................... 175
Geodetic Position Computation—Forward and Inverse ................ 175
  Puissant Forward (BK18) ............................................................. 176
  Puissant Inverse (BK19) ............................................................... 177
  Numerical Integration ................................................................... 178
  BK18: Forward .............................................................................. 178
  BK19: Inverse ............................................................................... 181
  Geodetic Position Computations Using State Plane Coordinates .... 185
  GSDM 3-D Geodetic Position Computations ............................ 186
  Forward—BK3 ............................................................................. 186
  Inverse—BK4 ............................................................................... 187
  GSDM Inverse Example: New Orleans to Chicago ................... 188
References ........................................................................................ 193


Chapter 8    Geodetic Datums .......................................................................... 195

  Introduction .................................................................................... 195
  Horizontal Datums ........................................................................ 196
    Brief History ............................................................................... 196
    North American Datum of 1927 (NAD 27) ............................... 198
    North American Datum of 1983 (NAD 83) ............................... 198
    World Geodetic System 1984 ..................................................... 199
    International Terrestrial Reference Frame ................................. 200
    High Accuracy Reference Network—HARN ............................ 202

Continuously Operating Reference Station—CORS..................204
NA2011..........................................................................................205
Vertical Datums .................................................................................205
Sea Level Datum of 1929 (now NGVD 29) ...............................205
International Great Lakes Datum ...............................................206
North American Vertical Datum of 1988—NAVD 88 .................206
Datum Transformations......................................................................207
NAD 27 to NAD 83 (1986) .......................................................208
NAD 83 (1986) to HPGN ..........................................................208
NAD 83 (xxxx) to NAD 83 (yyyy)..............................................208
NGVD 29 to NAVD 88...............................................................208
HTDP..........................................................................................208
Software Sources ......................................................................208
7-(14-) Parameter Transformation .............................................209
3-D Datums .......................................................................................209
References .........................................................................................210

**Chapter 9**    Physical Geodesy .............................................................................213

Introduction ......................................................................................213
Gravity...............................................................................................214
Definitions .........................................................................................215
Elevation (Generic)....................................................................216
Equipotential Surface ................................................................216
Level Surface .............................................................................216
Geoid ..........................................................................................216
Geopotential Number .................................................................217
Dynamic Height..........................................................................217
Orthometric Height.....................................................................217
Ellipsoid Height..........................................................................217
Geoid Height ..............................................................................217
Gravity and the Shape of the Geoid .................................................218
Laplace Correction ............................................................................219
Measurements and Computations .....................................................221
Interpolation and Extrapolation.................................................221
Gravity .......................................................................................222
Tide Readings.............................................................................223
Differential Levels .....................................................................223
Ellipsoid Heights .......................................................................224
Time............................................................................................225
Use of Ellipsoid Heights in Place of Orthometric Heights .............225
The Need for Geoid Modeling ...........................................................227
Geoid Modeling and the GSDM .......................................................231
Using a Geoid Model ........................................................................232
References .........................................................................................234

**Chapter 10**  Satellite Geodesy and Global Navigation Satellite Systems ............237

Introduction .................................................................................237
Brief History of Satellite Positioning .............................................240
Modes of Positioning ....................................................................244
    Elapsed Time ...........................................................................244
    Doppler Shift ...........................................................................244
    Interferometry..........................................................................246
Satellite Signals ...........................................................................247
    C/A Code ................................................................................249
    Carrier Phase ..........................................................................250
Differencing.................................................................................251
    Single Differencing..................................................................252
    Double Differencing ................................................................252
    Triple Differencing ..................................................................252
RINEX.........................................................................................252
Processing GNSS Data..................................................................253
    Spatial Data Types ..................................................................254
    Autonomous Processing ...........................................................255
    Vector Processing ...................................................................256
    Multiple Vectors......................................................................257
    Traditional Networks ...............................................................259
    Advanced Processing................................................................259
The Future of Survey Control Networks—Has It Arrived? .............262
References ....................................................................................265


**Chapter 11**  Map Projections and State Plane Coordinates................................267

Introduction: Round Earth—Flat Map ...........................................267
Projection Criteria ........................................................................268
Projection Figures ........................................................................270
Permissible Distortion and Area Covered .......................................273
U.S. State Plane Coordinate System (SPCS) ..................................274
    History ....................................................................................275
    Features...................................................................................275
    NAD 27 and NAD 83 ..............................................................277
    Current Status—NAD 83 SPCS ...............................................279
        Advantages.........................................................................280
        Disadvantages ....................................................................280
    Procedures ..............................................................................281
    Grid Azimuth ..........................................................................281
    Grid Distance ..........................................................................281
    Traverses.................................................................................284
        Loop Traverse ....................................................................285
        Point-to-Point Traverse .......................................................285

Algorithms for Traditional Map Projections ...................................285
  Lambert Conformal Conic Projection .........................................286
    BK10 Transformation for Lambert Conformal Conic
    Projection ..............................................................288
    BK11 Transformation for Lambert Conformal Conic
    Projection ..............................................................288
  Transverse Mercator Projection...............................................289
    BK10 Transformation for Transverse Mercator Projection ....292
    BK11 Transformation for Transverse Mercator Projection ....294
  Oblique Mercator Projection .................................................297
    BK10 Transformation for Oblique Mercator Projection ........299
    BK11 Transformation for Oblique Mercator Projection ........300
Low-Distortion Projection.......................................................302
References ......................................................................302

**Chapter 12** Spatial Data Accuracy ....................................................305

Introduction ...................................................................305
Forces Driving Change ...........................................................305
Transition.......................................................................306
Consequences ....................................................................308
Accuracy.........................................................................309
  Introduction ................................................................309
  Definitions .................................................................311
  Absolute and Relative Quantities ...........................................311
  Spatial Data Types and Their Accuracy.......................................313
    Accuracy Statements.......................................................313
    But Everything Moves .....................................................313
Observations, Measurements, and Error Propagation .....................315
  Finding the Uncertainty of Spatial Data Elements .....................315
  Using Points Stored in a *X/Y/Z* Database .................................317
Example..........................................................................319
  Control Values and Observed Vectors...........................................320
  Blunder Checks ..............................................................321
  Least Squares Solution ......................................................322
  Results .....................................................................323
  Network Accuracy and Local Accuracy ......................................323
References ......................................................................328

**Chapter 13** Using the GSDM to Compute a Linear Least Squares
                GNSS Network..........................................................329

Introduction ...................................................................329
Parameters and Linearization ....................................................329
Baselines and Vectors ...........................................................330

Observations and Measurements ......................................................330
Covariance Matrices and Weight Matrices ....................................331
Two Equivalent Adjustment Methods .............................................332
Formulations of Matrices—Indirect Observations ......................333
Example GNSS Network Project in Wisconsin ............................336
RINEX Data Used to Build the Wisconsin Network ....................338
Blunder Checks ................................................................................338
Building Matrices for a Linear Least Squares Solution ................341
    f Vector—n, 1 ............................................................................341
    B Matrix—n, u ...........................................................................343
    Q Matrix—n, n ..........................................................................343
Computer Printouts .........................................................................345
Notes Pertaining to Adjustment ......................................................364
References ........................................................................................364


**Chapter 14** Computing Network Accuracy and Local Accuracy
            Using the Global Spatial Data Model ....................................365

Introduction ....................................................................................365
Background ......................................................................................366
Summary of Pertinent Concepts .....................................................366
Detailed Example Based on Wisconsin Network ..........................368
Conclusion .......................................................................................376
References ........................................................................................376


**Chapter 15** Using the GSDM—Projects and Applications ....................379

Introduction ....................................................................................379
Features ...........................................................................................381
    The Functional Model ...............................................................381
    The Stochastic Model ...............................................................381
Database Issues ...............................................................................384
Implementation Issues ....................................................................385
Examples and Applications ............................................................387
    Example 1—Supplemental NMSU Campus Control Network ...387
    Example 2—Hypothesis Testing ..............................................400
    Example 3—Using Terrestrial Observations in the GSDM ........401
    Example 4—Using the GSDM to Develop a 2-D Survey Plat ....407
    Example 5—New Mexico Initial Point and Principal Meridian .....410
    Example 6—State Boundary between Texas and New
    Mexico along the Rio Grande River .........................................417
    Example 7 in Wisconsin—Leveling in the Context of the
    GSDM (Example in Wisconsin) ...............................................427
    Example 8—Determining the NAVD 88 Elevation of
    HARN Station REILLY .............................................................427

Example 9—Determining the Shadow Height at a Proposed
NEXRAD Installation ............................................................... 432
Example 10—Comparison of 3-D Computational Models ......... 434
Example 11—Underground Mapping ......................................... 438
Example 12—Laying Out a Parallel of Latitude Using the
GSDM ....................................................................................... 440
    Analogous to Solar Method ................................................... 441
    Analogous to Tangent-Offset Method .................................. 442
The Future Will Be What We Make It ............................................ 443
References ..................................................................................... 445

**Appendix A: Rotation Matrix Derivation** .......................................... 447

**Appendix B: 1983 State Plane Coordinate Zone Constants** ........................... 451

**Appendix C: 3-D Inverse with Statistics** ............................................. 459

**Appendix D: Development of the Global Spatial Data Model (GSDM)** ......... 461

**Appendix E: Evolution of Meaning for Terms: Network Accuracy
and Local Accuracy** ........................................................ 465

**Index** ................................................................................................ 473

# Preface to the Second Edition

This Preface describes what material is added and why a second edition is justi-fied. The big reason is *applications*. The second edition adds information about least squares adjustment and provides examples that highlight benefits associated with using a comprehensive model for three-dimensional (3-D) digital spatial data. A bonus take-away is demystifying the concepts of spatial data accuracy and pro-viding mathematical clarity to issues of network accuracy and local accuracy. As a result, critical decisions that rely on the quality of spatial data can be made with more confidence knowing "accuracy with respect to what."

Many disciplines have participated in the digital revolution and new spatial data industries have sprung up in the recent past. Productivity has been enhanced and efficiencies have been realized in many aspects of the spatial data user community as existing processes were computerized and automated. But, too often, policies and practices were carried forward without reexamining underlying assumptions and models. Although those efforts have been largely successful, there is a better way to exploit the characteristics of 3-D digital spatial data. Fundamental concepts of solid geometry and error propagation are added to the assumption of a single origin for 3-D geospatial data to form the foundation for modern practice. That is the context in which the characteristics of 3-D digital spatial data are examined with the global spatial data model (GSDM) emerging as the result. The GSDM is not unique to any one discipline but can be shared by all who work with digital spatial data. Benefits in addition to those described in this book will be realized as users worldwide come to rely on common definitions for 3-D spatial data and as common accuracy standards are developed and implemented.

Specific differences and additions in the second edition include the following:

- A new Foreword focused on policy considerations for implementation.
- Moving material from the previous Foreword into a new Chapter 2.
- Adding material related to BIG DATA and a grade report on the spatial data infrastructure to the same Chapter 2.
- Anticipating the new horizontal and vertical datum definitions to be intro-duced in 2022 by the National Geodetic Survey (NGS).
- Incorporating detailed examples of linear least squares adjustments.
- A new Chapter 14 that contains a comprehensive discussion of network accuracy and local accuracy and a comprehensive example showing how they are computed.
- A new Appendix E that describes chronological development of network accuracy and local accuracy concepts.
- The history of development of the GSDM. Although not "mission critical," details of that history dovetail with other aspects of the digital revolution, and omitting that material would leave an awkward gap in the story.

Another way of looking at the difference from the first edition to the second is to say that the first edition looks at the "how" while the second edition places more emphasis on the "why." Ultimately, it will be the reader who decides which view is more appropriate.

The Preface to the first edition is reprinted because those points remain valid.

*Note*: All web URLs were accessed and validated during the copyediting process (May 2017). A topical web engine search is recommended in the event a link is found to be defective.

# Preface to the First Edition

"The right tool for the job" is a simple phrase with profound implications. The proliferation of tools for handling spatial data is somewhat daunting as benefits associated with their use spawns development of even better tools. Although we are where we are because of where we came from, the path to the future should be viewed in terms of the analog-to-digital revolution. Given the many specializations associated with developing technology, it is difficult to write a comprehensive book about an umbrella topic like spatial data. Therefore, acknowledging that others will add details to illuminate the path ahead even better, this book is written to define and describe a global spatial data model (GSDM) that

- Is easy to use because it is based upon rules of solid geometry
- Is standard between disciplines and can be used all over the world
- Accommodates modern measurement and digital data storage technologies
- Supports both analog map plots and computer visualization of digital data
- Preserves geometrical integrity and does not distort physical measurements
- Combines horizontal and vertical data into a single three-dimensional database
- Facilitates rigorous error propagation and standard deviation computations
- Provides (and defines assumptions associated with) various choices with respect to spatial data accuracy

In a way, this book is organized backward. Chapter 1 contains the results and Chapter 2 justifies Chapter 1. Fundamental geometrical concepts are developed in terms of more traditional material in subsequent chapters. That is done to accommodate readers with various backgrounds. Managers and those with a strong technical background might concentrate only on the beginning chapters. Spatial data professionals at various levels who wish to gain a better understanding of geometrical relationships should start with the beginning chapters so they know where the rest of this book is going. Given that Chapters 1 and 2 are not easy reading, they should be read first as an overview. It is expected then, as the reader progresses through subsequent chapters, that Chapters 1 and 2 will be revisited as required to help refresh the focus on the overall objective of defining an appropriate spatial data model. For those just beginning to work with spatial data, serious reading and study should begin in Chapter 3. With that said, the plan for building a comprehensive spatial data model is to present fundamental mathematical concepts in Chapter 3 and to add concepts from surveying, geodesy, and cartography in subsequent chapters.

The material is presented as simply as possible without compromising technical rigor. Some readers will find the review of mathematical concepts redundant and some readers may never have occasion to use linear algebra, matrix manipulation, or error propagation. Acknowledging the certain diversity of readers, the goal is to provide a logical development of concepts for those who wish to follow the theory and to provide all readers a collection of tools that can be used to handle spatial data more efficiently.

Whether the reader is involved in technical applications, is making managerial and administrative decisions with regard to spatial data, or is a programmer writing software for handling spatial data, all should agree that the most appropriate tools for handling spatial data are those which are, at the same time, both simple and appropriate. The GSDM is simple because it uses existing practice and rules of solid geometry for manipulating spatial data. And, the GSDM is appropriate because it is built on local coordinate differences, preserves true three-dimensional geometrical integrity on a global scale, accommodates modern digital technology, handles error propagation with aplomb, and supports subsequent computation of complex geometrical relationships in geodesy, cartography and other sciences. In the past, spatial data models were selected by default as people (rightfully) focused on impressive gains in utility and productivity made possible by automating existing processes. But the GSDM is a result of examining those processes in terms of digital technology and fundamental geometrical concepts. With features of the various models described and compared, it is anticipated that spatial data analysts in various fields will, as a matter of conscious choice, begin using the GSDM because it establishes a common geometrical link between spatial data sets, applications, and disciplines and because it provides an efficient method of defining, tracking, and evaluating the accuracy of spatial data.

What does it mean to "think outside the box?" Is thinking outside the box something beneficial and desirable? Or is thinking outside the box to be avoided? What do elephant jokes have to do with boxes? Without answering those questions, consider the following: (To whom does one credit elephant jokes?)

1. How does one determine the number of elephants in the refrigerator? Answer: Count their tracks in the butter.
2. How does one kill a blue elephant? Answer: Shoot it with a blue elephant gun. OK, now the pattern is established and the reader is ready for whatever else comes along.
3. How does one kill a red elephant? No, you don't shoot it with a red elephant gun because you don't have one. The correct answer is, "Choke it until it turns blue, then shoot it with your blue elephant gun."
4. Elephant jokes may have no place in a rigorous technical book (except maybe in the preface), but these illustrate a very important point. Humans are very good at using whatever tools are available to do what needs to be done. Without being critical, many wonderful accomplishments have involved (figuratively) choking the elephants. But, everyone should be aware that sometimes it is better, easier, and more appropriate to look for a red elephant gun than it is to keep choking those red elephants. Most red elephant guns are found outside the box.

The GSDM is viewed as a red elephant gun for handling 3-D digital geospatial data.

# Acknowledgments

As with the first edition, I am indebted to many persons for assistance in writing this second edition. I retired from classroom teaching in 2010 and I miss the invigorating interchange of ideas with students. But interacting with former students (and others) in the professional arena continues to provide motivation and inspiration. I also acknowledge the importance of my family who graciously understood the time I spent on "the book." To them I owe a huge debt of gratitude—thank you!

The astute reader may notice an improvement in organization, style, and clarity of this edition over the first edition. For that I am indebted to Mr. Glen Schaefer, retired Geodetic Engineer from the Wisconsin Department of Transportation. He read each chapter for grammar, style, and technical content. While I did not incorporate every suggestion he made, our back-and-forth discussions are reflected in this "improved" second edition. However, I alone am responsible for any errors and all defects. I also need to acknowledge the contribution that Dr. Charles Ghilani made with respect to Hypothesis Testing, a very important tool for the spatial data analyst. Yes, an actual example is included, but conclusions are left to the reader.

Over the years, I have paid close attention to many conversations and correspondence with many professional colleagues who have offered support for the work I have done and those who have challenged various ideas. While I remain convinced that starting with the assumption of a single origin for 3-D geospatial data is legitimate, I have learned much from those who insist that my view of the world is only one of many. A partial list of revered colleagues includes, but is not limited to, Steve Frank, Kurt Wurm, Ahmed F. Elaksher, James P. Reilly, Alfred Leick, David Garber, Ronnie Taylor, Thomas Meyer, Tomás Soler, Dru Smith, Michael Dennis, Jan van Sickle, Bill Stone, Larry Hothem, Robert Merry, Conrad Keyes, Jr., Bill Hazelton, Tim Pitts, Scott Farnham, Doug Copeland, Ty Trammell, David Cooper, Glen Thurow, and Bob Green. I am grateful for the example and inspiration that each has provided. And, lastly, I must mention someone I've never met but whose work I have learned to appreciate—that is Owen Gingerich, author of *The Book Nobody Read: Chasing the Revolutions of Nicolaus Copernicus*. I suspect many readers may understand after reading (or browsing) that book.

# Author

A native Virginian, **Earl F. Burkholder** grew up in Virginia's Shenandoah Valley and graduated from Eastern Mennonite High School in 1964. He earned a BS in civil engineering from the University of Michigan, Ann Arbor, Michigan, in 1973 and an MS in civil engineering (Geodesy) in 1980 from Purdue University. From 1980 to 1993, he taught upper division surveying classes at the Oregon Institute of Technology, Klamath Falls, Oregon. After five years of being self-employed, he taught in the Surveying Engineering program at New Mexico State University from 1998 to his retirement in 2010.

His professional career began as a draftsman with Gould Engineering, Inc. of Flint, Michigan in 1968. Following graduating from University of Michigan, Ann Arbor, Michigan, he worked five years for Commonwealth Associates, Inc. of Jackson, Michigan, an international consulting firm for the utility industry. Assigned to the Transmission Line Engineering Division at Commonwealth, he was responsible for surveying related computations on projects in numerous states and was promoted to Survey Project Manager prior to leaving in 1978 to attend Purdue University.

While teaching at Oregon Tech, he became editor of the *ASCE Journal of Surveying Engineering* and served two separate four-year terms; 1985–1989 and 1993–1998. He also became involved in ABET accreditation activities while at Oregon Tech and went on to serve on the Engineering Related Accreditation Commission (now known as the Applied Science Accreditation Commission) culminating as chair of the RAC in 2000/2001.

While self-employed, he completed three major projects for the Southeastern Wisconsin Regional Planning Commission (SEWRPC), Waukesha, Wisconsin. The first project was to develop a reliable bidirectional algorithm for transforming data between the NAD 27 datum being used by SEWRPC and the new NAD 83 datum published by the National Geodetic Survey (NGS). Upon successful completion of the horizontal transformation project, the next project was a similar assignment for the bidirectional transformation of data between the NGVD 29 datum (again being used by SEWRPC) and the new NAVD 88 datum published by NGS.

Prior to beginning the horizontal transformation project, he suggested to Dr. Bauer, SEWRPC Executive Director, that the horizontal and vertical datum transformation challenges would be an excellent opportunity to combine the two databases into a single 3-D database. After a rather short deliberation, Dr. Bauer indicated that the 3-D proposal was untested, too radical, and not proven practical. However, upon completion of the first two projects, Dr. Bauer commissioned the preparation of a report outlining and defining how such an integrated model could be implemented. That report (link below) became the basis for the first edition of *The 3-D Global Spatial Data Model: Foundation of the Spatial Data Infrastructure* published by CRC Press in 2008. http://www.sewrpc.org/SEWRPCFiles/Publications/ppr/definition_three-dimensional_spatial_data_model_for_wi.pdf, accessed May 4, 2017.

# List of Abbreviations

| | |
|---|---|
| 1-D | One-dimensional |
| 2-D | Two-dimensional |
| 3-D | Three-dimensional |
| 4-D | Four-dimensional |
| AASHTO | American Association of State Highway and Transportation Officials |
| ACSM | American Congress on Surveying and Mapping |
| ALTA | American Land Title Association |
| ARP | Antenna Reference Point |
| A-S | Anti-Spoofing |
| ASCE | American Society of Civil Engineers |
| ASCII | American Standard Code for Information Interchange |
| ASME | American Society of Mechanical Engineers |
| ASPRS | American Society of Photogrammetry and Remote Sensing |
| BIPM | Bureau International des Poids et Mesures |
| BK1 thru BK22 | Labels assigned to routine procedures listed in Table 1.1 and identified in Figure 1.4 |
| BLM | Bureau of Land Management |
| BURKORD™ | Trademark for 3-D software and database |
| C/A code | Coarse Acquisition Code |
| COGO | Coalition of Geospatial Organizations |
| $c + r$ | Correction for curvature and refraction |
| CORPSCON | Coordinate conversion software by the U.S. Army Corps of Engineers |
| CORS | Continuously Operating Reference Station |
| CTP | Conventional Terrestrial Pole |
| DGNSS | Differential Global Network Satellite System |
| DGPS | Differential GPS |
| DMA | Defense Mapping Agency (no longer used—see NIMA) |
| DMD | Double-Meridian-Distance |
| $D°$ | Degree of curve |
| DOD | Department of Defense |
| DORIS | Doppler Orbitography and Radio Positioning Integrated by Satellite |
| DOT | Department of Transportation |
| $e/n/u$ | Local perspective right-handed rectangular coordinates |
| ECEF | Earth-centered Earth-fixed |
| EDM | Electronic Distance Measurement |
| EDMI | Electronic Distance Measuring Instrument |
| EVC | End of Vertical Curve |
| FAA | Federal Aviation Administration |
| FGCC | Federal Geodetic Control Committee |

| | |
|---|---|
| FGDC | Federal Geographic Data Committee |
| FOC | Full Operational Capability |
| FRN | Federal Register Notice |
| GALILEO | European satellite positioning system (similar to GPS owned by the United States) |
| GD&T | Geometric Dimensioning and Tolerancing |
| GEOID12B | Geoid model that supercedes previous geoid models |
| GIS | Geographic Information System |
| GLONASS | Globalnaya Navigazionnaya Sputnikovaya Sistema (similar to GPS owned by the United States) |
| GM | Geocentric Gravitational Constant |
| GMT | Greenwich Mean Time |
| GN | Geodetic Networks |
| GNSS | Global Navigation Satellite System |
| GPS | Global Positioning System |
| GRS 80 | Geodetic Reference System 1980 |
| GS | Geodetic Survey |
| GSDI | Global Spatial Data Infrastructure |
| GSDM | Global Spatial Data Model |
| GTS | Great Trigonometric Survey |
| HARN | High Accuracy Reference Network |
| HD | Horizontal Distance |
| HD(1) | Horizontal distance used in plane surveying |
| HI | Height of Instrument |
| HPGN | High Precision Geodetic Network |
| HTDP | Horizontal Time Dependent Positioning |
| IAG | International Association of Geodesy |
| IERS | International Earth Rotation Service |
| iFT | International Foot equal to 0.3048 meter exactly |
| IGLD | International Great Lakes Datum |
| IGLD 55 | International Great Lakes Datum of 1955 |
| IGLD 85 | International Great Lakes Datum of 1985 |
| IOC | Initial Operational Capability |
| IT | Information Technology |
| ITRF | International Terrestrial Reference Frame |
| ITRS | International Terrestrial Reference Service |
| LDP | Low-distortion projection |
| LiDAR | Light Detection and Ranging |
| LLR | Lunar Laser Ranging |
| MSL | Mean Sea Level |
| NA2011 | North American Datum of 1983 based upon general adjustment of 2011 |
| NAD | North American Datum |
| NAD 27 | North American Datum of 1927 |
| NAD 83 (xx) | North American Datum of 1983 (realized in 19xx) |
| NAD 83 (yyyy) | North American Datum of 1983 (realized in yyyy) |

| | |
|---|---|
| NAD 83 (CORS) | North American Datum of 1983 based upon NGS CORS |
| NAD 83 (NSRS2007) | North American Datum of 1983 based upon 2007 adjustment of the NSRS |
| NADCON | North American Datum Conversion software |
| NAPA | National Academic of Public Administration |
| NAVD 88 | North American Vertical Datum of 1988 |
| NAVSAT | Transit satellite system |
| NAVSTAR | *NAV*igation *S*atellite *T*iming *A*nd *R*anging |
| NGA | National Geospatial-Intelligence Agency (formerly NIMA) |
| NGRS | National Geodetic Reference System |
| NGS | National Geodetic Survey |
| NGVD 29 | National Geodetic Vertical Datum of 1929 |
| NILS | National Integrated Land System |
| NIMA | National Imagery and Mapping Agency (formerly DMA, now NGA) |
| NMAS | National Map Accuracy Standards |
| NMPM | New Mexico Principal Meridian |
| NMSU | New Mexico State University |
| NOAA | National Oceanic and Atmospheric Administration |
| NRC | National Research Council |
| NSDI | National Spatial Data Infrastructure |
| NSPS | National Society Professional Surveyors |
| NSRS | National Spatial Reference System |
| OGC | Open Geospatial Consortium |
| OPUS | Online Positioning User Service |
| OPUS-RS | OPUS–Rapid Static |
| PAGEOS | *PA*ssive *GEO*detic *S*atellite |
| PI | Point of Intersection |
| POB | Generic Point of Beginning—used in surveying |
| P.O.B. | Point of Beginning as used specifically in the GSDM |
| PC | Point of Curvature |
| P-code | Precision code |
| PPP | Precise Point Positioning |
| PPS | Precise Position Service |
| PT | Point Number for Survey Station |
| PZS | Pole-Zenith-Star |
| RFI | Request for Information |
| RINEX | Receiver Independent Exchange format |
| RSME | Root-Mean-Square-Error |
| RTK | Real-Time Kinematic |
| RTN | Real-Time Network |
| SA | Selective Availability (discontinued on May 1, 2000) |
| SEWRPC | Southeastern Wisconsin Regional Planning Commission |
| sFT | U.S. Survey Foot equals 12/39.37 meter exactly |
| SI | International System of Units |
| SLR | Satellite Laser Ranging |

| | |
|---|---|
| SPC | State Plane Coordinate |
| SPCS | State Plane Coordinate System |
| SPR | Seconds per Radian |
| SPS | Standard Positioning Service |
| TAI | International Atomic Time |
| TCT | Transcontinental Traverse |
| TS | Station at Transition |
| UAV | Unmanned Aerial Vehicle |
| USC&GS | U.S. Coast & Geodetic Survey (now NGS) |
| USFS | U.S. Forest Service |
| USPLSS | U.S. Public Land Survey System |
| UTC | Coordinated Universal Time |
| UTM | Universal Transverse Mercator |
| VERTCON | Vertical Datum Conversion software |
| VLBI | Very Long Baseline Interferometry |
| VRS | Virtual Reference System |
| WAAS | Wide Area Augmentation System |
| WADGPS | Wide Area Differential GPS |
| WISCORS | Statewide GNSS CORS Network operated by WisDOT |
| WisDOT | Wisconsin Department of Transportation |
| WGS 60 | World Geodetic System 1960 |
| WGS 84 | World Geodetic System 1984 |
| WGS 84 (Gxxxx) | World Geodetic System 1984 (specifies epoch of GPS operations) |
| WVD yyyy | World Vertical Datum yyyy |
| $X/Y/Z$ | Geocentric rectangular coordinates |

# Introduction

The purpose of this Foreword is to give the reader some insight as to the philosophy and inspirations guiding the author. The primary objective in writing the first edition was to serve the end user by identifying fundamental concepts underlying the use of spatial data—especially as related to the spatial data infrastructure. Given the success of the first edition, the concepts are extended in the second edition by applying well-known principles of solid geometry and information management to real-world applications. This second edition is offered with the hope that it will be useful to many users across the spatial data spectrum—from high-level scientific applications to the endless list of flat-Earth applications. The global spatial data model (GSDM) serves spatial data users all over the world and within the confines of near-space.

As described in the first edition, the GSDM includes functional model equations of solid geometry and stochastic model equations for error propagation. Building on those, this second edition includes additional information on least squares adjustment and applications of the GSDM. The focus of the least squares treatment is twofold—highlighting the use of linear least squares adjustments and being very specific on tools for handling both network accuracy and local accuracy. The impact of those concepts could be very far-reaching in the development of standards for monitoring the location of autonomous robots—be it driverless vehicles (land based) or drones (airborne). Knowledge of spatial data accuracy "with respect to what" is becoming absolutely essential.

Challenges to adopting and using the GSDM have been identified based upon feedback from first edition readers. It is not possible to address all challenges, but second edition readers can enjoy a "heads up" by familiarizing themselves with the following:

1. Geocentric $X/Y/Z$ coordinates are the primary definition for the location of any point worldwide. This includes all points in a point cloud.
2. Coordinate differences—$\Delta X/\Delta Y/\Delta Z$—define the 3-D vector between two points.
3. Using a rotation matrix, the same vector is viewed from the local perspective.
4. The origin of a local coordinate system is coincident with the tail of the vector.
5. A local tangent plane is perpendicular to the ellipsoid normal through the origin.
6. Except for orientation, the mathematical characteristics of a rotated vector are unchanged.
7. Resulting $\Delta e/\Delta n$ components are used for "flat-Earth" surveying computations.
8. Horizontal distance and azimuth "here" to "there" are computed from $\Delta e/\Delta n$ components.
9. The $\Delta u$ component is the perpendicular distance from the tangent plane to the forepoint.

10. Computations are performed in 3-D space when using the GSDM. There is no need to reduce measured distances to the ellipsoid or to a mapping surface. The GSDM does not distort distances or angles. Geometrical integrity is insured by basing derived quantities on $X/Y/Z$ values that are not changed by the model.

But, because humans stand erect and walk on a "flat" Earth, a bigger challenge arises from a mind-set that separates horizontal and vertical. A common reaction to the GSDM is "we don't do it that way." That challenge reflects pedagogical issues that also need to be considered. René Descartes formalized concepts of solid geometry in 1637 and those tools (especially plane geometry) have been used extensively ever since. But, with the advent of the digital revolution, Descartes' fundamental principles also need to be implemented in the 3-D environment.

Embracing competence and more efficient ways of handling 3-D digital spatial data will also require consideration of those pedagogical issues. Traditional discussions about the difference between training and education remain relevant and should be actively pursued. But "learning how to learn" is more fundamental and will pay huge dividends to those having the foresight to accept that challenge. Equipment manufacturers, vendors, and numerous consultants are bringing "traditional solutions" to the marketplace and spatial data users are faced with an ever-expanding list of options regarding purchase of equipment, software, data, and professional services. Marvelous accomplishments are possible and many good things are being achieved but, increasingly, a "black box" is replacing the critical analysis processes of the licensed professional. That is a two-edged sword. Many data collection businesses are both legitimate and profitable. Society benefits from increased productivity but, on the other hand, the role of too many licensed professionals is in danger of being relegated to that of a button-pushing technician.

The author's aspiration is that more efficient methods of handling 3-D spatial data will be embraced in the educational and learning processes for various spatial data professionals. The GSDM is viewed as a unifying concept for spatial data disciplines worldwide and should be included as a standard part of both undergraduate and graduate spatial data curricula.

**Earl F. Burkholder, PS, PE, F.ASCE**
*Emeritus Faculty New Mexico State University*
*President, Global COGO, Inc.*

# 1 The Global Spatial Data Model (GSDM) Defined

## INTRODUCTION

Geospatial data representing real-world locations are three dimensional (3-D), and modern measurement systems collect data in a physical 3-D environment. Time as the fourth dimension is acknowledged, but this book focuses on 3-D data. This chapter defines and describes the Global Spatial Data Model (GSDM) as a collection of mathematical concepts and procedures that can be used to collect, organize, store, process, manipulate, evaluate, and use 3-D spatial data. Although the terms "spatial data" and "geospatial data" are often used interchangeably, spatial data are those generic data that describe the size and shape of an object while geospatial data are referenced to planet Earth. It can be argued that geospatial data are a special category of spatial data. It can also be argued that spatial data are a subcategory of geospatial data. While the default context of the GSDM is geospatial data, the geometrical and mathematical properties of spatial data are essentially the same as for geospatial data.

Measurements of quantities such as angles, length, time, current, mass, and temperature are used with known physical and geometrical relationships to compute spatial data components that are stored for subsequent use and reuse. In the past, records of such measurements were written in field books, logs, or journals, and the spatial information was compiled into an analog map that typically served two purposes. The map was simultaneously the primary storage medium for the spatial information and the end product of the data collection process. Spatial data are now collected, stored, and manipulated digitally in an electronic environment, and the primary storage medium is rarely the end product. Instead, the same digital data file can be duplicated repeatedly and used to generate and/or support many different spatial data products. In either case, whether developing an analog or digital spatial data product, algorithms are the mathematical rules used to manipulate measurements and spatial data to obtain meaningful spatial information. In addition, the quality of spatial information is dependent upon the quality of the original measurement, completeness of the required information, and appropriateness of the algorithms used to manipulate the data.

The GSDM includes both the algorithms for processing spatial data and procedures that can be used to provide a defensible statistical description of spatial data quality. This means measurement professionals can focus on building and/or using systems that generate reliable spatial data components, and spatial data users in various disciplines can devote attention to using and interpreting the data with the assurance that all parties generating and/or using the data are "on the same page," that is, using a common spatial data model.

This chapter is a summary of the defining document for the GSDM (Burkholder 1997b). The intent is to cite primary works, because other people developed most of the concepts described herein. For example, Appendix C in Bomford (1971) is titled "Cartesian Coordinates in Three Dimensions." Leick (2004) defines the 3-D Geodetic Model of which the GSDM is a part, Mikhail (1976) provides a comprehensive discussion of functional and stochastic models, and, when discussing models, Moritz (1978) comments on the simplicity of using the basic global rectangular $X/Y/Z$ system without an ellipsoid. When the aforementioned concepts are combined in a systematic way with particular attention to the manner in which spatial data are used, the synergistic whole—the GSDM—appears to be greater than the sum of the parts.

Neither is the GSDM concept a new one. Seeber (1993) states that H. Burns proposed the concept of a global three-dimensional polyhedron network as early as 1878. The difference now is that the global navigation satellite system (GNSS) and other modern technologies have made a global network practical and that the polyhedron need not be limited to Earth-based points. The GSDM is an appropriate model for describing the "best" instantaneous positions of a global network of continuously operating reference stations (CORS) computed in real time with respect to the International Terrestrial Reference Frame (ITRF). An adopted mean position for each CORS may serve the needs of most users, but corrections for short-term variations caused by Earth tides, long-term continental drift velocities, and even catastrophic events such as earthquakes should be available to those needing them. It is readily acknowledged that such policies are already being used in the scientific community and that a space-fixed inertial reference system is more appropriate for describing the motion of Earth satellites. The GSDM should not be viewed as a prescriptive model, but as an inclusive model that accommodates the diverse practice of many spatial data users and provides an efficient bridge between local "flat-Earth" uses and rigorous scientific applications.

## THE GSDM

The GSDM is a collection of mathematical concepts and procedures that can be used to manage spatial data both locally and globally. It consists of a functional model that describes the geometrical relationships and a stochastic model that describes the probabilistic characteristics—statistical qualities—of spatial data. The functional part of the model includes equations of geometrical geodesy and rules of solid geometry as related to various coordinate systems and is intended to be consistent with the 3-D Geodetic Model described by Leick (2004) with the following exception: The GSDM, being strictly spatial, does not accommodate gravity measurements but presumes gravity affects are appropriately accommodated before data are entered into the spatial model. The stochastic portion of the GSDM is an application of concepts described by Mikhail (1976), Leick (2004), and Ghilani (2010).

Although the GSDM makes no attempt to accommodate non-Euclidean space or concepts, it does provide a simple universal foundation for many disparate coordinate systems used in various parts of the world and offers advantages of standardization for spatial data users in disciplines such as those listed in Figure 1.1. As such, the GSDM

**Global Spatial Data Model - GSDM**
**(A universal 3-D model for spatial data)**

The Global Spatial Data Model (GSDM) provides a simple, universal 3-D
mathematical foundation for the Global Spatial Data Infrastructure (GSDI)
which supports Geographic Information System (GIS) database applications
in disciplines such as:

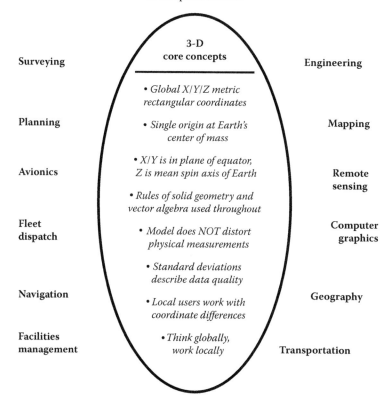

**3-D**
**core concepts**

- *Global X/Y/Z metric rectangular coordinates*
- *Single origin at Earth's center of mass*
- *X/Y is in plane of equator, Z is mean spin axis of Earth*
- *Rules of solid geometry and vector algebra used throughout*
- *Model does NOT distort physical measurements*
- *Standard deviations describe data quality*
- *Local users work with coordinate differences*
- *Think globally, work locally*

Surveying

Planning

Avionics

Fleet
dispatch

Navigation

Facilities
management

Engineering

Mapping

Remote
sensing

Computer
graphics

Geography

Transportation

**FIGURE 1.1** The Global Spatial Data Model

should be viewed as the geometrical portion of a larger concept being promoted as the
Global Spatial Data Infrastructure (GSDI) described by Holland et al. (1999) as, "The
policies, organizational remits, data, technologies, standards, delivery mechanisms,
and financial and human resources necessary to ensure that those working at the global
and regional scales are not impeded in meeting their objective." For more information
on the GSDI, see https://www.fgdc.gov/international/gsdi (accessed May 4, 2017).

## FUNCTIONAL MODEL COMPONENT

The functional model component of the GSDM is based upon a three-dimensional,
right-handed, rectangular Cartesian coordinate system with the origin located at the
center of mass of the Earth. The *X/Y* plane lies in the equatorial plane with the *X*-axis

at the 0° (Greenwich) meridian. The Z-axis coincides very nearly with the mean spin axis of the Earth, as defined by the Conventional Terrestrial Pole (Leick 2004). This geocentric coordinate system is called an Earth-centered Earth-fixed (ECEF) coordinate system by the U.S. National Imagery and Mapping Agency (NIMA 1997) and is widely used by many who work with GNSS and related data. Rules of solid geometry and vector algebra are universally applicable when working with ECEF coordinates and coordinate differences.

As shown in Figure 1.2, the unique 3-D position of any point on Earth or near space is equivalently defined by traditional latitude/longitude/ellipsoid height coordinates or by a triplet of $X/Y/Z$ coordinates expressed in meters. Due to the large distances involved, the $X/Y/Z$ coordinate values can be quite large, but most computational devices handle 15 significant digits routinely. Twelve significant digits will accommodate all ECEF coordinate values within the "birdcage" of satellites down to 0.1 mm. Some users may object to working with such large coordinate values, but, as shown in Figure 1.3, such objections will likely become inconsequential to the extent end user applications are designed to utilize coordinate differences (much smaller numbers and fewer digits).

Figure 1.4 is a schematic that illustrates relationships between the ECEF coordinate system and various other coordinate systems commonly used in connection with spatial data. A key feature on the diagram is a rotation matrix used to convert $\Delta X/\Delta Y/\Delta Z$ coordinate differences to local $\Delta e/\Delta n/\Delta u$ coordinate differences at any point (local origin) specified by the user. Since a vector in 3-D space is not altered by moving the origin or by changing the orientation of the reference coordinate system, a vector defined by its geocentric $\Delta X/\Delta Y/\Delta Z$ components is equivalently defined by local components, and the rotation matrix is the mechanism that efficiently transforms a global perspective into a local one. The transpose of the rotation matrix is used to transform local components of a space vector to corresponding geocentric components.

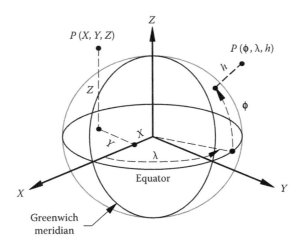

**FIGURE 1.2**   Geocentric $X/Y/Z$ and Geodetic $\phi/\lambda/h$ Coordinates

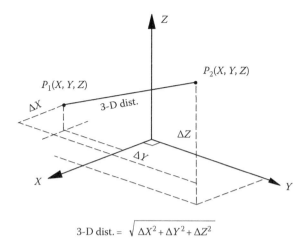

$$\text{3-D dist.} = \sqrt{\Delta X^2 + \Delta Y^2 + \Delta Z^2}$$

**FIGURE 1.3**   GNSS Technology Provides Precise $\Delta X/\Delta Y/\Delta Z$ Components

## COMPUTATIONAL DESIGNATIONS

With regard to Figure 1.4, the functional model includes equations for transforming spatial data described by coordinates in one numbered box to equivalent expressions in a different coordinate system. The contents of the numbered boxes are as follows:

Box 1.  Geocentric $X/Y/Z$ coordinates are the basis for all other coordinate values obtained from the GSDM. These are the primary defining values stored for each point in a digital spatial data file. Coordinate values in other coordinate systems are derived from the stored ECEF coordinates using algorithms that have been tested and proven for mathematical "exactness" and computational precision. This part of the GSDM features meter units, a linear adjustment model, and vector algebra along with universal rules of solid geometry.

Box 2.  Geodetic coordinates of latitude and longitude combined with ellipsoid height can define a three-dimensional position with the same precision and exactness as geocentric $X/Y/Z$ coordinates. Equations are listed in a subsequent section by which coordinate values in one box can be converted to equivalent values in another. Using both angular sexagesimal units (degrees, minutes, and seconds) on the ellipsoid and length units of meters for height makes traditional 3-D geodetic computations more complicated than when using ECEF rectangular coordinates.

Box 3.  GNSS technology has been a driving force behind use of 3-D spatial data and helps create demand for the GSDM. Although practice includes displaying coordinates in a defined system, the primary output of a GNSS survey historically has been a 3-D vector defined by its $\Delta X/\Delta Y/\Delta Z$ components. Because existing control stations were defined with geodetic coordinates of latitude and longitude (and other reasons), it was natural to

**The 3-D Global Spatial Data Model diagram**

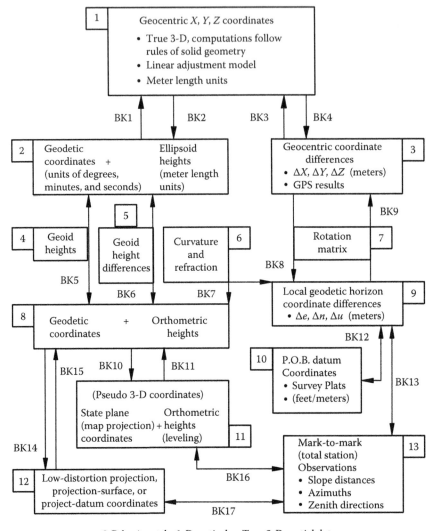

2-D horizontal + 1-D vertical or True 3-D spatial data

**FIGURE 1.4**  Diagram Showing Relationship of Coordinate Systems

continue building a two-dimensional network using 3-D measurements. And there certainly are cases where that practice can still be justified. But, the GSDM defines an environment in which the full value of 3-D data can be used to build high-quality 3-D networks without being encumbered by 2-D assumptions and complex equations found in classical geometrical geodesy. Another benefit of the GSDM is that the associated stochastic model lends itself to implementation in the rectangular 3-D environment more readily than in the latitude/longitude/height system.

Box 4. In practice, geoid height is taken to be the difference between ellipsoid height and orthometric height. With any two of the three elements known, the third can be found. If a reliable ellipsoid height for a point (from GNSS data) is combined with an appropriate geoid height (from geoid modeling), it is possible to obtain high-quality orthometric height (elevations). Appropriate use of standard deviations for the constituent components will provide a statistical assessment of the quality of such elevations.

Box 5. This is the same as Box 4 except that the computations are performed using differences. As explained in Chapter 9, modeled geoid height differences are often more reliable than modeled absolute geoid heights. This means better elevations can be computed by starting with a known high-quality bench mark elevation and combining observed ellipsoid height differences with modeled geoid height differences to compute the orthometric height difference.

Box 6. The $\Delta u$ component from Box 9 is the perpendicular distance of the forepoint from the tangent plane through the standpoint. An elevation difference from the standpoint to the forepoint includes the $\Delta u$ component plus the curvature and refraction $(c + r)$ correction. This $c + r$ procedure is based on the modeled distance between the horizontal plane and a level surface and does not include any geoid modeling between the standpoint and forepoint.

Box 7. Given that the statistical qualities of a vector in space are independent of the perspective from which it is viewed, the rotation matrix is a very efficient method for changing a global perspective (geocentric coordinate differences) into a local perspective (local "flat-Earth" components). Similarly, the transposed rotation matrix converts the local perspective into a global one.

Box 8. Historically, horizontal latitude and longitude coordinates have been combined with vertical elevations when mapping features on or near the Earth's surface. The generic zero reference surface for elevation has been the geoid (or mean sea level) that admits to a physical definition but, as it turns out, is very difficult, if not impossible, to find. As a result, an arbitrary reference surface that approximates, but does not define, mean sea level was selected for the North American Vertical Datum of 1988 (Zilkoski et al. 1992).

Box 9. The geodetic horizon (Leick et al. 2015, p. 167) is essentially the same as the local geodetic frame described by Soler and Hothem (1988) and shares many similarities with local plane surveying practice. The primary difference when using the GSDM is that "up" is defined by the ellipsoid normal instead of the plumb line. This difference is largely inconsequential except (1) where very high precision is required, (2) where the slope of the geoid (with respect to the normal) is severe, and (3) if the "up" component is quite large—for example, as encountered in underground mapping. Another difference for plane surveyors using the GSDM is that the origin moves with the observer because one is working with local coordinate differences with respect to the user-specified standpoint. See Box 10 for working with a traditional (or fixed) origin.

The $\Delta u$ component in Box 9 can be used as an approximate elevation difference because it does not include slope of the geoid, Earth curvature, or refraction $(c + r)$—all inconsequential for short lines. Although suggested as a secondary means for obtaining elevation differences, the standard $c + r$ correction can be combined with the $\Delta u$ component to obtain elevation differences between standpoint and forepoint. Understandably, the primary method for obtaining elevation differences still relies on differential leveling, accurate geoid heights and ellipsoid heights, or their differences. See Chapter 9 for more details.

Box 10. Point-of-Beginning (P.O.B.) datum coordinates is a feature within the GSDM that accommodates long-established local plane surveying practices without compromising geometrical integrity. P.O.B. coordinates permit the user to select any point in the database as an origin. The 3-D location of each additional point selected is listed with respect to the P.O.B. Admittedly, this practice makes little sense for large distances, but these local coordinate differences can be treated in the same manner as local plane coordinates and used on survey plats. Horizontal distances are in the tangent plane through the P.O.B., and azimuths are with respect to the meridian through the P.O.B. If surveys of adjacent tracts do not use the same P.O.B., there will be two azimuths for a common line (the difference is convergence of the meridians between the two P.O.B.s). However, if the P.O.B. is the same for both tracts, they will share a common basis of bearing—the geodetic meridian through the P.O.B.

Box 11. Map projections were invented to address the challenge of representing a curved Earth on a flat map. In particular, conformal projections have been used in surveying and mapping to define precisely a 2-D relationship between latitude/longitude positions on the Earth and equivalent plane coordinate positions on a flat map. Systematic use of map projections includes state plane coordinate systems as implemented in the United States and worldwide use of Universal Transverse Mercator (UTM) coordinates.

However, it is important to note that elevations combined with map projection $x/y$ (or *east/north*) plane coordinates is not an appropriate 3-D rectangular model for two reasons:

1. Conformal projections are well defined in two dimensions only. There is no mathematical definition of elevation in conformal mapping.
2. The zero reference surface for elevation (approximated by sea level) is a nonregular curved surface. Full 3-D integrity is preserved only to the extent a flat Earth can be safely assumed. Therefore, map projection coordinates combined with elevations are referred to as pseudo 3-D.

Box 12. An important consideration when using state plane coordinates is the relationship of the grid inverse distance to actual ground-level horizontal distance. In applications such as highway centerline stationing, the difference between grid and ground distance quickly becomes too great to ignore. Project datum coordinate systems were invented to accommodate this difference. Lack of standardization is an issue when considering project datum coordinates. For a summary of comments from 46 out of 50 state DOTs on the grid/ground distance difference, see Appendix III of Burkholder (1993a). On the other hand, states such as Wisconsin (Wisconsin 1995, 2009) and Minnesota (Whitehorn 1997) have formally defined countywide coordinate systems for local use.

When working with the $\Delta e/\Delta n$ components, the horizontal distance is in the tangent plane through the standpoint and is the same horizontal distance plane surveyors have been using for generations. It is also the same as $HD(1)$ as described in Burkholder (1991). Understandably, with a unique tangent plane at each standpoint, the tangent plane from Point A to Point B is slightly different than the tangent plane from Point B to Point A. But, geometrical integrity in three dimensions is preserved by the GSDM and underlying $X/Y/Z$ coordinates.

The 3-D azimuth from standpoint to forepoint obtained from arctan $(\Delta e/\Delta n)$ gives the correct azimuth between each pair of points. The forward azimuth of a line differs from the back azimuth of the same line due to convergence of the meridians between the two endpoints. The GSDM competently provides the correct answer in each case (Burkholder 1997a). The 3-D azimuth is defined simply and is easy to use. The azimuth of a geodetic line has a more complex definition and differs only slightly from the 3-D azimuth. The geodetic azimuth is "better" than the 3-D azimuth only in the most demanding cases. See Chapter 7 for more details.

Box 13. Spatial data measurements with conventional total station surveying instruments include slope distances, vertical (or zenith) angles, and determinations of bearings or azimuths. These measurements are used to compute local geodetic horizon coordinate differences of $\Delta e/\Delta n/\Delta u$. In reality, these measurements are referenced to the plumb line, while the GSDM presumes the results are normal based. The difference is small, but important. Current procedures for making Laplace corrections are still applicable and should be used as described in Chapter 9.

Equations for moving spatial data from one box to another have been given various names over the years. When used in context, there may be little confusion over what is a "forward" and what is an "inverse" computation. But, when brought together in a common collection, the duplication of conventional names can be confusing and misleading. Therefore, as a matter of convenience and in the interest of promoting unambiguous communication, the designations shown in Table 1.1 are used to describe the various computations and transformations. Many of them are illustrated in Figure 1.4.

**TABLE 1.1**

**Designation for Spatial Data Computations and Transformations**

| Name | Conventional Description |
|---|---|
| BK1 | Convert geodetic latitude/longitude/height coordinates to geocentric $X/Y/Z$ coordinates. |
| BK2 | Convert geocentric $X/Y/Z$ coordinates to geodetic latitude/longitude/height coordinates. |
| BK3 | 3-D geodetic forward computation using $\Delta X/\Delta Y/\Delta Z$. |
| BK4 | 3-D geodetic inverse computation using $\Delta X/\Delta Y/\Delta Z$. |
| BK5 | Any combination of using ellipsoid height, orthometric height (elevation), and geoid height. |
| BK6 | Any combination of using *differences* for ellipsoid height, orthometric height (elevation), and geoid height. |
| BK7 | Use curvature and refraction corrections to refine elevation difference computations—especially trig heights. |
| BK8 | Convert geocentric coordinate differences to local coordinate differences. |
| BK9 | Convert local coordinate differences to geocentric coordinate differences. |
| BK10 | Convert geodetic latitude/longitude to state plane or map projection coordinates (also known as "forward" computation). |
| BK11 | Convert state plane or map projection coordinates to latitude/longitude (also known as "inverse" computation). |
| BK12 | Use local coordinate differences to compute P.O.B. datum coordinates and vice versa. |
| BK13 | Use conventional total station observations to compute local coordinate differences (1-D, 2-D, or 3-D) and vice versa. |
| BK14 | Convert geodetic latitude/longitude to low distortion, project datum, or surface coordinates. |
| BK15 | Convert low distortion, project datum, or surface coordinates to latitude/longitude. |
| BK16 | 2-D COGO computations based upon named state plane coordinate system zone. |
| BK17 | 2-D COGO low distortion, surface, or project datum computations of designated (countywide) system. |
| BK18 | 2-D geodetic forward computation—not shown in Figure 1.4. |
| BK19 | 2-D geodetic inverse computation—not shown in Figure 1.4. |
| BK20 | Generic 2D COGO computations—not shown in Figure 1.4. |
| BK21 | Generic differential leveling—not shown in Figure 1.4. |
| BK22 | Generic trig height leveling—not shown in Figure 1.4. |

*Note:* BK5, BK6, and BK7 are quite similar, but having different designations will help avoid problems caused by the subtle differences.

## ALGORITHM FOR FUNCTIONAL MODEL

A more complete set of equations and derivations is provided in Chapter 7, but the following symbols are defined and used in this summary:

$X/Y/Z$ = geocentric right-handed rectangular coordinates.
$\Delta X/\Delta Y/\Delta Z$ = geocentric coordinate differences.

$\Delta e/\Delta n/\Delta u$ = local coordinate differences.
$\phi/\lambda/h$ = geodetic latitude/longitude (east) and ellipsoid height.
$a$ and $b$ = semimajor and semiminor axes of reference ellipsoid.
$f$ = flattening of reference ellipsoid.
$e^2$ = eccentricity squared of reference ellipsoid; $e^2 = 2f - f^2$.
$N$ = length of ellipsoid normal, also used for geoid height.
$S$ = spatial slope distance between standpoint and forepoint.
$\alpha$ = geodetic azimuth at standpoint to forepoint.
$z$ or $V$ = zenith direction or vertical angle to forepoint.
$H$ = orthometric height (elevation).
$HD(1)$ or $D$ = ground-level horizontal distance.

*Notes*:
1. All distances are in units of meters.
2. Where two points are concerned, the standpoint is indicated by subscript 1 while the forepoint is indicated by subscript 2.

The BK1 equations are

$$N = \frac{a}{\sqrt{1 - e^2 \sin^2 \phi}} \tag{1.1}$$

$$X = (N + h)\cos\phi\cos\lambda \tag{1.2}$$

$$Y = (N + h)\cos\phi\sin\lambda \tag{1.3}$$

$$Z = \left(N\left[1 - e^2\right] + h\right)\sin\phi \tag{1.4}$$

The BK2 equations are more difficult to use because iteration is normally required to solve them. Equation 1.5 is quite straightforward, but Equations 1.6 and 1.7 need to be iterated as explained in Chapter 7. An alternate (noniterative) method for performing the BK2 transformation is also given in Chapter 7.

$$\lambda = \tan^{-1}\left(\frac{Y}{X}\right) \tag{1.5}$$

$$\phi = \tan^{-1}\left[\frac{Z}{\sqrt{X^2 + Y^2}}\left(1 + \frac{e^2 N \sin\phi}{Z}\right)\right] \tag{1.6}$$

$$h = \frac{\sqrt{X^2 + Y^2}}{\cos\phi} - N \tag{1.7}$$

The BK3 and BK4 equations are also called 3-D "forward" and "inverse":

BK3—Forward

$$X_2 = X_1 + \Delta X \tag{1.8}$$

$$Y_2 = Y_1 + \Delta Y \tag{1.9}$$

$$Z_2 = Z_1 + \Delta Z \tag{1.10}$$

BK4—Inverse

$$\Delta X = X_2 - X_1 \tag{1.11}$$

$$\Delta Y = Y_2 - Y_1 \tag{1.12}$$

$$\Delta Z = Z_2 - Z_1 \tag{1.13}$$

The BK5 computation handles any combination of orthometric height ($H$), ellipsoid height ($h$), and geoid height ($N$) as follows:

$$N = h - H \tag{1.14}$$

$$H = h - N \tag{1.15}$$

$$h = H + N \tag{1.16}$$

The BK6 computation is the same as BK5 except that *differences* are used as follows:

$$\Delta N = \Delta h - \Delta H \tag{1.17}$$

$$\Delta H = \Delta h - \Delta N \tag{1.18}$$

$$\Delta h = \Delta H + \Delta N \tag{1.19}$$

Differences are important because geoid modeling provides better answers when using relative geoid height differences rather than absolute geoid heights. See Chapter 9 for more details.

The BK7 computation relies on the combined curvature and refraction ($c + r$) correction for the difference between a level surface and tangent plane surface. For modest precision over short distances, the $c + r$ correction can be used beneficially as (Davis et al. 1981, Equation 5.7)

$$H_2 = H_1 + \Delta u + 0.0675 \frac{D^2}{1,000,000} = H_1 + \Delta u + 0.0675 \frac{\Delta e^2 + \Delta n^2}{1,000,000} \tag{1.20}$$

The BK8 and BK9 transformations involve using a rotation matrix to convert geocentric differences to local differences and local differences to geocentric differences. See Appendix A for more details.

The BK8 transformation of geocentric differences to local differences is

$$
\begin{bmatrix} \Delta e \\ \Delta n \\ \Delta u \end{bmatrix} = \begin{bmatrix} -\sin\lambda & \cos\lambda & 0 \\ -\sin\phi\cos\lambda & -\sin\phi\sin\lambda & \cos\phi \\ \cos\phi\cos\lambda & \cos\phi\sin\lambda & \sin\phi \end{bmatrix} \begin{bmatrix} \Delta X \\ \Delta Y \\ \Delta Z \end{bmatrix} \tag{1.21}
$$

The BK9 transformation of local differences to geocentric differences is

$$
\begin{bmatrix} \Delta X \\ \Delta Y \\ \Delta Z \end{bmatrix} = \begin{bmatrix} -\sin\lambda & -\sin\phi\cos\lambda & \cos\phi\cos\lambda \\ \cos\lambda & -\sin\phi\sin\lambda & \cos\phi\sin\lambda \\ 0 & \cos\phi & \sin\phi \end{bmatrix} \begin{bmatrix} \Delta e \\ \Delta n \\ \Delta u \end{bmatrix} \tag{1.22}
$$

Equations 1.23 and 1.24 are not one of the BKX transformations, but they are used to obtain the local tangent horizontal distance and the true direction from the standpoint (PT 1) to the forepoint (PT 2):

$$
\text{Distance} = \sqrt{\Delta e^2 + \Delta n^2} \tag{1.23}
$$

$$
\tan\alpha = \left(\frac{\Delta e}{\Delta n}\right) \quad \text{with due regard to the quadrant} \tag{1.24}
$$

The BK10 and BK11 transformations are used to handle state plane coordinate transformations and are discussed in Chapter 11.

BK12 computations are used to develop local tangent plane coordinates with respect to any P.O.B. selected by the user. See 2-D plat example in Chapter 15.

BK13 transformations are used to convert terrestrial observations into local coordinate differences that can then be converted to geocentric differences using the BK9 transformation. Such observations may need to be corrected for instrument calibration, atmospheric conditions, polar motion, and local deflection-of-the-vertical.

$$
\Delta e = S\sin z\sin\alpha = HD(1)\sin\alpha \tag{1.25}
$$

$$
\Delta n = S\sin z\cos\alpha = HD(1)\cos\alpha \tag{1.26}
$$

$$
\Delta u = S\cos z \tag{1.27}
$$

There is no one correct set of equations for BK14 and BK15 computations. The primary force behind the use of project datum (or surface) coordinates is that an inverse between grid coordinates (grid distance) is not the same as the horizontal ground distance. Various methods for handling the grid/ground distance difference are described in Burkholder (1993a). Concise rigorous procedures for using local coordinate systems are provided in Burkholder (1993b), and some applications of project datum coordinates may continue to be justified. More recently, the concept of a low-distortion projection (LDP) has been proposed, and the advantages of central administration for LDPs certainly have merit. But the issue of using project datum coordinates or a LDP becomes moot when using the GSDM.

Computations BK16 and BK17 are the same traditional coordinate geometry computations with the following exception: BK16 computations involve formal use of state plane coordinates, and BK17 computations involve use of project datum coordinates. Being specific between these two is very prudent and can help avoid the frustration and wasted efforts as a result of unwittingly using one for the other.

BK18 through BK22 computations consist of traditional surveying practices and are not shown in Figure 1.4.

## STOCHASTIC MODEL COMPONENT

The stochastic component of the GSDM is devoted to answering the question, "accuracy with respect to what?" The stochastic model is based upon storing the covariance matrix associated with the geocentric $X/Y/Z$ rectangular coordinates that define the location of each stored point. A user can compute the local east/north/up covariance matrix of any point on an "as needed" basis using the standard covariance error propagation (this minimizes storage requirements). The same basic procedure is extended to other functional model computations and provides a statistically defensible method for tracking the influence of random errors to any derived quantity. In particular, the user can look at the standard deviation of a coordinate position (by individual component) in either the geocentric or local reference frame. The standard deviation of other derived quantities such as distance, azimuth, slope, area, or volume can be obtained using the same error propagation procedure with the appropriate functional model equations.

## THE GSDM COVARIANCE MATRICES

The functional component of the GSDM consists of geometrical equations that are used to manipulate $X/Y/Z$ geocentric coordinates defining the spatial position of each point. The stochastic component of the GSDM is an application of the laws of variance/covariance error propagation and utilizes the following matrix formulation (Mikhail 1976, Burkholder 1999, 2004):

$$\Sigma_{YY} = J_{YX}\Sigma_{XX}J_{XY}^{t} \qquad (1.28)$$

where:

$\Sigma_{YY}$ = covariance matrix of computed result.

$\Sigma_{XX}$ = covariance matrix of variables used in computation.

$J_{YX}$ = Jacobian matrix of partial derivatives of the result with respect to the variables.

The GSDM uses two covariance matrices for each point (but stores only one): the geocentric covariance matrix and the local covariance matrix. In particular, the following symbols and matrices are used in the stochastic model:

$\sigma_X^2 \sigma_Y^2 \sigma_Z^2$ = variances of geocentric coordinates for a point.

$\sigma_{XY} \sigma_{XZ} \sigma_{YZ}$ = covariances of geocentric coordinates for a point.

$\sigma_e^2 \sigma_n^2 \sigma_u^2$ = variances of a point in the local reference frame.

$\sigma_{en} \sigma_{eu} \sigma_{nu}$ = covariances of a point in the local reference frame.

$\sigma_{\Delta X}^2 \sigma_{\Delta Y}^2 \sigma_{\Delta Z}^2$ = variances of geocentric coordinate differences.

$\sigma_{\Delta X \Delta Y} \sigma_{\Delta X \Delta Z} \sigma_{\Delta Y \Delta Z}$ = covariances of geocentric coordinate differences.

$\sigma_{\Delta e}^2 \sigma_{\Delta n}^2 \sigma_{\Delta u}^2$ = variances of coordinate differences in local frame.

$\sigma_{\Delta e \Delta n} \sigma_{\Delta e \Delta u} \sigma_{\Delta n \Delta u}$ = covariances of coordinate differences in local frame.

$\sigma_S^2 \sigma_\alpha^2$ = variances of local horizontal distance and azimuth.

$\sigma_{S\alpha}$ = covariance of local horizontal distance with azimuth.

$\sigma_{X_1 X_2} \sigma_{Y_1 Y_2}$ = elements of point 1–point 2 correlation matrix.

Geocentric Covariance Matrix

$$\Sigma_{X/Y/Z} = \begin{bmatrix} \sigma_X^2 & \sigma_{X/Y} & \sigma_{X/Z} \\ \sigma_{X/Y} & \sigma_Y^2 & \sigma_{Y/Z} \\ \sigma_{X/Z} & \sigma_{Y/Z} & \sigma_Z^2 \end{bmatrix} \tag{1.29}$$

Local Covariance Matrix

$$\Sigma_{e/n/u} = \begin{bmatrix} \sigma_e^2 & \sigma_{e/n} & \sigma_{e/u} \\ \sigma_{e/n} & \sigma_n^2 & \sigma_{n/u} \\ \sigma_{e/u} & \sigma_{n/u} & \sigma_u^2 \end{bmatrix} \tag{1.30}$$

Notes on the individual point covariance matrices include the following:

1. Each covariance matrix is $3 \times 3$ and symmetric. Six numbers are required to store upper (or lower) triangular values.
2. The unit for each covariance matrix element is meters squared; the off-diagonal elements represent correlations; diagonal elements are called variances; and standard deviations are computed as the square root of the variances.
3. Each station covariance matrix (with its unique orientation) represents the accuracy of a point with respect to a defined reference frame (or to whatever control is held fixed by the user) and is designated *datum accuracy*.

The local covariance matrix and the geocentric covariance matrix are related to each other mathematically by a rotation matrix for the latitude/longitude position of a point computed from its X/Y/Z coordinates (Burkholder 1993a):

$$R = \begin{bmatrix} -\sin\lambda & \cos\lambda & 0 \\ -\sin\phi\cos\lambda & -\sin\phi\sin\lambda & \cos\phi \\ \cos\phi\cos\lambda & \cos\phi\sin\lambda & \sin\phi \end{bmatrix} \tag{1.31}$$

The relationship between the covariance matrices is

$$\Sigma_{e/n/u} = R\Sigma_{X/Y/Z}R^t \tag{1.32}$$

$$\Sigma_{X/Y/Z} = R^t\Sigma_{e/n/u}R \tag{1.33}$$

With regard to the rotation matrix in Equation 1.31, longitude is counted $0°–360°$ east from the Greenwich Meridian, west longitude is a negative value, and latitude is counted positive north of the equator and negative south of the equator.

## THE GSDM 3-D INVERSE

Given that point 1 is defined by $X_1/Y_1/Z_1$ and point 2 by $X_2/Y_2/Z_2$, matrix formulations of the 3-D geocentric coordinate inverse and covariance error propagation are

$$\begin{bmatrix} \Delta X \\ \Delta Y \\ \Delta Z \end{bmatrix} = \begin{bmatrix} -1 & 0 & 0 & 1 & 0 & 0 \\ 0 & -1 & 0 & 0 & 1 & 0 \\ 0 & 0 & -1 & 0 & 0 & 1 \end{bmatrix} \begin{bmatrix} X_1 \\ Y_1 \\ Z_1 \\ X_2 \\ Y_2 \\ Z_2 \end{bmatrix} \tag{1.34}$$

The matrix of coefficients to the variables is called the Jacobian matrix, and the general error propagation formulation in the form of Equation 12.1 is

$$\Sigma_\Delta = J\Sigma_{1\rightarrow2}J^t \tag{1.35}$$

The Jacobian matrix from Equation 1.34 and the general covariance error propagation procedure, Equation 1.35, are used to find the overall geocentric inverse covariance matrix as

$$\Sigma_\Delta = \begin{bmatrix} -1 & 0 & 0 & 1 & 0 & 0 \\ 0 & -1 & 0 & 0 & 1 & 0 \\ 0 & 0 & -1 & 0 & 0 & 1 \end{bmatrix} \begin{bmatrix} \begin{bmatrix} \sigma_{X_1}^2 & \sigma_{X_1Y_1} & \sigma_{X_1Z_1} \\ \sigma_{X_1Y_1} & \sigma_{Y_1}^2 & \sigma_{Y_1Z_1} \\ \sigma_{X_1Z_1} & \sigma_{Y_1Z_1} & \sigma_{Z_1}^2 \end{bmatrix} & \begin{bmatrix} \sigma_{X_1X_2} & \sigma_{X_1Y_2} & \sigma_{X_1Z_2} \\ \sigma_{Y_1X_2} & \sigma_{Y_1Y_2} & \sigma_{Y_1Z_2} \\ \sigma_{Z_1X_2} & \sigma_{Z_1Y_2} & \sigma_{Z_1Z_2} \end{bmatrix} \\ \begin{bmatrix} \sigma_{X_1X_2} & \sigma_{Y_1X_2} & \sigma_{Z_1X_2} \\ \sigma_{X_1Y_2} & \sigma_{Y_1Y_2} & \sigma_{Z_1Y_2} \\ \sigma_{X_1Z_2} & \sigma_{Y_1Z_2} & \sigma_{Z_1Z_2} \end{bmatrix} & \begin{bmatrix} \sigma_{X_2}^2 & \sigma_{X_2Y_2} & \sigma_{X_2Z_2} \\ \sigma_{Y_2X_2} & \sigma_{Y_2}^2 & \sigma_{Y_2Z_2} \\ \sigma_{Z_2X_2} & \sigma_{Z_2Y_2} & \sigma_{Z_2}^2 \end{bmatrix} \end{bmatrix} \begin{bmatrix} -1 & 0 & 0 \\ 0 & -1 & 0 \\ 0 & 0 & -1 \\ 1 & 0 & 0 \\ 0 & 1 & 0 \\ 0 & 0 & 1 \end{bmatrix} \tag{1.36}$$

Correlation between points 1 and 2 is described by the off-diagonal submatrices. Various accuracies are defined by a choice with regard to use of the covariance matrix in Equation 1.36. The matrix operation in Equation 1.36 can be used to compute

1. *Local accuracy* if the full covariance matrix is employed
   (relative accuracy based upon the quality of measurements connecting adjacent points)

2. *Network accuracy* if the correlation between points 1 and 2 is zero (relative accuracy based upon the combined quality of each point with respect to the network)
3. *P.O.B. accuracy* if the covariance matrix of point 2 is the only one used (relative accuracy based solely on the network quality of point 2)

Implementation issues related to naming the various accuracies defined by Equation 1.36 are discussed later. For the sake of completeness, the remaining inverse computations for local coordinate differences, directions, distances, and associated standard deviations are given next. The matrix formulation for computing local coordinate differences from geocentric coordinate differences is derived from Equation 1.21 as

$$
\begin{bmatrix} \Delta e \\ \Delta n \\ \Delta u \end{bmatrix} = \begin{bmatrix} -\sin\lambda & \cos\lambda & 0 \\ -\sin\phi\cos\lambda & -\sin\phi\sin\lambda & \cos\phi \\ \cos\phi\cos\lambda & \cos\phi\sin\lambda & \sin\phi \end{bmatrix} \begin{bmatrix} \Delta X \\ \Delta Y \\ \Delta Z \end{bmatrix} = R \begin{bmatrix} \Delta X \\ \Delta Y \\ \Delta Z \end{bmatrix} \tag{1.21}
$$

The Jacobian matrix ($R$) from Equations 1.21 and 1.31 is used with the general error propagation formulation to compute the covariance matrix of local coordinate differences as

$$
\Sigma_{3D\text{-}INV} = \begin{bmatrix} \sigma^2_{\Delta e} & \sigma_{\Delta e \Delta n} & \sigma_{\Delta e \Delta u} \\ \sigma_{\Delta e \Delta n} & \sigma^2_{\Delta n} & \sigma_{\Delta n \Delta u} \\ \sigma_{\Delta e \Delta u} & \sigma_{\Delta n \Delta u} & \sigma^2_{\Delta u} \end{bmatrix} = R \Sigma_\Lambda R^T = R \begin{bmatrix} \sigma^2_{\Delta X} & \sigma_{\Delta X \Delta Y} & \sigma_{\Delta X \Delta Z} \\ \sigma_{\Delta X \Delta Y} & \sigma^2_{\Delta Y} & \sigma_{\Delta Y \Delta Z} \\ \sigma_{\Delta X \Delta Z} & \sigma_{\Delta Y \Delta Z} & \sigma^2_{\Delta Z} \end{bmatrix} R^T
$$

$$(1.37)$$

The functional model equations for a 2-D local tangent plane horizontal distance and 3-D azimuth are given by Equations 1.23 and 1.24 as

$$
D = HD(1) = \sqrt{\Delta e^2 + \Delta n^2} \tag{1.23}
$$

$$
\alpha = \tan^{-1}\left(\frac{\Delta e}{\Delta n}\right) \tag{1.24}
$$

And the Jacobian matrix ($J_2$) of those partial derivatives is

$$
J_2 = \begin{bmatrix} \dfrac{\partial D}{\partial \Delta e} & \dfrac{\partial D}{\partial \Delta n} & \dfrac{\partial D}{\partial \Delta u} \\ \dfrac{\partial \alpha}{\partial \Delta e} & \dfrac{\partial \alpha}{\partial \Delta n} & \dfrac{\partial \alpha}{\partial \Delta u} \end{bmatrix} = \begin{bmatrix} \dfrac{\Delta e}{D} & \dfrac{\Delta n}{D} & 0 \\ \dfrac{\Delta n}{D^2} & \dfrac{\Delta e}{D^2} & 0 \end{bmatrix} \tag{1.38}
$$

Finally, using the covariance propagation formulation, the 2-D results are

$$
\Sigma_{2D\text{-}INV} = \begin{bmatrix} \sigma^2_D & \sigma_{D\alpha} \\ \sigma_{D\alpha} & \sigma^2_\alpha \end{bmatrix} = J_2 \Sigma_{3D\text{-}INV} J_2^T = J_2 \begin{bmatrix} \sigma^2_{\Delta e} & \sigma_{\Delta e \Delta n} & \sigma_{\Delta e \Delta u} \\ \sigma_{\Delta e \Delta n} & \sigma^2_{\Delta n} & \sigma_{\Delta n \Delta u} \\ \sigma_{\Delta e \Delta u} & \sigma_{\Delta n \Delta u} & \sigma^2_{\Delta u} \end{bmatrix} J_2^T \tag{1.39}
$$

Take the square root of the diagonal elements of Equation 1.39 to get standard deviations of the distance and the azimuth. Convert radians for standard deviation of azimuth to seconds of arc using 206,264.8062 seconds of arc per radian.

Equation 1.36 is really the heart of the stochastic model portion of the GSDM. Yes, Equations 1.32 and 1.33 can be used to convert absolute datum accuracy from one reference frame to another (geocentric and local), but when looking at one point with respect to another, Equation 1.36 offers important choices in answering the question, "accuracy with respect to what?" If a connecting measurement between two points has a smaller standard deviation than would be computed given no correlation between them, then the local accuracy, of one point with respect to the other, can be computed with statistical reliability. These tools give the spatial data user a number of choices and the option of computing the standard deviation of all subsequently derived quantities, such as distance, direction, height, volume, and area. More details on spatial data accuracy are included in Chapter 12.

## BURKORD™: SOFTWARE AND DATABASE

The mathematical concepts and equations described and used in formulating the Global Spatial Data Model are all in the public domain. The phrase "Global Spatial Data Model (GSDM)" is generic. The term "BURKORD" has been trademarked (1) as the name of a software package that performs 3-D coordinate geometry and error propagation computations as described in this chapter and (2) as the name of a 3-D database used by the BURKORD software. The end user is free to use the term "BURKORD™" as applied to the underlying database or to software obtained from Global COGO, Inc. However, anyone offering a product or service to others whose value relies upon or is enhanced by reference to or use of the BURKORD trademark will be expected to obtain appropriate permission. Inquiries related to using the term "BURKORD™" should be directed to Global COGO, Inc., P.O. Box 3162, Las Cruces, NM, 88003.

## SUMMARY

The GSDM gives each user both control and responsibility. If good or bad information is used inappropriately, unreliable answers can be obtained. However, the opposite case is the important one. The GSDM defines a model and computational environment that can be used to manage spatial data efficiently. Each user has the option of establishing criteria that must be met before spatial data can be used for a given purpose. The concept of metadata is important in establishing and preserving the credibility of spatial data, but standard deviation (in any or all components) is a very efficient method for evaluating the quality of spatial data. Once the $X/Y/Z$ position of a point is defined along with its variance/covariance matrix, the spatial data can be exchanged in a very compact format. The same solid geometry and error propagation equations for using such shared spatial data are equally applicable worldwide, and the mathematical procedures are already proven and individually implemented. Using the GSDM is primarily a matter of choosing to do so.

## REFERENCES

Bomford, G. 1971. Cartesian coordinates in three dimensions. In *Geodesy*, 3rd ed., Appendix C. Oxford, U.K.: Oxford University Press.

Burkholder, E. 1991. Computation of horizontal/level distances. *Journal of Surveying Engineering* 117 (3): 104–116.

Burkholder, E. 1993a. Using GPS results in true 3-D coordinate system. *Journal of Surveying Engineering* 119 (1): 1–21.

Burkholder, E. 1993b. Design of a local coordinate system for surveying, engineering, and LIS/GIS. *Surveying and Land Information Systems* 53 (1): 29–40.

Burkholder, E. 1997a. The 3-D azimuth of GPS vector. *Journal of Surveying Engineering* 123 (4): 139–146.

Burkholder, E. 1997b. *Definition and Description of a Global Spatial Data Model (GSDM)*. Washington, DC: U.S. Copyright Office.

Burkholder, E. 1999. Spatial data accuracy as defined by the GSDM. *Surveying and Land Information Systems* 59 (1): 26–30.

Burkholder, E. 2004. *Fundamentals of Spatial Data Accuracy and the Global Spatial Data Model (GSDM)*. Washington, DC: U.S. Copyright Office.

Davis, R., F. Foote, J. Anderson, and E. Mikhail. 1981. *Surveying: Theory and Practice*, 6th ed. New York: McGraw-Hill.

Ghilani, C.D. 2010. *Adjustment Computations: Spatial Data Analysis*, 5th ed. Hoboken, NJ: John Wiley & Sons.

Holland, P., M. Reichardt, D. Nebert, S. Blake, and D. Robertson. 1999. The Global Spatial Data Infrastructure initiative and its relationship to the vision of a digital Earth. Paper presented by Peter Holland *International Symposium on Digital Earth*, Beijing, China, November 29–December 2, 1999.

Leick, A. 2004. *GPS Satellite Surveying*, 3rd ed. Hoboken, NJ: John Wiley & Sons.

Leick, A., L. Rapoport, and D. Tatarnikov. 2015. *GPS Satellite Surveying*, 4th ed. Hoboken, NJ: John Wiley & Sons.

Mikhail, E. 1976. *Observations & Least Squares*. New York: Harper & Row.

Moritz, H. 1978. Definition of a geodetic datum. *Proceedings of the Second International Symposium on Problems Related to the Redefinition of North American Geodetic Networks*, Arlington, VA. Washington, DC: U.S. Department of Commerce, Department of Energy, Mines & Resources (Canada); and Danish Geodetic Institute.

National Imagery and Mapping Agency (NIMA). 1997. Department of defense world geodetic system 1984: Its definition and relationships with local geodetic systems. Technical Report 8350.2. Fairfax, VA: National Imagery and Mapping Agency.

Seeber, G. 1993. *Satellite Geodesy*. Berlin, Germany: Walter de Gruyter.

Soler, T. and L. Hothem. 1988. Coordinate systems used in geodesy: Basic definitions and concepts. *Journal of Surveying Engineering* 114 (2): 84–97.

Whitehorn, K.L. 1997. *The Minnesota County Coordinate System: A Handbook for Users*. St. Cloud, MN: Precision Measuring Systems.

Wisconsin. 1995. *Wisconsin Coordinate System*. Madison, WI: Wisconsin State Cartographer's Office.

Wisconsin. 2009. *Wisconsin Coordinate Systems*, 2nd ed. Madison, WI: Wisconsin State Cartographer's Office.

Zilkoski, D., J. Richards, and G. Young. 1992. Results of the general adjustment of the North American Vertical Datum of 1988. *Surveying and Land Information Systems* 52 (3): 133–149.

# 2 Featuring the 3-D Global Spatial Data Model

## INTRODUCTION

The Global Spatial Data Model (GSDM) is an arrangement of time-honored solid geometry equations and proven mathematical procedures. In that respect, it contains nothing new. But, the GSDM is built on the assumption of a single origin for 3-D geospatial data and formally defines procedures for handling spatial data that are consistent with digital technology and modern practice. In that respect, the GSDM is a newly defined model. The model for geodetic computations is an ellipsoid chosen for a global "best fit." The model for many mapping and cartographic applications is a map projection chosen as a trade-off between ease of use and standardization versus geometrical distortion. In each case, physical observations and measurements are modified to conform to the geometry imposed by the selected model. The GSDM accommodates measurements and observations that have been reduced to their rectangular components, and reliable geometrical computations are performed in 3-D space without being distorted by the model. *The model fits the measurements instead of forcing the measurements to fit the model.* The GSDM stores ECEF coordinates and (optionally) the standard deviations of each component in a BURKORD™ database, and those ECEF values can be transformed into any legitimate coordinate system for which bidirectional conversions exist. All users can enjoy the luxury of working in a well-understood and familiar environment with spatial data having a common heritage shared by other disciplines worldwide.

But, why another spatial data model? Compatibility is essential when persons from disparate disciplines share data. Success is assured to the extent fundamental concepts are clearly defined and basic procedures for data exchange are formalized. Such standardization and centralization provide economies of scale to the user community. Subsequently, additional benefits are derived through decentralization in which innovative applications expand upon the capabilities supported by the underlying standard. Such benefits, in turn, spawn new markets and applications. As the cycle continues, the underlying standards and the assumptions upon which they were established are reexamined and, if appropriate, updated. The telephone is an example. A centralized, regulated monopoly was largely responsible for placing a telephone in most homes in the United States. With the underlying infrastructure in place, additional benefits were realized as the industry was deregulated and competition between providers brought the consumer more options relative to telecommunication equipment and services. The twisted-pair analog standard upon which the telecommunications network was built has been replaced, and digital technology has been implemented to support significantly greater levels of service. Consumers can now select from an ever-expanding list of options for communication choices.

Geospatial data are another example. In the past, analog storage of geospatial data on a map was standard practice. Geographic coordinates provide global standardization, and derivative uses such as map projection (or state plane) coordinates are commonplace. The benefits of such centralization were a driving force in the early stages of building a geographic information system (GIS) as users and agencies needed to pool resources to achieve desired economies of scale. But spatial data users have experienced the same analog-to-digital transition as the telecommunications industry, and the underlying model needs to be reexamined. With the advent of affordable digital technologies, that is, GNSS and related microprocessor-based devices, the demand for spatial data products continues to grow. Enormous gains in productivity have been achieved by automating procedures for handling spatial data and by exploiting the advantages of digital spatial data storage. However, traditional (horizontal and vertical) spatial data models fail to exploit fully the wealth of information available. In a sense, the spatial data user community continues to "put new (digital) wine into old bottles." The GSDM is a new bottle model that preserves the integrity of 3-D spatial data while providing additional benefits, that is, simpler equations, worldwide standardization, and enhanced ability to establish and track spatial data accuracy. Of numerous relevant articles devoted to a better understanding of the characteristics and uses of spatial data, the following are particularly informative: Dobson (2016) and Coumans (2016).

## THE GSDM FACILITATES EXISTING INITIATIVES

Computer databases are digital. Analog maps are still used, but, increasingly, maps and data visualizations are generated upon demand from a digital database. Rarely is a map now used for primary spatial data storage. Spatial data are 3-D and maps are 2-D. Modern measurement systems collect 3-D data, yet some professional practices continue to handle spatial data separately as 2-D horizontal data and 1-D vertical (elevation) data. A better practice is to build and share a 3-D database that supports both 2-D and 3-D applications in either analog or digital mode.

### U.S. NATIONAL ACADEMY OF PUBLIC ADMINISTRATION REPORTS

Two documents prepared by the U.S. National Academy of Public Administration (NAPA) provide a backdrop for considering the advantages of using the GSDM:

1. "The Global Positioning System: Charting the Future" was prepared for the U.S. Congress and the U.S. Department of Defense (DOD) and published in 1995. It describes the history, performance, and future of GPS. This document is particularly important to those who build, operate, and utilize the systems that generate reliable geospatial data. The Executive Summary states, in part, "GPS is much more than a satellite system for positioning and navigation. It represents a stunning technological achievement that is

becoming a global utility with immense benefits for the U.S. Military, civil government, and commercial users and consumers worldwide."

2. "Geographic Information for the 21st Century" was prepared for the Bureau of Land Management, Forest Service, United States Geological Survey, and the National Ocean Service, and published in 1998. It describes the instrumental roles that agencies of the U.S. government have played in "surveying, mapping, and other geographic information functions since the beginning of the Republic." The report includes various excellent recommendations based upon use of a geographic information system (GIS) under the conceptual umbrella of the National Spatial Data Infrastructure (NSDI).

Many professionals are comfortable with both reports. GNSS professionals (system designers and builders) include highly technical specialists such as aerospace engineers, electrical engineers, geodesists, physicists, and photogrammetrists. Alternatively, the professional and technical focus of spatial data users tends toward administration, local government services, information technology (IT), civil engineering, surveying and mapping, planning, and business. In addition to those having a professional interest in generating and/or using spatial data, the number of consumers using GNSS and/or GIS data on a personal level is growing exponentially, a trend reasonably expected to continue.

Interoperability is the key. The GSDM is cloud compatible and builds a conceptual bridge between spatial data disciplines—including the two NAPA reports—by providing a consistent 3-D geometrical framework for both GNSS and GIS. The GSDM serves the scientific end of the spectrum without sacrificing technical rigor while simultaneously providing local spatial data users the opportunity to work with coordinate differences in a flat-Earth environment and to view the world from any location. In addition to examples of the "interoperability bridge" highlighted in later chapters, the GSDM is seen as contributing to success of the following initiatives.

## NATIONAL OCEANIC AND ATMOSPHERIC ADMINISTRATION

In March 2014, the National Oceanic and Atmospheric Administration (NOAA) issued a request for information (RFI) for advice on ways to make the data holdings of the agency more available to business sectors—energy/gas, insurance, finance, and agriculture/aquaculture (NOAA 2014). The RFI states that NOAA currently stores approximately 100 petabytes of environmental data with demand for data storage increasing about 30 petabytes per year. While not stated specifically in the RFI, "location" is an attribute common to many of those data. But, those data are obtained from various sensors under different (sometimes unique) circumstances, and establishing interoperability between data sets is a challenge to be addressed. The GSDM provides a way to "level the playing field" in that raw data can be preprocessed to a common geometrical standard by knowledgeable specialists in various disciplines. With geometrical interoperability established, the vision in the NOAA RFI has a better chance of being realized, and those vast spatial data holdings can be exploited for their commercial value by a multitude of users.

## COALITION OF GEOSPATIAL ORGANIZATIONS

In February 2015, the Coalition of Geospatial Organizations (COGO) published "A Report Card on the National Spatial Data Infrastructure" (COGO 2015). Every 4 years, the American Society of Civil Engineers (ASCE) publishes a report card on America's infrastructure that calls attention to the importance of spending priorities. Although the ASCE report card includes transportation, energy, water, and thirteen other categories, it does not include spatial data as a separate infrastructure category. As a coalition of thirteen different geospatial organizations, it is fortitudinous that COGO prepared and published a comprehensive geospatial data infrastructure report card.

The Executive Summary of the COGO report includes background information on the NSDI as created in 1994 and delegated to the Federal Geographic Data Committee (FGDC). The Executive Summary recognizes the excellent efforts of federal, state, regional, and local government agencies that have contributed to development of the NSDI as it currently exists. But, the COGO report also candidly points out some of the shortcomings of current efforts. Of the points made in the Executive Summary, some are as follows:

> The cornerstone of the program is a common digital base map that would aggregate the best representations of fundamental data from all levels of government. These Framework data layers are intended to serve as the unified foundation upon which all other geographic information could be created and shared. By maintaining a standardized high-quality series of Framework data, the Federal government would provide access to reliable, current data from all of the above partners, not just Federal agencies.

> While Framework data have been collected and made available for use over the past two decades, a digital geospatial Framework, that is national in scope, is not yet in place and may never exist.

The point here is that such a framework—the GSDM—does exist; the GSDM is already in place, and using the GSDM is a matter of choosing to do so. Without being critical of the COGO report, wording in the summary about a digital base map and framework data layers exposes a 2-D conceptual bias. The GSDM is global, 3-D, digital, cloud compatible, and fully supports those derivative "layer" applications.

The importance of spatial data accuracy should also be considered and is highlighted by publication of ASPRS Positional Accuracy Standards for Digital Geospatial Data (ASPRS 2014). Previous spatial data accuracy standards focused primarily on analog data and practices, but the 2014 update is a credible effort to recognize the digital nature of spatial data. Those digital standards are an important step in providing guidance for modern practice and will find beneficial application in many circumstances. However, the ASPRS standards are based largely on assumptions that treat horizontal and vertical components separately as opposed to capitalizing on and exploiting the characteristics of 3-D digital spatial data. These issues and others are addressed in this book by reexamining the underlying spatial data model and by designing spatial data collection, storage, manipulation, adjustment, and visualization procedures based upon assumptions of a single origin for 3-D digital geospatial data. Burkholder (2004) contains a discussion of fundamentals of spatial data accuracy and the GSDM.

## OTHER APPLICATIONS

### Dynamic Environments

The previous discussion refers specifically but not exclusively to static location and spatial data accuracy. The GSDM is also applicable when performing location and accuracy computations in a dynamic environment in which at least one (source or target) if not both (source and target) are moving relative to the Earth or relative to each other. Examples include intelligent transportation (driverless cars, fleet monitoring, collision avoidance, etc.); navigation (aircraft, missiles, ships, and drones—large/small/private/commercial/military); and real-time kinematic (RTK) surveying activities. There are many challenges and much research to be done in this arena—especially related to standardization, efficiency, reliability, and safety. Briefly, some of the issues (see more details in Chapter 14) are as follows:

- Is the separation distance measured directly (irrespective of a coordinate system) or computed from database coordinate positions?
- The quality of information (including datum) in the database of "mapped" features against which an observed dynamic position is to be compared is essential.
- The reference frame (datum) for observed (instantaneously computed) positions and on-board storage of relevant locations must be compatible with "mapped" features.
- In some cases, the use of network accuracy (relative to the datum being used) may be sufficient. In other cases, local accuracy (relative to the position of an adjacent object/sensor) is essential. In most cases, a reliable result (conclusion) involves a combination of network accuracy and local accuracy. Knowledge of "accuracy with respect to what" is critical to safety issues and to risk management.
- If the separation of sensor and target is not measured directly, detect-and-avoid computations and operations rely on datum compatibility between a source and a target (any combination of database and/or sensor) and use either or both network and local accuracies. The GSDM accommodates both network and local accuracies.

See Dobson (2016) for more considerations.

### Static Environments

In addition to supporting initiatives in a static environment, as discussed earlier in this chapter under the section "The GSDM Facilitates Existing Initiatives," Chapter 15 describes specific projects in which the GSDM was used beneficially. The following applications can also benefit from adopting the GSDM:

- The spatial data infrastructure on which "smart cities" are built (Coumans 2016).
- Reliable 3-D underground mapping for thousands of oil/gas wells (Burkholder 2014).
- Documentation for location of offshore energy exploration and production activities.
- Design and construction of civil facilities and infrastructure.
- Parcel-based land information systems and public works management.
- Geographic databases at all levels of practice for surveying, engineering, mapping, and photogrammetry.
- Local surveying and engineering practice.

## INFORMATION PROVIDED BY THE GSDM

Whether static or dynamic, the GSDM can provide a view of any or all other (even cloud-based) points in a data set from any point specified by the user. This gives the user complete flexibility with respect to answers desired. If appropriate covariance data are stored, the standard deviation of any computed quantities can also be determined. Whatever the choice of an origin, the following information (static or dynamic) is instantaneously available:

- 3-D slope distance between points (useful in detect-and-avoid computations).
- Standard deviation of any computed quantities.
- Tangent plane horizontal distance from the origin to the chosen point.
- HD(3) distance between points (Burkholder 2015).
- Ellipsoid distance between points—Chapter 7.
- True geodetic azimuth—also called the 3-D azimuth (Burkholder 1997)— from the selected origin to any other specified point.
- Local plane surveying latitudes/departures from the origin to any selected point. These local differences ($\Delta e/\Delta n/\Delta u$) can be used in plane surveying operations.
- Reliable height differences between points. These ellipsoid height differences are used with known (or modeled) geoid height differences to obtain orthometric height (elevation) differences.

## SUMMARY

Using the GSDM, spatial data computations are performed in 3-D space, and results are stored in terms of ECEF geocentric $X/Y/Z$ coordinates. Covariance data for each point and correlation between points can also be stored in a BURKORD™ database. One set of equations as given in Chapter 1 is applicable worldwide. Although ECEF values are stored for each point, the user selects any desired origin and views the relative location of any or all other points in terms of local plane surveying differences.

Geometrical integrity is preserved in the stored ECEF coordinate values because those values remain unchanged in subsequent derivative computations of geodetic and/or map projection coordinates. The user remains in complete control of the map products generated from the GSDM database. Maps can be plotted directly from local coordinate differences, derivative map projection coordinates (SPCS/UTM), or geodetic values. The 3-D GSDM equations are not as complex as geodesy equations or map projection equations for performing similar computations and can be readily programmed. Given that matrix manipulation tools are available, the error propagation and spatial accuracy computations can be used to provide either or both network and local accuracies as supported by information stored in the covariance matrices.

## REFERENCES

American Society of Photogrammetry & Remote Sensing (ASPRS). 2014. Positional accuracy standards for digital geospatial data. Edition 1, Version 1.0. Bethesda, MD: American Society of Photogrammetry & Remote Sensing. http://www.asprs.org/PAD-Division/ASPRS-POSITIONAL-ACCURACY-STANDARDS-FOR-DIGITAL-GEOSPATIAL-DATA.html (accessed May 4, 2017).

Burkholder, E.F. 1997. The 3-D azimuth of a GPS vector. *Journal of Surveying Engineering* 123 (4): 139–146.

Burkholder, E.F. 2004. *Fundamentals of Spatial Data Accuracy and the Global Spatial Data Model (GSDM)*. Washington, DC: U.S. Copyright Office. http://www.globalcogo.com/fsdagsdm.pdf (accessed May 4, 2017).

Burkholder, E.F. 2014. Underground (well) mapping revisited. *Proceedings of the ASCE Shale Energy Engineering 2014*, Pittsburgh, PA, July 20–23, 2014. http://www.globalcogo.com/underground-mapping.pdf (accessed May 4, 2017).

Burkholder, E.F. 2015. Comparison of geodetic, state plane, and geocentric computational models. *Surveying and Land Information Science* 74 (2): 53–59.

COGO. 2015. Report card on the U.S. National Spatial Data Infrastructure. Washington, DC: Coalition of Geospatial Organizations. http://www.cogo.pro/uploads/COGO-Report_Card_on_NSDI.pdf (accessed May 4, 2017).

Coumans, F. 2016. Spatial data and smart cities are interdependent. *GIM-International* 30 (10): 31–33. https://www.gim-international.com/content/article/spatial-data-and-smart-cities-are-interdependent (accessed May 4, 2017).

Dobson, M. 2016. More on spatial databases for autonomous vehicles. Exploring Local Blog. Laguna Hills, CA: TeleMapics, LLC. http://blog.telemapics.com/?p=628 (accessed May 4, 2017).

National Oceanic and Atmospheric Administration (NOAA). March 18, 2014. NOAA looks for advice to make its data easier to use. *EOS* 95 (11): 95. https://www.fbo.gov/index?s=opportunity&mode=form&id=d9844cb78b4527fb11a6ac6d2b80a742&tab=core&_cview=0 (accessed May 4, 2017).

# 3 Spatial Data and the Science of Measurement

## INTRODUCTION

Many disciplines work with spatial data and many people use a GIS to reference geospatial data. Starting with a concise definition of spatial data, this chapter describes how spatial data and their accuracy are related to the measurement process and one's choice of a measurement system. The goal is to describe how 3-D spatial data can be manipulated more efficiently and how spatial data accuracy can be established without ambiguity using the GSDM as the foundation for GISs and the GSDI.

Modern practice and instruments are used to collect and record spatial measurements. These data are processed electronically and digital results are stored in computer files. Paper maps are inherently two dimensional (they flatten the Earth), and humans traditionally view spatial relationships in terms of "horizontal" and "vertical." Computer graphics and data visualization procedures offer an endless array of display options. Although not exhaustive, this chapter summarizes characteristics of pertinent coordinate systems, defines spatial data, and looks at measurement processes by which spatial data are generated. Today, 3-D digital spatial data are more appropriately stored in a database that combines horizontal and vertical into a single database. And, as discussed later in this chapter, differences between how spatial and geospatial data are used may become significant. The accuracy of spatial data is also considered, and an important distinction is made between primary and derived spatial data.

## SPATIAL DATA DEFINED

Use of the GSDM can foster greater insight into the relationships between coordinate systems and how they are used to handle spatial data. Spatial data are *described* as those numerical values that represent the location, size, and shape of objects found in the physical world. Examples include points, lines, directions, planes, surfaces, and objects. For purposes of this book, spatial data are *defined* as the distance between endpoints of a line in Euclidean space (see the definition of a point in Chapter 4). The endpoints may be nonphysical entities such as an origin or a specific location on the axis of a coordinate system. An endpoint may also represent the location of some physical feature such as a survey monument, building corner, bench mark, or other object. Geometrical elements such as planes, surfaces, and other objects formed by the movement and aggregation of distances also qualify as spatial data. Although straight-line distances are generally presumed, the measure of a distance can also be along a curved line, in either linear or angular units, without violating the definition. As used here, the definition of spatial data also includes, but is not limited to geospatial data.

## COORDINATE SYSTEMS GIVE MEANING TO SPATIAL DATA

When working with spatial data, assumptions are made about the underlying coordinate system. Since each reader deserves to know at all times, "with respect to what," an attempt is made to be very specific about the underlying coordinate system and whether the spatial data are absolute or relative. As a matter of convention, *absolute* spatial data are taken to be with respect to a defined coordinate system while *relative* spatial data are taken to be the difference between two absolute values in the same system. A coordinate is an absolute distance with respect to the defined coordinate system. An azimuth is an absolute direction with respect to the zero reference. Spatial data components arc coordinate differences (in the same system) and used as relative values. An angle, defined as the difference between two directions, is also a relative value. Absolute data are often used to store spatial information while relative data are more often associated with measurements.

Admitting the use of undefined terms, relying upon prior knowledge, and acknowledging a difference between a reference system and a reference frame, the information presented in this chapter is intended to be consistent with current definitions of coordinate systems, such as those described by Hothem and Soler (1988). Three coordinate systems are an integral part of the GSDM:

1. *ECEF*: A foundation 3-D geocentric coordinate system for spatial data is called the Earth-centered Earth-fixed (ECEF) rectangular Cartesian coordinate system and defined by the National Geospatial-Intelligence Agency (NGA 2014). See Figure 3.1. With its origin at the center of mass of the

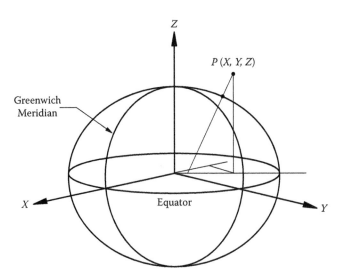

**FIGURE 3.1** Geocentric ECEF Coordinate System

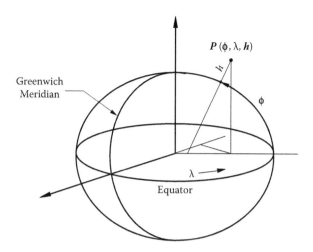

$P(\phi, \lambda, h)$

Greenwich
Meridian

$h$

$\phi$

$\lambda$

Equator

**FIGURE 3.2**   Geodetic Coordinate System

Earth, the *X/Y* plane is coincident with the equator and the *Z*-axis is defined by the location of the Conventional Terrestrial Pole (CTP). The *X*-axis is defined by the arbitrarily fixed location of the Greenwich Meridian, and the *Y*-axis is at longitude 90° east, giving a right-handed coordinate system.

2. *Geodetic*: A geodetic coordinate system, Figure 3.2, is used to reference spatial data by geodetic positions on the ellipsoid, a mathematical approximation of the surface of the Earth. Position is defined in the north-south direction by angular units (degrees, minutes, and seconds) of latitude and in the east-west direction by angular units of longitude. Lines of equal latitude are called parallels and lines of equal longitude are called meridians. The sign convention for latitude is positive north of the equator and negative south of the equator. The sign convention for longitude is positive eastward for a full circle from 0° on the Greenwich Meridian to 360° (arriving again on the Greenwich Meridian). A west longitude, as commonly used in the western hemisphere (Howse 1980, p. 144), is mathematically compatible if used as a negative value.

   Geodetic latitude and longitude are 2-D curvilinear coordinates given in angular units. The third dimension, ellipsoid height, in this worldwide coordinate system is the distance above or below the mathematical ellipsoid and is measured in length units, meters being the international standard. With the conceptual separation of horizontal and vertical, this system of geodetic coordinates more closely matches physical reality in a global sense than does the ECEF system and remains very useful for cartographic visualizations. But, the geodetic coordinate system is computationally more complex and more cumbersome to use than rectangular components when working with 3-D spatial data.

3. *Local (flat-Earth)*: Local coordinate systems (Figure 3.3) portray the location of spatial data with respect to some user specified reference and/or origin.

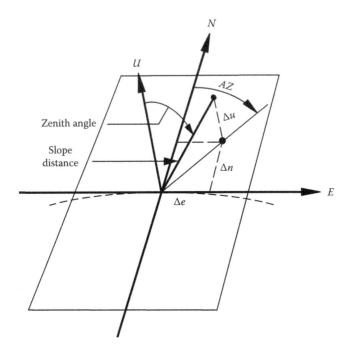

**FIGURE 3.3**   Local Coordinate System

A local coordinate system can be defined in such a way that horizontal and vertical relationships are both accurately portrayed and 3-D relationships are preserved. However, many local coordinate systems enjoy true 3-D geometrical integrity only to the extent that a flat Earth can be assumed. If spatial data issues are addressed strictly on a local basis, the error caused by such flat-Earth assumptions can be negligible. However, as one works over larger areas, needs greater precision in small areas, or needs to establish compatibility between local coordinate systems, the flat-Earth model is not adequate for referencing spatial data. But, when used as a component of the GSDM, the local flat-Earth model can support visualization and use of 3-D data without being adversely affected by the underlying curved-Earth distortions. That means local rectangular (flat-Earth) relationships can be utilized in a global environment without compromising the geometrical integrity of spatial data.

## Spatial Data Types

Given descriptions of the geocentric ECEF coordinate system, the geodetic coordinate system, and a local coordinate system, the following spatial data types are listed:

1. Absolute geocentric $X/Y/Z$ coordinates are perpendicular distances in meter units from the respective axes of an ECEF reference system.

2. Absolute geodetic coordinates of latitude/longitude/height are derived and computed from ECEF coordinates with respect to some named model (geodetic datum).
3. Relative geocentric coordinate differences, $\Delta X/\Delta Y/\Delta Z$, are obtained by differencing compatible geocentric $X/Y/Z$ coordinate values.
4. Relative geodetic coordinate differences, $\Delta\phi/\Delta\lambda/\Delta h$, are obtained as the difference of compatible (common datum) geodetic coordinates.
5. Relative local coordinate differences, $\Delta e/\Delta n/\Delta u$, are local components of a space vector defined by relative geocentric coordinate differences.
6. Absolute local coordinates, $e/n/u$, are distances from some origin whose definition may be mathematically sufficient in 3-D, 2-D, or 1-D. Examples are as follows:
   a. Point-of-Beginning (P.O.B.) datum coordinates as defined in Chapter 1. These derived coordinates enjoy full mathematical definition in 3-D, suffer no loss of geometrical integrity in the GSDM, and serve the local needs of many spatial data users.
   b. Map projection (state plane) coordinates, which are well defined in 2-D with respect to some named origin and geodetic datum.
   c. Elevations, which are 1-D distances above or below some named reference equipotential surface. In the past mean sea level was assumed to be acceptable as a vertical reference but, due to the difficulty of finding mean sea level precisely, modern vertical datums are referenced to an arbitrary equipotential reference surface (Zilkoski et al. 1992).
7. Arbitrary local coordinates may be 1-D (assumed elevations), 2-D (assumed plane coordinates), or 3-D (spatial objects, rectangular coordinates, or assumed elevations and plane coordinates). Although useful in some applications, arbitrary local coordinates are generally not compatible with other local systems and have limited value in the broader context of georeferencing. Many computer graphics and data visualization programs use arbitrary local coordinates.

The GSDM efficiently handles the spatial data types that fall into categories 1, 3, and 5 (absolute geocentric coordinates, relative geocentric coordinate differences, and relative local coordinate differences). Spatial information is stored most efficiently using digital geocentric coordinates, manipulated most readily using geocentric coordinate differences, and displayed for human visualization and analysis using relative local coordinate differences. Spatial data consisting of geodetic coordinates and geodetic coordinate differences (categories 2 and 4) are useful for cartographic portrayal and, to the extent they can be competently related to category 1, generally are not a problem. Category 6 spatial data (local coordinate differences) can be incorporated into the GSDM if and only if they enjoy full 3-D mathematical definition. Without additional survey measurements, attempts to incorporate category 7 data into the GSDM are not viewed as fruitful. This is where the difference between spatial and geospatial data definitions may become significant.

## SPATIAL DATA VISUALIZATION IS WELL DEFINED

Spatial data are used extensively in computer graphics, visualization programs, computer-aided design and drafting, and in the manipulation of spatial objects. The GSDM provides for the connection of spatial data to the physical Earth, but otherwise makes no attempt to impose conditions on the use of spatial data. It is anticipated that the scope, utility, and value of many spatial data manipulation, visualization, and 3-D coordinate geometry (COGO) programs may be enhanced by taking advantage of the physical Earth connection as defined by the GSDM.

## DIRECT AND INDIRECT MEASUREMENTS CONTAIN UNCERTAINTY

Spatial data are created by measurement and no measurement is perfect. In a simple case, a distance is determined by direct comparison of some unknown length with a standard such as a ruler, steel tape, or wavelength. Whether the distance is horizontal or vertical is a condition noted by the person recording the observation. More often, however, spatial data are obtained as the result of an indirect measurement in which one or more spatial data components are computed from the observations as is the case when a slope distance is resolved into its horizontal and vertical components. In other cases, some physical quantity is observed and a distance is computed using known mathematical relationships (a model). An example is computing a distance from a voltage that represents the phase shift of a sine wave signal in an electronic distance measuring instrument (EDMI). Restating, spatial data measurements may be the result of a direct comparison or, more often, they are computed indirectly from observations of various fundamental physical quantities—for example, photogrammetry, remote sensing, GNSS, and/or LiDAR.

### FUNDAMENTAL PHYSICAL CONSTANTS ARE HELD EXACT

Fundamental physical quantities as expressed in the International System (SI) are as follows:

| | |
|---|---|
| Length—meter | Temperature—kelvin |
| Time—second | Luminous intensity—candela |
| Mass—kilogram | Amount of substance—mole |
| Current—ampere | |

Derived physical quantities include the following (there are others—Nelson (1999)):

| | |
|---|---|
| Frequency—hertz | Electric charge—coulomb |
| Force—newton | Electric potential—volt |
| Pressure—pascal | Plane angle—radian |
| Energy—joule | Solid angle—steradian |
| Power—watt | |

## MEASUREMENTS/OBSERVATIONS CONTAIN ERRORS

Spatial data are created by measurement of some combination of physical quantities, and those measurements are used in models that relate the observed quantity to a physical distance (spatial data) relative to one of the three coordinate systems listed earlier. The accuracy of such spatial data is dependent upon (1) the quality and sufficiency of the measurements, (2) the appropriateness of the models used to compute the spatial data components, and (3) error propagation computations. The GSDM accommodates all three considerations.

# MEASUREMENTS USED TO CREATE SPATIAL DATA

## TAPING

A calibrated tape is laid flat on a horizontal surface at some specified tension and temperature. The measurement involves a visual comparison of the unknown length with uniform markings on the tape (a fundamental physical quantity). The observation is recorded as a measurement. If the temperature (another physical quantity) is different than the specified calibration temperature or if the tension (a derived physical quantity) is not what it should be, these other measurement conditions must also be noted. Using this additional information and appropriate equations, corrections to the taped distance are computed and applied to this otherwise direct measurement. Whether the computed distance is a direct or an indirect measurement is left to the reader.

## LEVELING

A level rod with graduations marked on it is held erect in the field of view of an observer looking through the telescope of an automatic (or tilting) level, and the distance from the bottom of the rod to the cross-hair intercept is read and recorded. Separate readings are made with the rod resting on other objects. In this case, the difference of two direct readings provides an indirect determination of the relative height of the two objects. Among others, the accuracy of such an indirect measurement is affected by (1) whether the line-of-sight is perpendicular to the plumb line, (2) the presence of parallax, (3) whether or not a vernier was used to refine the reading, (4) plumbness of the level rod when the readings were made, (5) and by the distance from the instrument to the rod (curvature and refraction correction). Modern practice includes use of bar-scale reading instruments in which automated observations are recorded digitally.

# EDMI

An EDMI emits electromagnetic radiation that is modulated with a known frequency (giving a known wavelength). The signal is returned by a retro-reflector from the forepoint end of a line, and the phase of the returned waveform is electronically compared to that of the transmitted signal. The measurement of phase differences on several modulated frequencies provides information used to compute the distance between the EDMI and a reflector. Other quantities such as temperature and barometric pressure are also measured to determine corrections that account for the signal traveling through the atmosphere between the standpoint and forepoint at some speed slower than it would have traveled through a vacuum. The point is that several physical quantities are measured and that the physical environment is modeled with equations before a collection of observations can be converted into spatial data.

With later generation pulse laser technology, the physical distance between the EDMI and target object is determined using the time interval required for a pulse to travel from the EDMI to the object and back. In some cases, the return signal comes back from a target having sufficient reflectance—known as a reflectorless measurement. Of course, atmospheric delay must be modeled and direction to the target must be known before a slope distance can be resolved into rectangular components. See a later section in this chapter on LiDAR measurements.

## ANGLES

Although not a fundamental physical quantity, angles and directions are commonly measured and used in computing spatial data components. Two examples are (1) using a vertical angle to resolve a slope distance into horizontal and vertical rectangular components and (2) using the bearing of a line to find the latitudes and departures of a traverse course. Looking beyond the obvious, where an angle is measured directly with a protractor on paper or on the ground using a total station surveying instrument, angles are also measured indirectly as the difference of two directions such as might be observed with a compass, a gyroscope, or with GNSS. Whether an angle was measured in the horizontal, vertical, or some other plane is also an important consideration, especially when using angles to resolve the hypotenuse of a triangle into its rectangular components. One example is resolving slope distances into horizontal and vertical components. Another is resolving traverse courses into latitudes and departures.

## GPS AND GNSS

The global positioning system (GPS) was developed by the U.S. government as part of the space program and is used worldwide for a variety of positioning and timing applications. Other nations and organizations have also developed similar positioning systems that both compete with and that supplement GPS. Therefore, the term global navigation satellite system (GNSS) is recognized as an appropriate generic descriptor. All GNSS satellites orbiting the Earth broadcast signals that are received by sensors enjoying electronic line-of-sight with the satellite. Conventionally, the user sensing unit is passive in that it does not send signals back to the orbiting GNSS satellites. However, depending upon design and manufacturer, a GNSS unit might, in fact, transmit signals to other sensors/devices. Operators of the various GNSSs routinely upload information to their orbiting satellites for rebroadcast to the Earth.

Physical concepts used by GNSSs used to compute spatial data components and location include (1) distance as the product of velocity (of signal) and elapsed time (of transmission), (2) the Doppler shift of a frequency recorded by a sensor as compared to the frequency transmitted by the satellite, and (3) interferometric inter-action of a signal as recorded simultaneously at two different sensors. Sophisticated electronic signal processing techniques have been developed to determine spatial and geospatial data components from the GNSS signals. In the early years of GNSS two distinct modes of positioning were popular—*absolute* positioning at one point using code phase instruments receiving signals from multiple satellites and *relative* positioning based upon carrier phase interferometric interaction of the same signal as received simultaneously at two separate locations. In each case, a minimum of four visible satellites was required to obtain a position fix although better solutions could be obtained if using signals from more satellites. In practice, an approximate *absolute* location (5–10 meters worldwide) obtained from a code phase instrument served the needs of many applications. However, the *relative* precision of one point with respect to another (ultimately within millimeters over distances of 20 km or more) obtainable from carrier phase instruments made GNSS technology especially valuable in applications such as surveying, mapping, precision agriculture, and oth-ers. The disadvantages being that carrier phase equipment is more expensive, at least two receivers are required to obtain data for one baseline, and to obtain the best results from a static survey, each data set needed to be an hour long or more. Those disadvantages are mitigated by more recent technological advances. Operating in an area covered by a virtual reference system (VRS), it is now possible for a user to take one instrument to the field and to obtain survey grade results in a matter of minutes.

Generally, code phase instruments provided approximate absolute positions rather quickly while carrier phase instruments required longer observing times to obtain precise relative positions. However, over time, more sophisticated algorithms, bet-ter equipment, and operational procedures were developed that blurred the distinc-tion between results from code phase instruments and carrier phase instruments. For example, although now phased out, the U.S. Coast Guard operated a differential code phase system (DGPS) to support navigation nationwide. Possible improvement in accuracy went from about 10 meters to less than 0.5 meter. Likewise real-time kine-matic (RTK) procedures were developed for carrier phase instruments that enable a user to set up a master base unit and, using a second receiver, "rove" from point to point to achieve centimeter-level results in much less time than required for a static baseline survey. Current state-of-the-art GNSS positioning includes the VRS systems as mentioned previously.

GNSS technology has invaded the mass consumer market to the point GNSS chips are embedded in automobiles, cell phones, tracking devices, and others. Conflicting criteria for GNSS manufacturers include time-to-fix, accuracy of position, and cost to consumer. Various combinations and many options are available for many applica-tions. In addition to the mass consumer market, research and development within the GNSS industry for high-end applications currently includes precise point positioning (PPP) in which an affordable instrument can be used anywhere worldwide to deter-mine a single position within centimeters within minutes.

As will be discussed in later chapters, the GSDM can serve the needs of GNSS users worldwide in all operational modes.

## REMOTE SENSING

ASPRS (1984) describes remote sensing as the process of gathering information about an object without touching or disturbing it. Photogrammetry is an example of remote sensing, and ray tracing based upon stereo photographs of a common image is very geometrical. Bethel (1995) also discusses remote sensing and describes interpretative (less metrical) applications of remote sensing that include use of invisible portions of the electromagnetic spectrum to record the response of an object or organism to stimuli from a distant source. Sensors include infrared film, digital cameras, radar, satellite imagery, etc., and information is stored pixel by pixel in a raster format. Determining the unique spatial location represented by each pixel can be a daunting challenge and typically requires enormous digital storage capacity.

## PHOTOGRAMMETRY

Relative spatial data, both local and geocentric, can be determined efficiently and precisely using geometrical relationships reconstructed from stereoscopic photographs of a common image. A photogrammetric measurement is the relative location of an identifiable feature on a photographic plate with respect to fiducial marks on the same plate as determined with a comparator. A more complex measurement of 3-D spatial relationships based upon principles of photogrammetry requires mechanical reconstruction of the stereoscopic image by achieving the proper relative and absolute orientation of the stereo photographs in a mechanical stereo plotter. A 3-D contour map of the ground surface is the end result of the plotting operation. That traditional photogrammetric mapping process has been computerized and automated (bypassing mechanical reconstruction) and now comes under the banner of softcopy photogrammetry. The end result of the modern computerized process is a 3-D digital model of the terrain. Hardcopy maps, computer displays, and other products, both digital and analog, are made from a common digital spatial data file.

## LiDAR

In addition to photogrammetry, Light Detection and Ranging (LiDAR) also comes under the category of remote sensing in that LiDAR gathers information about an object without touching or disturbing it. Photogrammetry relies upon ambient light to illuminate the target and the photogrammetric image captures the relative direction of light rays from the camera to the object. Stereo images are used to generate a 3-D spatial model. LiDAR is different in that laser ranging is used to determine the distance (very quickly—at the speed of light) from the sensor to the object. When combined with directions determined by the same sensing unit, the 3-D position of each point becomes part of a scanned "point cloud."

## LOGISTICS

A wide variety of sensors and platforms can be used to support both static and kinematic spatial data collection. For example, these include fixed and rotary wing aircraft, tripods, mobile vehicles, robots, balloons, and UAVs. Areas with difficult access can often be scanned with handheld scanning devices. In all cases, the laser (which requires line-of-sight to the object) is used to determine the distance or range to the object. If the location and orientation of the sensor are known, then the 3-D spatial coordinates can be determined for each measurement.

The result of scanning the field of view is a cloud of 3-D points—commonly referred to as a point cloud. These point clouds can be georeferenced to the desired coordinate system by including targets at known locations that can be identified in the scan data. In some cases, the GNSS position of the scanner becomes part of the georeferencing process. Capable of operating indoors or out, scanners can be versatile 3-D measuring devices.

Other measurement methods are also used to create spatial data. But, regardless of the technology used to measure fundamental physical quantities, the GSDM provides a common universal foundation for expressing fundamental spatial relationships. Various equations (models) are used to convert observations into spatial data components that are then used as measurements in subsequent operations such as traverse computations, network adjustments, plotting maps, etc. The GSDM also accommodates fundamental error propagation in all cases by storing that information in the covariance matrix for each point. From there each user can make informed decisions about whether or not the spatial data accuracy is sufficient to support a given application.

## Errorless Spatial Data

Several cases exist in which spatial data are considered errorless. Examples include (1) spatial data created during the design process, (2) physical dimensions (such as the width of a street right-of-way) defined by ordinance or statute, and (3) spatial data whose standard deviations are small enough to be judged insignificant for a given application. In the case of a proposed development such as a highway, bridge, skyscraper, or other civil works project, the planned location of a feature and the numbers representing the size and shape of each feature qualify as spatial data. But, they are the result of a design decision instead of a measurement process. Such design dimensions are without error until they are laid out. After being laid out and constructed, the location of the feature or object is determined by measurement and typically recorded on as-built drawings or in project files. Considered this way, the perfection of a design dimension is transitory and ceases to exist when laid out during construction.

An exception to the transitory nature of an errorless design dimension exists when a dimension is established by ordinance or statute. Such a dimension may be fixed by law, but the physical realization of that dimension is still subject to the procedures and quality of measurements used to create it. Under ideal conditions, there will be no conflict between a statutory dimension and the subsequent as-built measurement to document its location. For example, a 100-foot right-of-way may have been monumented very carefully and current measurements between the monuments are all 100.00 feet, plus or minus 0.005 foot. In that case, the right-of-way width can be shown as 100.00 feet, measured and recorded. Under less-than-ideal conditions, several possible dilemmas are as follows:

1. The right-of-way monuments really are separated by 100.00 feet at ground level, but the survey is based upon a state plane coordinate grid and the calculated grid distance is 99.97 feet, plus or minus 0.005 foot (possible at elevations over 4,200 feet). Understandably, the monuments are not to be moved so they are separated by a grid distance of 100.00 feet, but some users may not be willing to accept the implication that a foot is not really a foot. The apparent discrepancy arises from the use of two different definitions for horizontal distance, local tangent plane distance or grid distance (Burkholder 1991).

2. The right-of-way monuments appear stable and undisturbed but the measurement between them is consistently 99.96 feet, plus or minus 0.005 foot (it could happen if the monument locations were staked during cold weather and no temperature corrections were applied to the steel tape measurements at the time). The conflicting principles are that the statutory dimension (of 100.00 feet) must be honored and that the original undisturbed monument controls, even if originally located with faulty measurements.

The intent here is not to solve those problems, but to acknowledge the potential for conflicting principles when working with so-called errorless spatial data. In addition to references such as ASPRS, ACSM, and ASCE (1994), other authors have written entire textbooks devoted to survey law, evaluation of evidence, and analysis of survey measurements. The point here is that the GSDM offers a consistent standard environment in which to make comparisons between conflicting data. The GSDM does not distort horizontal distances as does the use of map projections and/or state plane coordinates.

When the coordinates of any point are held "fixed," the result is the same as assuming the standard deviations associated with the coordinates are very small or are zero. In many cases, such an assumption is reasonable and defensible because the standard deviations of a point are small enough to be insignificant and the implied statement "with respect to existing control" is acceptable. However, each spatial data user making decisions about which control points are held "fixed" should document such decisions specifically so that subsequent users may always know, "with respect to what." With the accuracy of spatial data becoming ever more important, criteria for judging the quality of spatial data should be unambiguous and easy to understand. The stochastic model portion of the GSDM uses 3-D standard deviations to describe the accuracy of spatial data and accommodates errorless spatial data as those data having zero standard deviations (Burkholder 1999). An answer to the question "accuracy with respect to what?" is determined by (1) the control points used as primary data, (2) the 3-D standard deviations of those control points, (3) the quality of measurements used to establish new positions, (4) competent determination of the covariance matrix of each new point, and (5) the manner in which Equation 1.36 is used in subsequent computations.

## SOURCES OF PRIMARY SPATIAL DATA

Earlier, spatial data were defined as distances. Spatial data types were also listed as distances represented by coordinates or coordinate differences in one of several coordinate systems. And, unless attempting to convert from one datum to another (see Chapter 8), it should be understood that equations for converting spatial data from one coordinate system to another should have little or no mathematical uncertainty associated with them. Any uncertainty should be the result of an imperfect measurement, not a defective model, equation, or algorithm. With that said, primary spatial data are defined as geocentric $X/Y/Z$ coordinates, their associated covariance matrices, and point-pair correlation matrices. Primary spatial data are created by a specific measurement process or determined on the basis of some prescribed geometry. Measurements have standard deviations and covariances associated with them while errorless quantities have zero standard deviations. The GSDM accommodates both measurements and errorless quantities by using standard deviations of all three components at each point, covariances between components, and correlations between points.

### OBSERVATIONS AND MEASUREMENTS

Mikhail (1976) describes how measurement and observation are very similar and, in fact, used interchangeably. A mathematical distinction made here is that observations are independent while measurements may be correlated. Stated differently, an observation (whether it is the process or the numerical outcome) is taken to be the actual comparison of some quantity with a standard while a measurement is taken to be either the same as an observation or as a subsequently computed quantity after corrections are applied as dictated by observation conditions. For example, a horizontal distance is said to be measured by an EDMI. Actually, an EDMI uses (1) the observed phase difference of two electromagnetic signals on several frequencies (the transmitted/received signal and an internal reference signal), (2) the estimation (observation) of air temperature and barometric pressure for the atmospheric correction, and (3) measurement of the vertical (or zenith) angle. These observations are used to compute the horizontal distance, which is called a measurement, when, in fact, physical quantities in addition to length were observed. Also, note that the same observations are used to compute the vertical component of the slope distance. If one of the observations is changed, it may affect both computed values. Hence, the horizontal and vertical measurements are correlated and not independent. Slope distance and zenith directions are the independent observations.

Having made a distinction between observations and measurements, several other points also need to be addressed:

- In the strictest sense, primary spatial data should include only errorless quantities and independent observations. However, given the multitude of sensors used to make observations and the number of steps often needed to convert observations into spatial data components (measurements), it would be onerous indeed for each spatial data user to assume responsibility for the integrity of his/her data all the way back to the observation. It is hereby suggested the GSDM will conveniently serve two distinct groups: those responsible for generating quality spatial data and those who use spatial data. The work of scientists, physicists, electrical engineers, and programmers is completed upon delivering a measurement system that can be used to generate quality spatial data. If cartographers, geographers, planners, and other spatial data users know they can rely on the quality of data provided, they need not be so concerned with the science of measurement, computations, and adjustments but are free to focus their energies on spatial analysis and other chosen applications. The geomatician (geodesist, surveying engineer, photogrammetrist, etc.) provides a valuable service to society by interacting with and serving both groups.

- All primary spatial data have covariance matrices associated with them. In the case of errorless quantities, the covariance matrix is filled with zeros. Otherwise, the covariance matrices are obtained by formal error propagation from basic observations through a competent network adjustment.
- Computation of measurements often results in correlation between computed spatial data components. That correlation is defined and determined by the error propagation computation procedure applied to independent observations and the mathematical equations used to obtain the spatial data components. For that reason, it is necessary to store the full (3 × 3 symmetric, 6 unique values) covariance matrix along with the X/Y/Z coordinates of each point.

### ERRORLESS QUANTITIES

As used here, errorless quantities and unadjusted measurements are the basis of primary spatial data. But, in reality, primary spatial data are the X/Y/Z coordinates and associated covariance values stored following rigorous network adjustments and successful application of appropriate quality control measures.

A statement of the obvious is that primary spatial data having small standard deviations are more valuable than primary spatial data with large standard deviations. Whether a standard deviation is large or small is dependent upon the measurements made and the correct propagation of the measurement errors to the spatial data components. The GSDM handles 3-D spatial data the same way, component by component, regardless of the magnitude of the standard deviations, and each user has the option of deciding what level of uncertainty is acceptable for a given application. Additionally, the GSDM is strictly three dimensional and makes no mathematical distinction between horizontal and vertical data. But, the GSDM readily provides local $\Delta e/\Delta n/\Delta u$ components that can be used locally as flat-Earth (local tangent plane) distances.

## DERIVED SPATIAL DATA ARE COMPUTED
## FROM PRIMARY SPATIAL DATA

Spatial data that owe their existence to mathematical manipulation of existing primary spatial data are considered to be derived spatial data. Derived spatial data include geodetic coordinates, UTM coordinates, state plane coordinates, project datum coordinates, and coordinates in other mathematically defined systems. Derived spatial data also includes inversed bearings and distances (as shown on survey plats and subdivision maps), areas, volumes, and elevations. In each case, the accuracy of each derived quantity is computed from the standard deviations (and covariances) of the underlying primary spatial data on which they are based.

A clear distinction between primary spatial data and derived spatial data is critical to efficient collection, storage, management, and use of spatial data. Primary spatial data require measurement of physical quantities and computation of spatial data components according to very specific procedures. For example, taping corrections are needed to determine a precise horizontal distance measured with a steel tape, and an EDMI measurement needs to be corrected for reflector offset, atmospheric conditions, and slope if the endpoints are at different elevations. The cost of acquiring primary spatial data is still prohibitive in some cases but, due to automation, computerization, and other technical developments (such as GNSS), spatial data are less expensive to obtain now than in the past. Even so, by comparison, derived spatial data are generally less expensive than primary spatial data. Derived data can be computed, used, and discarded without detrimental economic consequence. Other than the effort required to assemble the needed primary spatial data and to make the computations, derived spatial data can be generated as often as needed based upon a prescribed algorithm. The challenge is to archive primary spatial data efficiently and to make sure other users know specifically what algorithms were used in generating the derived quantities. The practice of storing derived spatial data is potentially wasteful.

## ESTABLISHING AND PRESERVING THE VALUE OF SPATIAL DATA

Establishing the value and integrity of spatial data is not a trivial undertaking. An oversimplified statement is that the right measurement needs to be made with the correct equipment under well-documented conditions and appropriate equations must be used to compute primary spatial data components. Once the components are run through an appropriate least squares network adjustment, they are attached to the chosen 3-D datum and the resulting geocentric $X/Y/Z$ coordinates (and covariances) become the primary spatial data. A thorough treatise on the science of measurement and subsequent computation of spatial data components would fill an entire book. The goal of this chapter is to establish a connection between measurements and spatial data components with the idea of showing how both can be handled more efficiently in the context of the GSDM.

In addition to the points made here about measurements and computations, another question also needs to be asked: "What makes spatial data lose their value?" Often attention is focused on doing whatever is necessary to get the most data for the least cost, but avoiding the cost of replacement or the inconvenience of not having the needed data also needs to be considered. Therefore, efforts made to preserve the value of spatial data may be efforts well spent. Although less directly, the GSDM also facilitates those efforts by providing a simple standard model that can be used worldwide by all spatial data disciplines and users. The key concepts are data compatibility and interoperability.

Specifically, spatial data lose their value if

- Potential users do not know they exist or that they are available
- They are incomplete, incompatible, or of dubious quality
- They are in the wrong format or stored in the wrong location
- A user does not know what to do with them
- They are replaced by data having smaller standard deviations

## SUMMARY

This chapter

- Defines spatial data as the distance between endpoints of a line
- Describes three coordinate systems used to reference spatial data
- Lists spatial data types as coordinates and coordinate differences
- Acknowledges that spatial data visualization is a solved problem
- Admits that spatial data created by measurements contain uncertainties
- Gives examples of nontrivial spatial data measurements
- Describes the role of errorless spatial data
- Defines primary spatial data and derived spatial data
- Suggests ways to preserve the value of spatial data

The remainder of this book

- Reviews mathematical concepts needed for working with spatial data
- Describes existing geometrical models used for spatial data computations
- Develops geometrical relationships in the context of
  - Geometrical geodesy
  - Physical geodesy
  - Satellite geodesy and GNSS surveying
  - Geodetic datums
  - Map projections and state plane coordinate systems
  - How spatial data are used
- Shows how a linear least squares adjustment can be used in the context of the GSDM
- Includes a number of example projects that show how the GSDM accommodates 3-D coordinate geometry and error propagation

## REFERENCES

American Society of Photogrammetry & Remote Sensing (ASPRS). 1984. *Multilingual Dictionary of Remote Sensing and Photogrammetry*. Bethesda, MD: American Society Photogrammetry and Remote Sensing.

American Society of Photogrammetry & Remote Sensing (ASPRS), American Congress on Surveying & Mapping (ACSM), American Society of Civil Engineers (ASCE). 1994. *Glossary of Mapping Sciences*. Bethesda, MD: American Society of Photogrammetry and Remote Sensing.

Bethel, J.S. 1995. Chapter 53: Surveying engineering. In *Civil Engineering Handbook*. Boca Raton, FL: CRC Press.

Burkholder, E.F. 1991. Computation of level/horizontal distance. *Journal of Surveying Engineering* 117 (3): 104–116.

Burkholder, E.F. 1999. Spatial data accuracy as defined by the GSDM. *Journal of Surveying and Land Information Systems* 59 (1): 26–30.

Howse, D. 1980. *Greenwich Time and the Discovery of Longitude*. Oxford, U.K.: Oxford University Press.

Mikhail, E.M. 1976. *Observations and Least Squares*. New York: Harper & Row.

National Geo-spatial Intelligence Agency (NGA). 2014. Department of defense world geodetic system 1984: Its definition and relationships with local geodetic systems, Version 1.0.0. Arnold, MO: National Geo-spatial Intelligence Agency.

Nelson, R.A. August 1999. Guide for metric practice. *Physics Today* 52 BG11–BG12.

Soler, T. and Hothem, L. 1988. Coordinate systems used in geodesy: Basic definitions and concepts. *Journal of Surveying Engineering* 114 (2): 84–97.

Zilkoski, D., Richards, J., and Young, G. 1992. Results of the general adjustment of the North American Datum of 1988. *Surveying and Land Information Systems* 52 (3): 133–149.

# 4 Summary of Mathematical Concepts

## INTRODUCTION

The term "mathematics" is difficult to define, in part, because it includes so many concepts. Even so, the primary goal of this book is to organize mathematical concepts and geometrical relationships for the convenience of spatial data users. The approach is to start with simple, well-defined ideas and add understandable pieces as needed to develop tools for handling spatial data more efficiently. Mathematics has been concisely defined as the study of quantities and relationships through the use of numbers and symbols. The terms "quantities" and "relationships" may be somewhat abstract but their meaning should become clearer with use. "Numbers" includes the set of all real values from negative infinity to positive infinity and "symbols" includes letters of the alphabet (English, Greek, or otherwise) used to represent numerical values. Symbols also include other markings to indicate mathematical operations such as addition, subtraction, multiplication, division, square root, and others. As illustrated by the definition and the two following examples, the goal of presenting simple well-defined concepts and keeping a focus on practical applications for spatial data will not be easy to achieve. Understandably, some readers will be distracted by temptations to pursue peripheral interests. Although that is acceptable and encouraged, space and print limitations do not permit joint excursions. The reader is always welcome back.

- With respect to numbers, few people can comprehend the vastness of infinity, yet reference is made to negative infinity and positive infinity with the implication that they might somehow be the same size (but in opposite directions on the real number line). The point is not whether that implication is true, but to note instead that there are just as many numbers between zero and one (0 and 1) as there are numbers greater than one. That statement is proved by taking the reciprocal of any number greater than 1.
- Symbols for mathematical operations such as $+$, $-$, $\times$, $\div$, and $\sqrt{}$ are simple and used the world over. Symbols also include letters that are used to represent certain numerical values. Perhaps the most common mathematical symbol is the Greek letter pi ($\pi$) used to represent the ratio of the circumference of a circle divided by its diameter. The definition is simple and that ratio finds many applications when working with spatial data. However, mathematicians (in what could be called esoteric pursuits) have spent years chasing an increasing number of digits for pi (Beckman 1971). Access to millions of digits for $\pi$ is now as simple as typing "pi" into a World Wide Web search engine.

## CONVENTIONS

If you went looking for $\pi$, welcome back. In an effort to be specific about concepts and to communicate clearly using unambiguous symbols, the following conventions are identified and used throughout this book as consistently as possible even though they may differ from one discipline to the next or from one culture to another.

### NUMBERS

Integers are whole numbers, used to count things, and are a subset of real numbers. A line such as that shown in Figure 4.1 can be used to represent all numbers—both integer and real. If some point on the line is assigned the value "zero," points to the left of zero are negative numbers, and points to the right of zero are positive numbers. Numbers are composed of the Arabic digits 0, 1, 2, 3, 4, 5, 6, 7, 8, and 9 and the Hindu-Arabic number system used in modern mathematics is decimally constructed by decades where the column of a digit implies its value times 1, 10, 100, 1,000, etc.

### FRACTIONS

A ratio of one integer divided by any other (except 0) is known as a fraction. Many examples exist but, successive division by 2 is an intuitive example that serves very well when cutting a pie or a pizza. Fractions of 1/2, 1/4, 1/8, etc., are familiar to everyone and are used, for example, when driving a car and judging the amount of fuel remaining in the tank. Successive division by 2 is also appropriate in other cases but, carried too far, it becomes cumbersome and using decimal equivalents is easier. In the case of the U.S. Public Land Survey System (USPLSS), it is no coincidence that 1 square mile nominally contains 640 acres. Successive division of area by quarters (two divisions of length by two) is convenient down to a parcel of 10 acres. Although more could be said about repetitive division, it is noted that some disciplines in the United States still use fractions (e.g., architects, carpenters, millwrights, and ironworkers).

### DECIMAL

Another prevalent practice for counting objects utilizes decades of 10, presumably based upon prehistoric man having 10 fingers. With the invention of "zero" by mathematicians in India about AD 600, the decimal system was developed in its

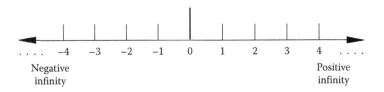

**FIGURE 4.1**    The Real Number Line

**TABLE 4.1**
**Prefixes for Numbers**

| 1,000,000,000,000,000. | $10^{15}$ | peta- | One quadrillion |
|---|---|---|---|
| 1,000,000,000,000. | $10^{12}$ | tera- | One trillion |
| 1,000,000,000. | $10^9$ | giga- | One billion |
| 1,000,000. | $10^6$ | mega- | One million |
| 1,000. | $10^3$ | kilo- | One thousand |
| 1. | $10^0$ | unit | One |
| 0.001 | $10^{-3}$ | milli- | One thousandth |
| 0.000 001 | $10^{-6}$ | micro- | One millionth |
| 0.000 000 001 | $10^{-9}$ | nano- | One billionth |
| 0.000 000 000 001 | $10^{-12}$ | pico- | One trillionth |
| 0.000 000 000 000 001 | $10^{-15}$ | femto- | One quadrillionth |

present form, borrowed by the Arabs in about AD 700 and subsequently adopted by European merchants. The decimal form conveniently handles both positive and negative real numbers of any size from the very large to the very small and is used worldwide.

Spatial data users are probably more interested in development of the meter as a decimal standard of length. The goal of the French Academy of Science in the 1790s was to devise a standard of length that could be duplicated in nature and which was decimally divided. The arc distance from the equator of the Earth to the North Pole was determined as accurately as possible by means of a geodetic survey and the result was 5,130,766 toises. This equator to pole distance was then defined to be exactly 10 million meters (Smith 1986, Alder 2002).

The decimal system is used in the International System of Units (SI) adopted by *the 11th General Conference on Weights and Measures* in 1960 (Chen 1995, p. 2455). The SI is a coherent system of units included in, or derived from, the seven independent SI base units of the meter, kilogram, second, ampere, degrees Kelvin, mole, and candela. One advantage of the decimal system is names for units that differ by a magnitude of 1,000 as shown in Table 4.1. Perhaps the names are best recognized when referring to computer speeds (megahertz), data storage (gigabytes) or very short periods of time (nanoseconds). While the SI defines decimal subdivision for length (meters), time (seconds), and angles (radian), standard practice in many parts of the world still uses sexagesimal division of angles (degrees, minutes, and seconds) and time (hours, minutes, and seconds).

## Radian

One radian is the angle formed by an arc whose length is the same as the radius of the circle. The definition of radian angular measure is the ratio of the circumference of a circle divided by diameter of the same circle. Since diameter is twice the radius, there are $2\pi$ radians in a full circle of 360°, and it is common to use the relationship 1 radian = $180°/\pi$ or $57°\ 17'\ 44.''806247$. Another useful relationship is derived from the fact that there are 1,296,000 seconds of arc in a full circle. Dividing that number of seconds by $2\pi$, there are 206,264.806247+ seconds of arc per radian.

## SEXAGESIMAL

The sexagesimal system is used in both angular measure and for measuring time. About 5,000 years ago, the Babylonians related 360 degrees in a circle to 365 days in a year. Given that six circles will fit exactly around a seventh circle, all of the same diameter, a subdivision of 360 degrees into six sectors of 60 degrees each is plausible. Tooley ([1949]1990) credits the Babylonians with subdividing both the sky into degrees and the day into hours. The sexagesimal system of minutes and seconds was applied to each, allowing stars of the night sky to be plotted in a consistent proportional manner. Wilford (1981) credits Ptolemy with subdividing the degree into 60 minutes and each minute into 60 seconds while Smith (1986) credits the Chinese with first using zero and a circle sexagesimally divided into degrees, minutes, and seconds. Regardless of the origin of the practice, the sexagesimal second is now the defining unit of time in addition to being used extensively as an angular measure. The sexagesimal system of units is used for time worldwide (60 seconds equals 1 minute, 60 minutes equals 1 hour, and 24 hours equals 1 day). The sexagesimal system is also widely used in angular measurement where a full circle of 360 degrees is equivalent to one 24 hour day—giving a conversion of 15 degrees per hour. But, the radian is the defining unit for rotation in the International System of Units.

## BINARY

Computer science professionals use combinations of zeros and ones (0s and 1s) to represent numbers, letters, and other symbols within a computer's memory. A string of 8 bits (0s and 1s) is called a byte and can be used to represent up to 256 different items. The American Standard Code for Information Interchange (ASCII) uses combinations of 7 of the 8 bits to represent upper- and lowercase letters of the English alphabet, digits 0–9, and other symbols as text characters. If ASCII characters were used to represent real numbers in a computer, it would be quite costly in terms of memory requirements. Therefore, numeric data are stored using a base 2 (binary) system that accommodates real numbers and integers, both positive and negative. Rather than exploring details of the binary system, the point here is to recognize that numerical values, whether integer or real, are coded differently than text. In most cases, the user need not be concerned with computer binary operations because numeric input in decimal form is immediately converted to binary and numeric output, unless specified otherwise by the user, is displayed or printed in conventional decimal format. It is interesting to note, however, that there are similarities between fractions (dividing by two) and binary arithmetic. For an interesting tongue-in-check discussion of the advantages of binary computations for survey measurements, see Stanfel (1994).

## UNIT CONVERSIONS

The ability to use numbers is important, but the real meaning of any mathematical operation comes from knowing what physical quantities—that is, units—are associated with the numbers. And, it is important to understand the use of ratios where the units cancel out and the relative size of the number is really the issue. The value of $\pi$ is one example, trigonometric ratios is another. The solution of most problems involves a reasonable number of something (units) that can be understood. Conversion is the process whereby equivalency is established between seemingly unrelated physical quantities. Mathematical operations also apply to the units. An easy example might be area (m²) = length (m) times width (m). Not as obvious, but not really that unusual, the volume of concrete in a hypothetical sidewalk is length (30 meters) times width (5 feet) times thickness (4 inches). Unit conversions are "exact" and used as ratios to find the desired answer. Note how units cancel in the separate computations and make it easy to find the volume of concrete in either cubic yards or cubic meters:

$$\text{Vol yd}^3 = (30 \text{ m})\left(\frac{39.37 \text{ ft}}{12.00 \text{ m}}\right)(5 \text{ ft})(4 \text{ inches})\left(\frac{1 \text{ ft}}{12 \text{ inches}}\right)\left(\frac{1 \text{ yd}^3}{27 \text{ ft}^3}\right) = 6.1 \text{ yd}^3$$

$$\text{Vol m}^3 = (30 \text{ m})(5 \text{ ft})\left(\frac{12.00 \text{ m}}{39.37 \text{ ft}}\right)(4 \text{ inches})\left(\frac{1 \text{ m}}{39.37 \text{ inches}}\right) = 4.6 \text{ m}^3$$

## COORDINATE SYSTEMS

An origin and three mutually perpendicular axes that intersect at the origin define a generic Cartesian coordinate system as shown in Figure 4.2. When studying two-dimensional phenomena, the X-axis and Y-axis are the ones most commonly used.

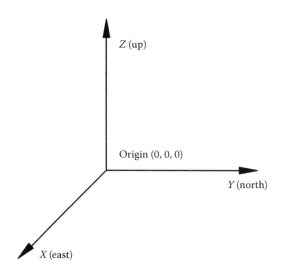

**FIGURE 4.2** Three-Dimensional Rectangular Coordinate System

In a three-dimensional system, the direction of the positive Z-axis for a right-handed coordinate system is given by the direction of the thumb on one's right hand when rotating the positive X-axis (one's index finger) toward the Y-axis (one's palm). An example of a left-handed coordinate system is given as *north/east/up*. The convention in this book is to use a right-handed rectangular Cartesian coordinate system whenever possible. That includes both *X/Y/Z* and *east/north/up*. Note that while labeling the axes of a coordinate system is really the prerogative of the user, the GSDM uses *X/Y/Z* for geocentric ECEF coordinates and uses *e/n/u* for local perspective coordinates.

## Significant Digits

In some cases, the terms significant digits and significant figures are used interchangeably. The rules of significant digits are not arbitrary, but based upon the principles of error propagation. If the validity of an answer obtained using significant digits is in doubt, the uncertainty of any answer can be verified using the standard deviations of measured quantities and error propagation computations.

- Integer values may have an infinite number of significant digits. For example, when dividing the area of a rectangle in half, 2 is an exact number. But, physical operations are not so precise. If one of two siblings cuts a piece of candy in half, the astute parent permits the second sibling to have first choice of the two pieces (in case one piece is larger).
- Conversions that are exact ($12'' = 1'$ or $27 \text{ ft}^3 = 1 \text{ yd}^3$) contain an infinite number of significant digits when used as a ratio.
- With regard to controlling round-off error, it is recommended to carry at least one more digit in the computations than can be justified by the original data. The final answer of any computation should reflect the user's judgment with respect to significant digits.

### Addition and Subtraction

The column (decade) of the least accurate number in a sum (addition) or in a difference (subtraction) determines the number of significant digits in the answer. A zero listed after the decimal point is usually counted as significant. But, zeros that serve only to position the decimal point are generally not significant. Table 4.2 shows

## TABLE 4.2
## Significant Digit Examples Using Addition and Subtraction

| Addition | Addition | Addition | Subtraction | Subtraction |
|---|---|---|---|---|
| 120. | 120.00 | 0.0023 435 | 1.00000000 | $5.44536724 \times 10^9$ |
| +13.42 | 13.42 | 0.101 | −0.00676865 | $-5.43407673 \times 10^9$ |
| 130. | 133.42 | 0.103 | 0.99323135 | 11,290,510 |
| 2 s.d. | 5 s.d. | 3 s.d. | 8 s.d. | 7 s.d. |

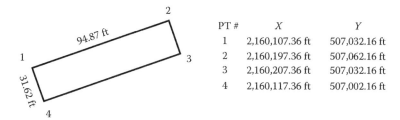

| PT # | X | Y |
|------|---|---|
| 1 | 2,160,107.36 ft | 507,032.16 ft |
| 2 | 2,160,197.36 ft | 507,062.16 ft |
| 3 | 2,160,207.36 ft | 507,032.16 ft |
| 4 | 2,160,117.36 ft | 507,002.16 ft |

**FIGURE 4.3** Rectangle for Area Computation

examples for addition and subtraction. For an exception, see the area computation in Figure 4.3 where 3,00$\overline{0}$ ft$^2$ has four significant digits, not one. In such cases, some authors place a bar over the last significant digit.

### Multiplication and Division

The number of significant digits in the product or quotient is determined by the term used in the operation containing the least number of significant digits. A product or quotient does not contain more significant digits than either term used to compute it. A simple example is area computed as length ($L = 94.87'$) times width ($W = 31.62'$), which gives area = 2,999.7894 ft$^2$. An appropriate answer to four significant digits is 3,00$\overline{0}$ ft$^2$.

Area for the same rectangle, Figure 4.3, can be computed using the area-by-coordinates (cross multiplication) equation and the listed state plane coordinates of its corners. A common form of the area-by-coordinates equation is

$$A = \frac{1}{2}\left[ \begin{array}{l} (Y_1X_2 + Y_2X_3 + Y_3X_4 + Y_4X_1) \\ -(X_1Y_2 + X_2Y_3 + X_3Y_4 + X_4Y_1) \end{array} \right] \tag{4.1}$$

| **Sum of Negative Products** | **Sum of Positive Products** |
|---|---|
| $X_1Y_2 = 1.095308704 \times 10^{12}$ ft$^2$ | $Y_1X_2 = 1.095289533 \times 10^{12}$ ft$^2$ |
| $X_2Y_3 = 1.095289533$  " | $Y_2X_3 = 1.095359410$  " |
| $X_3Y_2 = 1.095229798$  " | $Y_3X_4 = 1.095248971$  " |
| $X_4Y_3 = 1.095248971$  " | $Y_4X_1 = 1.095179097$  " |
| $Sum_1 = 4.381077006 \times 10^{12}$ ft$^2$ | $Sum_2 = 4.381077011 \times 10^{12}$ ft$^2$ |

Area is one half the difference of the two sums:

$$Sum_2 = 4,381,077,011,000 \ \text{ft}^2$$
$$Sum_1 = 4,381,077,006,000 \ \text{ft}^2$$
$$Difference = \qquad\qquad 5,000 \ \text{ft}^2$$

$Difference/2 = 2,500$ ft$^2$ Not Good! The correct answer is 3,00$\overline{0}$ ft$^2$.

What happened? The area equation is correct. The coordinates are good. A ten-digit calculator was used to compute the answer. But, the answer is obviously in error by 500 ft². The problem is one of significant digits. Two issues (innocent mistakes) are:

- The $Y$ coordinates contain only eight significant figures, yet each $XY$ product lists ten significant digits. This is mistake number one and invalidates the computation.
- But, a second mistake (and separate issue) is that significant digits are lost in taking the difference of two large numbers of similar magnitudes. In this case, only one significant digit remains after computing the difference of the two sums of products.

### Avoid Mistakes by Working with Coordinate Differences

By finding a coordinate difference first, the problem of working with large coordinate values is avoided because the cross products are formed using much smaller numbers. A derivation of the improved area Equation 4.2 is given in Burkholder (1982) but it can also be obtained from Equation 4.1 by moving the coordinate system origin to the first point of the figure (a fair amount of algebraic manipulation is required). Another advantage of Equation 4.2 is that it can be extended easily for any number of points. A minimum of 3 points is required in the first line of Equation 4.2. Beyond that, point 4 is brought into line 2, point 5 is brought into line 3, etc. Note that with the orderly progression of subscripts, a program need only be written for the first line and used in a loop with updated subscripts until all points around the figure are used. Points used must be in graphical sequence to form a closed figure and no line crossovers are permitted.

$$2A = (X_2 - X_1)(Y_3 - Y_1) - (Y_2 - Y_1)(X_3 - X_1)$$
$$+ (X_3 - X_1)(Y_4 - Y_1) - (Y_3 - Y_1)(X_4 - X_1)$$
$$+ (X_4 - X_1)(Y_5 - Y_1) - (Y_4 - Y_1)(X_5 - X_1)$$
$$+ \cdots \text{for any number of points.} \qquad (4.2)$$

## LOGIC

The following 13 items on logic are adapted from Chapter 1, Bumby and Klutch (1982):

1. A *statement*, also called an assertion, is any sentence that is true or false, but not both. The truth or falsity of a statement is called its "truth value."
2. Placeholders in mathematical sentences are called *variables*. An open sentence contains one or more placeholders. Variables are selected from a collection of possibilities called the *domain* and the collection of all variables from the domain that make an open sentence true is called the *solution set*.
3. If a statement is represented by "*p*," then "*not p*" is the *negation* of that statement. A negation of a negation is same as the original. That is, "*p*" equals "*not (not p)*."
4. A compound statement formed by joining two statements with the word "and" is called a *conjunction*. Each of the statements is called a *conjunct*.

5. A compound statement formed by joining two statements with the word "or" is called a *disjunction*. Each statement is called a *disjunct*.

6. A compound statement formed by joining two statements with the words "if...then" is called a *conditional*.

7. In logic, the symbol "→" is used to represent a conditional. Therefore, the conditional if $p$ then $q$ can be written as "$p \rightarrow q$." Statement $p$ is called the *antecedent* of the conditional and the statement $q$ is called the *consequent*.

8. A *tautology* is a compound statement that is true regardless of the truth values of the statements of which it is composed.

9. The *converse* of a conditional is formed by switching the order of $p$ and $q$ of the original conditional.

10. The *inverse* of a conditional is formed by negating both the antecedent and the consequence.

11. The *contrapositive* of a conditional is formed by the inverse of the converse, that is, by negating both $p$ and $q$ and reversing their order.

12. A statement formed by the conjunction of the conditionals $p \rightarrow q$ and $q \rightarrow p$ is a *bi-conditional*. The phrases "necessary and sufficient" and "if and only if" are bi-conditionals.

13. Whenever two statements are always either both true or both false, the two statements are *equivalent*.

The following rules of logic, still applicable in problem solving and computer programming, were adopted by the French mathematician and philosopher, Rene Descartes (Russell 1959):

- Never accept anything but clear distinct ideas.
- Divide each problem into as many parts as are required to solve it.
- Thoughts must follow an order from the simple to the complex. Where there is no order, one must be assumed.
- One should check thoroughly to assure no detail has been overlooked.

## ARITHMETIC

Arithmetic consists of the manipulation of numbers by addition, subtraction, multiplication, division, and square roots to solve problems. Simple examples are:

Addition: $8 + 4 = 12$ and $8 + (-4) = 4$
Subtraction: $8 - 4 = 4$ and $8 - (-4) = 12$
Multiplication: $8 \times 4 = 32$ and $8 \times (-4) = -32$
Division: $8 \div 4 = 2$ and $8 \div (-4) = -2$
Square root: $\sqrt{(64)} = 8$ and $\sqrt{(-64)} = $ undefined

## ALGEBRA

Algebra is an extension of arithmetic that includes the use of letters to represent unknown numerical values. This permits a problem to be solved in a general form and for the solution of a specific problem to be obtained more efficiently by substituting the variables into an algebraic solution as opposed to resolving the problem with different numbers on every occasion.

Rules associated with the properties and manipulation of algebraic quantities (Dolciani et al. 1967) are as follows.

### AXIOMS OF EQUALITY (FOR REAL NUMBERS *A*, *B*, AND *C*)

Reflexive property: $A = A$.
Symmetric property: If $A = B$, then $B = A$.
Transitive property: If $A = B$ and $B = C$, then $A = C$.

### AXIOMS OF ADDITION (FOR REAL NUMBERS *A*, *B*, AND *C*)

Associative rule: $(A + B) + C = A + (B + C)$.
Existence of identity: There is a unique number zero such that $0 + A = A$ and $A + 0 = A$.
Existence of inverses: For each real number $A$ there is a number $-A$ such that $A + (-A) = 0$ and $(-A) + A = 0$.
Commutatively: For all real numbers $A$ and $B$, $A + B = B + A$.

### AXIOMS OF MULTIPLICATION (FOR REAL NUMBERS *A*, *B*, AND *C*)

Associative rule: $(A \times B) \times C = A \times (B \times C)$.
Existence of identity: There exists a unique number 1 such that for every real number $A$, $1 \times A = A$ and $A \times 1 = A$.
Existence of inverse*: For every real number $A$ (except 0), there is an element $1/A$ such that $A \times (1/A) = 1$ and $(1/A) \times A = 1$.
Commutativity: $A \times B = B \times A$.
Distributive: $A \times (B + C) = (A \times B) + (A \times C)$ and $(B + C) \times A = (B \times A) + (C \times A)$.

### BOOLEAN ALGEBRA

Boolean algebra involves the use of logic conditions in which the result of some operation takes on only one of two values, true or false. In terms of binary computer code, the conditions equate to 1 for true, 0 for false. Computer programmers make extensive use of Boolean algebra in writing computer programs and testing for differing conditions that may exist at the time a given part of the program is executed. Boolean algebra is beyond the scope of this book.

## GEOMETRY

This book is written to define a model for spatial data on a global scale. Geometry is fundamental to that mission and includes the study of points, lines, circles, curves, planes, triangles, rectangles, cubes, spheres, and other objects. In Chapter 2, spatial

---

* This property is the basis of the rule that division by zero is undefined.

data were defined as the distance between endpoints of a line in Euclidean space. In order to provide additional clarification, the following elements are described.

## POINT

A point is a dimensionless quantity that occupies a unique location. The irony is that location cannot be defined without reference to (distance from) some other point. And, what does it mean? "A point has no dimension."

## DISTANCE

Distance is defined as the spatial separation of two objects (points) and has length as an attribute—dimension. Note that this is an example of circular logic because the concept of a point is used to define distance and the concept of distance (reference to axes) is used to define a point.

## DIMENSION

Dimension, units of the quantity being measured, is another intuitive concept that seems to defy definition. A point has no dimension. A line has one dimension, surface area has two dimensions, volume has three dimensions, and space-time is generally considered to have four dimensions. According to Hawking (1988), anything more than four dimensions is a part of science fiction. But, he also describes "string theories" that accommodate many dimensions.

## LINE

As stated earlier, a line has length and can be described as the path of a moving point. Strictly speaking, a straight line is infinitely long but a line segment has two end points. A line can also be straight or curved. A straight line is the shortest distance between two points in Euclidian space while a curved line has a radius associated with it. Understandably, a straight line can also be defined as a curve with an infinite radius. The slope-intercept form for the equation of a line is $y = ax + b$ where $a$ is the slope and $b$ is the intercept on the $Y$-axis (that place where the $x$ value is zero).

## PLANE

A plane is two-dimensional (2-D), contains something called area, and is formed by the lateral movement of a straight line. More precisely, a plane is a flat surface defined by three noncollinear points. In 3-D space, a plane is also a flat surface that is perpendicular to a given straight line. Spatial data users should relate to the last definition because a horizontal plane is defined as perpendicular to the plumb line at a point. That is, humans stand erect and the perception is that we walk on a flat Earth. That is important because, in the GSDM, the origin moves with the observer and the model provides a view of all other points from that occupied or specified by the user.

## ANGLE

Given that straight lines are infinitely long, it is said for Euclidean geometry that two lines lying in the same plane are parallel if and only if they never intersect (i.e., the distance between them never changes). If two lines in a plane are not parallel, they will intersect. An angle is defined as the geometrical shape formed by the intersection of two lines in a plane. Later, an angle will also be defined as the difference between two directions. If the four angles formed by the intersection of two lines are all the same size, it is said the two lines are perpendicular to each other and the four angles are all called right angles (i.e., one-fourth of a circle).

## CIRCLE

A circle is a closed figure formed by a uniformly curved line lying in a 2-D plane. Being uniformly curved, all points on the circle are the same radius distance from a common center point. The maximum distance from one side of the circle to the other is the diameter or two radii. The angular measure of a full circle is one revolution, four right angles, $2\pi$ radians, or 360°.

## ELLIPSE

An ellipse is a continuous closed plane curve having a major axis dimension longer than its minor axis dimension. A unique characteristic of an ellipse is that the sum of distances from any point on the ellipse to each of the two foci located on the major axis is a constant whose value is twice the length of the semimajor axis. The ellipse becomes more nearly a circle as the distance between the two foci becomes smaller and smaller.

## TRIANGLE

A triangle is a three-sided figure in a 2-D plane. Said differently, a triangle is a closed figure in a plane formed by three straight line segments. The sum of the interior angles of any plane triangle is always 180°.

1. A right triangle has one right (90°) angle. In all seriousness, a student once asked, "What is a left triangle?" How should such a question be answered?
2. An equilateral triangle has three 60° angles.
3. An isosceles triangle has two equal angles. The length of sides opposite those angles are also equal.
4. An acute triangle is a triangle having three angles with no angle over 90°.
5. An obtuse triangle is a three-sided figure having one angle greater than 90°.

## QUADRILATERAL

A quadrilateral is any four-sided figure in a plane bounded by straight lines. If additional conditions are imposed, special names can be used.

### RECTANGLE

A rectangle is a quadrilateral having four right (90°) angles. That means opposite sides are parallel and the same length. Any rectangle divided by a line through opposite corners gives two right triangles.

### SQUARE

A square is a rectangle with all sides being the same length.

### TRAPEZOID

A trapezoid is any quadrilateral having only two parallel sides.

### PARALLELOGRAM

A parallelogram is a four-sided figure with two pairs of parallel sides.

### POLYGON

A polygon is a closed figure in a plane having many (any number of) sides. There is no restriction on lengths or angles except that the figure must be closed. A regular polygon is one in which all sides (regardless of the number) are of equal length and all deflection angles are equal. A regular polygon having an infinite number of sides is also the same as a circle.

### PYTHAGOREAN THEOREM

With regard to Figure 4.4, the area of the total figure is the product of two sides, $c$ times $c$, or $c^2$. If the area is also computed as the sum of four triangles plus the smaller square inside the figure, proof of the Pythagorean Theorem can be written as follows:

$$c^2 = 4\frac{ab}{2} + (a-b)(a-b) = 2ab + a^2 - 2ab + b^2 = a^2 + b^2 \tag{4.3}$$

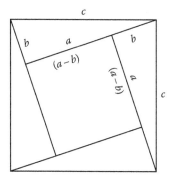

**FIGURE 4.4**   Theorem of Pythagoras

## SOLID GEOMETRY

The rules of solid geometry apply to 3-D objects and to the position(s) of other geometrical elements in 3-D space. Although an entire textbook could be written about solid geometry, only a brief summary is included here.

### SPHERE

A sphere, also known as a ball, is a closed, uniformly curving, 3-D surface all points on which are the same distance from an interior point, the center.

### ELLIPSOID

An ellipsoid is a solid object formed by rotating an ellipse about one of its two axes. An ellipse rotated about its minor axis is used to approximate the size and shape of the Earth.

### CUBE

A cube is a solid figure having six identically square faces (sides).

### POLYHEDRON

A polyhedron is a solid object bounded by plane surfaces—usually more than six.

### TETRAHEDRON

A tetrahedron is a regular solid whose sides consist of four equilateral triangles. It is a figure formed by the minimum number of plane sides.

### PYRAMID

A pyramid is a solid object whose base is a polygon and whose triangular sides meet at a common point. Most pyramids are five-sided, having a square base and four triangular sides.

### EQUATION OF A PLANE IN SPACE

A plane in space is described by a first-order (containing only powers of 1 for each variable) equation using variables *X/Y/Z* as

$$aX + bY + cZ + d = 0 \qquad (4.4)$$

The letters $a$, $b$, and $c$ are coefficients of the variables and $d$ is a constant, all of them real numbers. Note that the entire equation could be divided by $d$, giving Equation 4.5

having only three independent coefficients. That makes sense, because it takes three points in space to determine a plane.

$$(a/d)X + (b/d)Y + (c/d)Z + 1 = 0 = a'X + b'Y + c'Z + 1 \qquad (4.5)$$

## EQUATION OF A SPHERE IN SPACE

A sphere in space is described by a second-order (contains powers of 2 on one or more variables) equation using variables $X/Y/Z$ as

$$(X-a)^2 + (Y-b)^2 + (Z-c)^2 = R^2 \qquad (4.6)$$

where:
   The letters $a$, $b$, and $c$ are the $X/Y/Z$ coordinates of the center of the sphere.
   $R$ is the radius of the sphere.

If $a$, $b$, and $c$ are all zero, the center of the sphere lies at the origin of the coordinate system.

## EQUATION OF AN ELLIPSOID CENTERED ON THE ORIGIN

A 2-D ellipse rotated about its major axis forms a football-shaped object. A 2-D ellipse rotated about it minor axis is used to approximate the size and shape of the Earth. The North Pole and South Pole are on the minor axis of the ellipsoid, which is also the spin axis of the Earth.

$$X^2/a^2 + Y^2/a^2 + Z^2/b^2 = 1 \qquad (4.7)$$

where:
   The letter $a$ is used as the semimajor axis.
   $b$ is the semiminor axis of the ellipse.

The equatorial plane goes through the origin and is 90° from each pole. The equator forms a circle having $a$ as its radius. Each meridian section is perpendicular to the equator and forms an ellipse defined by the letters $a$ and $b$.

## CONIC SECTIONS

A cone is a triangular shaped solid whose base is a closed curve. A right circular cone is one whose base is a circle that is perpendicular to the cone axis. Conic sections are 2-D shapes obtained by intersecting a cone with a plane at different orientations.

1. A circle is formed by the intersection of a cone with a plane perpendicular to the axis of the cone. If the intersection occurs at the vertex, the circle is reduced to a point.
2. An ellipse is formed by the intersection of a cone with a plane that is not perpendicular to the axis of the cone. The intersection is a closed figure.

3. A parabola is formed by the intersection of a cone with a plane that is parallel with the opposite side of the cone. The intersection is not a closed figure.
4. A hyperbola is formed by the intersection of a cone with a plane that is parallel with the axis of the cone. The intersection is not a closed figure.

Conic sections can all be derived from the general second degree polynomial equation by appropriate selection of coefficients, $A$, $B$, $C$, $D$, $E$, and $F$. It is to be understood that the $X/Y$ coordinate system used in Equation 4.8 lies in the plane intersecting the cone.

$$AX^2 + BXY + CY^2 + DX + EY + F = 0 \qquad (4.8)$$

1. For a line: $A = B = C = 0$.
2. For a circle: $A = C$ and $B = D = E = 0$.
3. For an ellipse: $A \neq C$ and $B = D = E = 0$.
4. For a parabola: $B = C = 0$.
5. For a hyperbola: $A = -C$ and $B = D = E = 0$.

## VECTORS

A vector is a directed line segment in space. In terms of a right-handed coordinate system, a vector is composed of signed components in each of the **i/j/k** directions. The length of a vector is called its magnitude and is computed as the square root of the sum of the components squared (a three-dimensional hypotenuse). A unit vector has a length of 1.0 and is obtained by dividing each component by the vector magnitude. The resulting rectangular components of a unit vector are its direction cosines. The underlying **i/j/k** orientation of a vector can be rotated to any $X/Y/Z$ coordinate system without changing the length or statistical qualities of the vector. The direction cosines do change.

Vectors can be added and subtracted component by component (subtraction is same as addition given that the sign is changed for each component of the second vector).

A vector can be multiplied by a scalar that has the effect of changing only the magnitude of the vector. The direction cosines remain the same.

Vector multiplication takes two forms. First, a dot (inner) product of two vectors is a number (scalar) which can be used to find the angle between two vectors in a plane. Second, the cross product of two vectors is a third vector perpendicular to the plane common to the first two.

## TRIGONOMETRY

Trigonometry is the study of triangles and relationships between the various sides and angles. The trigonometric functions are defined as ratios of the sides of a right triangle. Once an angle is identified, such as angle $\theta$ in Figure 4.5, the defining trigonometric ratios are (*opp* = opposite, *adj* = adjacent, and *hyp* = hypotenuse).

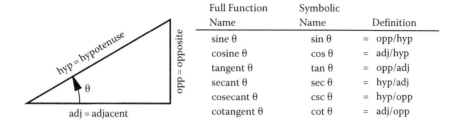

| Full Function Name | Symbolic Name | Definition |
|---|---|---|
| sine θ | sin θ | = opp/hyp |
| cosine θ | cos θ | = adj/hyp |
| tangent θ | tan θ | = opp/adj |
| secant θ | sec θ | = hyp/adj |
| cosecant θ | csc θ | = hyp/opp |
| cotangent θ | cot θ | = adj/opp |

**FIGURE 4.5** Trigonometric Definitions

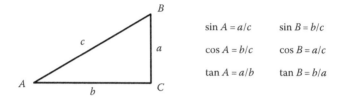

$$\sin A = a/c \qquad \sin B = b/c$$
$$\cos A = b/c \qquad \cos B = a/c$$
$$\tan A = a/b \qquad \tan B = b/a$$

**FIGURE 4.6** Right Triangle Relationships

Standard abbreviations for the six ratios are sin, cos, tan, sec, csc, and cot. The first three ratios are used extensively and the second three, while just as valid, are used much less. Generic nomenclature for a triangle is to label each of the vertex angles with capital letters $A$, $B$, and $C$. The sides opposite the vertex angles are labeled with lowercase letters $a$, $b$, and $c$. Using the generic labeling for a right triangle with the right angle being C and noting that the sum of angles $A$ and $B$ is 90°, the following relationships in Figure 4.6 are fundamental.

## TRIGONOMETRIC IDENTITIES

The following trigonometric identities can be proved using the Pythagorean Theorem, fundamental relationships, and referring to Figures 4.6 through 4.8.

$$\sin^2 \theta + \cos^2 \theta = 1 \Rightarrow \sin^2 \theta = 1 - \cos^2 \theta \quad \text{and} \quad \cos^2 \theta = 1 - \sin^2 \theta$$

$$\sin \theta = -\cos\left(\theta + 90^\circ\right) = \cos\left(90^\circ - \theta\right) = \cos\left(\theta - 90^\circ\right) = -\sin\left(\theta + 180^\circ\right)$$
$$= \sin\left(180^\circ - \theta\right)$$

$$\cos \theta = \sin\left(\theta + 90^\circ\right) = \sin\left(90^\circ - \theta\right) = -\sin\left(\theta - 90^\circ\right) = -\cos\left(\theta + 180^\circ\right)$$
$$= -\cos\left(180^\circ - \theta\right)$$

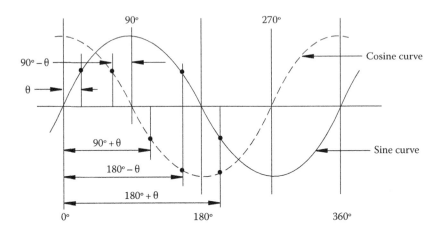

**FIGURE 4.7**   Plot of Sine and Cosine Functions

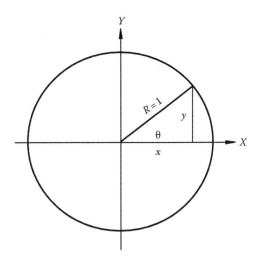

**FIGURE 4.8**   Circular Trigonometric Functions

## LAW OF SINES

Figure 4.9 is a standard generic triangle with extension lines added to make two right triangles; note that $\sin A = y/c$ and that $\sin(180° - C) = y/a$ from which $y = a \sin(180° - C)$. Next, use the trigonometric identity, $\sin(180° - C) = \sin C$, to get $y = a \sin C$. Substituting those in the original equation, $\sin A = (a \sin C)/c$, from which $(\sin A)/a = (\sin C)/c$. Similarly, with regard to the same Figure 4.9, $\sin B = z/c$ and $\sin(180° - C) = z/b$ from which $z = b \sin(180° - C)$. Use the same trigonometric identity again, $\sin(180° - C) = \sin C$. Making substitution for these two equalities, the original equation now becomes $\sin B = (b \sin C)/c$, from which $(\sin B)/b = (\sin C)/c$. Since it's

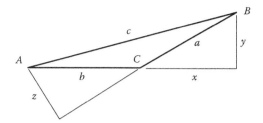

**FIGURE 4.9** Generic Triangle

been shown the two quantities are both equal to (sin C)/c, they are equal to each other and the Law of sines is

$$\frac{\sin A}{a} = \frac{\sin B}{b} = \frac{\sin C}{c} \tag{4.9}$$

## LAW OF COSINES

The same Figure 4.9 is also used to write a relationship for the law of cosines. Using the Pythagorean Theorem on the two right triangles, write $c^2 = (b + x)^2 + y^2$ and $a^2 = x^2 + y^2$ (or $y^2 = a^2 - x^2$). Also note that $\cos(180° - C) = x/a$ from which $x = a \cos(180° - C)$. But a useful trigonometric identity is $\cos(180° - C) = - \cos C$. Putting these together

$$
\begin{aligned}
c^2 = \left(b + x\right)^2 + a^2 - x^2 &= b^2 + 2bx + x^2 + a^2 - x^2 \\
&= a^2 + b^2 + 2ab\cos\left(180° - C\right) \\
&= a^2 + b^2 - 2ab\cos C
\end{aligned}
\tag{4.10}
$$

An alternate version of the law of cosines giving an angle in terms of the sides is

$$\cos C = \frac{a^2 + b^2 - c^2}{2ab} \tag{4.11}$$

## SPHERICAL TRIGONOMETRY

The rules of spherical trigonometry can be used to solve for the great circle arc distance between latitude/longitude points on a spherical Earth or to solve triangles on the celestial sphere when determining the astronomical azimuth of a line on the ground. Two important considerations are as follows:

- The Earth is slightly flattened at the poles so the spherical great circle distance will be somewhat longer than the actual ellipsoidal distance. For example, the ellipsoidal (GRS 80) distance between points in New Orleans and Chicago is 1,354 kilometers and the great circle distance between the same two points is 1,359 kilometers. If the accuracy of the distance

New Orleans to Chicago is acceptable to two or three significant digits, the spherical triangle solution may be appropriate. The spherical triangle solution is certainly simpler and easier to find than the ellipsoidal distance using a geodetic inverse (BK19) computation.

• The radius of the celestial sphere is considered to be infinitely large, which means there is no similar spherical/ellipsoidal approximation when spherical trigonometry is applied to solving the pole-zenith-star (*PZS*) triangle. The equations and procedures for determining an astronomical azimuth are not covered here but can be found in texts such as Moffitt and Bossler (1998), Davis et al. (1981), or Ghilani and Wolf (2012).

As shown in Figure 4.10, a spherical triangle has three vertex angles and three sides. All six elements are given in terms of angles because changing the size of the sphere does not change the angular relationships. The three vertex angles are on the surface of the sphere and are labeled with capital letters *A*, *B*, and *C* while the sides opposite each vertex angle are labeled with lowercase letters *a*, *b*, and *c*. Each side is the arc of an angle subtended at the center of the sphere and the value listed for each side is really the value of the corresponding subtended angle.

Equations for solving a spherical triangle include the spherical law of sines and two forms of the spherical law of cosines. In order to solve a spherical triangle, three of the six angles must be known. If two vertex angles and the side opposite one of them are known or if two sides and the vertex angle opposite one of them are known, the spherical law of sines can be used. The spherical law of sines is

$$\frac{\sin A}{\sin a} = \frac{\sin B}{\sin b} = \frac{\sin C}{\sin c} \tag{4.12}$$

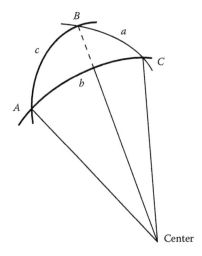

**FIGURE 4.10**   Spherical Triangle

The spherical law of cosines has two forms; the first form solves for a vertex angle if the sides are all known and the second form solves for a side if all the vertex angles are known. They are

$$\cos A = \frac{\cos a - \cos b \cos c}{\sin b \sin c} \qquad (4.13)$$

$$\cos a = \frac{\cos A + \cos B \cos C}{\sin B \sin C} \qquad (4.14)$$

Note that Equations 4.13 and 4.14 are cyclic in that the symbols for the vertices and the sides can be relabeled on the diagram or that equations for vertex angles $B$ and $C$ can be found by rearranging the sides in Equation 4.13. Similarly, equations for sides $b$ and $c$ can be found by rearranging vertex angles in Equation 4.14. For example

$$\cos B = \frac{\cos b - \cos c \cos a}{\sin c \sin a} \quad \text{while} \quad \cos b = \frac{\cos B + \cos C \cos A}{\sin C \sin A}$$

Equation 4.13 is the form used to solve for great circle arc distances. Given that the radius of the Earth is 6,372 km and that the latitude and longitude for the two cities are as follows:

New Orleans: Latitude = 30° 02′ 17″, Longitude = 90° 09′ 56″ W
Chicago: Latitude = 42° 07′ 39″, Longitude = 87° 55′ 12″ W

Note that vertex angle $A$ is the longitude difference at the North Pole, vertex angle $B$ is at Chicago, and vertex angle $C$ is at New Orleans.
    Find the great circle distance between the two cities.

Side $b$ = 90° − 30° 02′ 17″ = 59° 57′ 43″
Side $c$ = 90° − 42° 07′ 39″ = 47° 52′ 21″
Angle $A$ = 90° 09′ 56″ − 87° 55′ 12″ = 2° 14′ 44″

Rewriting Equation 4.13 to solve for side $a$ and using the values for $b$, $c$, and $A$

$$\cos a = \sin b \sin c \cos A + \cos b \cos c$$

$$\cos a = \sin\left(59° \ 57′ \ 43″\right)\sin\left(47° \ 52′ \ 21″\right)\cos\left(2° \ 14′ \ 44″\right) \\ + \cos\left(59° \ 57′ \ 43″\right)\cos\left(47° \ 52′ \ 21″\right)$$

$$\cos a = 0.9773288; \quad a = 12° \ 13′ \ 25″ = 0.21334 \text{ radians} \left(\text{only 5 s.f.}\right)$$

$$\text{Distance} = Ra\left(\text{in radians}\right) = 1,359.4 \text{ km} = 844.70 \text{ miles}$$

Remember, this computation assumes the Earth is a sphere. Computing the ellipsoid arc distance between the same two points will be covered in Chapter 7.

## CALCULUS

Calculus is a valuable mathematical tool that can be described as a study of rates of change or, said differently, cause and effect. Calculus has an undeserved reputation of being difficult to learn. That may be true for some but consider that most people who receive a paycheck for wages are already experts at calculus. That is, upon being notified of a 5 percent increase in wages they will quickly determine how the different pay rate affects their take-home pay. In a more formalized manner, the procedures of *differential* calculus apply known rules to smaller and smaller increments so that the instantaneous rate of change at any point can be computed. Called a derivative and given the symbol *dy/dx*, the graphical representation of *dy/dx* is the slope of a line in the *X/Y* plane.

---

### Convention

Three different ways of showing similar values are
   Deltas: $\Delta x$, $\Delta y$, when using small or specific numerical values
   Derivative: $dx$, $dy$, infinitesimally small components of calculus
   Partial: $\partial x$, $\partial y$, same as a derivative, but the derivative of the computed result is
      taken with respect to one variable at a time

---

In another application of calculus, small pieces of a quantity (such as arc length or area) are computed according to some equation. *Integral* calculus uses the anti-derivative of *dy/dx* to perform an infinite number of summations in order to find the grand total result (see computation of the meridian arc in Chapter 7). In cases where a given mathematical equation cannot be integrated, a fallback procedure is to use numerical integration to compute an approximation based upon intervals ($\Delta X$s) selected by the user. With a computer programmed to do the repetitive calculations, a numerical integration can be made to be as accurate as desired (within reason) by adding up more and more ever smaller pieces. Computing area under the standard error bell curve is an example of numerical integration.

Many students learn calculus by memorizing the rules of manipulation and, with continued use, the concepts become more understandable. Therefore, the goal here is to include several simple examples along with a summary of the fundamental rules of manipulation. It is hoped that the examples will lead toward greater understanding of specific applications involving spatial data manipulation.

### EXAMPLE

The volume of a cylindrical storage tank (see Figure 4.11) is computed from the radius (*R*) and height (*h*) of the tank; see Equation 4.15. Two questions are as follows:

   1. If the height of the tank is changed, how will that affect the total volume?
   2. If the radius of the tank is changed, how will that affect the total volume?

$$V = \pi R^2 h \qquad (4.15)$$

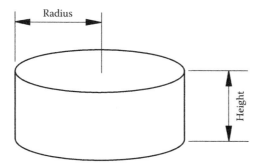

**FIGURE 4.11**   Volume of Tank

Certainly, one could compute the volume for specific incremental values of radius
and height. The change is then found by taking the difference of various answers.
Calculus is not needed for that. But, by looking at the rate of change given by the
derivative of the equation used to compute the volume, extra calculations can be
avoided. And, different answers are obtained depending upon whether the height
changes or the radius changes. Using calculus, height and radius variables are handled separately and the user has the option of using either, neither, or both depending
upon the question to be answered.

   The relationship between tank volume, height, and radius is illustrated in
Figure 4.12. Note that volume increases linearly with height, but that the volume of
the tank increases exponentially with radius. These changes are handled with partial
derivatives, one variable at a time. Note too that when taking the volume partial
derivative with respect to height, the answer is the slope of the line in the volume/
height plane. The radius is treated as a constant and the slope (derivative) is constant.
In the radius/volume plane, the slope of the curve increases exponentially with radius
and is not linear—the larger the radius, the larger the slope.

   Partial derivative of volume with respect to height is $\partial V/\partial h = \pi R^2$
   Partial derivative of volume with respect to radius is $\partial V/\partial R = 2\pi Rh$

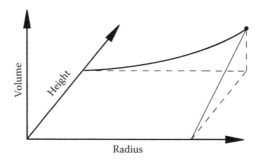

**FIGURE 4.12**   Plot of Volume

In order to find the combined impact of changing both radius and height, the two expressions are combined. The symbols are changed to the "delta" nomenclature to indicate that one is expected to use actual differences to compute a change.

$$\Delta V = \pi R^2 \Delta h + 2\pi R h \Delta R \qquad (4.16)$$

Equation 4.16 is quite useful but, because $R^2$ is nonlinear, Equation 4.16 is accurate only for "small" values of $\Delta R$. As $\Delta R$ grows larger, the accuracy of Equation 4.16 becomes unacceptable. The problem will be addressed in the section on "Error Propagation" where $\Delta h$ and $\Delta R$ are taken to be standard deviations of those dimensions. Using standard deviations of $R$ and $h$, partial derivatives of the volume with respect to $R$ and $h$, and formal error propagation procedures, the standard deviation of the volume can be determined with statistical reliability.

## DIFFERENTIAL CALCULUS EQUATIONS

The following is a brief list of derivatives. These rules presume $x$ is the independent variable, $u$ and $v$ are intermediate variables, $y$ is the computed result, and the derivative is $dy/dx$. Appropriate substitutions for $y$ are listed in each case.

| | |
|---|---|
| $y$ = constant, $a$ | $d(\text{constant})/dx = 0$ |
| $y$ = constant × variable, $a \times u$ | $d(au)/dx = a\, du/dx$ |
| $y$ = sum of variables $u$ and $v$ | $d(u + v)/dx = du/dx + dv/dx$ |
| $y$ = product of variables $u$ and $v$ | $d(uv)/dx = u\, dv/dx + v\, du/dx$ |
| $y$ = quotient of variables $u$ and $v$ | $d(u/v)/dx = (v\, du/dx - u\, dv/dx)/v^2$ |
| $y$ = variable raised to power $n$ | $d(un)/dx = n\, un^{-1}\, du/dx$ |
| $y = \sin u$ | $d(\sin u)/dx = \cos u\, du/dx$ |
| $y = \cos u$ | $d(\cos u)/dx = -\sin u\, du/dx$ |
| $y = \tan u = \sin u/\cos u$ | $d(\tan u)/dx = \sec^2 u\, du/dx$ |

## INTEGRAL CALCULUS EQUATIONS

Using the same conventions as in the previous section, the following is a brief summary of integrals. When evaluating integrals between stipulated limits, the constant of integration ($C$ in the following equations) cancels out.

$$\int du = u + C$$

$$\int a\, du = a \int du = au + C$$

$$\int (du + dv) = \int du + \int dv = u + v + C$$

$$\int u^n\, du = \left[ u^{n+1}/(n+1) \right] + C$$

$$\int du/u = \ln|u| + C$$

$$\int \cos u \, du = \sin u + C$$

$$\int \sin u \, du = -\cos u + C$$

$$\int \sec^2 u \, du = \tan u + C$$

## PROBABILITY AND STATISTICS

### INTRODUCTION

The fields of probability and statistics are distinct disciplines—each deserving more coverage than given here. Since knowledge of underlying mathematical principles is essential to understanding the importance of the contribution of each discipline to spatial data, the reader is referred to a variety of sources for more information. The following references are examples that have a focus:

- Ghilani (2010)—for geomatics professionals and other spatial data users
- Dwass (1970)—on the mathematical theory and principles, for example, an undergraduate text for math majors
- Wine (1964)—on applications in various fields, for example, mathematical, not behavioral, sciences
- Cressie (1993)—to explore specific spatial data applications

The goal in this book is to utilize fundamental principles from each discipline and to organize them as needed to describe efficient procedures by which 3-D spatial data accuracy (in this case, standard deviations) can be reliably established, stored, tracked, and used—component by component. Other more sophisticated tools such as kriging are also useful when analyzing spatial data. Although not included here, such tools are viewed as being compatible with the underlying GSDM.

Concepts of probability and statistics overlap each other when applied to spatial data. Beyond the definitions, little effort is made here to preserve the distinction of various concepts. Each of the following two references is quite specific on the definition of probability but each also includes significant discussion of using statistics when working with spatial data.

- Probability is the ratio of the number of times that an event should occur divided by the total number of possibilities (Ghilani 2010).
- Mikhail (1976) defines probability as the limit of the relative frequency of occurrences of a random event.

A simple example of probability is the results one would expect when tossing a coin. It should come up heads half the time and tails the other half. If the coin is tossed seven times, the ratio of heads to the number of tosses will not be 0.5 for several reasons: (1) because of the odd number of tosses and (2) because of the outcome of a random event is never certain. It would be possible, but not probable, for eight persons to toss a coin seven times each and for each person to obtain different results ranging from zero heads to seven heads. However, if a very large number of tosses is used, the limit of relative frequency should be the same as the (expected) ratio of "heads up" to the total number of tosses. Two important characteristics of probability are that a zero probability is associated with an event that will not happen while a 100 percent probability is associated with an event that is certain to happen. Therefore, a number between zero and one gives an estimate of uncertainty associated with some event or statement.

Statistics is the science of decision-making in the face of uncertainty and should be thought of as "both a pure and an applied science which is involved in creating, developing, and applying procedures in such a way that the uncertainty of inferences may be evaluated in terms of probability" (Wine 1964). Another book begins by saying that statistics, the science of uncertainty, attempts to model order in disorder (Cressie 1993). Incorporating those definitions and concepts into the goal of this book, the GSDM defines an environment and computational procedures by which the standard deviation of each 3-D spatial data component can be determined, enabling better decisions to be made regarding use of the data.

## STANDARD DEVIATION

The standard deviation of a distance is used to describe its uncertainty at some level of confidence. Convention often equates one sigma with one standard deviation. Procedures for computing standard deviation have been formalized, are well documented under the umbrella of error propagation, and are summarized later in this chapter. The confidence level at which decisions are made regarding use of standard deviations is selected by the user. One standard deviation is associated with a confidence level of 68 percent, two standard deviations correspond to a 95 percent confidence level, and virtual certainty (99.7 percent) is achieved at three standard deviations (Figure 4.13).

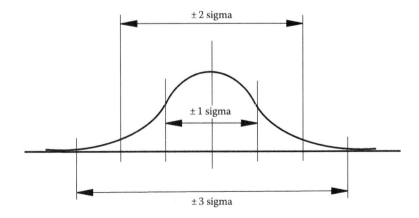

**FIGURE 4.13**  Normal Distribution Curve Showing Standard Deviations in Terms of Sigmas

Standard deviations are not all that exact. Although they can be computed by very specific equations, they are acknowledged to be approximations rarely having more than two significant figures. That means standard deviation can, at times, also be assigned on the basis of seasoned judgment. For example, the standard deviation of a GNSS vector might be assigned a standard deviation of 5 mm, a distance measured by an EDMI may have a standard deviation of 0.01 foot and a single code-phase GNSS position may have a standard deviation of 100 meters, 10 meters, 5 meters, 1 meter, or even less based upon circumstances of the measurement. Such assignments, while subjective, are based upon prior experience and can be quite valid. But, more specifically, standard deviations are determined from repetitious measurements made under known conditions and/or computed via error propagation of such measurements.

## MEASUREMENT

In a sense, measurement and observation are both the results of comparing some unknown quantity with a standard. Accepting a graduated meter (yard) stick as a standard, the distance between points is measured by aligning graduations on the scale with the end points of the distance being measured. The process utilizes both the human eye and judgment. The measurement is the observation and the observation is the measurement. If the same distance is measured by GNSS, the result may be more precise but the concept of observation is not so clear because the comparison was computed rather than viewed by human eye. Admittedly, care and judgment are involved in positioning the GNSS antenna precisely over a mark but the concept remains that the measurement involved computations rather than a direct scale comparison. Chapter 3 describes the difference between a direct measurement and an indirect one. Chapter 3 also notes that observation and measurement are essentially the same with one subtle distinction: observations are taken to be independent while measurements may be correlated.

## ERRORS

Another fundamental principle is that no measurement is perfect. Here a distinction is made between a count and a measurement. An integer count of pencils can be exact but repeated measurements of the length of a pencil, when compared carefully to a finely graduated scale, will invariably yield different results. The standard deviation of a measurement is determined from those variations. Admitting that a true length can never be found, the goal in making a measurement is to find an acceptable estimate and to have some knowledge of the uncertainty of the estimate. The mean (or average) of a group of measurements is taken to be the estimate and the standard deviation of the mean provides a measure of confidence. Results of measuring the length of a pencil could be reported as 181.3 mm ± 0.4 mm.

The word "error" is used when referring to variability of results within a set of measurements. When associated with a mistake or blunder, the word "error" has a justifiably bad connotation. But, given the impossibility of making a perfect measurement, errors are not necessarily bad but regularly occur with predictable behavior and are categorized as systematic errors or random errors.

## Blunders

Blunders are mistakes and are not considered legitimate observations. Any measurement containing a blunder should be discarded and not used. A blunder is the responsibility of the person making the measurement and is eliminated by exercising care, checking one's work, and making redundant observations. There is no mathematical magic for accommodating blunders in a set of data.

## Systematic Errors

Systematic errors arise from a mismatch between ideal (assumed) conditions and actual conditions of a measurement. A systematic error is characterized by its predictability and the fact that it always occurs with a given set of measurement conditions. Bias is another word sometimes used to describe systematic errors. Two examples of systematic error are (1) measuring a distance with a steel tape which has been foreshortened by cold temperature and (2) measuring a distance with an EDMI without specifying the appropriate correction for given temperature and pressure conditions.

## Random Errors

Random errors are the result of imperfect observations. Even though the goal is to use well-calibrated equipment according to proper procedures in order that random errors are kept as small as practical, they do occur and their predictable behavior has been well documented. The Normal Distribution Curve shown in Figure 4.13 illustrates the collective characteristics of random errors.

The characteristics of random error are as follows:

1. Small random errors occur more frequently than large ones.
2. Positive and negative random errors occur with similar frequency.
3. Very large random errors do not occur. If they do, they are taken to be blunders and discarded.

## ERROR SOURCES

Knowledge of possible error sources is important, both to those making measurements (collecting spatial data) and those responsible for determining the circumstances under which data are collected (buying equipment and/or writing specifications). Three general error sources are discussed next.

## Personal

All three types of errors can be attributed to a person making a measurement. It is common for a careless person to make mistakes and blunders. Even the most conscientious person will make random errors and systematic errors can arise from choices made about how computations are made or how data are used. For example, the difference between grid area and ground area for a parcel is a correction that can be computed and applied. As such it is counted as a personal systematic error if the person responsible for making that computation chooses not to make the correction. If the difference is small, the error may be inconsequential. However, such a difference could also be counted as a blunder if the difference is significant and the person responsible for the computation is oblivious to the need for such.

### Environmental

Environmental conditions give rise to both systematic errors and random errors. If one determines the temperature of a steel tape, it is possible to compute a temperature correction for a measured distance. By applying the temperature correction to the measurement, it is possible to minimize that systematic error source. Imperfect knowledge of the actual temperature leads to an imperfect correction. That difference is most likely a random error. If a cold front moves through during GNSS data collection, the results of the survey may be affected by the changing weather conditions. Depending upon the severity of the weather changes and upon availability of reliable meteorological data, the differences of a computed GNSS position could be a combination of systematic and random errors.

### Instrumental

If a piece of equipment malfunctions, one could call it a blunder. More often, instrumental errors are systematic and due to the physical construction of the instrument. An example is a steel tape whose length at standard temperature and tension differs from its nominal length. Random error for the same tape could be represented by scale graduations that are not perfectly spaced. And, gradual change of calibration parameters with respect to the performance of an EDMI give rise to random errors until such time as the EDMI is calibrated. With the calibration parameters known, that part of the error becomes systematic and an appropriate correction should be applied.

### ACCURACY AND PRECISION

Accuracy and precision are also related to errors. Blunders are mistakes related to one's level of diligence and professionalism. Systematic errors are related to the concept of accuracy, while random errors are related to the concept of precision.

Accuracy is a measure of absolute nearness of some measurement to the true value (the true value is never known, but an estimate is used in its place). A data set containing little or no systematic error is said to be accurate. Due to random error, there may be significant variation among different measurements of the same quantity, but the mean of the data set will be quite close to the true value.

Precision is a measure of consistency (or repeatability) within a given data set. A data set containing small random errors is said to be precise. However, if systematic error (a bias) is also present, it is possible to be precisely wrong.

Examples of accuracy and precision are often given as the result of shooting a gun at a target. Figure 4.14 shows four different results:

1. Both accurate and precise: There is a small grouping of holes located near the center of the target. This represents the desired result.
2. Precise, but not accurate: There is a small grouping of holes but they are obviously not located near the center of the target. Such a result indicates

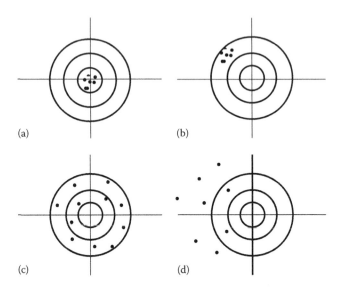

**FIGURE 4.14** Examples of Accuracy and Precision: (a) Both Accurate and Precise; (b) Precise, but Not Accurate; (c) Accurate, but Not Precise; and (d) Neither Accurate Nor Precise

    the presence of systematic error and/or the need to remove a bias by adjusting the cross-hairs in the telescope of the rifle.

3. Accurate, but not precise: There is a wide dispersion of holes over the target with no obvious grouping. Taken as a whole, the results can be said to be accurate, but the randomness in the grouping indicates the need for more practice, a steady rest, or some other factor to assure greater consistency.

4. Neither accurate nor precise: When some of the shots obviously miss the target, the marksman is neither accurate nor precise. The first step to improvement might be to eliminate the blunders (misses) after which decisions can be made as to the need for greater improvement.

## Computing Standard Deviations

What is the definition of "small" when dealing with random errors? The relative magnitude of a random error is defined mathematically by its standard deviation. Whether the standard deviation is large or small, the shape of the distribution is that given in Figure 4.13. However, whether the range covered by the random error is large or small is illustrated in Figure 4.14. In either case, the Greek letter sigma, $\sigma$, is used to denote standard deviation, which is computed using Equation 4.17.

## Standard Deviation of the Mean

Equation 4.17 gives the standard deviation associated with an individual measurement in the data set. That is an appropriate value to use when making comparisons between different methods of measurement or evaluating equipment. But, when using the mean of a set of observations in subsequent computations, the computed mean has a greater chance of being close to the true value than do any of the individual measurements. Specifically, in subsequent computations, the standard deviation of the mean, Equation 4.18, is needed rather than the standard deviation of a measurement. The standard deviation of the mean is the standard deviation of the measurement divided by the square root of the number of measurements. In the example, the standard deviation of a single 1000-meter measurement is 0.032 meter, but the standard deviation of the mean is 0.014 meter:

$$\sigma_i = \frac{\sqrt{\sum_{i=1}^{n} \left( Mean - x_i \right)^2}}{n-1} \tag{4.17}$$

$$\sigma_{mean} = \frac{\sqrt{\sum_{i=1}^{n} \left( Mean - x_i \right)^2}}{n(n-1)} = \frac{\sigma_i}{\sqrt{n}} \tag{4.18}$$

where:

$\sigma_i$ = Greek letter, sigma, for standard deviation of data set.
$\sigma_{mean}$ = standard deviation of the mean.
*Mean* = average of data set.
$x_i$ = single measurement.
$n$ = number of observations.

| No. | Observation (m) | Mean (m) | Difference (m) | Difference Squared |
|------|-----------------|----------|----------------|--------------------|
| 1. | 999.98 | 999.994 | 0.014 | 0.000196 |
| 2. | 1,000.02 | 999.994 | −0.026 | 0.000676 |
| 3. | 999.95 | 999.994 | 0.044 | 0.001936 |
| 4. | 1,000.03 | 999.994 | −0.036 | 0.001296 |
| 5. | 999.99 | 999.994 | 0.004 | 0.000016 |
| Total | 4,999.97 | | Sum | = 0.004120 |

Results: $\sigma_i = 0.032$ meter and $\sigma_{mean} = \sigma_i/\sqrt{n} = 0.014$ meter.

## CONFIDENCE INTERVALS

The standard deviation as computed in the example corresponds to a confidence level of 68 percent. As applied to the computed mean, there is a 68 percent chance the true value of the distance, whatever it is, lies between 999.980 meters and 1,000.008 meters. If one is not comfortable at the 68 percent level of confidence, it is routine to quote results at the 95 percent confidence level. For 95 percent confidence, the range quoted is the mean plus/minus two standard deviations. In this case, there is a 95 percent probability that the true value lies between 999.966 meters and 1,000.022 meters. The level of confidence is greater, but the quoted interval of uncertainty is also larger. Detail: Two standard deviations correspond to 95.5 percent and 95 percent corresponds to 1.96 standard deviations. Common practice is to use 2 standard deviations with 95 percent.

A limit of three standard deviations is associated with a confidence level of 99.7 percent. This criterion is often used to judge whether or not a given observation should be discarded as a blunder. As applied to the standard deviation of each observation, if the difference of an individual observation from the mean is greater than 3 standard deviations of the data set, standard practice is to eliminate that observation from the data set and recompute the mean and standard deviations with a smaller data set. Strictly speaking, a difference could exceed 3 standard deviations 3 times out of 1000 observations and not be a blunder but, generally, little harm is done by rejecting an observation that lies more than 3 standard deviations from the mean of the data set.

## HYPOTHESIS TESTING

Extending the concept of confidence intervals involves hypothesis testing. When comparing the results (statistics) of one data set with another, one question to ask is whether the data sets are from the same population of data or has something changed and thus the data sets are from two different populations of data. For example, is it really proper to make the comparisons between them and what is the likelihood of drawing conclusions based upon comparisons of unlike data (apples and oranges)? A brief discussion is included here, but a comprehensive discussion of hypothesis testing is beyond the scope of this book. Those principles are well formulated in books such as (Ghilani 2010) and constitute additional valuable tools for the spatial data analyst.

The basic elements of a hypothesis test are as follows:

1. Null hypothesis, $H_0$, which is the statement that compares one population statistic against its sample statistic or two sample statistics against each other.
2. Alternative hypothesis, $H_a$, which is the statement that is accepted if the null hypothesis is rejected by the test.
3. The test statistic, which is a statistic computed from the sample of data. This value is used to reject the null hypothesis.
4. The rejection region, which is the mathematical expression that is used to determine if the null hypothesis should be rejected.

Two errors can occur when performing a hypothesis test. The first error is known as the *Type I* error, which occurs when the $H_0$ is rejected as false when in fact it is true. This error will occur at the level of significance of the test. For example, assume that the test is performed at a level of significance of $\alpha$ percent, which means a $(1 - \alpha)$ percent confidence level. The Type I error will occur $\alpha$ percent of the time. Thus a hypothesis test, which is performed at a 5 percent level of significance or 95 percent confidence level will indicate that $H_0$ is false when, in fact, it is true 5 percent of the time. In essence, the results of the test are going to be incorrect a known 5 percent of the time.

The other error that can occur is known as the Type II error. This error occurs when the $H_0$ statement is false but the hypothesis test does not reject it. This error happens when the two samples of data are, in fact, from two different populations of data, and thus the sample statistics are not related. Unfortunately, nothing is often known about the population of the alternative hypothesis and thus the percentage of time that a Type II error will occur is often unknown. For this reason, the statement in $H_0$ can never be accepted. It can only be stated that there is no reason to believe that statement in $H_0$ is not true (Ghilani 2016).

Station BROMILOW is a GNSS-surveyed point on the NMSU campus and embedded in a section of sidewalk near Goddard Hall. In April 2008, NMSU facilities personnel replaced that section of sidewalk—removing the tablet marking the surveyed location of BROMILOW in the process. The construction manager insisted that the survey mark would be "replaced exactly where it was." A discussion of details supporting a hypothesis test of the resurveyed location is included in Chapter 15.

## Matrix Algebra

Matrix algebra is a compact way of representing and manipulating systems of linear equations. Compact representation helps humans grasp and discuss overall concepts without getting bogged down in details. Compact manipulation makes it possible to utilize standard programming procedures more efficiently in computerized solutions. Rules of matrix manipulation are developed in many math books and included, often as an appendix, in many texts devoted to survey computations or data adjustment. In particular, Ghilani (2010) covers matrix theory and applications for spatial data.

A matrix is a rectangular array of rows and columns and denoted by bold-faced type when written in text or in equations. Each subscripted position in the array is assigned a real number (as opposed to the cell of a spreadsheet which may contain text or equations in addition to numbers). The number of rows and columns are the dimensions of a matrix and are, at times, written as subscripts to the bold-faced matrix symbol. Individual elements within a matrix are often given by lowercase subscripted variables. An example of a simple 2 rows by 3 columns matrix is

$$A_{2,3} = \begin{bmatrix} a_{1,1} & a_{1,2} & a_{1,3} \\ a_{2,1} & a_{2,2} & a_{2,3} \end{bmatrix} = \begin{bmatrix} 2 & 1 & -1 \\ 4 & 0 & -2 \end{bmatrix} \tag{4.19}$$

Several important matrix concepts are as follows:

1. A square matrix has the same number of rows and columns.
2. The diagonal of a square matrix contains elements in those positions having identical row and column numbers. A diagonal matrix is one in which any nondiagonal element is zero.
3. An identity matrix is a diagonal matrix with 1s on the diagonal.
4. A symmetrical matrix has a mirror image with respect to the diagonal.
5. The transpose of a matrix (indicated by the superscript $t$) is obtained by switching the rows and columns of the parent matrix. The dimensions are switched accordingly. The transpose of $A_{2,4}$ is $A_{4,2}^t$.
6. A vector is a matrix having only one row or one column. A matrix with only one row and one column contains a single element, a real number.

7. Matrix addition is defined, for matrices having compatible dimensions, as a matrix containing the sum of the corresponding elements in the two parent matrices.

8. Likewise, matrix subtraction is defined, for matrices having compatible dimensions, as the difference of corresponding elements in the stipulated order.

9. Matrix multiplication is defined for matrices having compatible dimensions. That is, the number of rows in the second matrix must be the same as the number of columns in the first matrix. Each element in the product matrix is obtained as the sum of the products of row/column elements of the matrices being multiplied as shown in Equation 4.20.

$$
\begin{bmatrix} a_{1,1} & a_{1,2} \\ a_{2,1} & a_{2,2} \end{bmatrix} \times \begin{bmatrix} b_{1,1} & b_{1,2} \\ b_{2,1} & b_{2,2} \end{bmatrix} = \begin{bmatrix} a_{1,1}b_{1,1} + a_{1,2}b_{2,1} & a_{1,1}b_{1,2} + a_{1,2}b_{2,2} \\ a_{2,1}b_{1,1} + a_{2,2}b_{2,1} & a_{2,1}b_{1,2} + a_{2,2}b_{2,2} \end{bmatrix}
$$
$$
\begin{bmatrix} 1 & 2 \\ 2 & 3 \end{bmatrix} \times \begin{bmatrix} -2 & 4 \\ -1 & 2 \end{bmatrix} = \begin{bmatrix} -4 & 8 \\ -8 & 14 \end{bmatrix}
$$
(4.20)

10. Matrix division is not defined. Instead, the matrix inverse is an alternate procedure that is used and produces the same result. Given matrix $A$, the inverse is $A^{-1}$. In regular algebra $A \times (1/A) = 1$. In matrix algebra $A \times A^{-1}$ gives the identity matrix, $I$. Details for computing a matrix inverse are somewhat involved and not included here.

## MODELS

Models provide a connection between abstract concepts and human experience. In the context of spatial data, models give relevance and meaning to the concepts of location and geometrical relationships. Two kinds of models are used in this book, functional models and stochastic models.

### FUNCTIONAL

Functional models consist of physical, geometrical, mechanical, electrical, and other relationships that exist with respect to the cause and effect between observed fundamental quantities (e.g., length, time, temperature, and current) and computed results such as spatial data components. Using a model consists of writing and solving equations that describe the problem being considered and interpreting the results in terms of how well they agree with the original assumptions and observations.

### STOCHASTIC

A stochastic model describes the probabilistic characteristics of various elements of the functional model. Whether a quantity is fixed by law, determined by repeated measurements, or assigned on the basis of personal judgment, the stochastic model represents the "totality of the assumptions on the statistical properties of the variables involved" (Mikhail, 1976). The standard deviation of any quantity is a statistical measure of its quality. Statistical interaction between variables is known as correlation and—along with standard deviations—is captured in the appropriate variance/covariance matrix. With regard to 3-D coordinates representing spatial data, the following variance/covariance matrix represents the probabilistic characteristics of the defined point:

$$\Sigma_{X/Y/Z} = \begin{bmatrix} \sigma_X^2 & \sigma_{X/Y} & \sigma_{X/Z} \\ \sigma_{Y/X} & \sigma_Y^2 & \sigma_{Y/Z} \\ \sigma_{Z/X} & \sigma_{Z/Y} & \sigma_Z^2 \end{bmatrix} \qquad (4.21)$$

*Notes*:

1. The diagonal elements are called variances and the off-diagonal elements are called covariances. It is proper to refer to the entire matrix as a covariance matrix even though it contains both variances and covariances.
2. The standard deviation of each respective $X/Y/Z$ coordinate is the square root of the variance.
3. Correlation between variables (coordinates) is a number between $-1.0$ and $1.0$ and is obtained from elements in the covariance matrix. Since the correlation of $X$ with respect to $Y$ is the same as the correlation of $Y$ with respect to $X$, the covariance matrix is symmetric. Quantities that are statistically independent have zero correlations. Correlation is mathematically defined as

$$\rho_{XY} = \frac{\text{cov}(XY)}{\sigma_X \sigma_Y} \qquad (4.22)$$

where:
$\text{cov}(XY) = $ covariance value for element $X, Y$.
$\sigma_x$ and $\sigma_y = $ standard deviations of $X$ and $Y$.

## ERROR PROPAGATION

The theory of error propagation is derived in books such as Mikhail (1976) and Ghilani (2010) and presented concisely in matrix form as

$$\Sigma_{YY} = J_{YX} \Sigma_{XX} J_{XY}^t \qquad (4.23)$$

where:
$\Sigma_{YY} = $ covariance matrix of computed result.
$\Sigma_{XX} = $ covariance matrix of variables used in computation.
$J_{YX} = $ Jacobian matrix of partial derivatives of the result with respect to the variables.

Error propagation involves calculus and is used to answer the question, "If something is computed on the basis of a measurement and the measurement contains uncertainty, how is the computed quantity affected?" In a trivial case ($Y = X$), there is a direct correspondence between the two and the error in the result is the same as the error in the measurement. In another simple case—the volume of a tank and its height—the relationship between the measurement and the computed result is linear: volume = (area of base) × (height). Other cases—the volume of a tank and its radius—are more complex: volume = ($\pi R^2$) × (height). Here the relationship between the volume and radius is exponential. However, even more complexity arises when several measurements contribute simultaneously to the quantity being computed. Equation 4.23 handles all cases from the trivial to the most complex.

Equation 4.16 could be used to compute approximate changes in volume based upon changes in radius and height, but that approach is somewhat limited. Equation 4.23 is used to answer the specific question, "What is the standard deviation of the volume if the standard deviation of the radius and the standard deviation of the height are both known?" Admitting that standard deviations are estimates, the answer will still be an estimate. But, unlike using Equation 4.16 (which is accurate only for "small" values of $\Delta h$ and $\Delta R$), Equation 4.23 is a definitive procedure that is statistically reliable and the approximation is in the standard deviation of the measurements (the user's responsibility) and not in the equation. A simplified list of steps for performing error propagation is as follows:

1. Identify the variables (i.e., the measurements) and determine their standard deviations on the basis of repeated measurements, computations, or professional judgment.

    Measurements and standard deviations in the tank example are

$$R = 50.0 \text{ meters} \pm 0.07 \text{ m}$$

$$h = 10.00 \text{ meters} \pm 0.02 \text{ m}$$

2. Formulate the equations that will be used to compute the result:

$$V = \pi R^2 h = 78,539.82 \text{ m}^3 \tag{4.24}$$

3. Take the partial derivatives, one variable at a time:

$$\partial V / \partial h = \pi R^2 = 7,853.98 \text{ m}^2 \tag{4.25}$$

$$\partial V / \partial R = 2\pi R h = 3,141.59 \text{ m}^2 \tag{4.26}$$

4. Build the matrices as shown in Equation 4.23:

$$\Sigma_{XX} = \Sigma_{RH} = \begin{bmatrix} \sigma_R^2 & \sigma_{Rh} \\ \sigma_{Rh} & \sigma_h^2 \end{bmatrix} = \begin{bmatrix} 0.0049 & 0.0000 \\ 0.0000 & 0.0004 \end{bmatrix} \tag{4.27}$$

$$J_{YX} = \left[\frac{\partial V}{\partial R} \quad \frac{\partial V}{\partial h}\right] = \left[3,141.59 \quad 7,853.98\right] \qquad (4.28)$$

5. Perform the matrix operations. Computers make this task much easier.

$$\Sigma_{YY} = \left[3,141.59 \quad 7,853.98\right]\begin{bmatrix} 0.0049 & 0.0000 \\ 0.0000 & 0.0004 \end{bmatrix}\begin{bmatrix} 3,141.59 \\ 7,853.98 \end{bmatrix}$$

$$\Sigma_{YY} = \text{Variance of volume} = 73,034.98 \ \text{m}^6 \qquad (4.29)$$

6. Interpret the results:
   a. The standard deviation of the volume = square root of the variance = 270.25 m³. Realistically, this answer has no more than two significant figures. At the 68 percent confidence level, the standard deviation of the volume is 270 m³, and at the 95 percent confidence level, the standard deviation of the tank volume is 540.5 m³ (540 m³).
   b. The answer in step two really has only three significant figures and could be reported as 78,500 m³ ± 270 m³ (or, at 2 sigma, ± 540 m³).
   c. In this case, there is no correlation between the measurements of height and radius. Had there been, the $\Sigma_{XX}$ matrix off-diagonals would be nonzero.

What happens if correlation is present? The following example may help. A total station surveying instrument was used to measure the size of the tank, as shown in Figure 4.15. Admittedly, this might not be the best way to measure a tank but this procedure was chosen to show how correlation is included. Measurement of the radius and height is derived from independent observations of slope distances and zenith directions. This example assumes a standard deviation of 0.10 meter for each

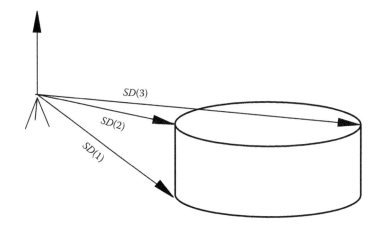

**FIGURE 4.15**   Survey Measurements of a Tank

slope distance and 20 seconds of arc for each zenith direction. It could also be appropriate to make other assumptions about standard deviations for the observations.

$$SD_1 = 101.119 \pm 0.10\,\text{m} \quad Z_1 = 98°\,31'\,51'' \pm 20''$$

$$SD_2 = 100.125 \pm 0.10\,\text{m} \quad Z_2 = 92°\,51'\,45'' \pm 20''$$

$$SD_3 = 200.062 \pm 0.10\,\text{m} \quad Z_3 = 91°\,25'\,55'' \pm 20''$$

To find radius and height from the field measurements, the functional model equations are

$$R = 0.5\left(SD_3 \sin Z_3 - SD_2 \sin Z_2\right) \tag{4.30}$$

$$H = SD_1 \cos Z_1 - SD_2 \cos Z_2 \tag{4.31}$$

The partial derivatives of the radius are

$$\partial R / \partial SD_1 = 0 = 0.0 \tag{4.32}$$

$$\partial R / \partial SD_2 = -0.5 \sin Z_2 = -0.499376165 \tag{4.33}$$

$$\partial R / \partial SD_3 = 0.5 \sin Z_3 = 0.499843856 \tag{4.34}$$

$$\partial R / \partial Z_1 = 0 = 0.0 \tag{4.35}$$

$$\partial R / \partial Z_2 = -0.5 \cos Z_2 = 2.500011935 \tag{4.36}$$

$$\partial R / \partial Z_3 = 0.5 \cos Z_3 = -2.499971439 \tag{4.37}$$

The partial derivatives of the height are

$$\partial H / \partial SD_1 = \cos Z_1 = -0.148340663 \tag{4.38}$$

$$\partial H / \partial SD_2 = -\cos Z_2 = 0.049937816 \tag{4.39}$$

$$\partial H / \partial SD_3 = 0.0 = 0.0 \tag{4.40}$$

$$\partial H / \partial Z_1 = -SD_1 \sin Z_1 = -100.0002519 \tag{4.41}$$

$$\partial H / \partial Z_2 = SD_2 \sin Z_2 = 100.0000769 \tag{4.42}$$

$$\partial H / \partial Z_3 = 0.0 = 0.0 \tag{4.43}$$

The Jacobian matrix (transposed for ease of printing) of partial derivatives is

$$
J^T = \begin{bmatrix}
0 & -0.148340663 \\
-0.499376165 & 0.049937816 \\
0.499843856 & 0 \\
0 & -100.0002519 \\
2.500011935 & 100.0000769 \\
-2.499971439 & 0
\end{bmatrix}
\tag{4.44}
$$

The covariance matrix of the observations is

$$
\Sigma_{observations} = \begin{bmatrix}
\sigma^2_{SD_1} & 0 & 0 & 0 & 0 & 0 \\
0 & \sigma^2_{SD_2} & 0 & 0 & 0 & 0 \\
0 & 0 & \sigma^2_{SD_3} & 0 & 0 & 0 \\
0 & 0 & 0 & \sigma^2_{Z_1} & 0 & 0 \\
0 & 0 & 0 & 0 & \sigma^2_{Z_2} & 0 \\
0 & 0 & 0 & 0 & 0 & \sigma^2_{Z_3}
\end{bmatrix}
\tag{4.45}
$$

Note that each slope distance was assumed to have a standard deviation of 0.10 meter and that each zenith direction has a standard deviation of 20 seconds of arc. Given that 20 seconds squared in radians is $9.401755 \times 10^{-9}$, the elements of the observation covariance matrix are

$$
\Sigma_{obs} = \begin{bmatrix}
0.01 & 0 & 0 & 0 & 0 & 0 \\
0 & 0.01 & 0 & 0 & 0 & 0 \\
0 & 0 & 0.01 & 0 & 0 & 0 \\
0 & 0 & 0 & 9.401755E-9 & 0 & 0 \\
0 & 0 & 0 & 0 & 9.401755E-9 & 0 \\
0 & 0 & 0 & 0 & 0 & 9.401755E-9
\end{bmatrix}
\tag{4.46}
$$

The covariance matrix of the derived radius and height is then computed as

$$
\Sigma_{RH} = J\Sigma_{obs}J^T = \begin{bmatrix}
0.0049923218667 & -0.000247027986 \\
-0.000247027986 & 0.00043302309592
\end{bmatrix}
\tag{4.47}
$$

Note that the covariance matrix in Equation 4.47 is almost the same as in Equation 4.27 except that Equation 4.47 contains correlation data for the derived measurements of radius and height. The standard deviation of the radius is 0.0707 meter (not 0.07 meter) and the standard deviation of the height is 0.0208 meter (not 0.02 meter). In the first example, radius and height were independent measurements (observations). In the second example, radius and height were both computed from the independent

slope distance and zenith direction observations. Radius and height are not independent and the covariance matrix contains correlation. The variance of the computed
volume is now computed (the same procedure as in Equation 4.28) as

$$\Sigma_{YY} = \begin{bmatrix} 3,141.59 & 7,853.98 \end{bmatrix} \begin{bmatrix} 0.0049923219 & -0.000247027 \\ -0.000247027 & 0.0004330231 \end{bmatrix} \begin{bmatrix} 3,141.59 \\ 7,853.98 \end{bmatrix}$$

$$\Sigma_{YY} = \text{Variance of volume} = 63{,}792.91 \text{ m}^6 \tag{4.48}$$

The correlated standard deviation of the computed volume is 252.57 m³.

When the results in Equation 4.48 are compared to those in Equation 4.28 the correlated results are somewhat smaller. However, in this case, when significant digits
are taken into account, the overall reported answer is the same in each case. The tank
volume is 78,500 ± 250 m³.

When does correlation make a significant difference? Each user must answer that
question for himself/herself. As technology permits observations to be made with
greater and greater precision and as smaller tolerances are imposed upon the computed result, correlated measurements will need to be considered. The important fact
here is that the independent observations must be identified and that the equations
(models) used to compute spatial data components will need to be used properly to
compute the correlated covariance matrices.

As shown in the tank example, Equation 4.23 is very powerful. Specifically,
matrix tools were used to illustrate using both correlated and uncorrelated measurements. Many surveying measurements are uncorrelated and, as shown here, even correlated measurements may give the same answer. In the past, the nonmatrix form of
Equation 4.23 has been quoted as the special law of propagation of variance, which
is used without correlations as

$$\sigma_U^2 = \left( \frac{\partial U}{\partial X} \right)^2 \sigma_X^2 + \left( \frac{\partial U}{\partial Y} \right)^2 \sigma_Y^2 + \left( \frac{\partial U}{\partial Z} \right)^2 \sigma_Z^2 + \cdots \tag{4.49}$$

where $U = f(X, Y, Z, \ldots)$ and $X/Y/Z$ are independent variables.

If the variables really are independent, Equation 4.49 can be applied to simple
equations to give error propagation equations listed in various textbooks and memo-
rized by many as

$$U = \text{Sum} = A + B \quad \sigma_{A+B} = \sqrt{\sigma_A^2 + \sigma_B^2} \tag{4.50}$$

$$U = \text{Difference} = A - B \quad \sigma_{A-B} = \sqrt{\sigma_A^2 + \sigma_B^2} \tag{4.51}$$

$$U = \text{Product} = A \times B \quad \sigma_{A \times B} = \sqrt{A^2 \sigma_B^2 + B^2 \sigma_A^2} \tag{4.52}$$

$$U = \text{Quotient} = A/B \quad \sigma_{A/B} = \sqrt{\frac{\sigma_A^2}{B^2} + \left( \frac{A}{B} \right)^2 \frac{\sigma_B^2}{B^2}} \tag{4.53}$$

Even when used without correlations, the error propagation equation is a powerful tool that has been underutilized. But, the matrix form of the error propagation, Equation 4.23, is even more powerful in that it utilizes the power of matrices to handle systems of complex equations and it handles any and all correlations that may be part of a problem, simple or complex.

## ERROR ELLIPSES

Error ellipses are a graphical tool used to illustrate the pair-wise correlation that exists between computed values. Using 2-D plane coordinates as an example, if the correlation between the computed coordinates is zero, then the orientation of the error ellipse corresponds to that of the host coordinate system. Special case: if the correlation is zero and the standard deviations are the same for both coordinates, then the error ellipse is a circle (and orientation is immaterial). However, if the standard deviation is the same for both coordinates and the correlation is not zero, then the maximum and minimum standard deviations will occur with some other orientation.

The general case, illustrated in Figure 4.16, accommodates standard deviations of coordinates that are *not* the same with respect to the *x* and *y* axes and in which correlation *does* exist. The respective *X/Y* standard deviations and orientation of the ellipse major axis are obtained from the point covariance matrix. Additional material on error ellipses is given in Ghilani (2010) and should be studied carefully in order to fully understand their usefulness. Error ellipses provide excellent visualization of correlation in two dimensions, typically in the horizontal plane. Visualization of 3-D error ellipsoids needs more study and discussion.

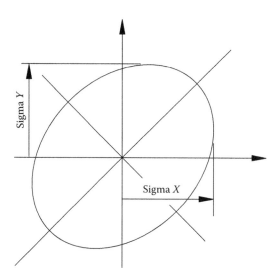

**FIGURE 4.16**   General Error Ellipse

## LEAST SQUARES

The principle of least squares states that the sum of the squares of the residuals—multiplied by their appropriate weights—will be a minimum for the set of answers (parameters) that has the greatest probability of being correct. The concept is simultaneously simple and complex because it applies with equal validity to computing a simple mean of two equally weighted measurements as well as adjustment of the most complex problem that can be described with functional model equations. Although it is correct to say that there is no known method proven to be better than a least squares solution, it is also true to say, within reason, that least squares can be used to obtain any desired answer. The difference lies in selection of weights, which is the responsibility of the user. The least squares procedure is specific and proven but least squares can also be abused, sometimes unwittingly, to an extreme where a solution has questionable, marginal, little, or no value at all. The challenge in using least squares is to select the appropriate model, to write the equations (observation and/or condition) correctly, and to assign legitimate weights to the observations. Generating the solution could be a challenge (if done long-hand), but computers are programmed to handle the matrices and to crunch the numbers needed to find the most probable solution to the problem. Having done all that, one could still argue that the most challenging part of using least squares is interpreting the results - such as using the covariance matrices to track 3-D spatial data accuracy.

Given one measurement of a distance, there is no basis for adjustment. In order to use least squares, there must be "extra" measurements. When tying in side shots to a survey traverse, there is no "extra" measurement to the radial point and no adjustment of that point is possible. But when computing around a closed loop traverse, the coordinates of the end point must be the same as used for the beginning point. Once coordinates of each traverse point are computed, least squares is an appropriate procedure by which to find the most probable coordinate values of the surveyed points. Of course, if the side shot points are tied in a second time from a separate survey point, redundancy does then exist and an adjustment of such redundant positions is possible.

In order to use least squares competently, the user must decide upon the appropriate model (is the coordinate system local, state plane, UTM, geodetic, or 3-D) and write equations for the computations that utilize the measurements. As stated above, if there is no redundancy in the measurements, no adjustment is possible. But, given a loop traverse and redundant measurements, the equations used in the computations must be consistent with the model being used. In most cases, a slope distance must be reduced to horizontal and the horizontal distance must be reduced to the ellipsoid if using geodetic coordinates or to the state plane coordinate grid (in the appropriate zone) if using state plane coordinates. Least squares cannot be used to correct errors caused by using the wrong model. Often, the model is implicit and defined by the context. For example, when measuring a distance with a plumb bob and steel tape, the implied model is horizontal distance and everyone knows what that is. But, if the distance is measured with EDMI or with GNSS, it becomes more important to be specific about the definition of horizontal. For example, six different definitions of horizontal are given by Burkholder (1991), each being more precise (specific) than the previous one. It is the user's responsibility to assure compatibility between the measurements, the model, and the solution obtained from a least squares adjustment.

In addition to providing the best possible geometrical answer to a network of redundant observations, a least squares adjustment can also be used to determine the statistical properties (standard deviations) of the answers obtained from the adjustment. This provides the spatial data analyst with valuable (error propagation) tools for making decisions about how the answers are used or interpreted. For example, elevations of points on a building are determined very carefully and compared with elevations determined, say 6 months, earlier. Did the building move during that time interval? If the building did move, how much did it move? If the differences are small, does that mean the building really did move? Or, is it possible that the observed difference is the result of accumulation of random errors in the measurements? Least squares, error propagation, error ellipses, and hypothesis testing are tools that can be used to make statements based upon inferences having a rigorous statistical foundation at a level of confidence chosen by the user.

## LINEARIZATION

Taken by itself, the concept of least squares represents too much computational effort to be practical. However, if the least squares process is combined with matrices (for computational compactness) and with computers (for processing speed), a least squares solution becomes feasible for larger and larger problems. A second drawback to using least squares is that matrices are valid only for systems of linear equations and many spatial data computations involve nonlinear geometrical relationships. That obstacle is overcome using a process called linearization in which nonlinear equations are replaced by their Taylor series approximation.

When using least squares to solve a linear problem, the solution is obtained on the basis of a single iteration. However, successive iteration is generally needed when using least squares to solve nonlinear problems. Since only the first two terms of the Taylor series are used in the matrix formulation (point-slope form of a line), the solution (a set of corrections to the previously adopted values) of a set of equations will be only an approximation. Stated differently, a consequence of linearization is that the answer being sought changes from "the actual numerical value" to "what correction(s) to the previous approximate value(s) will provide a better solution?" (For linear problems, the correction is the value between zero and the correct answer.) The most important question is, "What is an acceptable answer?" An acceptable answer is one that fulfills the original (nonlinear) conditions within some tolerance selected by the user. That puts the user in control.

A corollary to the previous question might be, "How small must the corrections be to be acceptable?" Answering this question also keeps the user in control. But, achieving the goal of finding the right answer or making the corrections go to zero (within some tolerance) typically requires an enormous amount of number crunching—feasible only when done on a computer.

PROCEDURE FOR NONLINEAR SOLUTION

Very briefly, the overall process for solving a nonlinear problem is as follows:

1. Identify the geometry of the problem and write the appropriate equations.
2. Linearize the equations and take partial derivatives to be used in the matrix formulation.
3. Establish some initial value for each unknown parameter as being reasonably close to the final answer.
4. Run the least squares adjustment to find corrections to the initial estimates.
5. Look at the results. Are the corrections small enough to quit? If so, quit. If not, update the previous estimate using current corrections and run the adjustment again. Are the corrections smaller and are they small enough?

Two important concepts described above are iteration and convergence. Iteration is the process of using results from a previous solution to solve the problem again. Convergence is the desirable condition realized when each successive correction is smaller than the previous one. If a solution converges slowly, it may take many iterations to solve a problem. Given computers are programmed to do the number crunching, the time and effort required may or may not be an issue. If solutions are being generated in real time, rapid convergence is preferable and linear models that do not require iteration are even more desirable. When using the GSDM, network adjustments can be formulated as a linear model.

## APPLICATIONS TO THE GSDM

The First Edition of this book was primarily devoted to describing the GSDM. This Second Edition includes an additional discussion of linear least squares in Chapter 13, expands coverage on network and local accuracy in Chapter 14, and gives numerous examples of practical applications in Chapter 15. In looking ahead, a summary of concepts includes the following:

- Observations are independent measurements of fundamental physical quantities.
- Observations are manipulated in conformance with physical laws and geometrical relationships to obtain spatial data components. For example, carrier phase GPS observations are processed to obtain geocentric components while total-station survey measurements provide local components. Each can be used and/or combined with the other when using the GSDM.
- The standard deviation of each observation is propagated to the corresponding spatial data component. Correlation data are tracked in the associated covariance matrix. Correlation is largely responsible for the difference between network accuracy and local accuracy from one point to another (Burkholder 1999, 2004).
- Spatial data components (indirect measurements) are combined into networks according to existing geometrical configurations and evaluated for statistical reliability.

- Once spatial data components pass rigorous quality control criteria, they are combined with existing primary database points in a least squares adjustment.
- Stored values include the geocentric $X/Y/Z$ values of each point, the associated covariance values, and the point-correlation values.
- When drawn from the 3-D database, the spatial accuracy of each point is given by its standard deviation, component by component. The value or utility of each database point is determined by whether it passes a tolerance filter (for each component) as selected by the user.
- Standard deviations of any/all derived quantities are available using proven standard error propagation computations.
- Both local and network accuracies can be computed using the full covariance matrix in Equation 1.36. That matrix formulation was first given in Burkholder (1999). The example in Chapter 14 shows computation of both network and local accuracies for points that were directly connected as well as points that were not directly connected.

## REFERENCES

Alder, K. 2002. *The Measure of All Things: The Seven Year Odyssey and Hidden Error that Transformed the World.* New York: The Free Press.

Beckman, P. 1971. *A History of PI* New York: St. Martin's.

Bumby, D.R. and Klutch, R.J. 1982. *Mathematics: A Topical Approach.* Columbus, OH: Merrell.

Burkholder, E.F. 1982. Algorithm for accurate area computations. *Surveyor Calculations Journal* 92 (7): 9–10 (Joe Bell, Publisher, San Bernardino, CA 92412).

Burkholder, E.F. 1991. Level/horizontal distances. *Journal of Surveying Engineering* 117 (3): 104–116.

Burkholder, E.F. 1999. Spatial data accuracy as defined by the GSDM. *Surveying and Land Information Systems* 59 (1): 26–30.

Burkholder, E.F. 2004. *Fundamentals of Spatial Data Accuracy and the Global Spatial Data Model (GSDM).* Washington, DC: Filed with U.S. Copyright Office.

Chen, W.F., (ed.) 1995. *The Civil Engineering Handbook.* Boca Raton, FL: CRC Press.

Cressie, N.A.C. 1993. *Statistics for Spatial Data* New York: John Wiley.

Davis, R., Foote, F., Anderson, J., and Mikhail, E. 1981. *Surveying: Theory & Practice.* New York: McGraw-Hill.

Dolciani, M.P., Beckenbach, E.F., Donnelly, A., Jurgensen, R., and Wooton, W. 1967. *Modern Introductory Analysis.* Boston, MA: Houghton Mifflin.

Dwass, M. 1970. *Probability: Theory & Applications.* New York: W.A. Benjamin.

Ghilani, C. 2016. Personal communication.

Ghilani, C. and P.R. Wolf. 2012. *Elementary Surveying: An Introduction to Geomatics*, 13th ed. Upper Saddle River, NJ: Pearson Education.

Ghilani, C.D. 2010. *Adjustments Computations: Spatial Data Analysis*, 5th ed. Hoboken, NJ: John Wiley.

Hawking, S.W. 1988. *A Brief History of Time.* New York: Bantam.

Mikhail, E. 1976. *Observations and Least Squares.* New York: Donnelly.

Moffitt, F. and J. Bossler. 1998. *Surveying*, 10th ed. New York: Addison-Wesley.

Russell, B. 1959. *Wisdom of the West.* London, U.K.: Crescent Book.

Smith, J.R. 1986. *From Plane to Spheroid: Determining the Figure of the Earth from 3000 B.C. to the 18th Century Lapland and Peruvian Survey Expeditions.* Rancho Cordova, CA: Landmark Enterprises.

Stanfel, L. 1994. Survey distance units: A better way. *Journal of Surveying Engineering* 120 (3): 130–132 (See also subsequent discussion in Vol. 122 (1): 41–44).
Tooley, R.V. [1949] 1990. *Maps and Mapmakers*. New York: Dorset.
Wilford, J.N. 1981. *The Mapmakers*. New York: Vintage.
Wine, R.L. 1964. *Statistics for Scientists and Engineers*. Englewood Cliffs, NJ: Prentice-Hall.

# 5 Geometrical Models for Spatial Data Computations

## INTRODUCTION

As previously defined, a mathematical model is a set of rules used to make a conceptual connection between abstract concepts and human experience. A model is judged "good" to the extent it is both simple and appropriate. When working with spatial data, the simplest model is a one-dimensional (1-D) distance. Models of increasing complexity include a 2-D plane coordinate system formed by the perpendicular intersection of the $X$ and $Y$ axes, a generic 3-D $X/Y/Z$ rectangular Cartesian coordinate system having three mutually perpendicular axes, a spherical Earth model, and finally, the ellipsoidal Earth model. Figure 5.1a shows the standard 2-D $X/Y$ coordinate system, Figure 5.1b shows a right-handed $X/Y/Z$ coordinate system, and Figure 5.1c illustrates the sexagesimal coordinate system of latitude and longitude used to describe the geodetic location of points on an ellipsoidal surface representing a mathematical approximation of the surface of the Earth. Other choices will be discussed later, but the goal at this point is to identify a variety of geometrical model choices. With regard to working with spatial data, considerations include, but are not necessarily limited to, the following:

- Are the observations or subsequently computed measurements 1-D, 2-D, or 3-D?
- Is a 1-D or 2-D model sufficient, or is a 3-D model required?
- Is the extent of a given project small enough to use "flat-Earth" relationships or is a different model needed to accommodate curvature of the Earth? Is a spherical Earth model appropriate or is the ellipsoidal Earth model required?
- Is the project of such a nature that a local coordinate system is sufficient or should the data be referenced to the National Spatial Reference System (NSRS)?
- What issues of compatibility (e.g., units of feet or meters) must be addressed? What is required for new measurements to be compatible with and/or added to the value of existing data?
- Is there a spatial data model that accommodates all computational concerns? If so, what is it? That decision should be documented specifically for each project. Otherwise subsequent users are forced to infer the model from the way spatial data are used. For example, project datum (also called surface) coordinates often resemble state plane coordinates and serious problems may result if project datum coordinates are used as if they were state

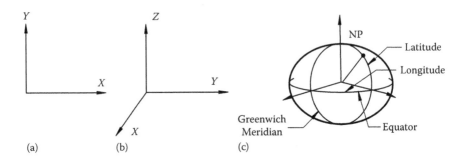

**FIGURE 5.1**  (a) 2-D Rectangular System, (b) 3-D Rectangular System, and (c) the Geodetic Curvilinear System

plane coordinates. A low-distortion projection (LDP) is becoming popular for some applications. Those too can be used successfully if documented and administered properly. See Chapter 11 for more information on use of 2-D map projections.

## CONVENTIONS

With regard to the use of various models for handling spatial data, the following conventions, although contradictory at times, are commonly used:

- Standard computational practice in scientific, engineering, and mathematical disciplines employs the right-handed rectangular 3-D coordinate system described earlier and shown in Figure 5.1b. The conventional 2-D $X/Y$ rectangular coordinate system is a subset of the right-handed 3-D $X/Y/Z$ system.
- Conventional surveying and mapping practice includes use of latitudes and departures, northings and eastings, and $X/Y$ coordinates in the 2-D plane. The goal here is to accommodate existing practice to the extent possible. However, the convention of east/north/up (not north/east/up) will be used for local perspective coordinates because it is right-handed and compatible with the underlying geocentric ECEF system described in Chapter 3. Regretfully, clockwise azimuth is a left-handed convention.
- Coordinate differences between points are denoted as "$\Delta$" and computed as forepoint (point 2) minus standpoint (point 1). Given that computations are performed with respect to the standpoint and that another point of interest is the forepoint, then rectangular spatial data (vector) components are computed as

$$\Delta x = x_2 - x_1 \qquad (5.1)$$

$$\Delta y = y_2 - y_1 \qquad (5.2)$$

$$\Delta z = z_2 - z_1 \qquad (5.3)$$

- Directions are given in sexagesimal units as bearings or azimuths. An azimuth is counted from north clockwise through a complete circle of 360°. Negative azimuths imply a counterclockwise (angle left) rotation. A negative azimuth can be changed to a positive azimuth by adding 360°. Bearings are quadrant based and related to azimuths as follows:
  - Quadrant I: A northeast (NE) bearing has the same value as an azimuth.
  - Quadrant II: A southeast (SE) bearing is 180° minus the azimuth of the line.
  - Quadrant III: A southwest (SW) bearing is the azimuth of the line minus 180°.
  - Quadrant IV: A northwest (NW) bearing is 360° minus the azimuth of the line.
- Bearings and azimuths are distinguished by the meridian to which they are referenced. Common examples are as follows:
  - Magnetic bearings are referenced to lines of the Earth's magnetic field that converge at the magnetic poles. Given that the magnetic pole is not coincident with the true pole (as defined by the Earth's spin axis), magnetic declination is the difference between magnetic north and true north.
  - An astronomic azimuth is determined by observing the sun or stars using a transit or theodolite leveled with respect to the local plumb line (vertical). An astronomical meridian goes to the Earth's instantaneous spin axis. This pole position changes slowly and differs slightly from the Conventional Terrestrial Pole (CTP), the mathematical North Pole, adopted by international agreement.
  - Geodetic azimuth is referenced to the meridian to the CTP with respect to the ellipsoid normal (instead of the vertical) and is used in geodetic position computations. A geodetic azimuth differs from an astronomic azimuth due to (1) the difference between the direction of the ellipsoid normal and the direction of the vertical (deflection-of-the-vertical) and (2) due to the difference between the instantaneous North Pole and the CTP (polar wandering). See Chapter 7 for more detail on each difference.
  - Grid azimuths are commonly encountered when working with state plane (or map projection) coordinates. At the center of each map projection zone, the central meridian coincides with the true geodetic meridian. All geodetic meridians converge to the CTP but grid meridians are generally parallel on the projection surface. Convergence is the difference between a geodetic azimuth and its corresponding grid azimuth from one point to another. Convergence is zero on the central meridian of a map projection.
  - A 3-D azimuth is very nearly the same as a geodetic azimuth but is much easier to compute. The 3-D azimuth is computed as $\tan^{-1}(\Delta e/\Delta n)$ where $\Delta e$ and $\Delta n$ are the local tangent plane components of a 3-D vector. See Burkholder (1997) or Chapter 7 for more details.

- An assumed azimuth, encountered in many places, is whatever the user declares it to be.
- Radian mode is the standard convention for computing trigonometric functions on a computer or spreadsheet. However, many calculators operate in the decimal degree mode when computing trigonometric functions. It is important to convert sexagesimal degrees-minutes-seconds to decimal degrees before computing a trigonometric function. Time is also expressed in sexagesimal units and some calculators provide a hardwired function for converting hours-minutes-seconds to decimal hours. It is the same sexagesimal conversion. Look for D-M-S to DD or H-M-S to HR for converting sexagesimal to decimal. Or, it can be done easily on the keyboard as

$$D.D \left( \text{decimal degrees} \right) = Degrees + Minutes/60 + Seconds/3600$$

The opposite computation is also important. When computing inverse trigonometric functions such as $\tan^{-1}(\Delta x/\Delta y)$ for azimuth, the answer is displayed by many calculators as decimal degrees. The function to convert decimal degrees to degrees-minutes-seconds is hardwired in many calculators as *HR* to *H-M-S* (or *DD* to *D-M-S*). If not hardwired in the calculator, the conversion can also be done on the keyboard.

The keyboard procedure for converting decimal degrees to degrees/minutes/seconds is

*Degrees* = integer portion of *DD.DDDDDDDDD*
*Minutes* = integer portion of ([*DD.DDDDDDD* − *degrees*] × 60)
*Seconds* = *DD.DDDDDDDD* × 3600 − *degrees* × 3600 − *minutes* × 60

Some calculators will show symbols for degrees-minutes-seconds. An alternative formatting convention for showing angular units in a calculator display is

Decimal degrees: *DD.DDDDDDDDDDDD*
Degrees/minutes/seconds: *DD.MMSSSSSSS* (with an implied decimal point in the seconds as SS.SSSS)

Another practice used by some software packages includes

Decimal degrees: *DD.DDDDDDDD*
Degrees/minutes/seconds: *DDMMSS.SSSSSSS*

Standard degree-minute-second symbols, use of zeros for placeholders, and placement of the decimal point are illustrated as

Latitude = 038° 00′ 03.″44563 where the seconds symbol (″) is either over or follows the decimal point in the seconds.

Of course, a calculator may also be set to operate in radian units, in which case the conversions are

$$\text{Radians} = DD \times \left(\pi/180°\right) \quad \text{and} \quad DD = \text{Radians} \times \left(180°/\pi\right) \quad \text{or}$$
$$\text{Radians} = DD \times 3600/SPR \quad \text{and} \quad DD = \text{Radians} \times SPR/3600 \quad \text{or}$$
$$\text{Radians} = \text{Seconds}/SPR \quad \text{and} \quad \text{Seconds} = \text{Radians} \times SPR$$

where $SPR = 206{,}264.806247096$ seconds per radian.

- When geodetic latitude or longitude is given in sexagesimal units, it is customary to show five decimal places of seconds for control point positions. If the Earth were a sphere with a radius of 6,372,000 meters, 1 second of arc would correspond to a distance of 30.89 meters on the surface of the sphere. So, $0.''00001$ of arc corresponds to $0.000309$ meter (submillimeter accuracy).
- It is also understood that latitude is used as a positive value (0° to 90°) in the northern hemisphere and as a negative value (0° to −90°) in the southern hemisphere.
- Longitude starts with 0° at the Greenwich Meridian and is used 0°–360° eastward. However, at the October 1884 International Meridian Conference held in Washington, DC, it was agreed that longitude would be counted both east and west from the Greenwich Meridian up to 180° (Howse 1980). Therefore, west longitude is the standard practice for geographic locations in the western hemisphere. Using a west longitude as a negative number is compatible with the mathematically unambiguous practice of using 0°–360° eastward.
- For many spatial data applications, horizontal distance is computed from plane surveying latitudes and departures. As such, it is the hypotenuse of a plane right triangle. But, when working with different elevations, horizontal distance is also taken to be the right-triangle component of a slope distance. Depending upon the coordinate system being used, other definitions of horizontal distance include the geodetic distance on the ellipsoid, the grid distance as used in a state plane coordinate system, or a local plane coordinate system.

Conventions used by the GSDM include horizontal distance as the local tangent-plane right-triangle component of a GNSS vector (same as used in plane surveying) and the 3-D azimuth (in degrees-minutes-seconds) as computed from the latitude and departure components of the same horizontal distance. To the extent possible, the GSDM includes relevant plane surveying practices without sacrificing advantages of a rigorous connection to the NSRS.

## TWO-DIMENSIONAL CARTESIAN MODELS

The standard 2-D rectangular coordinate system has an origin formed by the perpendicular intersection of the abscissa and the ordinate. Two systems, called the Math/Science Reference System and the Engineering/Surveying Reference System,

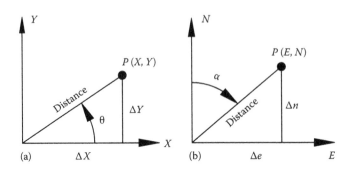

**FIGURE 5.2**   (a) Math/Science and (b) Engineering/Surveying Coordinate Systems

are quite similar but, from one system to the other, the reference axes are different and the direction of positive rotation is reversed; see Figure 5.2a and b.

### MATH/SCIENCE REFERENCE SYSTEM

The 2-D coordinate system commonly used by mathematicians and scientists labels the abscissa as the $X$-axis and the ordinate as the $Y$-axis. The positive $X$-axis is considered to be the reference for angles and rotation is counted positive counterclockwise. In this system, the $\Delta x$ and $\Delta y$ components of any directed line segment (vector) are

$$\Delta x = d \cos \theta \tag{5.4}$$

$$\Delta y = d \sin \theta \tag{5.5}$$

where:
  $d$ = length (distance) of the directed line segment.
  $\theta$ = direction of the line segment as measured counterclockwise with respect to positive $X$-axis (consistent with right-handed rule).

### ENGINEERING/SURVEYING REFERENCE SYSTEM

Surveyors, engineers, and others who work with mapping data often use a 2-D rectangular coordinate system, which is similar to the math/science system except that cardinal directions of north-south and east-west are superimposed upon the two axes and rotation is counterclockwise from north. In this system, the $\Delta x$ (*easting*) and $\Delta y$ (*northing*) components of any directed line segment are

$$\Delta x = \Delta e = d \sin \alpha \tag{5.6}$$

$$\Delta y = \Delta n = d \cos \alpha \tag{5.7}$$

where:
  $d$ = length (distance) of the directed line segment.
  $\alpha$ = azimuth of the line segment as measured clockwise with respect to north, the positive $Y$-axis (contrary to right-handed rule).

The specific relationship between the math/science system and the engineering/ surveying system is that, in all cases, $\alpha + \theta = 90°$, which also means $\alpha = 90° - \theta$ or $\theta = 90° - \alpha$. Being aware of these similarities and differences helps spatial data users make greater use of the polar/rectangular conversions hardwired into many calculators.

For example, most scientific calculators are hardwired according to the math/science system. To compute rectangular components of a line 100.000 meters long having an azimuth of 30°, the calculator will show 50.000 meters as being the north/south component when it is really the east/west component. Similarly, because it is using the math/science convention, the calculator will show 86.602 meters as being the east/west component when it really is the north/south component (in the engineering/surveying system). The only thing the user needs to do is switch the label of the computed components. A word of caution: Each user should practice and initially work with known quantities to make sure the procedure being used is giving correct answers in the intended system.

In each case, the 2-D math/science reference system and the 2-D engineering/ surveying reference system can be expanded into a 3-D system by adding a Z-axis. The convention adopted in this book is to use $e/n/u$ for the local reference frame because it is right-handed and $\Delta e/\Delta n/\Delta u$ can be conveniently rotated into the ECEF right-handed $X/Y/Z$ reference frame. Regretfully, azimuth in the engineering/ surveying system is not consistent with the right-hand rule.

## COORDINATE GEOMETRY

Once rectangular components are obtained in either the math/science or engineering/ surveying system, standard coordinate geometry (often referred to as COGO) operations are the same. Admittedly, it becomes a challenge to keep track of various conventions employing $X/Y$ coordinates and *eastings/northings* in the same rectangular system, but the COGO procedures are the same in each case. Derivations for the following 2-D COGO operations are given in Burkholder (1984).

1. Forward (traverse to new point)
2. Inverse (find direction and distance from standpoint to forepoint)
3. Line-line (also called bearing-bearing) intersection
4. Line-circle (also called bearing-distance) intersection
5. Circle-circle (also called distance-distance) intersection
6. Perpendicular offset distance from a line to a point

### FORWARD

Given:
   $e_1$ and $n_1$ = coordinates at standpoint.
   $d$ = distance from standpoint to forepoint.
   $\alpha$ = azimuth from standpoint to forepoint.

Compute:
   $\Delta e$ and $\Delta n$ = rectangular components.
   $e_2$ and $n_2$ = coordinates of forepoint.

Solution:

$$e_2 = e_1 + \Delta e = e_1 + d \sin \alpha \qquad (5.8)$$

$$n_2 = n_1 + \Delta n = n_1 + d \cos \alpha \qquad (5.9)$$

## INVERSE

Given:
$e_1$ and $n_1$ = coordinates at standpoint.
$e_2$ and $n_2$ = coordinates at forepoint.

Compute:
$\Delta e$ and $\Delta n$ = rectangular components.
$d$ = distance between standpoint and forepoint.
$\alpha$ = azimuth from standpoint to forepoint.

Solution:

$$\Delta e = e_2 - e_1 \qquad (5.1)$$

and

$$\Delta n = n_2 - n_1 \qquad (5.2)$$

$$d = \sqrt{\left( \Delta e^2 + \Delta n^2 \right)} \qquad (5.10)$$

$$\tan \alpha = \Delta e / \Delta n$$

$$\left( \text{use signs of } \Delta e \text{ and } \Delta n \text{ to determine the proper quadrant} \right)$$

$$\text{If } \Delta n \text{ is } -, \quad \alpha = 180° + \arctan\left( \Delta e / \Delta n \right) \qquad (5.11)$$

$$\text{If } \Delta n \text{ is } + \text{ and } \Delta e \text{ is } -, \quad \alpha = 360° + \arctan\left( \Delta e / \Delta n \right) \qquad (5.12)$$

$$\text{If } \Delta n \text{ and } \Delta e \text{ are both } +, \quad \alpha = \arctan\left( \Delta x / \Delta y \right) \qquad (5.13)$$

## INTERSECTIONS

When performing design or other COGO computations, it is often necessary to determine where lines intersect; see Figure 5.3. Three methods are

1. Line-line: In this case, points 1 and 2 are given. The direction from point 1 and the direction to point 2 are also given. The problem is to compute the coordinates of the intersection point. Line-line is also called a bearing-bearing intersection.

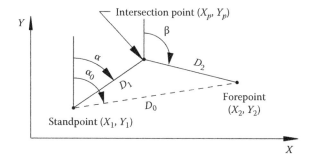

**FIGURE 5.3** Geometry of Intersections

2. Line-circle: In this case, points 1 and 2 are given. The direction from point 1 and the distance from the intersection point to point 2 are given. The problem is to compute the coordinates of the intersection point. Line-circle is also called a bearing-distance intersection.

3. Circle-circle: In this case, points 1 and 2 are given. The distance from point 1 to the intersection point and the distance from the intersection point to point 2 are given. The problem is to compute the coordinates for the intersection point. Circle-circle is also called a distance-distance intersection.

In each case, the computation starts at the standpoint and ends at the forepoint while establishing the location of an intermediate intersection point. The rules to be used in establishing the computed intersection point vary according to the type of intersection, but they fall into one of the three categories listed. Symbols and conventions for the computations are as follows:

$(e_1, n_1)$ = coordinates of standpoint (given).
$(e_2, n_2)$ = coordinates of forepoint (given).
$(e_p, n_p)$ = coordinates of intersection point (computed result).
$d_0$ = distance from standpoint to forepoint.
$d_1$ = distance from standpoint to intersection point.
$d_2$ = distance from intersection point to forepoint.
$\alpha_0$ = azimuth from standpoint to forepoint.
$\alpha$ = azimuth from standpoint to intersection point.
$\beta$ = azimuth from intersection point to forepoint.

In each case, the solution starts by computing $\Delta e = e_2 - e_1$ and $\Delta n = n_2 - n_1$. Also note that no solution exists if, in the first case, the lines are parallel; in the second case, the line does not intersect the circle; and lastly, the circles do not intersect. A different mathematical impossibility is encountered in each case:

Line-line: If the lines are parallel, azimuths $\alpha$ and $\beta$ are the same and the denominator of Equation 5.14 goes to zero. As a result, the distance $d_1$ is undefined. There is no intersection. See Figure 5.4a.

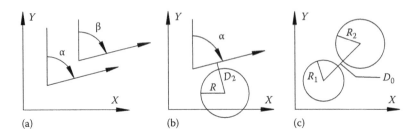

(a)                 (b)                 (c)

**FIGURE 5.4** Examples of Intersections That Fail. (a) Parallel Lines Do Not Intersect, (b) Line Does Not Intersect the Circle, and (c) Circles Do Not Intersect

Line-circle: If the line does not intersect the circle, the perpendicular offset distance from the line to the circle is greater than the radius of the circle. If that happens, the quantity under the radical in Equation 5.15 is negative. Since it is not possible to take the square root of a negative number, $d_1$ is undefined and there is no intersection. See Figure 5.4b.

Circle-circle: If the two circles do not intersect, it is not possible to find the angle $\gamma$ using Equation 5.16 because, in one case, the value of cos $\gamma$ is greater than 1.0 and, in the other case (one circle entirely within the other), the value of cos $\gamma$ is less than −1.0. See Figure 5.4c.

Other than the nonintersection cases described, the following formulas provide very efficient procedures for computing intersections in a two-dimensional plane.

**Line-Line (One Solution or No Solution If Lines Are Parallel)**

Given:

$$e_1, n_1, e_2, n_2, \alpha, \text{ and } \beta \left( \Delta e = e_2 - e_1 \text{ and } \Delta n = n_2 - n_1 \right)$$

Compute:

$$d_1 = \left( \Delta e \cos \beta - \Delta n \sin \beta \right)/\sin \left( \alpha - \beta \right) \quad \left( d_1 \text{ may be either a positive or negative value.} \right) \tag{5.14}$$

Solution:

$$e_p = e_1 + d_1 \sin \alpha \tag{5.8}$$

$$n_p = n_1 + d_1 \cos \alpha \tag{5.9}$$

Check: Inverse intersection point to forepoint, compare computed direction $\beta$ with given direction $\beta$. If they are not the same, a mistake was made.

**Line-Circle (May Have Two Solutions, One Solution, or No Solution)**

Given:

$$e_1, n_1, e_2, n_2, \alpha, \text{ and } d_2$$

Compute:

$$d_1 = \Delta e \sin \alpha + \Delta n \cos \alpha \pm \sqrt{d_2^2 - \left[\Delta e \cos \alpha - \Delta n \sin \alpha\right]^2}$$

$$\left(d_1 \text{ normally has two values, one for each solution.}\right)$$

(5.15)

Solution:

$$e_p = e_1 + d_1 \sin \alpha$$

(5.8)

$$n_p = n_1 + d_1 \cos \alpha$$

(5.9)

Check: Inverse intersection point to forepoint, compare computed distance $d_2$ with given distance $d_2$. They should be the same. Also make sure the solution obtained is the one desired. Depending upon where the line intersects the circle, the values of $d_1$ could both be positive, one positive and one negative, or both negative.

## Circle-Circle (May Have Two Solutions, One Solution, or No Solution)

Given:

$$e_1, n_1, e_2, n_2, d_1 \quad \text{and} \quad d_2$$

Compute:

$$d_0 = \sqrt{\left(\Delta e^2 + \Delta n^2\right)}$$

(5.10)

$$\alpha_0 = \tan^{-1}\left(\Delta e / \Delta n\right)$$

(5.11, 5.12, or 5.13)

$$\gamma = \frac{\cos^{-1}\left(d_1^2 + d_0^2 - d_2^2\right)}{2 d_1 d_0}$$

(5.16)

$$\alpha = \alpha_0 + \gamma$$

(5.17)

$$\alpha = \alpha_0 - \gamma$$

(5.18)

$$\left(\text{two solutions}\right)$$

Solution:

$$e_p = e_1 + d_1 \sin \alpha$$

(5.8)

$$n_p = n_1 + d_1 \cos \alpha$$

(5.9)

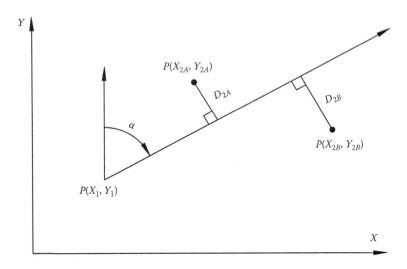

**FIGURE 5.5**  Perpendicular Offset to a Line. *Note*: $D_2$ Will Be Positive for Points Right of the Line but Negative for Points Left of the Line

Check: Inverse intersection point to forepoint, compare computed distance $d_2$ with given distance $d_2$. They should be the same. Also make sure the solution obtained is the one desired. Several solutions may exist. Also be aware that Equation 5.16 has no solution if the two circles do not intersect.

## PERPENDICULAR OFFSET

In many situations, it is desirable to find the perpendicular distance from a line to a point. Given one is on a standpoint $P(e_1, n_1)$ and looking in the direction of a line, $\alpha$, the problem, as illustrated in Figure 5.5, is to find the perpendicular offset distance right (positive) or left (negative) from the line to the point specified as the forepoint $P(e_2, n_2)$. Note that the azimuth, $\alpha$, may be any azimuth 0° to 360°. Using the conventions given earlier, the perpendicular offset distance is the distance $d_2$ computed as

$$\Delta e = e_2 - e_1 \quad \text{and} \quad \Delta n = n_2 - n_1$$

$$d_2 = \Delta e \cos \alpha - \Delta n \sin \alpha \tag{5.19}$$

## AREA BY COORDINATES

Area is the product of length multiplied by width and generally is presumed to be computed on a flat surface. Computing area on a spherical or ellipsoidal surface is more of a challenge and is addressed in Chapter 7. The method generally used for computing the area of an irregular tract is area-by-coordinates as used in Chapter 4. Area by double-meridian-distance (DMD), useful if working with latitudes and departures, is described in many surveying texts and is not presented here.

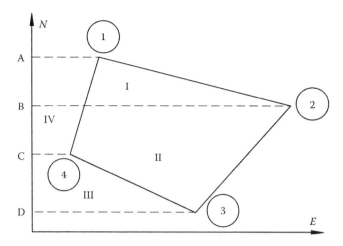

**FIGURE 5.6** Area by Coordinates

Development of the area equation (4.1) involves adding and subtracting trapezoids as shown in Figure 5.6.

$$\text{Area} = \text{Trapezoid I} + \text{Trapezoid II} - \text{Trapezoid III} - \text{Trapezoid IV}$$

| | |
|---|---|
| Trapezoid I: A-1-2-B-A | Area $= 0.5(e_1 + e_2)(n_1 - n_2)$ |
| Trapezoid II: B-2-3-D-B | Area $= 0.5(e_2 + e_3)(n_2 - n_3)$ |
| Trapezoid III: C-4-3-D-C | Area $= 0.5(e_3 + e_4)(n_3 - n_4)$ |
| Trapezoid IV: A-1-4-C-A | Area $= 0.5(e_4 + e_1)(n_4 - n_1)$ |

Combining the four trapezoids into one equation and multiplying by two gives

$$2A = (e_1 + e_2)(n_1 - n_2) + (e_2 + e_3)(n_2 - n_3) - (e_3 + e_4)(n_4 - n_3) - (e_4 + e_1)(n_1 - n_4)$$

Considerable algebraic manipulation and combination of terms are needed to get

$$2A = n_1 e_2 + n_2 e_3 + n_3 e_4 + n_4 e_1 - (e_1 n_2 + e_2 n_3 + e_3 n_4 + e_4 n_1) \qquad (5.20)$$

Comments about Equation 5.20 are as follows.

The coordinates are often arranged in tabular form with the beginning values listed again at the end. The computation is then illustrated by showing accumulation of cross products and area is computed as half the difference of the two accumulated sums. See example in Chapter 4 using Equation 4.1.

- If points around a figure are chosen in a clockwise sequence as was done in the derivation, a positive area is computed. If a counterclockwise sequence is used, a negative area will be the result. Although each user is free to "traverse" a parcel in either direction, the convention is to use positive area.

- Care must be exercised to assure that area is being computed for a legitimate figure. If one line of the figure crosses another (e.g., due to erroneous point sequence input), the equation will provide an answer that reflects removal of the crossover area.
- Significant digits may become an issue if Equation 5.20 is used with large (i.e., state plane) coordinate values. See the example in Chapter 4. Equation 4.2 is recommended as the preferred area-by-coordinates formula.

## CIRCULAR CURVES

Circular logic defines a curve as a line that is not straight and a straight line as one that is not curved. Separating circular geometry from circular logic, a circle is a curve that has completed one cycle or, said differently, a curve is a portion of a circle. There are two fundamental definitions that relate a circle to the value of pi ($\pi$) and to the measure of an angle.

### DEFINITIONS

$$\text{Definition } 1 : \pi \equiv \text{Circumference/Diameter} = C/2R \tag{5.21}$$

$$\text{Definition } 2 : \text{Angle in radians} \equiv \text{Arc length/Radius} = L/R \tag{5.22}$$

Using the two definitions, it is quickly established that there are $2\pi$ radians in a complete circle (360°) and that any arc length is the product of the radius times the subtended angle in radians. One of the simplest, most powerful equations available to the spatial data user is

$$L = R\theta = R\Delta°\left(\pi/180°\right) \tag{5.23}$$

where:
  $L$ = arc length.
  $R$ = radius of curve.
  $\theta$ = a subtended angle in radians.
  $\Delta°$ = a subtended angle in decimal degrees.
  $\pi$ = the value of pi.

### DEGREE OF CURVE

Circular curves are frequently encountered when working with road, street, or highway data. Distance along a road centerline is often measured in stationing (increments of 100 feet or other units), and straight centerline segments between the angle points are called tangents. The angle point in a centerline is called a point of intersection (PI). A curve is used at each PI to provide for a gradual, rather than an abrupt, change in direction. Historically, the sharpness of a curve has been defined by

**TABLE 5.1**

**Comparison of Degree of Curve Radii**

| Degree of Curve | Highway Definition | Railroad Definition |
|---|---|---|
| 1° | 5,729.578' | 5,729.651' |
| 2° | 2,864.789' | 2,864.934' |
| 5° | 1,145.915' | 1,146.279' |
| 10° | 572.958' | 573.686' |
| 20° | 286.479' | 287.939' |
| 30° | 190.986' | 193.185' |
| 45° | 127.324' | 130.656' |
| 90° | 63.662' | 70.711' |

degree-of-curve ($D°$) according to one of two definitions; one definition for railroad work and one for highway work. The $D°$ definitions and the equations for computing radius from each of them are as follows:

*Degree of curve* (highway) $\equiv$ Central angle subtended by an arc of 100 feet

$$Radius = 100/\left[D°\left(\pi/180°\right)\right] = 5,729.57795/D° \qquad (5.24)$$

$$D° = 5,729.57795/Radius \qquad (5.25)$$

*Degree of curve* (railroad) $\equiv$ Central angle subtended by a chord of 100 feet

$$Radius = \left(100/2\right)/\sin\left(D°/2\right) = 50/\sin 0.5D° \qquad (5.26)$$

$$D° = 2\sin^{-1}\left(50/Radius\right) \qquad (5.27)$$

Table 5.1 compares values of radii for several values of $D°$ by each of the two definitions. Note that values of radius decrease linearly for the highway definition, but not for the railroad definition.

## ELEMENTS AND EQUATIONS

Historically, degree-of-curve definitions are used with the English (foot) system. Although a metric degree-of-curve can be defined, modern practice tends to use the radius and another element to define a circular curve. As shown in Figure 5.7, there are five primary circular curve elements routinely encountered when working with spatial data involving circular curves. Of the ten possible combinations of five elements taken two at a time, the combinations of tangent/length and tangent/long chord are rarely used because of their weak geometry and computational complexity

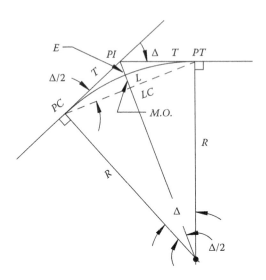

**FIGURE 5.7**   Elements of a Circular Curve

(see, for example, Thompson (1974)). Given any other pair of primary circular curve elements, the solution can be found using some combination of Equations 5.28 through 5.33.

$R$ = radius.
$T$ = tangent.
$LC$ = long chord.
$L$ = length.
$\Delta°$ = central angle.

Secondary curve elements also labeled in Figure 5.7 include

$PI$ = point of intersection, intersection of straight line tangents.
$PC$ = point of curvature, centerline station at beginning of curve.
$PT$ = point of tangency, centerline station at end of curve.
$\Delta°$ = intersection angle, same as the central angle of the curve.
$\Delta°/2$ = deflection angle, half the central angle.
$M.O.$ = middle ordinate, radial distance from chord to arc.
$E$ = external, radial distance from arc of curve to $PI$.

With reference to Figure 5.7, the following points are noted:

- The word "tangent" is used twice. The straight centerline segment between $PI$'s is called a tangent. And, one of the curve elements, the distance from the $PC$ to the $PI$, is called the tangent (of the curve). Added to those two uses, the word "tangent" is also associated with the use of a trigonometric function. The context of usage generally dictates which meaning is intended.

- The radius of the curve meets the centerline (both tangents) at a right angle.
- The angle of intersection is the same as the central angle of the curve.
- The deflection angle is half the central angle.
- The diagram is symmetric. The distance $PC$ to $PI$ equals the distance $PI$ to $PT$ and a radial line to the $PI$ is perpendicular to the long chord. This right angle statement should not be accepted at face value but each reader should prove it to himself/herself.

Using $L = R\theta$, the definitions of trigonometric ratios, the labels assigned to the curve elements, and the diagram in Figure 5.7, it is possible to write the following relationships ($\Rightarrow$ means "implies that"):

$$L = R\Delta°\left(\pi/180°\right) \Rightarrow R = L180°/\left(\pi\Delta°\right) \tag{5.28}$$

$$\Delta° = \left(L/R\right)\left(180°/\pi\right) \tag{5.29}$$

$$\tan \Delta/2 = T/R \Rightarrow T = R\tan \Delta/2 \tag{5.30}$$

$$R = T/\tan \Delta/2 \tag{5.31}$$

$$\sin \Delta/2 = LC/\left(2R\right) \Rightarrow LC = 2R\sin \Delta/2 \tag{5.32}$$

$$R = LC/\left(2\sin \Delta/2\right) \tag{5.33}$$

$$M.O. = R - R\cos \Delta/2 = R\left(1 - \cos \Delta/2\right) \tag{5.34}$$

$$E = R/\cos \Delta/2 - R = R\left(1/\cos \Delta/2 - 1\right) \tag{5.35}$$

## STATIONING

When designing and building a road (or other centerline referenced project), horizontal distance along centerline (or reference line) is the basis of stationing. An arbitrary value such as 0+00 or 100+00 is assigned to an initial point near the beginning of a project and points on the centerline are stationed as $XX+XX.XX$ (the value of $XX+XX.XX$ being the accumulated centerline distance from the initial point). Difference in stationing is the horizontal distance along the centerline between stationing values—even along curved portions of the centerline. For example, the distance from station 132+16.58 to station 163+45.32 is 3,128.74 feet. Discontinuities in stationing (either gaps or overlaps) are handled with a station equation assigned to some centerline point. A common format for a station equation is "$XXX+XX.XX$ Back = $YYY+YY.YY$ Ahead".

Station equation policies vary from one organization to another but, if changes are made to the centerline alignment or if curves are added after centerline stationing is assigned, stationing for the entire project might be reassigned. Since restationing

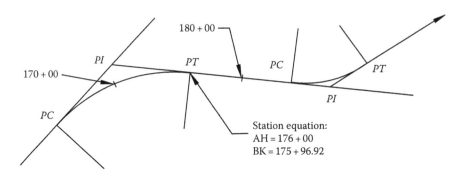

**FIGURE 5.8**   Station Equation Example

is often not practical, a station equation is used to account for gaps and overlaps. In the case of adding a curve, a station equation is used at the end of the curve (*PT*). The "ahead" station is found by adding the curve tangent distance to the *PI* station and the "back" station for the same point is found by adding the curve length to the *PC* station. Used in reverse, distance along centerline is the difference in stationing—except when a station equation is encountered. Then, distance along centerline is computed separately on two sides of the station equation. For example, a station equation (175+96.92 BK = 176+00 AH) is used at the end of the curve shown in Figure 5.8. What is the centerline distance between station 170+00 on the curve and station 180+00 on the tangent?

   Distance along curved portion = 175+96.92 − 170+00 = 596.92 feet
   Distance along tangent portion = 180+00 − 176+00 = 400.00 feet
   Total distance = 996.92 feet

Of course, other changes to the centerline alignment also require a station equation to accommodate the change in length along the revised route. Some changes will lengthen the centerline causing an overlap at the station equation. Other changes will shorten the centerline and cause a gap between the "back" and "ahead" stations at a point.

## Metric Considerations

Stationing is used to identify cross-section locations on road or similar projects. Cross-section spacing of each station or half-station are common. There appears to be no set standard when using stationing on metric-based projects. Some use a 100-meter stationing with cross-section intervals of 20 or 30 meters. Others use 1000 meters (1 kilometer) as a station length with cross-section intervals every 100 meters. Therefore, each user is encouraged to confirm and/or be specific about stationing policies on any metric project. Possible sources of information include any state Department of Transportation (DOT) office or the American Association of State Highway Transportation Officials (AASHTO).

## AREA FORMED BY CURVES

The area of a rectangular shaped figure is length multiplied by width. The area of a circle is developed using small rectangles and tools of calculus. As shown in Figure 5.9, a differential element of area is the length (circumference at that radius distance from the center) times an infinitesimally small width ($dR$). Written in calculus as a summation of an infinite number of small rings, the area of a circle with radius $R$ is computed as

$$dA = 2\pi R\, dR$$

$$\text{Area} = \int_0^R 2\pi R\, dR = \pi R^2 \tag{5.36}$$

A *sector* of a curve is defined as the "pie-shaped portion" of a circle; see Figure 5.10. Area of a curve sector is linearly proportional to the total curve area in

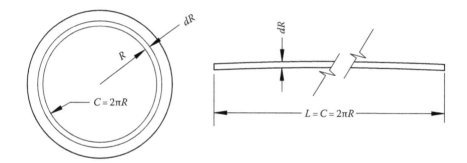

**FIGURE 5.9**   Area of a Circle

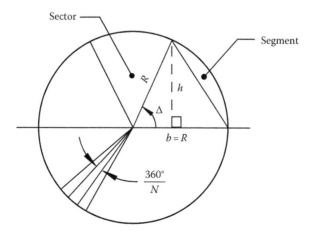

**FIGURE 5.10**   Area of a Sector and Area of a Segment

the same manner as central angle is proportional to a complete circle. If the central angle is 1/4th of 360°, then the area of the sector is 1/4th the area of the circle, or if the central angle is 0.4475 of 360°, then the area of the sector is $0.4475 \pi R^2$. For any sector, the area of the sector is

$$\text{Sector area} = \left(\frac{\Delta}{360}\right)\pi R^2 \tag{5.37}$$

where $\Delta$ = central angle in decimal degrees.

A segment of a curve is defined as the area between the arc and the chord of the curve; see Figure 5.10. The area of a segment is computed as the remainder left when the area of the inscribed triangle is subtracted from the area of the corresponding sector as shown in Figure 5.10. The area of a triangle is ½ the base multiplied by the height. Using the radius as the base and the radius multiplied by the *sin* of the central angle as the height of the inscribed triangle, the segment area is computed as

$$\text{Area of segment} = \frac{\Delta}{360}\pi R^2 - \frac{1}{2}R^2 \sin \Delta$$

And, with a bit of algebraic manipulation

$$\text{Area of segment} = \frac{R^2}{2}\left(\frac{\pi}{180}\Delta - \sin \Delta\right) \tag{5.38}$$

Note that the last expression in Equation 5.38 is really $\Delta$ in radians minus the sin of $\Delta$.

## AREA OF UNIT CIRCLE

What is the area of a unit circle ($R = 1$)? The answer is $\pi$. If the area of a unit circle is computed as the sum of a large number of small triangles, the answer will be in error by the accumulated segment area between the arc and the chord. As a larger number of triangles is used, the error becomes very small. As illustrated in Figure 5.10, Equation 5.39 approximates the area of a unit circle and permits the user to choose any large value of N, the number of triangles.

$$\text{Area} = \pi \approx \frac{N}{2}\left(\sin\frac{360}{N}\right) \quad \text{for large values of } N. \tag{5.39}$$

Table 5.2 shows a summary of answers for various values of N. The last two lines are the values of $\pi$ computed using a spreadsheet, $\pi = 4 \arctan(1.0)$, and the first 20 digits of $\pi$ (Beckmann 1971).

Comment: Mathematicians have devised better ways of computing $\pi$, but this technique shows an interesting connection between the value of $\pi$ and the area of a unit circle while illustrating the concept of limits.

**TABLE 5.2**

**Approximations for π Based on Area of Unit Circle**

| Value of N | Area of Corresponding Unit Circle |
|---|---|
| | (to 15 significant digits) |
| 100 | 3.13952597646567 |
| 1,000 | 3.14157198277948 |
| 10,000 | 3.14159244688129 |
| 100,000 | 3.14159265152271 |
| 1,000,000 | 3.14159265356913 |
| 10,000,000 | 3.14159265358959 |

*Notes:* Spreadsheet computation: $4 \times \text{atan}(1) = 3.14159265358979$.
First 20 digits of π: 3.1415926535897932385.

## SPIRAL CURVES

A spiral is defined as a curve whose radius is inversely proportional to its length. Various kinds of spirals can be developed from the same definition. Mathematicians often view spirals from the perspective of an origin located at that point where the radius approaches zero. Spirals used in geomatics applications share the definition but are viewed from a different perspective (i.e., beginning at that point where the spiral length is zero and the radius is infinitely long). This origin is selected because spirals are used to provide a rigorous gradual transition from traveling in a straight line along a tangent to traveling along a circular curve having a constant radius. Spirals are used extensively on railroad layout but they are also used in some highway applications.

Typically, spirals are used in pairs. Starting on a straight-line tangent, a spiral is used to make the transition from traveling in a straight line to traversing a circular curve. At the end of the circular curve, another spiral is used to transition back to traveling in a straight line on the next tangent. Although the entrance spiral and exit spiral could have different lengths, standard practice is for them to be symmetrical. Therefore the discussion here will focus only on the geometry of the entrance spiral.

### SPIRAL GEOMETRY

In some cases, spirals have been avoided because of their computational complexity. But, even though the equations have not changed, modern computers, coordinate geometry routines, and radial surveying techniques have eased the burden of using spirals. It is intended for the following to provide a comprehensive computational algorithm that can be completed using only a simple scientific calculator. Admittedly, the formulas given here do not lend themselves to easy long-hand solutions but will find ready applications in a spreadsheet solution or computer program. The equations presented here will likely be most useful to those writing coordinate geometry routines (for computers and/or data collectors).

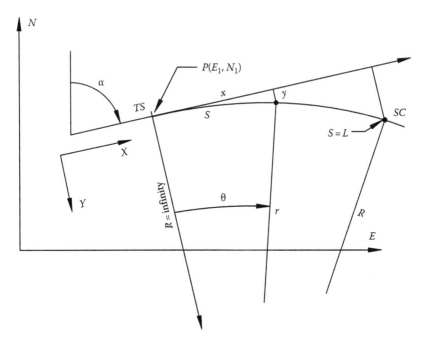

**FIGURE 5.11**  Spiral Elements and Geometry

Symbols for the spiral illustrated in Figure 5.11 are

$\alpha$ = azimuth of beginning tangent.
$TS$ = station at transition tangent to spiral.
$SC$ = station at transition spiral to circular curve.
$L$ = total length of spiral. Difference of stationing, $SC - TS$.
$R$ = radius of circular curve at end of the spiral.
$K$ = spiral constant = $1/(2RL)$.
$S$ = Capital letter $S$ is used when $s = L$.
$s$ = distance along spiral from TS to specified point on spiral.
$r$ = instantaneous radius of spiral at location defined by $s$.
$\theta$ = angular difference between spiral radius vectors at two points. Typically one point is at the beginning of the spiral and the second is any point on the spiral at a distance "$s$" from the beginning.
$x$ = tangent component of the distance from $TS$ to any point on spiral.
$y$ = perpendicular distance from the tangent to any point on the spiral. An appropriate sign convention is for "$y$" to be positive if spiral is to the right and negative if spiral curves to the left when standing at $TS$ and looking along the tangent adjacent to the spiral.

Equations for computing coordinates of a point on a spiral are as follows:

Given:
$e_1, n_1$ = local plane coordinates of tangent to spiral, $TS$.
$\alpha$ = local azimuth of straight line tangent at $TS$.

$R$ = radius of circular curve at the end of the spiral.
$L$ = total length of the spiral.
$s$ = centerline distance from $TS$ to point on the spiral.

Store: Constants common to spiral solutions.

| | |
|---|---|
| $C_1 = 1/3$ | $C_2 = -1/10$ |
| $C_3 = -1/42$ | $C_4 = 1/216$ |
| $C_5 = 1/1,320$ | $C_6 = -1/9,360$ |
| $C_7 = -1/75,600$ | $C_8 = 1/685,440$ |

Compute:

$$\theta = s^2/(2RL) = Ks^2. \tag{5.40}$$

$$x = s\left(1 + C_2\theta^2 + C_4\theta^4 + C_6\theta^6 + C_8\theta^8 + \cdots\right)$$
$$= s\left(1 + \theta^2\left(C_2 + \theta^2\left(C_4 + \theta^2\left(C_6 + C_8\theta^2\right)\right)\right)\right) \tag{5.41}$$

$$y = s\left(C_1\theta + C_3\theta^3 + C_5\theta^5 + C_7\theta^7 + \cdots\right)$$
$$= s\theta\left(C_1 + \theta^2\left(C_3 + \theta^2\left(C_5 + C_7\theta^2\right)\right)\right) \tag{5.42}$$

Solution: Coordinates for a point on spiral are computed using the COGO forward computation Equations 5.8 and 5.9 as (sign convention: $y$ is negative for spiral to left.)

$$e_2 = e_1 + x\sin\alpha + y\sin(\alpha + 90°)$$
$$= e_1 + x\sin\alpha + y\cos\alpha \tag{5.43}$$

$$n_2 = n_1 + x\cos\alpha + y\cos(\alpha + 90°)$$
$$= n_1 + x\cos\alpha - y\sin\alpha \tag{5.44}$$

A line parallel to a spiral is not a spiral. But, points on an offset to a spiral, $(e_3, n_3)$, can be computed using Equations 5.45 and 5.46. To compute coordinates of points lying to the right of the spiral, use a plus offset distance. Points to the left of the spiral are computed using the offset distance as a negative value. The azimuth of the line (radius vector) perpendicular to the line tangent to the spiral at a point is $(\alpha + 90° + \theta)$.

$$e_3 = e_2 \pm \text{Offset distance} \times \sin(\alpha + 90° + \theta) \tag{5.45}$$

$$n_3 = n_2 \pm \text{Offset distance} \times \cos(\alpha + 90° + \theta) \tag{5.46}$$

With spiral points and offset line points computed according to Equations 5.43 through 5.46, these points can be used along with other project points according to radial surveying techniques and standard coordinate geometry procedures available in most field computers and/or data collectors.

### INTERSECTING A LINE WITH A SPIRAL

Computing the position of the intersection of a straight line with a spiral is not encountered very often. Possibilities are no intersection (trivial), one intersection (covered here), or, in the extremely rare case, two intersections. This section looks at the single intersection case. As shown in Figure 5.12, the key to finding the solution is determining a correct value of $s$, the distance from the $TS$ along the spiral centerline to the intersection. Once the value of $s$ is known, the $x$ and $y$ spiral components are computed according to Equations 5.41 and 5.42. Then the $e$ and $n$ coordinates of the intersection point are computed using Equations 5.43 and 5.44. The value of $s$ is determined using an iterative process.

Given:
$e_1, n_1$ = local plane coordinates of tangent to spiral, $TS$.
$\alpha$ = local azimuth of straight line tangent at $TS$.
$e_2, n_2$ = local plane coordinates of any point on line.
$\beta$ = local azimuth from point on line to spiral intersection.
$R$ = radius of circular curve at the end of the spiral.
$L$ = total length of the spiral.

Compute:
$K = 1/(2RL)$.
$M$ = line-line intersection distance along original tangent. Use Equation 5.14 and distance $d_1$ as $M$. $M$ does not change.
$s_1$ = value of $M$. The value of $s$ will be improved.

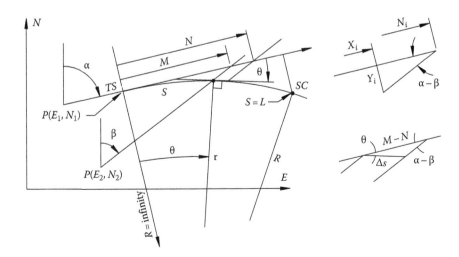

**FIGURE 5.12**   Spiral Intersection Elements and Geometry

Iterate: Start with $i = 1$, continue incrementing $i$ by 1 until tolerance is met.

$\theta_i$ = angle (in radians) to initial trial point on spiral = $Ks_i^2$.

$x_i$ = trial distance along spiral tangent—Equation 5.32.

$y_i$ = trial distance perpendicular to spiral tangent—Equation 5.33.

$N_i$ = check distance; $N_i = x_i + y_i \cot(\alpha - \beta)$.

$Tol$ = absolute value $(M - N)$. Is it small enough? If yes, quit.

If not:

$\Delta s_i$ = Correction to $s$ $\Delta s = \dfrac{(M - N)\sin(\alpha - \beta)}{\sin(\alpha - \beta + \theta_i)}$

$s_{i+1} = s_i + \Delta s_i$ ($\Delta s_i$ should get smaller and smaller.)

Increment $i$ by 1.

Return to beginning of iteration using new value of $i$.

Solution:

When the tolerance is met, $s_i$ is the correct value.

The values of $x_i$ and $y_i$ have already been computed.

Use Equations 5.43 and 5.44 to compute $e$ and $n$ at intersection.

As a final check, inverse from $e_2$ and $n_2$ to $e$ and $y$. The computed direction should be same as $\beta$.

## COMPUTING AREA ADJACENT TO A SPIRAL

If a spiral is part of a boundary, computing the area adjacent to a spiral becomes an issue. The method presented here allows the user to compute the area between the original tangent and the spiral to any precision desired by choosing smaller and smaller increments of $\Delta s$ and accumulating the area by numerical integration. Without a computer programmed to perform the repetitive calculations, the method loses its practicality. As shown in Figure 5.13, the area is accumulated from numerous

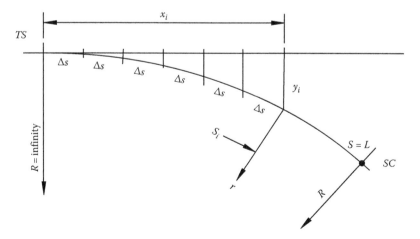

**FIGURE 5.13** Area Adjacent to a Spiral

trapezoids formed by the original tangent and perpendicular lines to the spiral. The separation of construction lines perpendicular to the tangent is not constant, but determined by equal values of $\Delta s$ on the spiral centerline.

Given:
  $R$ = radius of circular curve at end of spiral.
  $L$ = total length of spiral.
  $s = L$ or some portion thereof.

Find: Area bounded by original tangent and spiral up to distance $s$.

Solution:
  $K = 1/(2RL)$. Spiral constant.
  $\Delta s$ = Increment chosen by user. $\Delta s = s/n$. User chooses $n$.
  $s_0 = 0.0$.
  $x_0 = 0.0$.
  $y_0 = 0.0$.

Loop: for $i = 1$ to $n$
  $s_i$ = Distance to point $i$. $s_i = s_{i-1} + \Delta s$.
  $\theta_i = Ks_i^2$. Angle in radians to point on spiral.
  $x_i$ = Tangent component distance to point on spiral. See Equation 5.41.
  $y_i$ = Perpendicular distance tangent to spiral. See Equation 5.42.

End of loop.

Area by trapezoids:

$$\text{Area} = (x_1 - x_0)(y_1 + y_0) + (x_2 - x_1)(y_2 + y_1) + \cdots + (x_n - x_{n-1})(y_n + y_{n-1})$$

which, after considerable algebraic manipulation, reduces to

$$2\,\text{Area} = x_2 y_1 - x_1 y_2 + x_3 y_2 - x_2 y_3 \cdots + x_n y_{n-1} - x_{n-1} y_n + x_n y_n \qquad (5.47)$$

Or, the area adjacent to a spiral can also be written as

$$2A = \sum_{i=2}^{n} x_i y_{i-1} - \sum_{i=2}^{n} x_{i-1} y_i + x_n y_n \qquad (5.48)$$

## RADIAL SURVEYING

The following discussion relates specifically to conventional total-station surveying procedures, but it is also somewhat applicable to using GNSS equipment and procedures. Radial surveying is a term used to describe the practice of measuring an angle and distance from a known point to locate another point (determine coordinates for its location) or for setting out a point at a predetermined location. The procedures are well defined and simple to perform, but they have the disadvantage of no redundancy.

If a blunder is made in measuring the angle, measuring the distance, or setting up over the wrong point, there is no built-in check. One way to check the position of such open side shots is to set up over a different point, locate all side shot points a second time and compare the results. Another method is to measure, say with a steel tape, distances between the side shot points. The inverse distance in each case should compare favorably with the taped distance. Lack of expected consistency in the results using either technique is an indication of a blunder or uncontrolled errors. A high degree of consistency in such checks is an indication, but not a guarantee, that there are no blunders and that random and systematic errors have been controlled at an acceptable level. Given that many production measurements may be checked in this manner, it must be understood that such checks are a poor substitute for carefully designed redundant measurements.

Radial surveying methods can be accomplished using various equipment combinations, but they are ideally suited for the modern total station instrument used in conjunction with a data collector. Radial surveying techniques are also suited for the one-person crew using a robotic total station. Each point is identified by number in the data file and instructions are given the instrument in terms of commands, point numbers, and in some cases, attributes. When collecting data for a topographic site plan, the point numbers may be assigned sequentially by default. In the layout mode, the operator specifies the points to be used or staked in any order desired. In either case, collecting data or laying out points, the point occupied by the instrument and the backsight point must both be specified by the user.

Radial surveying has two primary advantages:

1. The geometry of points to be staked may involve curves, spirals, offsets, or other complex geometrical relationships but, in the field, the solution boils down to an angle from a known backsight line and a distance from the instrument.
2. Decisions about which point to occupy with the instrument and which point to use as a backsight can be deferred to the responsible person in the field. Intervisibility between points is essential for line-of-sight equipment but, whether tying in points or staking locations, the logistical operations are made easier by the flexibility of the method.

When performing radial stakeout, the computations are performed by computer and details are rarely of concern to the end user. However, should it be necessary to perform angle to the right and distance computations by hand, the following procedure as illustrated in Figure 5.14 may be helpful. The conventional procedure is to perform two separate inverse computations and to use the two computed azimuths to find the appropriate angle to the right. Using a trigonometric identity, the procedure can be shortened and done directly on any scientific calculator.

The following procedure is moot if using GNSS to stake out points. If using a total station for layout, three points are required to perform an angle to the right and distance layout computation; the standpoint, a backsight point, and the forepoint. The solution is the angle to the right (at the standpoint) from the backsight point to the forepoint and the distance from the standpoint to the forepoint.

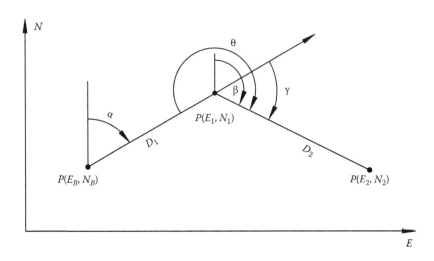

**FIGURE 5.14**  Angle-Right on Radial Stakeout

Symbols:

$e_B$, $n_B$ = backsight point coordinates.

$e_1$, $n_1$ = standpoint coordinates.

$e_2$, $n_2$ = foresight coordinates.

$\alpha$ = azimuth from backsight point to standpoint.

$\beta$ = azimuth from standpoint to forepoint.

$\gamma$ = deflection angle, $\beta - \alpha$.

$\theta$ = angle-right, $\gamma + 180°$.

$D_1$ = distance backsight to standpoint.

$D_2$ = distance standpoint to forepoint.

Trigonometric identity:

$$\tan(B - A) = \frac{\tan B - \tan A}{1 + \tan A \tan B} \qquad (5.49)$$

Compute:

$$\Delta e_1 = e_1 - e_B, \quad \Delta e_2 = e_2 - e_1$$

$$\Delta n_1 = n_1 - n_B, \quad \Delta n_2 = n_2 - n_1$$

$$\tan \alpha = \Delta e_1 / \Delta n_1, \quad \tan \beta = \Delta e_2 / \Delta n_2$$

$$D_1 = \sqrt{\Delta e_1^2 + \Delta n_1^2}, \quad D_2 = \sqrt{\Delta e_2^2 + \Delta n_2^2}$$

Solution (first find deflection angle, then angle to the right):

$$\tan \gamma = \tan(\beta - \alpha) = \frac{\tan \beta - \tan \alpha}{1 + \tan \alpha \tan \beta}$$

Substitute values for tan α and tan β from above:

$$\tan \gamma = \frac{\left(\dfrac{\Delta e_2}{\Delta n_2}\right) - \left(\dfrac{\Delta e_1}{\Delta n_1}\right)}{1 + \left(\dfrac{\Delta e_1}{\Delta n_1}\right)\left(\dfrac{\Delta e_2}{\Delta n_2}\right)} = \frac{\Delta e_2 \Delta n_1 - \Delta n_2 \Delta e_1}{\Delta n_1 \Delta n_2 + \Delta e_1 \Delta e_2} \tag{5.50}$$

$$\text{Angle right} = \theta = \gamma + 180° \left(\text{subtract } 360° \text{ if required}\right) \tag{5.51}$$

*Notes*:

1. The distance $D_1$ is not needed, but can be used to check distance to back-sight point.
2. Equation 5.50 can be solved using the rectangular/polar key of a scientific calculator by inputting the numerator and denominator separately. The result will be the deflection angle (add 180° for angle to the right) and distance $D_2$.
3. If programming Equation 5.50, include provision for a zero denominator. (Most arctan 2 functions will accommodate a zero denominator). If the denominator is zero, the angle to the right is 90° or −90° (270°).
4. Angle-right can also be computed by taking the difference of two azimuths as obtained from separate coordinate inverse computations; Equations 5.11 through 5.13.

## VERTICAL CURVES

The coordinate geometry tools discussed so far have looked at lines, curves, and spirals in the horizontal plane. This section looks at use of a parabolic equation in the vertical plane (profile) as used to provide gradual changes in grade on a road, street, or highway. Commonly called a vertical curve, the parabolic equation is one of the conic sections and is obtained from the general polynomial equation (4.8) by setting the $B$ and $C$ coefficients equal to zero. The result is an equation of the general mathematical form

$$y = ax^2 + bx + c \tag{5.52}$$

The grade (slope) of most highways is gradual and expressed in percent. A 2 percent grade means a uniform rise of 2 feet vertically per 100 feet horizontally (per station if stations are spaced at 100 feet). A −4 percent slope means a uniform drop of 4 feet per station. Sight distances across the crest of a hill and passenger comfort for those traveling on high-speed interstate roadways are related to the rate of change of grade when making a transition from one grade to another. This rate of change of grade per station is a design parameter that varies according to intended use of the travelway.

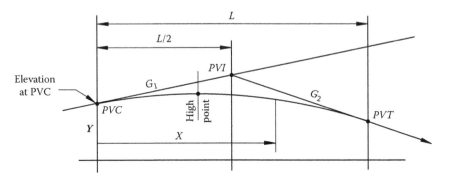

**FIGURE 5.15**   Vertical Curve Geometry

Symbols used in Equation 5.54 and Figure 5.15 are

$PVI$ = point of vertical intersection.
$PVC$ = point of vertical curve (also called the $BVC$, beginning of vertical curve).
$PVT$ = point of vertical tangency (also called $EVC$, end of vertical curve).
$L$ = length of vertical curve computed as difference in stationing (Station at $EVC$ − Station at $BVC$. $L$ is a horizontal distance).
$G_1$ = grade of roadway along the centerline coming into vertical curve.
$G_2$ = grade of roadway along the centerline leaving the vertical curve.
$y$ = elevation of point on vertical curve at station $XXX + XX$.
$x$ = distance of station $XXX + XX$ from $PVC$. $0 \le x \le L$.
$r$ = rate of grade change per station = $(G_2 − G_1)/L$, grade in percent and $L$ in stations.

Notes about derivation and solution of vertical curve problems:

1. The first derivative of the parabolic equation is the slope of the tangent to the vertical curve at that point.
2. The slope of the vertical curve is zero at the high point (summit) or low point (sag) of the curve.
3. The slope of the curve is $G_1$ at $BVC$ (at $x = 0$) and $G_2$ at $EVC$ ($x = L$).
4. The vertical curve is symmetrical with respect to length. That is, the $PVI$ is halfway between the $BVC$ and the $EVC$. That's why it is called an equal-tangent-length vertical curve. Note that an unequal-tangent-length vertical curve is solved by establishing secondary PVIs at the midpoints of the $BVC$ to $PVI$ and $PVI$ to $EVC$ and treating the result as two abutting equal-tangent-length vertical curves.
5. Units: Two consistent methods will each give good results. Mixing the two conventions may result in bad answers and lots of frustration. And, be careful when using metric stationing.
    a.  Grade expressed as slope (2 percent = 0.02) is consistent with the $x$ distance used as feet or meters.
    b.  Grade expressed as percent is consistent with $x$ used in stations.

The $a$, $b$, and $c$ coefficients must be found before Equation 5.52 can be used to solve for elevations at each station. Note that at $x = 0$, the first two terms of Equation 5.52 drop out leaving $c$ = elevation of vertical curve at $BVC$ ($x = 0$). To find coefficients $a$ and $b$, we need to take the derivative of Equation 5.52 as

$$dy/dx = 2ax + b$$

At $x = 0$, the slope is $G_1$ and $dy/dx = b$. Therefore, $b = G_1$.
At $x = L$, the slope is $G_2$ and $dy/dx = 2ax + G_1$. Therefore, $a = (G_2 - G_1)/(2L)$.

And, finally, if the first derivative is set to 0, the result is an expression that can be solved for $x$, the distance from the $BVC$ to the high point or low point of the vertical curve. Using values of $a$ and $b$ as found above

$$0 = 2 \times (G_2 - G_1)/(2L) + G_1 \Rightarrow x = G_1 L/(G_1 - G_2) \tag{5.53}$$

Note that if Equation 5.53 gives a value of $x$ less than zero or greater than $L$, it means there is no sag or summit between the $BVC$ and $EVC$. That will happen if both grades are positive or if both grades are negative. In that case, the highest point or the lowest point (not the sag or the summit) will be either at the $BVC$ or $EVC$.

The vertical curve equation is

$$y = \frac{G_2 - G_1}{2L} x^2 + G_1 x + \text{Elev at } BVC \tag{5.54}$$

**Example:** Compute elevations at each station and half station (every 50 feet) for the following vertical curve. Find the station and elevation at the low point (sag).

PVI station = 172+00
Elev at PVI = 2,300.00 feet
Length of curve = 400.00 feet
$G_1$ and $G_2$ = −4 percent and 1 percent, respectively

Solution:

1. Find BVC station and elevation:

Station at $BVC$ = Station at $PVI - L/2 = 170 + 00$

Elevation at $BVC$ = Elev at $PVI - L/2 \times G_1 = 2,308.00$ feet

2. Find EVC station and elevation:

Station at $EVC$ = Station at $PVI + L/2 = 174 + 00$ (also same as $BVC + L$)

Elevation at $EVC$ = Elev at $PVI + L/2 \times G_2 = 2,302.00$ feet

3. Find station of low point:

$$x(\text{low}) = G_1 \times L/(G_1 - G_2) = 320 \text{ feet}; \quad \text{Sta} = 173 + 20$$

4. Find coefficients a and b:

$$a = (G_2 - G_1)/(2L) = (0.01 - 0.04)/800 = 0.000062500$$

$$b = G_1 = -0.04$$

5. Use Equation 5.54 to find each elevation as

| Station | x | Elevation | Station | x | Elevation |
|---------|------|-----------|---------|------|-----------|
| 170+00 | 0 | 2,308.00' | 170+50 | 50' | 2,306.16' |
| 171+00 | 100' | 2,304.63' | 171+50 | 150' | 2,303.41' |
| 172+00 | 200' | 2,302.50' | 172+50 | 250' | 2,301.91' |
| 173+00 | 300' | 2,301.63' | 173+50 | 350' | 2,301.66' |
| 174+00 | 400' | 2,302.00' | | | |
| 173+20 | 320' | 2,301.60' | Low point of vertical curve | | |

## THREE-DIMENSIONAL MODELS FOR SPATIAL DATA

So far this chapter has looked specifically at COGO models used for routine 2-D computations. More complex models include three dimensions (3-D) and are needed to compute volumes. Several fundamental examples are as follows.

### VOLUME OF A RECTANGULAR SOLID

Perhaps the easiest volume to find is that of a rectangular solid computed as length multiplied by width multiplied by height. The same concept will be used along with tools of integral calculus to compute the volume of other shapes.

### VOLUME OF A SPHERE

To compute the volume of a sphere, an elemental surface area (length multiplied by width) is multiplied by an elemental increment of radius (height). The elemental surface area, as shown in Figure 5.16, is $Rd\phi$ in the north-south direction and $R\cos\phi d\lambda$ in the east-west direction. The surface area of a sphere in terms of the radius is obtained by performing a double integration on variables of longitude, 0 to $2\pi$, and latitude, $-\pi$ to $\pi$ (radian units). Surface area on a sphere bounded by latitude/longitude limits is

$$dA = Rd\phi R\cos\phi d\lambda$$

$$\text{Surface area on a sphere} = R^2 \int_{\lambda_1}^{\lambda_2} \int_{\phi_1}^{\phi_2} \cos\phi d\phi d\lambda \tag{5.55}$$

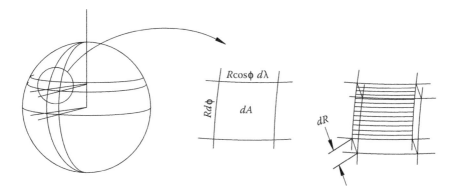

**FIGURE 5.16** Volume of a Sphere

To obtain total surface area, choose limits (in radians) of $\lambda = 0$ to $2\pi$ and $\phi = -\pi$ to $\pi$. Then total surface area is computed as

$$A_{Sphere} = R^2 \int_{0}^{2\pi}\int_{-\pi}^{\pi} \cos\phi\, d\phi\, d\lambda = 2\pi R^2 \int_{-\pi}^{\pi} \cos\phi\, d\phi = 2\pi R^2 - \left(-2\pi R^2\right) = 4\pi R^2$$

Building on that, the volume of a sphere as illustrated in Figure 5.16 is computed as

$$dV = dA \times dR = R^2 \cos\phi\, d\lambda\, d\phi\, dR$$

$$\text{Volume of a sphere} = \int_{0}^{R}\int_{-\pi}^{\pi}\int_{0}^{2\pi} R^2 \cos\phi\, d\lambda\, dR = 4\pi \int_{0}^{R} R^2 dR = \frac{4\pi R^3}{3} \qquad (5.56)$$

## VOLUME OF CONE

The (truncated) cone is another fundamental shape for which volume needs to be computed. For surveyors and engineers, it often comes disguised as the prismoidal formula for computing earthwork volumes. Referring to Figure 5.17a, calculus is also used to compute the volume of the cone shown as a cross-section area (a circle) times a differential height. For the cone shown, there is a linear relationship, $k$, between the height and the radius at that height. Note that at $h = 0$, $R = 0$ and at $h = h$, $R = kh$. The volume of the cone is computed as

$$dV = \text{Area} \times dh = \pi R^2 dh = \pi k^2 h^2 dh$$

$$V = \int_{0}^{h} \pi k^2 h^2 dh = \frac{\pi k^2 h^3}{3} = \frac{\pi R^2 h}{3} \qquad (5.57)$$

Note that there is no requirement that the cone axis be at a right angle to the base. In Figure 5.17b the volume of the cone is still computed as the area of the base times the height measured perpendicular to the base.

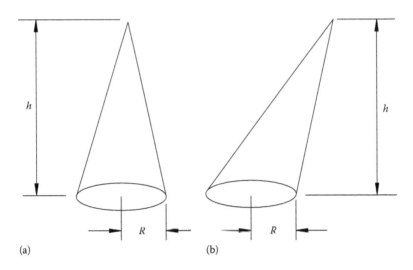

**FIGURE 5.17**    Volume of a Cone. (a) Right Angle Cone and (b) Oblique Angle Cone

## PRISMOIDAL FORMULA

Often earthwork volumes on road or highway construction are computed using the average-end-area method where volume of a given section is computed as the section length times the mean of the cross-section areas at the two ends. It is a good approximation, but based on a false assumption. The average end area method assumes that area varies linearly from one cross section to another. As will be shown using the example of the cone, the radius may vary linearly, but the cross sectional area varies exponentially. Consider the two-station length as illustrated in Figure 5.18. The shape is that of a cone for which volume can be computed without approximation. The volume contained between sections C and E in Figure 5.18 is the difference of cone AE minus the cone AC. Given that the radius varies linearly from A to E, the area at each cross section is computed accordingly.

The total "AE" cone volume is

$$V = \frac{\pi (2R)^2 2L}{3} = \frac{8\pi R^2 L}{3} \tag{5.58}$$

The volume of the small "AC" cone is

$$V = \frac{\pi R^2 L}{3}$$

The volume of the remaining "CE" interval is the difference and computed as

$$V = \frac{8\pi R^2 L}{3} - \frac{\pi R^2 L}{3} = \frac{L}{6}\left(14\pi R^2\right) \tag{5.59}$$

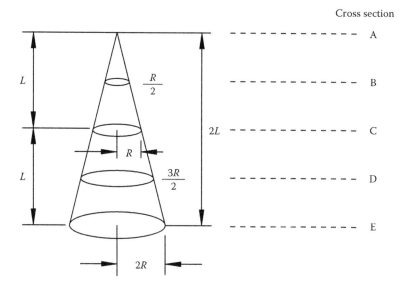

Cross section

FIGURE 5.18   Derivation of Prismoidal Formula

But, note that the cross-section areas at sections C, D, and E are as follows:

Cross-section area at C = $\pi R^2$
Cross-section area at D = $\pi(3R/2)^2 = 9\pi R^2/4$
Cross-section area at E = $\pi(2R)^2 = 4\pi R^2$

For the volume "CE," consider Section C to be the beginning, Section D to be the middle, and Section E to be the end. The prismoidal formula as derived by others computes prism volume as $L/6 \times$ (beginning cross-section area + 4 × cross-section area at midpoint + ending cross-section area) or, using the labeling on our example

$$V = \frac{L}{6}\left(A_{beginning} + 4A_{middle} + A_{end}\right) = \frac{L}{6}\left(\pi R^2 + 4\times\frac{9\pi R^2}{4} + 4\pi R^2\right) = \frac{L}{6}\left(14\pi R^2\right) \quad (5.60)$$

Note that the results of Equations 5.59 and 5.60 are identical. In a backhanded way, we have just derived the prismoidal formula.

The average-end-area method remains an excellent approximation in many cases. Most of the time "fluffing" will increase the volume of excavated material or loss of volume by subsequent compaction will far exceed the error caused by not using the prismoidal method. Lack of good cross-section data may also be a contributing factor to inaccuracies in computed earthwork volumes. But, the fact remains that the average-end-area method is based upon a false assumption and it is left up to the user to judge if or when the prismoidal formula should be used instead of the average-end-area method.

## TRADITIONAL 3-D SPATIAL DATA MODELS

The volume computations just considered are 3-D, but they really don't address the larger (global) issues facing spatial data users. Expanding attention from the local 2-D and 3-D considerations, traditional spatial data models include a variety of 2-D and 3-D options. Some are more complex than others. Examples along with various measurement units include the following:

1. Flat-Earth model used for local mapping, site development, and plane surveying:
   a. Units as selected by the user
   b. Two-dimensional $X/Y$ (or northing/easting) tangent plane coordinates
   c. In the third dimension
      i. Surveyors and engineers use profiles to show grades in terms of centerline stationing and elevation
      ii. Architects show elevation perspectives
      iii. Mappers use hachures and contour lines
2. Spherical Earth model used in geography and navigation:
   a. Mixed mode units: sexagesimal for horizontal, length for vertical.
   b. Two-dimensional curvilinear spherical Earth latitude/longitude positions.
   c. Third dimension is elevation or altitude in terms of length units.
3. Ellipsoidal Earth model used in geodesy, cartography, and geophysics:
   a. Mixed mode units: sexagesimal for horizontal, meters for vertical
   b. Two-dimensional curvilinear ellipsoidal Earth latitude/longitude positions
   c. Third dimension based upon
      i. Distance from geoid, orthometric height
      ii. Distance from ellipsoid, ellipsoid height
4. Map projection (state plane) and elevation spatial data model:
   a. Units are 2-D grid distances; true distances are distorted.
   b. North American Datum of 1927; $X/Y$ coordinates are U.S. Survey Feet.
   c. North American Datum of 1983; $N/E$ coordinates, units vary by state.
      i. NGS publishes all state plane coordinates in meters.
      ii. Some state statutes specify use of the U.S. Survey Foot.
      iii. Some state statutes specify use of the International Foot.
   d. Elevation, although referenced to a curved surface (approximately mean sea level), is typically given by the $Z$ coordinate value.

## THE 3-D GSDM

Many users prefer to work with a rectangular Cartesian coordinate system that defines location of a point as being the perpendicular distances from mutually orthogonal axes. Such a model is 2-D, simple, and appropriate for small areas when the Earth is essentially flat for the entire area. This chapter includes extensive discussions of 2-D COGO in the contexts encountered by many spatial data users. But spatial data are 3-D and spatial data users need simple, standard, reliable, tools and methodology for handling 3-D spatial data. This last section has identified the standard historical methods and coordinate systems that have been used for 3-D data.

Although the equations used in the GSDM are included as a part of the traditional 3-D models, the GSDM is different in that the arrangement of equations and the computations are based on the assumption of a single origin for 3-D spatial data. The consequence is that computations are performed in 3-D space—thus avoiding more complex equations and procedures needed to perform computations on an ellipsoid surface or on a map projection. 3-D geometrical integrity is established and preserved using standard long-proven rules of solid geometry. But, in those cases where geodetic latitude/longitude/ellipsoid height or map projection values are needed, they are readily determined from the primary $X/Y/Z$ coordinate definition of location.

Of the models listed, the state plane coordinate system model is called pseudo 3-D by this author because the third dimension is referenced to a nonregular curved surface. In spite of that deficiency, state plane coordinates combined with elevation data have been beneficially used in many applications. But, the fact remains, when performing 3-D computations using state plane coordinates and elevation, equations of solid geometry and vector algebra are not valid (in a strict sense) because horizontal and vertical distances do not enjoy a common origin.

Some will choose to use spherical geographic coordinates of latitude and longitude. Others will choose to use the more precise geodetic version of latitude and longitude based upon an ellipsoidal Earth. That choice is related to whether or not one anticipates that a spherical model is sufficiently accurate or whether one needs to use the ellipsoidal Earth model to preserve the geometrical integrity of the data being used. In either case, a study of geodesy supports an understanding of the underlying foundation of each model.

The goal of this book is to highlight features and characteristics of the 3-D GSDM that accommodate existing models, modern measurement technology, digital data storage, and spatial data manipulation practices common to various disciplines. A computational example comparing three different models— geodetic, cartographic, and 3-D— for the same simple 3-D position computation is contained in an article by Burkholder (2015). But, before the 2-D concepts, as discussed here, are extended to 3-D, it is appropriate to consider geodesy and other 3-D relationships of spatial data.

## REFERENCES

Beckman, P. 1971. *A History of PI*. New York: St. Martin's.
Burkholder, E.F. 1984. Coordinates, calculators, and intersections. *Surveying and Mapping* 46 (1): 29–39.
Burkholder, E.F. 1997. Three-dimensional azimuth of GPS vector. *Journal of Surveying Engineering* 123 (4): 139–146.
Burkholder, E.F. 2015. Comparison of geodetic, state plane, and geocentric computational models. *Surveying and Land Information Science* 74 (2): 53–59.
Howse, D. 1980. *Greenwich Time and the Discovery of Longitude*. Oxford, U.K.: Oxford University Press.
Thompson, W.C. 1974. The TH function. *Surveying and Mapping* 34 (2): 151–153 (and discussion in subsequent issues).

# 6 Overview of Geodesy

## INTRODUCTION: SCIENCE AND ART

Geomatics is a fairly new umbrella term being used to describe both a body of knowledge and the scope of professional activities having to do with generation, manipulation, storage, and use of spatial data. In a nonexclusive way, geomatics includes traditional disciplines such as surveying, mapping, geodesy, and photogrammetry. It also overlaps with other newer disciplines such as remote sensing, imaging, and information sciences. Of all such disciplines, geodesy provides the geometrical foundation for the rest. A literal meaning of the word geodesy is "dividing the Earth" and geodesy, as a discipline of inquiry, has been concerned with learning more about the size and shape of the Earth since the dawn of civilization.

The scope of geodesy includes both science and art and involves much more content than is summarized here. Without being restricted to either category, the science of geodesy includes issues such as (1) determining the size and shape of the Earth, (2) defining and quantifying the Earth's gravity field, and (3) defining reference systems and coordinate frames. The art and practice of geodesy includes using scientific data and various measuring systems to determine the location of points with respect to a defined geodetic framework. The point here is less about classifying an activity as being science or art but more about recognizing a mutual interdependence between the two.

## FIELDS OF GEODESY

Distinctions between the fields of geodesy are largely artificial and should not limit one's consideration. But historically the following categories have been used:

- Geometrical geodesy encompasses the 3-D geometrical elements of the ellipsoidal model of the Earth and the location of points relative to that model. Traditional geometrical geodesy includes separate horizontal and vertical datums while modern practice combines them into a single 3-D database.
- Physical geodesy involves the study of gravity and considers issues such as the distribution of mass within the Earth, the Earth's external gravity field, equipotential surfaces, and the cause-and-effect relationships between them (as evidenced, for example, by the behavior of a pendulum, the path of artificial satellites, and the shape of the geoid).
- Satellite geodesy deals with the trajectory of missiles in flight and satellites orbiting the Earth. It also includes processing signals received from various orbiting satellites for the purpose of determining the location and/or movement of the receiver.

- Geodetic astronomy has been used extensively to compute positions and directions on the Earth based upon optical (and radio) observations of stars in the sky. With the advent of GNSS surveying, geodetic astronomy has lost much of the relevance it once enjoyed.

Given the development of computers and convergence of other electronic technologies, this book considers how geometrical geodesy supports GIS and spatial data referencing, how physical geodesy relates to geoid modeling and computation of elevations, and how satellite geodesy contributes to an understanding of GNSS surveying and use of GNSS data. The GSDM incorporates concepts from various fields of geodesy as needed to support development of a comprehensive model for location and spatial referencing.

## GOALS OF GEODESY

Traditional goals of geodesy include determining the size and shape of the Earth, describing the gravity field associated with the Earth, and providing a means of locating points on or near the surface of the Earth. Geodetic scientists have defined reference frames and coordinate systems used for referencing spatial data and have determined both the size and shape of the Earth within impressive tolerances. They have also developed sophisticated geoid modeling procedures that can be used to compute both the shape of the geoid and, to a lesser extent, its precise location. Given those definitions and tools, geodetic surveyors and others using GNSS and other positioning technologies, are busy collecting, processing, and using spatial data to determine the location of points and to document the movement of objects on/near the Earth. Based upon what is being accomplished, it could be argued that the traditional goals of geodesy have been met.

The goals of modern geodesy are certainly more comprehensive than the traditional ones summarized here. Two publications by the National Research Council (NRC 1978, 1990) contain an informative overview of the field of geodesy and a description of future challenges and applications. With the advent of computers, satellites, and other modern technology, the fundamental goals of geodesy are being extended to the oceans, the moon, and the planets. With respect to geodetic networks, the 1978 report states: "The ultimate goal is a global geodetic system providing horizontal and vertical, or three dimensional, coordinates for national and international mapping and charting programs with the confidence that there will be no inconsistencies between the networks produced by individual countries." The ECEF system implemented by the DOD for GNSS lays the foundation for meeting that goal. The GSDM builds on that foundation.

Although the overall perspective of the 1978 report is excellent, its "crystal-ball" view didn't really anticipate the enormous impact of the digital revolution and makes little mention of the challenges associated with systematically describing spatial data accuracy. The final chapter of the 1978 report discusses future committee activities and includes seven specific important questions. Even though the view-of-today view enjoys the luxury of 30-plus years of hindsight, those questions remain pertinent and should be considered carefully by anyone attempting to articulate the goals of modern geodesy.

The National Research Council also published a follow-up study (NRC 1990) entitled, *Geodesy in the Year 2000*. Published only 12 years after the previous study, the 1990 report is more up-beat about the role of geodesy and begins by noting, "We stand on the threshold of a technological revolution in geodesy. With the introduction of space-based observational techniques over the past two decades, geodesy has undergone and continues to undergo profound changes." The preface of the 1990 report declares, "Geodesy is becoming a truly global science... And, since the Earth's topography and gravity field are continuous across all political and geographic boundaries, mapping them requires careful integration of data collected by various survey techniques in different countries and physiographic regions."

The 1990 NRC report also discusses, among other topics, how geodesy will use gravity and precise measurements to provide a broader understanding of earthquakes with the idea of better predicting their occurrence to mitigate their impact. And, it discusses the importance of oceanographic research with particular focus on the need for accurate measurements of the geoid and sea level surface. The report also details the need for measurements to show space-time departures from average plate motions predicted by the global rigid-body motion models. The importance of spatial data accuracy is not ignored, but the report fails to anticipate the need for and benefits of using a universal, concise, rigorous stochastic model for spatial data. The report includes five specific recommendations for priorities to "...be established in support of scientific and technological opportunities in geodesy for the year 2000." Of the five recommendations, number 3 states, "A global topographic data set should be acquired with a vertical accuracy of about 1 m, at a horizontal resolution of about 100 m." The "Overview and Recommendations" section closes with, "These data [measurements] would also support establishment and maintenance of a conventional terrestrial coordinate system, which is necessary for comparison of results obtained by different space geodetic technologies."

Since 1990, various agencies and organizations have collected an enormous amount of geospatial data and former Vice President Al Gore (1998) noted in a speech given in January 1998 at the California Science Center that "The hard part of taking advantage of this flood of geospatial information will be making sense of it – turning raw data into understandable information." The National Spatial Data Infrastructure (NSDI) concept was developed to meet that challenge and conceptualizers of the NSDI deserve credit for the help it provides. However, at a more fundamental level, the GSDM provides an underlying definition of spatial data and its accuracy. While the NSDI provides technical, organizational, and operational guidelines for using spatial data, the GSDM provides a specific definition of 3-D geometrical relationships and spatial data accuracy. In the rectangular ECEF environment, the rules of solid geometry and vector algebra are universal throughout and the rules of error propagation can be applied without ambiguity. Issues of data compatibility and interoperability are handled efficiently in the context of the GSDM. From there, derived uses of spatial data are the prerogative of each user. Using the NSDI and GSDM in concert will be more efficient than using them separately. Implementation of an integrated model will not occur immediately but, in the long run, their combined use will help address the challenge of making sense of the enormous quantities of spatial data being collected.

In the meantime, various initiatives (federal and otherwise) are addressing those challenges in a seemingly fragmented manner. Of many that could be cited, a partial list of organizations, efforts, and relevant web addresses include but are not limited to the following:

1. Former President Clinton's 1994 Executive Order established the National Spatial Data Infrastructure (NSDI).
   Use internet search for "Clinton Executive Order NSDI."
2. Former Vice President Al Gore's 1998 speech, "The Digital Earth: Understanding our planet in the 21st Century" addressed the issue.
   Use internet search for "Gore Digital Earth 1998."
3. The Federal Geographic Data Committee coordinated development of NSDI and develops standards for metadata (data about data).
   http://www.fgdc.gov/metadata/ (accessed May 4, 2017)
4. The Bureau of Land Management and U.S. Forest Service initiated the "National Integrated Land System (NILS)" is an effort to establish a common data model and software tools for collecting, processing, managing, and using spatial data.
   https://www.fgdc.gov/geospatial-lob/documents/BLM_NILS_geocommunicator_factsheet.pdf (accessed May 4, 2017).
5. The Global Spatial Data Infrastructure Association (GSDI) was formed as "an inclusive organization of academic and research institutions, government agencies, commercial firms, NGOs, and individuals from around the world.
   http://www.gsdi.org/ (accessed May 4, 2017)
6. The Open Geospatial Consortium (OGC) is an international not-for-profit organization committed to making quality open standards for the global community.
   http://www.opengeospatial.org (accessed May 4, 2017)
7. As described in Chapter 2, the National Oceanic and Atmospheric Administration (NOAA) solicited ideas via a request for information (RFI) in March 2014 for extracting commercial value from their vast data holdings. Having those data referenced to a common integrated 3-D spatial data model could add enormous value to that effort.
   https://www.fbo.gov/index?s=opportunity&mode=form&id=d9844cb78 b4527fb11a6ac6d2b80a742&tab=core&_cview=0 (accessed May 4, 2017).
8. In February 2015, the Coalition of Geospatial Organizations (COGO) published "A Report Card on the National Spatial Data Infrastructure" (also described in Chapter 2). A key finding in that report notes that a "digital geospatial Framework, that is national in scope, is not yet in place, and may never exist." The GSDM exists, is already in place, and using it is a matter of deciding to do so.
   http://www.cogo.pro/uploads/COGO-Report_Card_on_NSDI.pdf (accessed May 4, 2017)

9. The stated mission of the National Geodetic Survey (NGS) is to define, maintain, and provide access to the National Spatial Reference System (NSRS) to meet the economic, social, and environmental needs of the nation. A link to the NGS Ten-Year Strategic Plan 2013–2023 to meet that challenge is provided below. The GSDM is viewed as being compatible with and supportive of the goals and objectives of the NGS.

   http://www.ngs.noaa.gov/web/news/Ten_Year_Plan_2013-2023.pdf (accessed May 4, 2017)

Modern geodesy will certainly continue making contributions to the challenges of making sense of spatial data. But, the scope of geodesy includes much more than that. A statement of the goals for modern geodesy is left to the geodesists and other scientists. For example, the mission of the International Association of Geodesy (IAG) (http://www.iag-aig.org (accessed May 4, 2017)) is the advancement of geodesy. "The IAG implements its mission by furthering geodetic theory through research and teaching, by collecting, analyzing, modeling and interpreting observational data, by stimulating technological development, and by providing a consistent representation of the figure, rotation, and gravity field of the Earth and planets, and their temporal variations."

Recent research and development activities have produced efficient reliable positioning tools that can be used to solve most location problems. And, it is a self-feeding cycle in that more tools and better tools are used by people who dream up even more applications. In turn, they develop even more elaborate tools. Modern society owes much to geodesy, including both the geodetic scientists and the practitioners. Understandably, the collective appetite for technical miracles continues. For example, spatial data users ask for GNSS that works under heavy canopy and everyone wants a geoid model that can be used to convert ellipsoid height to reliable orthometric heights on a worldwide basis. Likewise, the goal of a single affordable autonomous GNSS instrument (precise point positioning—PPP) capable of providing centimeter-level positions anywhere in the world within minutes is a distinct possibility. Significant progress on various fronts has been and continues to be made.

Prior to the twenty-first century, geometrical geodesy activities were generally separated into considerations of horizontal and vertical. The primary reference for horizontal location was a geodetic datum of latitude and longitude positions monumented at points on the surface of the Earth. Elevation, the third dimension, was referenced to the geoid (or to an arbitrary level surface) according to a named vertical datum. Although horizontal and vertical datums both enjoy concise physical definition, they are incompatible in that there is no unambiguous geometrical relationship between mean sea level and the center of mass of the Earth. So, before goals of defining the size and shape of the Earth and locating points on or near the surface of the Earth can be mutually fulfilled, the incompatibility of horizontal and vertical datums must be resolved. In place of two datums having separate reference points, a combined 3-D datum having a single origin is required. The implication of a single origin is that elevation will be a derived quantity instead of an observed quantity. If spatial data are attached to a well-defined reference frame such as the NAD 83 or a named epoch of the ITRF, the GSDM provides a computational environment that has geometrical consistency for any and all points within the birdcage of orbiting GNSS satellites. More details are given in Chapter 8.

The shift to using a 3-D datum and a single origin for spatial data may be analogous to the way time is measured and the need for the Equation-of-Time. John Flamsteed was the first Royal Astronomer of the Greenwich Observatory that was built in 1675. The Greeks recognized the existence of the Equation-of-Time nearly 2000 years earlier, but it was not until reliable clocks were available and Flamsteed made the necessary observations that the Equation-of-Time was quantified. Even with the Equation-of-Time known, most people still reckoned time from the transit of the sun at noon each day and each railroad station had its own version of the correct time. In the United States, the problem was solved in 1883 by adopting a system of standard time zones as devised by Charles F. Dowd of Saratoga Springs, New York (Howse 1980). The system was adopted for use worldwide at *the International Meridian Conference* in Washington, DC, in 1884 and Greenwich Mean Time (GMT) became the world standard. Prior to 1972, time was called GMT but is now referred to as Coordinated Universal Time or Universal Time Coordinated (UTC). It is a coordinated time scale, maintained by the Bureau International des Poids et Mesures (BIPM). It is also known as "Z time" or "Zulu Time." Now, most peoples of the world take standard time for granted but, for scientific purposes, the Equation-of-Time and other time scale differences are known, documented, and used by those for whom the difference matters.

Sea level is an intuitive physical reference for elevation and has served well. But, on a global basis, locating the geoid precisely remains a challenge and will be the subject of research for years to come. Even in the United States where geoid modeling has progressed dramatically in the past 30 years, spatial data users remain dissatisfied when absolute geoid heights are not reliable at the millimeter or centimeter level. Are elevations being done backward? Using time as an analogy, it is like each person with a watch resets it at local noon each day then looks for the appropriate Equation-of-Time model as a means of finding the corresponding standard time. This is not how time is determined. Current practice is to use standard time routinely and the fact that the sun crosses the meridian before or after noon is of little consequence to most persons. Formal arguments for using ellipsoid height for elevation (in lieu of orthometric heights) are given in Burkholder (2002, 2006).

Reversing the process with regard to elevations means that ellipsoid height will be used routinely and that geoid height will be used to find orthometric height when needed (this part is already being done—see Meyer et al. 2004). Making that change will involve some or all of the following:

1. The word "elevation" connotes a generic meaning for most people. This is important. World Vertical Datum yyyy (WVD yyyy) is a designation that could satisfy most spatial data users. In the United States, it could be the vertical component of a 3-D geometrical datum to be published in 2022.

Great effort has been devoted to defining and determining orthometric heights that will be related to the 2022 vertical datum to be published by NGS. Precise orthometric heights are very important for scientific applications and will be used by those needing same.

2. Most people will remain oblivious to a change in definition that WVD yyyy elevation is the distance from the reference ellipsoid instead of sea level. In fact, a change in definition similar to that already occurred in 1973 with renaming the Sea Level Datum of 1929 as the National Geodetic Vertical Datum of 1929 (NGVD 29). Switching from Sea Level datum of 1929 to NGVD 29 was a change of name only and did not involve changing any numbers published for a bench mark (Zilkoski 1992).

3. To make a clean break, a specific word or phrase is needed. It will be similar to switching from NGVD 29 to the North American Vertical Datum of 1988 (NAVD 88)—see Chapter 8. In that case, the change in datums did involve different numbers on each bench mark due to readjustment. However, this time the change will be due to redefinition of the reference surface to be compatible with using a single origin for 3-D spatial data. It will also provide an opportunity to update orthometric heights based upon the 2022 vertical datum.

4. With adoption of the WVD yyyy, elevation will be the distance in meters from the reference ellipsoid. For the most part, elevation differences obtained from GNSS will be used without the need for geoid modeling. Elevation differences from differential leveling (gravity based) can be used as WVD yyyy elevation differences EXCEPT in those cases where deflection-of-the-vertical is large enough to make a difference at the level of accuracy required on the project. Examples include those persons using dynamic heights to compute hydraulic head in the Great Lakes System and physicists who deal with beam alignments in particle accelerators. Trigonometric height computations will no longer involve curvature of the Earth because that is handled implicitly by the GSDM. Vertical refraction of line-of-sight is still critical and will need to be considered where the required accuracy warrants it.

## HISTORICAL PERSPECTIVE

Many authors, for example, Berthon and Robinson (1991), generalize about humankind's collective sense of wonder and asking questions about things observed in nature. When did people first question the shape of the Earth? Hasn't it always been flat? Did Columbus (or was it just his crew) really believe that they would fall off the edge of the world if they sailed too far west? Civilizations develop as subsequent generations build on the body of knowledge developed by their ancestors. Books, libraries, the internet, and interaction with living professionals are all revered as storehouses of knowledge, and education is the process of drawing from that storehouse to gain a greater understanding of our physical world and the actions of humankind.

## RELIGION, SCIENCE, AND GEODESY

Scholars and philosophers certainly have much more to say about the interaction of science and religion but, for purposes here, consider the following:

- Religion
  - Truth is something that can't be refuted.
  - Faith is accepting as true something that can't be proved.
  - Perception is an understanding based upon experience and evidence available. A person with normal eyesight perceives an elephant differently than a blind person who bases their perception only on touching various parts.
  - Reality is based upon a collective evaluation of all the evidence, at least all that shared by others. An unsettling realization is that one can never be sure they have all the evidence. None of us really knows what it is we don't know.
  - Belief is a conviction one holds based upon experience, training, education, inquiry, and insight. One of the fundamental tenets of democracy is that each person is individually responsible for what they believe as well as subsequent decisions and actions. It is impossible to force anyone to believe something. They can only be invited to consider the evidence.
  - Dogma is characterized by a refusal to consider new or additional evidence. "My mind is made up. Do not confuse me with the facts."

- Science
  - Science can be defined as the systematic arrangement of knowledge in a logical order in which conclusions are consistent with beginning assumptions and subsequent observations.
    1. Physical science is concerned with physical matter, forces, and objects.
    2. Social science deals with the reasons for and consequences of decisions and actions made by humans (and other life-forms).
  - Whether physical or social, science is also categorized according to method of inquiry, theoretical or applied.
    1. Theoretical (also called pure) scientific research is conducted for the expressed purpose of gaining a better understanding of the matter, objects, or process of inquiry.
    2. Applied science (also called engineering) is conducted for the purpose of finding or documenting an arrangement of elements or sequence of events that will produce a desired outcome.

The historical development of geodesy could be written as a story of the development of science. Over the centuries, many intelligent persons have contributed to a better understanding of the size and shape of the Earth and people living today have the luxury of sharing innumerable triumphs of the human spirit. Measurements were made on the physical Earth and dutifully recorded. Observations were checked against prevailing thought and accepted, challenged, or ignored. Where warranted, old models were cast aside and new models were proposed and tested. In hindsight, one can see where progress was made and where progress was thwarted for various reasons (often having to do with prevailing religious attitudes). Russell (1959) states that one of the most difficult accomplishments for a person is to hold an idea with conviction and detachment at the same time. With regard to the development of science, that story undoubtedly has a thousand variations. Some variations are related by Alder (2002), Bell (1937), Carta (1962), Sobel (1999), Smith (1986, 1999), and Wilford (1981). Many such stories are related to geodesy and only a few of them are summarized here.

Learned persons have known since before the time of Christ that the Earth is not flat. Pythagoras (born 582 BC) declared the Earth to be a globe. Aristotle (384–322 BC), upon viewing the shadow of the Earth cast upon the moon during an eclipse, concluded the Earth must be spherical. Others pointed out that a curved Earth is apparent from watching a ship approach a port because the top of the mast is readily visible before the hull is visible. Admittedly, the local perspective is that the Earth is flat and that perspective is appropriate for many activities. But, when warranted the curvature of the Earth must be accommodated.

## DEGREE MEASUREMENT

At a fundamental level, the process of determining the size of the Earth requires two pieces of information from which the third is inferred using the well-known equation $L = R\theta$. Although not always the same, the process has been called degree measurement and consists of measuring an arc on the surface of the Earth along with the corresponding subtended angle. From that length-per-degree measurement, the circumference or radius is readily computed. For example, if $L = 500$ miles and the subtended angle is 1/50th of a circle, the circumference is immediately computed as 25,000 miles and the radius of the Earth is found to be about 4,000 miles.

## ERATOSTHENES

Eratosthenes lived approximately 276–195 BC in the Mediterranean port city of Alexandria near the mouth of the Nile River in Egypt. He is given credit for first determining the size of the Earth using the degree measurement method just described. He measured the length of a shadow cast by an obelisk at noon on the longest day of the year in Alexandria. He also knew that on the longest day of the year, the sun shown directly to the bottom of a well located on the Tropic of Cancer near the city of Syene (not far from the present site of the Aswan Dam). From that, he deduced the angle subtended at the center of the Earth to be 1/50th of a circle; see Figure 6.1. Researchers do not agree on how the distance between Alexandria and Syene was measured, but the consensus is the distance was 5,000 stadia and that there are about 185 meters per stadia. This being the case, the computed circumference is 46,250,000 meters—only about 16 percent larger than the currently known size of the Earth. It is conceded there may have been some compensating errors because Alexandria is not directly north of Syene as assumed by Eratosthenes, the quality of his arc distance is questionable, and the well was probably not located directly on the Tropic of Cancer

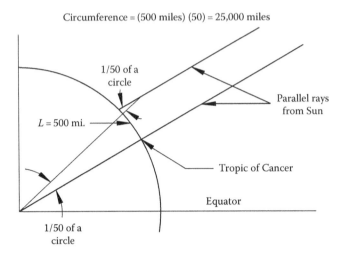

**FIGURE 6.1**   Erathosthenes' Measurement of the Earth

(Tompkins 1971). But, his results are a good approximation and show that humans have known for several millennia that the Earth is not flat.

## POSEIDONIUS

Poseidonius was a Greek astronomer, 135 to 51 BC, who also determined the size of the Earth using degree measurements. Alexandria, on the south shore of the Mediterranean, was the south end of his line. The north end was at Rhodes, on an island on the north side of the Mediterranean Sea about 20 kilometers from Turkey. The arc distance was based upon the sailing time of a ship and the angle subtended at the center of the Earth was deduced from the difference in vertical angle at the two ends of the line to the star Canopus as it crossed the meridian. His results were about 10 percent too big.

## CALIPH ABDULLAH AL MAMUN

Arabian efforts included a measurement on the plains near Baghdad by Caliph Abdullah al Mamun about AD 827 that yielded an answer only about 3.6 percent too big. Subsequent re-evaluation (Carta 1962) of the unit conversions from ells to barleycorns and Rhineland feet gave another value that was about 10 percent too big. Lack of length standardization still plagues modern interpretation of early work.

## GERARDUS MERCATOR

Navigation from the European continent flourished during the thirteenth and fourteenth century, and mapmaking became a valuable occupational talent. The person, shipping company, or sovereign possessing the better map enjoyed an enormous advantage. Stories of intrigue abound. Mercator was one of the most

famous mapmakers because he published a map of the world in 1569 having parallels of latitude on the map spaced so that one could sail from one port to another
on a constant bearing. Later, Mercator's map was shown to be what is now called
a conformal projection—the scale distortion at a given point is the same in all
directions.

## WILLEBRORD SNELLIUS

In 1615, a Dutchman, Willebrord Snellis, measured an arc more than 80 miles long
using a series of 33 triangles. His computation of the size of the Earth was too small
by about 3.4 percent—not bad for using a telescope with no cross hairs in it and
measuring baseline distances with an odometer connected to his carriage wheel. For
perspective, Galileo invented the telescope in 1611, but the Gunter's chain (used in
land surveying for several hundred years) was not invented until about 1620.

## JEAN PICARD

The French Academy of Science was founded in 1666 and sponsored, among others, geodetic surveying activities designed to answer questions about the size and
shape of the Earth. During 1669–1670, Jean Picard measured an arc of triangulation
from Paris to Amiens using a telescope containing cross hairs and "well-seasoned
varnished wooden rods" for measuring baseline lengths. Conventional wisdom at the
time presumed the Earth to be spherical.

## ISAAC NEWTON

Following up on Galileo's work on the pendulum, Newton formulated his theory of
universal gravitation during the mid-1660s. Using the commonly accepted values
for the size of the Earth, Newton was frustrated that the evidence was not consistent with his theory and laid the work aside. It wasn't until after he incorporated
Picard's results, that Newton's theories were validated and published in *Principia
Mathematica* in 1687.

   While his universal gravitational theory was laid aside, Newton became convinced
the Earth is flattened at the poles due to the vector sum of two forces as shown in
Figure 6.2. Gravitational attraction pulls an object toward the center of mass of the
Earth and centrifugal force acts on the same object perpendicular to the spin axis of
the Earth. An object can't differentiate between the two forces but responds only to
the vector sum of the two forces. Newton also noted that a level surface (sea level)
is always perpendicular to the plumb line and that the plumb line will point to the
center of the Earth only if one is at the equator or at one of the poles. At the equator, gravitational attraction and centrifugal force are colinear and centrifugal force
counteracts (reduces) gravitational attraction. At the pole, centrifugal force is zero.
Newton argued that an Earth flattened at the poles is the only shape that is consistent
with those conditions. The irony is that Newton needed Picard's work to verify correctness of his own theories, but Picard and the French Academy refused to accept
Newton's theory of a flattened Earth.

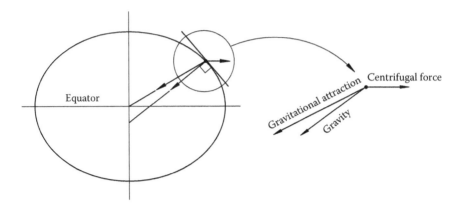

**FIGURE 6.2**    Newton's Logic for a Flattened Earth

## JEAN-DOMINIQUE AND JACQUES CASSINI

Jean-Dominique Cassini was a brilliant scientist who was lured away from Italy and service to the Pope to become the first Director of the Paris Observatory, built between 1667 and 1672. In the early 1680s, the King of France announced Picard's work would be extended from Dunkirk in the north to Collioure in the south. Since Picard had died, Jean-Dominique Cassini directed the survey work. As the work progressed, it became apparent, based upon computations performed separately on that part north of Paris and that part south of Paris, that the radius of the Earth for the northern portion was shorter than the radius for the southern portion. The implication was that the Earth is elongated at the poles, not spherical as had been believed and not oblate (flattened at the poles) as postulated by Newton.

It became a national debate, Newton and Huygens (British) arguing the Earth is oblate or flattened at the poles and the Cassinis (French) arguing the Earth is prolate or elongated at the poles. After Jean's death in 1718, his son, Jacques Cassini, carried on the work and insisted that the only way to settle the argument was to conduct decisive tests near the equator and near the Arctic Circle. Incidentally, the post of Director of the Paris Observatory was held by four generations of Cassinis (Smith 1986).

## FRENCH ACADEMY OF SCIENCE

In 1734, King Louis XV directed more geodetic measurements be taken in France. When those measurements failed to settle the argument, the French Academy of Science sponsored two geodetic surveying expeditions as recommended by Jacques Cassini. In 1735, one expedition left for Peru and another left for Lapland in 1736. Details of those two expeditions are included in Smith (1986). When the leader of the Lapland expedition returned in 1737 with conclusive proof that Newton was right, Voltaire, a French philosopher and social critic, wrote, "You have flattened the Earth and the Cassinis." When the more strenuous Peru expedition returned in 1741, Voltaire commented, "You have found by prolonged toil what Newton learned without leaving home."

## METER

An idea promoted by Picard was the need for a decimally divided standard of length based upon some natural quantity or simple physical observation. He proposed using the length of a pendulum having a period of 1 second as being the standard of length. His idea was not accepted, in part, because the period of a pendulum changes with both latitude and altitude. But, by 1790 the idea of a decimal length standard had caught on to the extent that the French Academy of Science sponsored yet another geodetic survey in France, the purpose of which was to determine as precisely as possible the arc distance from the equator to the pole. The French Revolution notwithstanding, the work was completed during the 1790s by Jean Baptiste Joseph Delambre (1749–1822) and Pierre Méchain (1744–1804). Their resulting distance from the equator to the pole was 5,130,740 toises and was set to be exactly 10,000,000 meters. Given further that 1 toise = 864 Paris lines, one can compute 1 meter = 443.296 Paris lines. "The Measure of All Things" (Alder 2002) is a postscript to this section and includes a fascinating account of the measurement-of-the-meter by Delambre and Méchain.

## DEVELOPMENTS DURING THE NINETEENTH AND TWENTIETH CENTURIES

By the end of the eighteenth century, the United States had won its independence, Thomas Jefferson had served as Minister to France, the French revolution was over, and the decimal meter was defined. A brief summary of geodetic developments during the next 200 years includes:

- Least Squares: Given various determinations for the size and shape of the Earth, the theory of least squares, published in 1806, was developed by Adrien-Marie Legendre (1752–1833) to give proper weight to the various determinations for the size and shape of the Earth. Credit for inventing least squares is also shared with Carl Friedrich Gauss (1777–1855) who, at the age of 18, independently invented the technique of least squares in 1795.
- The United States purchased the Louisiana Territory from France in 1803, and during most of the nineteenth century the rectangular U.S. Public Land Survey System was systematically and permanently etched on the curved surface of the North American continent.
- In 1807, President Thomas Jefferson established the Survey of the Coast and Ferdinand Hassler was hired as the first Director. Ferdinand Hassler (1770–1843) was a Swiss scientist who was well versed in making geodetic surveys and brought an impressive technical library and scientific instruments with him to America (including a standard meter bar and three standard toise bars). One of his first tasks as Director of the Survey of the Coast was to make a trip to Europe to acquire (additional) appropriate survey instruments. Over the next 150 years, the Survey of the Coast and its successor, the U.S. Coast & Geodetic Survey, observed high-order triangulation over the entire United States.

- In 1799, Captain William Lambton (1756–1823) "…drew up a project for a mathematical and geographical survey that would extend… from the southern tip to the northern extreme" of India. Formal orders for the survey were issued in 1800 and work commenced with measurement of a base in 1802. George Everest (1790–1866) joined the Great Trigonometric Survey (GTS) in 1819 and quickly established himself as a capable assistant to Superintendent Lambton who died in 1823. Everest carried on the work until he returned to England in November 1825 on sick leave. In 1829, Everest was named Surveyor General of India. He returned to India in October 1830 where he supervised the work until he retired in December 1843, on the pension of a full colonel. Everest then returned to England where, at the age of 56, he married a 23-year-old woman whose father was 6 years younger than himself. In all his years in India, George Everest never laid eyes on the highest mountain in the world that was named in his honor in August 1856 (Smith 1999).
- The U.S. Coast & Geodetic Survey (a part of which is now known as the National Geodetic Survey) completed triangulation arcs and computed massive network adjustments in 1927 and again in 1986 that provide local users reliable geodetic positions for monumented points on the surface of the Earth. These network adjustments were known respectively as the North American Datum of 1927 (NAD 27) and North American Datum of 1983 (NAD 83). Leveling networks were also observed and adjusted and vertical datum elevations were published.
- In October 1957, the Russians launched the first artificial satellite to orbit the Earth, and by 1964, the United States had developed the Transit satellite system (NAVSAT) that was used by the U.S. military for positioning Polaris submarines anywhere on the oceans (except at the North Pole), rain or shine. The first NAVSTAR global positioning system (GPS) satellites were launched in 1978, and by 1985, GNSS was recognized worldwide as the premier satellite positioning system. The full constellation of satellites was subsequently completed, and the GPS was declared fully operational in July 1995.
- The transistor and electronic computer were invented in the middle of the twentieth century and, with regard to surveying and mapping, the electronic distance measuring instrument, photogrammetry, GNSS, and remote sensing have revolutionized the use of spatial data. These developments are all considered part of the digital revolution. The internet and cellular telephone technologies are currently riding the wave of the digital revolution.
- The Global Spatial Data Model (GSDM) was first proposed by Burkholder (1997)—see Appendix D—and is described in a report prepared for the Southeastern Wisconsin Regional Planning Commission. The GSDM is intended to be compatible with the use of digital spatial data; the technologies employed for the generation, storage, manipulation, and use of spatial data; and the way humans perceive the world and use spatial relationships.

## FORECAST FOR THE TWENTY-FIRST CENTURY

With dawn of the new millennium, the planet Earth and civilization stand at the threshold of innumerable opportunities. With regard to spatial data and the evolution of geomatics as an umbrella discipline, a few forecasts include the following:

- A 3-D global datum will emerge as a global standard for data exchange.
- Compatibility issues related to horizontal and vertical datums will be addressed in terms of how each is related to the standard global model.
- Spatial data accuracy with respect to a datum will be assessed two ways:
  A. A collection of points or features (a spatial data set) will have one accuracy ranking based upon the metadata describing the modes of collection and the models used in processing.
  B. Each point location will have a covariance matrix associated with it that provides statistically reliable standard deviations in three dimensions. Some data sets will also include time as the fourth dimension.
- Statistical correlation between points will be stored and used as the basis for describing local and network accuracies.
- At the conceptual level, solid geometry, matrices, and vector algebra will be emphasized as appropriate tools for handling spatial data. In the applications arena, spatial data users will be able to work with simple local rectangular coordinate differences that reflect a local perception of a flat Earth. True geometrical integrity over the curved Earth will be preserved as a consequence of using the 3-D GSDM.
- WVD yyyy elevation (or some other name) will be a derived quantity computed as the distance above or below a named reference ellipsoid. Geodetic scientists will still be concerned about the difference between the ellipsoid and geoid and provide that information to researchers who need to find the geoid. But in local practice, except for specialized applications, the ellipsoid height difference will be used as if it were an elevation difference. Before adopting such a practice, the impact of making the change will need to be studied carefully, especially in areas where the deflection-of-the-vertical is large or changes rapidly. Otherwise, the need for elaborate geoid modeling will be drastically reduced.
- Persons using cellular telephones and the internet for communication along with GNSS (and other technology) for position will provide a wide range of location-based services for an increasing segment of the population on a global basis. Data compatibility and interoperability between disciplines are critical.
- A global 3-D spatial database provides the framework within which spatial data from any source can be freely exchanged. Decentralized use of spatial data is supported in that each user is free to manipulate data according to local practice and specifications. Rules for decentralized data manipulation are the responsibility of each user. At the foundation level, the functional and stochastic components of the GSDM stipulate the requirements for meeting any level of accuracy specified by the user.

## REFERENCES

Alder, K. 2002. *The Measure of All Things—The Seven-Year Odyssey and Hidden Error That Transformed the World*. New York: Free Press.

Bell, E.T. 1937. *Men of Mathematics*. New York: Simon & Schuster.

Berthon, S. and Robinson, A. 1991. *The Shape of the World*. Chicago, IL: Rand McNally.

Burkholder, E.F. 1997. *Definition of a Three-Dimensional Spatial Data Model for Southeastern Wisconsin*. Waukesha, WI: Southeastern Wisconsin Regional Planning Commission. http://www.sewrpc.org/SEWRPCFiles/Publications/ppr/definition_three-dimensional_spatial_data_model_for_wi.pdf (accessed May 4, 2017).

Burkholder, E.F. 2002. Elevations and the global spatial data model (GSDM). Paper presented at the *58th Annual Meeting of the Institute of Navigation*, Albuquerque, NM, June 24–26, 2002.

Burkholder, E.F. September 2006. The digital revolution: Whiter now? *GIM International* 20 (9): 25–27.

Carta, M., trans. 1962. *Jordan's Handbook of Geodesy*. Washington, DC: U.S. Army Corps of Engineers, Army Map Service.

Gore, A. January 1998. The Digital Earth: Understanding our planet in the 21st Century. Los Angeles, CA: California Science Center.

Howse, D. 1980. *Greenwich Time and the Discovery of Longitude*. Oxford, U.K.: Oxford University Press.

Meyer, T., Roman, D., and Zilkoski, D. 2004. What does height really mean? Part I: Introduction. *Surveying and Land Information Science* 64 (4): 223–233.

National Research Council (NRC). 1978. *Geodesy: Trends and Prospects*. Washington, DC: National Research Council, National Academy of Sciences.

National Research Council (NRC). 1990. *Geodesy in the Year 2000*. Washington, DC: National Research Council, National Academy Press.

Russell, B. 1959. *Wisdom of the West*. London, U.K.: Rathbone.

Smith, J.R. 1986. *From Plane to Spheroid: Determining the Figure of the Earth from 3000 B.C. to the 18th Century Lapland and Peruvian Survey Expeditions*. Rancho Cordova, CA: Landmark Enterprises.

Smith, J.R. 1999. *Everest: The Man and the Mountain*. Caithness, Scotland: Whittles.

Sobel, D. 1999. *Galileo's Daughter*. New York: Walker.

Tompkins, P. 1971. *Secrets of the Great Pyramids*. New York: Harper & Row.

Wilford, J.N. 1981. *The Mapmakers*. New York: Vintage.

Zilkoski, D. 1992. North American vertical datum and International Great Lakes Datum: They are one and the same. *Proceedings of the U.S. Hydrographic Conference '92*, Baltimore, MD.

# 7 Geometrical Geodesy

## INTRODUCTION

Geometrical geodesy is the branch of science that deals with the size and shape of the Earth. Viewed from a large distance, the Earth appears to be nearly spherical and, by comparison, the Earth is smoother than an orange—even when counting the highest mountains. If planet Earth were reduced to a globe having a diameter of 1.0000 meter at the equator, the length of the spin axis would be 0.99665 meter, only 3.35 millimeters less. As postulated by Newton, this flattening at the poles is due primarily to the Earth spinning on its axis. The sea-level shape of the Earth is a continuous equipotential surface called the geoid. As such, it could be viewed as the surface of the ocean at rest (no currents, winds, or tides) and extending coast-to-coast in a large transcontinental canal. An ellipse rotated about its minor axis generates a 3-D mathematical surface called the ellipsoid. Geodetic scientists select parameters of an ellipsoid that best approximate the geoidal shape of the Earth. The distance between the ellipsoid and geoid is called geoid height and varies, plus or minus, up to about 100 meters worldwide. Figure 7.1 is a meridian section of the Earth showing the poles, the equator, the spin axis, the mathematical ellipsoid, the geoid, the normal, and the vertical. Note that the ellipsoid-geoid separation and the flattening of the Earth are both exaggerated in Figure 7.1.

An important concept in physical geodesy is that a level surface is perpendicular to the plumb line at every point. The plumb line in Figure 7.1 defines the vertical. In geometrical geodesy, the ellipsoid normal is always perpendicular to the tangent to the ellipse. The angular difference between the normal and the vertical at a point is called the deflection-of-the-vertical. In this chapter, the deflection-of-the-vertical is taken to be zero. Chapter 9 contains information on the deflection-of-the-vertical in those cases where it is not zero.

As described by Newton and discussed in Chapter 6, the physical shape of the geoid is determined primarily by gravitational attraction and centrifugal force. The centrifugal force at a point is constant and can be computed with a high degree of certainty. But, due to the irregular distribution of mass and variations of density within the Earth, the force of gravity is not so uniform. Variations in the shape of the geoid caused by gravity anomalies are also discussed in Chapter 9.

An ellipsoid is the mathematical basis of geometrical geodesy. From a local perspective, and for many uses, it can be said that the Earth is flat. In that context, horizontal dimensions are flat and vertical dimensions are up. In many ways, such a flat Earth model is easier to use than the ellipsoid model, and many spatial data users prefer working with plane rectangular coordinates. But, as has been known for centuries, the Earth is not flat and numerous applications arise in which the flat-Earth model is not adequate. A more complex model—better for specifying location on the

147

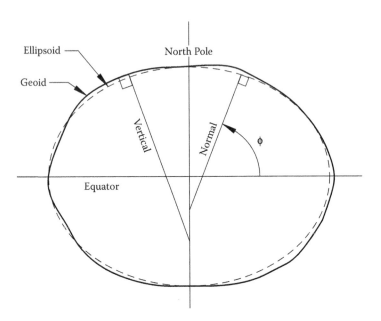

**FIGURE 7.1**  Ellipsoid and Geoid

Earth—is needed. One could say that the spatial data (location) model has evolved from saying the Earth is flat, to using a sphere to approximate the size and shape of the Earth, to using an ellipsoid chosen for a regional best fit, to using an ellipsoid selected for a global best fit.

So, even though the spatial data community may prefer using a flat-Earth model that accommodates a local perspective, a better understanding of basic spatial data concepts comes from learning the geometry of a rotational ellipsoid that

- Has its origin at (or near) the center of mass of the Earth
- Shares its minor axis with the spin axis of the Earth
- Has its major axis in the equatorial plane of the Earth
- Has its zero meridian coincident with the Greenwich Meridian
- Uses sexagesimal latitude and longitude coordinates on the ellipsoid surface
- Uses elevation (and/or ellipsoid height) in meters for the third (up) dimension

Esoteric issues (such as polar wandering and the aforementioned deflection-of-the-vertical) need more explanation than is offered in this chapter. But the emphasis here is that the GSDM is a mechanism that permits spatial data users to work with local rectangular (flat-Earth) differences while the underlying model appropriately accommodates the ellipsoidal shape of the Earth. Not only that, with careful planning, those esoteric issues, which are still a legitimate concern to geodesists, will have little or no impact on the local spatial data user. Additionally, the GSDM offers well-defined mathematical procedures by which the 3-D accuracy of spatial data can be established, tracked, and utilized—even from a local flat-Earth perspective.

## THE TWO-DIMENSIONAL ELLIPSE

The 3-D geometry of the ellipsoid begins with the meridian section of a 2-D ellipse. Figure 7.2 shows a meridian section in which the ellipse major axis is coincident with the equator of the Earth and the minor axis is essentially coincident with the spin axis. The length of the semimajor axis is denoted as $a$ and the length of the semiminor axis is denoted $b$. Equation 7.1 is the equation of a 2-D ellipse in the $X/Z$ plane whose size and shape are defined by parameters $a$ and $b$.

$$\frac{X^2}{a^2} + \frac{Z^2}{b^2} = 1 \tag{7.1}$$

Note that, for the Earth, $a$ is greater than $b$ and that the flattening of the ellipse is expressed several ways.

Flattening is defined as

$$f \equiv \frac{a-b}{a} = 1 - \frac{b}{a} \tag{7.2}$$

Eccentricity squared is

$$e^2 \equiv \frac{a^2 - b^2}{a^2} = 1 - \frac{b^2}{a^2} \tag{7.3}$$

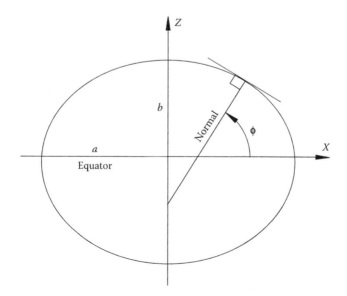

**FIGURE 7.2** Two-Dimensional Ellipse

Second eccentricity squared is

$$e'^2 \equiv \frac{a^2 - b^2}{b^2} = \frac{a^2}{b^2} - 1 \tag{7.4}$$

The polar radius of curvature is useful when making comparisons between ellipsoids. The polar radius of curvature is

$$c = \frac{a^2}{b} = \frac{a}{\sqrt{1 - e^2}} \tag{7.5}$$

Additional geometrical relationships derived from the definitions above include

$$e^2 = 2f - f^2 \tag{7.6}$$

$$b = a(1 - f) \tag{7.7}$$

$$\frac{b^2}{a^2} = 1 - e^2 \tag{7.8}$$

$$e'^2 = \frac{e^2}{1 - e^2} \tag{7.9}$$

Several observations with regard to the Earth ellipsoid, $a$, $b$, $f$, $e^2$, and $e'^2$ are as follows:

1. If $b = a$, the meridian ellipse is really a circle and the ellipsoid is a sphere. The flattening, the eccentricity, and the second eccentricity are all zero.
2. As $b$ approaches 0, the ellipse degenerates into a line and the ellipsoid becomes a flat plane. In that case, the values of $f$, $e^2$, and $e'^2$ are all exactly 1.
3. For the Earth, values of $f$, $e^2$, and $e'^2$ are all near zero and the ellipsoid is nearly spherical.
4. Some derivations in geodesy are accomplished more efficiently using the second eccentricity instead of the eccentricity. Although equally legitimate, the second eccentricity is used in this book only as required to be consistent with referenced material.

Two geometrical elements are required to define an ellipse. The most obvious pair of elements is the semimajor and semiminor axes. However, over the years the following conventions have been adopted and are used as noted. In each case, the ellipsoid is obtained by rotating the 2-D ellipse about its minor axis.

- $a$ and $b$: The semimajor axis and the semiminor axis. The Clarke Spheroid 1866 is defined by $a$ and $b$ and used in North America for the North American Datum of 1927 (NAD 27). See Table 7.1.

## TABLE 7.1
### Selected Geometrical Geodesy Ellipsoids

| Ellipsoid | Defined by | Derived Values |
|---|---|---|
| Clarke Spheroid 1866 (used for NAD 27) | $a = 6{,}378{,}206.4$ m<br>$b = 6{,}356{,}583.8$ m | $1/f = 294.978698214$<br>$e^2 = 0.0067686579973$<br>$c = 6{,}399{,}902.5516$ m |
| Geodetic Reference System 1980 (used for NAD 83) | $a = 6{,}378{,}137.000$ m<br>$1/f = 298.257222101$ | $b = 6{,}356{,}752.3141$ m<br>$e^2 = 0.0066943800229$<br>$c = 6{,}399{,}593.6259$ m |
| World Geodetic System 1984 (used in GNSS positioning) | $a = 6{,}378{,}137.000$ m<br>$1/f = 298.257223563$ | $b = 6{,}356{,}752.3142$ m<br>$e^2 = 0.0066943799902$<br>$c = 6{,}399{,}593.6258$ m |

- *a and* $1/f$: The semimajor axis and the reciprocal flattening. The Geodetic Reference System 1980 (GRS 80) and World Geodetic System 1984 (WGS 84) as used in geometrical geodesy are both defined by $a$ and $1/f$.
- *a and* $e^2$: The semimajor axis and the eccentricity squared. Most geodetic computations are arranged with the goal of preserving computational strength and efficiency. Although eccentricity squared is not a defining parameter, note that the quantity $(1 - e^2)$ is used in many geodesy equations and contains two more significant figures than does $e^2$ by itself.

Note that values of $c$, the radii of curvature at the poles, for the GRS 80 and the WGS 84, are within 0.1 mm of being identical even though the $1/f$ (and $e^2$) values are significantly different. That agreement supports the suggestion that the GRS 80 and WGS 84 ellipsoids can be used interchangeably. An important distinction is that, while they may be interchangeable as ellipsoids, WGS 84 datum coordinates are not interchangeable with the NAD 83 datum coordinates. See Chapter 8 for more details.

An ellipsoid was defined previously as that figure generated by rotating an ellipse about its minor axis. That same definition was used in the past to describe a spheroid, and the Clarke Spheroid 1866 still carries the word "spheroid" in its name. Although "ellipsoid" and "spheroid," have essentially the same definition in geometrical geodesy and are often used interchangeably, a distinction can be made. A spheroid is generically defined as a body that is nearly spherical, but not quite. That definition permits, but does not require, that the surface be rigorously defined mathematically. The ellipsoid enjoys rigorous mathematical definition in all cases. The convention in this book is to use the word "ellipsoid" when referring to a mathematical approximation of the size and shape of the Earth.

The ellipse shown in Figure 7.2 is taken through the Greenwich Meridian and shows the ellipse in terms of the $X/Z$ plane. Figure 7.3 shows the rectangular $X/Y/Z$ geocentric ECEF coordinate system superimposed upon the ellipsoid. The $X$ and $Y$ distances are in the plane of the equator; the Greenwich Meridian is in the $X/Z$ plane; the $Z$-axis represents the spin axis of the Earth; and the $X$-axis pierces the

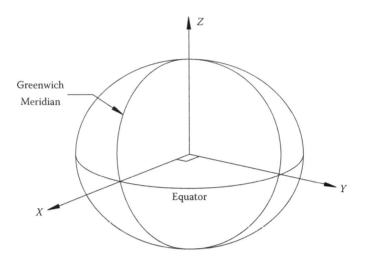

**FIGURE 7.3**   Three-Dimensional Ellipsoid

equator at the Greenwich Meridian. Figure 7.2 also shows a line tangent to the meridian ellipse. The line perpendicular to the tangent is the ellipsoid normal ($N$) and goes from the ellipse tangent to the spin axis. Geodetic latitude is the angle the normal makes with the equator and is denoted by the Greek letter phi ($\phi$). The geodetic latitude ranges from $-90°$ at the South Pole to $0°$ on the equator to $+90°$ at the North Pole.

With respect to Equation 7.1 and Figure 7.2, expressions for $X$ and $Z$ (in terms of $a$, $e^2$, and $\phi$) are found by equating the slope of the tangent to the ellipse with the first derivative of Equation 7.1 at the same point. The slope of the tangent to the ellipse is given by the trigonometric tangent of the angle $(90° + \phi)$. Using trigonometric identities

$$\tan(90+\phi) = \frac{\sin(90+\phi)}{\cos(90+\phi)} = \frac{\cos\phi}{-\sin\phi} = \frac{-1}{\tan\phi} \tag{7.10}$$

By algebraic manipulation, the derivative of Equation 7.1 is

$$\frac{2X}{a^2}dX + \frac{2Z}{b^2}dZ = 0 \quad \text{or} \quad \frac{dZ}{dX} = -\frac{2b^2Z}{2a^2X} \tag{7.11}$$

Equating Equations 7.10 and 7.11, substituting for $b^2/a^2$, and solving for $Z$, we get the following:

$$\frac{-1}{\tan\phi} = \frac{-b^2}{a^2}\frac{X}{Z} \quad \text{implies that} \quad Z = \left(1-e^2\right)\frac{\sin\phi}{\cos\phi}X \quad \text{and} \quad Z^2 = \left(1-e^2\right)^2\frac{\sin^2\phi}{\cos^2\phi}X^2$$

$$\tag{7.12}$$

Next, solve Equation 7.1 for $Z^2$, equate to Equation 7.12, and solve for $X$:

$$\frac{X^2}{a^2} + \frac{Z^2}{b^2} = 1 \text{ implies that } Z^2 = b^2\left(1 - \frac{X^2}{a^2}\right) = \left(1 - e^2\right)\left(a^2 - X^2\right) = \left(1 - e^2\right)^2 \frac{\sin^2\phi}{\cos^2\phi} X^2$$

(7.13)

Use algebraic manipulation on the last two terms in Equation 7.13 to solve for $X$ as

$$\cos^2\phi\left(a^2 - X^2\right) = X^2\left(1 - e^2\right)\sin^2\phi$$
$$a^2\cos^2\phi - X^2\cos^2\phi = X^2\sin^2\phi - X^2e^2\sin^2\phi$$
$$a^2\cos^2\phi = X^2\left(\cos^2\phi + \sin^2\phi\right) - X^2e^2\sin^2\phi = X^2\left(1 - e^2\sin^2\phi\right)$$

$$X^2 = \frac{a^2\cos^2\phi}{1 - e^2\sin^2\phi} \quad \text{and} \quad X = \frac{a\cos\phi}{\sqrt{1 - e^2\sin^2\phi}} \tag{7.14}$$

Now substitute Equation 7.14 into Equation 7.12 to solve for $Z$:

$$Z = \left(1 - e^2\right)\frac{\sin\phi}{\cos\phi}\left(\frac{a\cos\phi}{\sqrt{1 - e^2\sin^2\phi}}\right) = \frac{a\left(1 - e^2\right)\sin\phi}{\sqrt{1 - e^2\sin^2\phi}} \tag{7.15}$$

Note further in Figure 7.2 that the *normal* is the hypotenuse of a right triangle whose base is $X$. However, the usual statement, also apparent from Figure 7.2, is that

$$N\cos\phi = X \quad \text{from which } N = \frac{a}{\sqrt{1 - e^2\sin^2\phi}} \tag{7.16}$$

Because the Earth is flattened at the poles, the instantaneous radius of curvature in the north-south direction increases as one moves from the equator toward either pole. This is consistent with Newton's oblate Earth theory that was proven conclusively in the mid-eighteenth century by geodetic surveying expeditions to Lapland and Peru (modern day Finland and Ecuador). The instantaneous radius of curvature, $M$, in the north-south direction on the ellipsoid at a specified latitude is given by the general mathematical equation for curvature of a function as

$$M = \frac{-\left[1 + \left(\dfrac{dZ}{dX}\right)^2\right]^{3/2}}{\dfrac{d^2Z}{dX^2}} \tag{7.17}$$

Starting with Equation 7.1, taking the first and second derivatives, and substituting them into Equation 7.17 is tedious and involves much algebraic manipulation, but the result is stated concisely (the negative sign is dropped by convention) as

$$M = \frac{a\left(1 - e^2\right)}{\left(1 - e^2 \sin^2 \phi\right)^{3/2}} \tag{7.18}$$

## THE THREE-DIMENSIONAL ELLIPSOID

Equations 7.1 through 7.18 are presented in terms of the Greenwich Meridian section. Now, consider that the ellipse is rotated about the minor (Z) axis and that the foregoing equations are applicable in any meridian. And, since we are no longer restricted to the meridian ellipse, it is appropriate to speak in terms of the 3-D ellipsoid as depicted in Figure 7.3.

### ELLIPSOID RADII OF CURVATURE

Parallels of latitude are circles on the ellipsoid in planes parallel to the equator. The location of any parallel on the ellipsoid is measured in degrees, minutes, and seconds north or south from the equator. At each parallel of latitude, the underlying ellipsoid has a radius of curvature in the north-south direction (that is $M$ as discussed previously). At the same location, the underlying ellipsoid also has a radius of curvature, $N$, in the east-west direction (perpendicular to the meridian). The plane perpendicular to the meridian is called the prime vertical and contains the ellipsoid normal. At any point, the radius of curvature in the plane of the prime vertical is $N$, the length of the ellipsoid normal. Remember, if the Earth were perfectly spherical, the radius would be the same at all points. And, at any given point, the radius of curvature of the underlying sphere would be the same in any direction. But, the Earth is flattened at the poles and the ellipsoid radius of curvature changes with latitude and with direction of a line. At any point on a given parallel of latitude, $M$ is the radius of curvature of the underlying ellipsoid in the north-south direction and $N$ is the radius of curvature at the same point in the east-west direction. $M$ and $N$ are collinear, but are not the same length (except at the poles).

Table 7.2 shows the results of tabulating values for both $M$ and $N$ at the equator and at the poles. Note that the value of $N$ at the equator is $a$, the semimajor axis. The value of $M$

---

**TABLE 7.2**

**Comparisons of Radii of Curvature—$M$ and $N$**

| | | |
|---|---|---|
| *Equation:* | $M = \dfrac{a\left(1 - e^2\right)}{\left(1 - e^2 \sin^2 \phi\right)^{3/2}}$ | $N = \dfrac{a}{\sqrt{1 - e^2 \sin^2 \phi}}$ |
| *Results at two latitudes:* | | |
| Equator ($\phi = 0°$) | $M = a(1 - e^2)$ | $N = a$ |
| Pole ($\phi = 90°$) | $M = \dfrac{a}{\sqrt{1 - e^2}} = c$ | $N = \dfrac{a}{\sqrt{1 - e^2}} = c$ |

---

at the equator is somewhat shorter. Except at the poles, the ellipsoid radius of curvature at any point is shorter in the N-S direction than in the E-W direction. At the poles, values for $M$ and $N$ are identical and called $c$, the polar radius of curvature, Equation 7.5. The reason is that the prime vertical at the pole (90° to a meridian) is itself another meridian.

## NORMAL SECTION RADIUS OF CURVATURE

The normal section is defined as the intersection of a plane containing the ellipsoid normal and the ellipsoid surface. Given that the deflection-of-the-vertical is zero, the normal section lies in the plane shown by the vertical cross-hair of a carefully leveled transit, theodolite, or total station. There is an infinite number of normal sections radiating from a point, and each one has a unique azimuth with respect to the meridian through that point.

Previously the values of $M$ and $N$ were given as radii of curvature in the north-south and east-west directions. The radius of curvature, $R_\alpha$, in the plane of a normal section at any azimuth, $\alpha$, is given by Euler's equation as

$$R_\alpha = \frac{MN}{M \sin^2 \alpha + N \cos^2 \alpha} \quad \text{or} \quad \frac{1}{R_\alpha} = \frac{\cos^2 \alpha}{M} + \frac{\sin^2 \alpha}{N} \qquad (7.19)$$

## GEOMETRICAL MEAN RADIUS

Equation 7.19 is as good as it gets and can be used to compute the ellipsoid radius of curvature at a given latitude in a particular direction. But, in some cases, such as reducing horizontal distance to sea level, it is common to use an approximate spherical Earth radius of 6,372,000 m (or 20,906,000 feet). If, for whatever reason, a spherical radius is not good enough, then Equation 7.19 could be used. But there are also times when a spherical radius is not good enough and Equation 7.19 represents overkill. A radius of intermediate accuracy is the geometrical mean radius, $R_{mean}$, which is computed for a given ellipsoid at a given latitude. The azimuth of the normal section at the point is not needed or used. Rapp (1991) calls $R_{mean}$ the Gauss mean radius and derives it as

$$R_{mean} = \sqrt{MN} = \frac{a\sqrt{1 - e^2}}{\left(1 - e^2 \sin^2 \phi\right)} \qquad (7.20)$$

# ROTATIONAL ELLIPSOID

## EQUATION OF ELLIPSOID

Given that the Greenwich Meridian ellipse is rotated about its minor axis, the resulting figure is an ellipsoid used to approximate the size and shape of the Earth. In terms of the geocentric ECEF rectangular coordinate system, the equation of the ellipsoid is given by

$$\frac{X^2}{a^2} + \frac{Y^2}{a^2} + \frac{Z^2}{b^2} = 1 \qquad (7.21)$$

Note that if $Z = 0$, Equation 7.21 reduces to a circle in the plane of the equator and, separately, if $Y = 0$, Equation 7.21 reduces to an ellipse as given in Equation 7.1. Geodetic latitude and longitude are 2-D coordinates used to represent location on the 3-D ellipsoid surface. Ellipsoid height, the third dimension, is used to specify the distance of a point above or below the ellipsoid.

## GEOCENTRIC AND GEODETIC COORDINATES

Geodetic coordinates of latitude, longitude, and height have been the computational standard for many years. However, with the advent of GNSS technology, the use of the ECEF rectangular geocentric coordinates has much to offer with respect to modern technology and working with digital spatial data. This section first looks at the equations used to convert latitude, longitude, and height to ECEF geocentric $X/Y/Z$ coordinates. Then, the inverse computation—starting with $X/Y/Z$ and computing latitude, longitude, and height coordinates—is considered. With reference to the 3-D GSDM diagram shown in Figure 1.4 and strictly as a matter of convenience, the transformation of latitude, longitude, and height to $X/Y/Z$ coordinates is called a BK1 transformation and transforming $X/Y/Z$ coordinates to latitude, longitude, and height is called a BK2 transformation.

Figure 7.4 shows three views of the Earth.

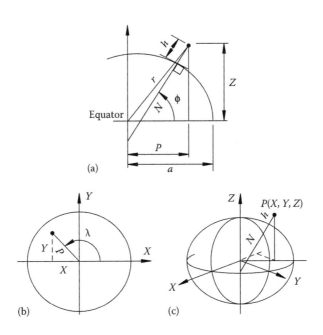

**FIGURE 7.4** Geometry and Symbols Used in BK1 and BK2 Transformations. (a) the Meridian Section Quadrant, (b) the Plane of the Equator, and (c) a 3-D View of the Ellipsoid

The following symbols are used as appropriate in the three views:

$X/Y/Z$ = geocentric ECEF rectangular coordinates—meters.
$\phi$ and $\lambda$ = geodetic latitude and longitude—sexagesimal units.
$h$ = ellipsoid height—meters.
$N$ = length of the ellipsoid normal—meters.
$P$ = projected distance of point from spin axis in equatorial plane.
$r$ = spatial distance from coordinate origin to point.
$a$ and $b$ = parameters of underlying ellipsoid.

Symbols used in the Vincenty BK2 transformation, but not appearing in Figure 7.4 include

$a'$ and $b'$ = parameters of an auxiliary ellipsoid.
$h'$ = approximate ellipsoid height.
$\phi'$ = approximate intermediate geodetic latitude.
$T$ and $U$ = intermediate values used for computational convenience.

## BK1 TRANSFORMATION

A BK1 transformation uses Equations 7.22 through 7.24. The relationships for those equations are illustrated in Figure 7.4a and b. The distance $Z$ in Figure 7.4a is given by Equation 7.25, but it is not obvious from the diagram that the factor $(1 - e^2)$ is responsible for removing that part of the normal lying below the equator. The equation for $Z$ was derived previously and given as Equation 7.15. Note too that Equations 7.23 through 7.25 give $X$, $Y$, and $Z$ values for any point at ellipsoid height, $h$, but that Equations 7.14 and 7.15 are specifically for $X$ and $Z$ on the ellipsoid surface in the Greenwich Meridian.

$$P = (N + h)\cos\phi \tag{7.22}$$

$$X = P\cos\lambda = (N + h)\cos\phi\cos\lambda \tag{7.23}$$

$$Y = P\sin\lambda = (N + h)\cos\phi\sin\lambda \tag{7.24}$$

$$Z = \left[N(1 - e^2) + h\right]\sin\phi \tag{7.25}$$

These BK1 equations are used to compute the ECEF geocentric coordinates for any point whose position is defined by latitude, longitude, and height on a specified datum or ellipsoid—typically NAD 83 geodetic coordinates or some other datum such as the WGS 84 or ITRF Reference Frame.

## BK2 TRANSFORMATION

The BK2 transformation uses geocentric $X/Y/Z$ coordinates to compute latitude, longitude, and height referenced to a named Earth-centered ellipsoid. Although a BK2 transformation is not as easy as a BK1 transformation, there are several ways it can be done. The longitude part of a BK2 computation is straight-forward and, with reference to Figure 7.4b, uses Equation 7.26 with due regard to the quadrant as described in Equations 5.11 through 5.13. Note, however, that the sign convention for longitude in the plane of the equator is consistent with the math/science convention. As such, $Y$ is in the numerator while $X$ is in the denominator when using the inverse tangent function:

$$\tan \lambda = \frac{Y}{X} \tag{7.26}$$

### ITERATION

Possibly the best approach to a BK2 transformation is to iterate on the geodetic latitude. Once latitude is computed, the ellipsoid height computation is routine. Leick (2004) recommends using Equation 7.27 for the iteration. An initial approximation for geodetic latitude is given by Equation 7.28. Equation 7.29 is used to compute the ellipsoid normal for the first iteration. Equation 7.30 is used to compute the next latitude approximation based upon previous values of latitude and ellipsoid normal. Values of geodetic latitude and ellipsoid normal are computed repeatedly until the change in successive values is negligible. After the geodetic latitude is found, Equation 7.32 can be used to compute the ellipsoid height. Given the iteration is programmed into a computer, the effort expended in the solution is minimal. A longhand iteration solution can become rather tedious.

$$\tan \phi = \frac{Z}{P}\left(1 + \frac{e^2 N \sin \phi}{Z}\right) \quad \text{where } P = \sqrt{X^2 + Y^2} \tag{7.27}$$

$$\phi_0 = \arctan\left(\frac{Z}{P\left(1 - e^2\right)}\right) \tag{7.28}$$

$$N_0 = \frac{a}{\sqrt{1 - e^2 \sin^2 \phi_0}} \tag{7.29}$$

$$\phi_i = \arctan\left[\frac{Z}{P}\left(1 + \frac{e^2 N_{i-1} \sin \phi_{i-1}}{Z}\right)\right] \tag{7.30}$$

$$N_i = \frac{a}{\sqrt{1 - e^2 \sin^2 \phi_i}} \tag{7.31}$$

$$h = \frac{P}{\cos \phi} - N \tag{7.32}$$

## Noniterative (Vincenty) Method

Another approach to the BK2 transformation is to use a "once-through" procedure devised by Vincenty (1980). In his method, the iteration is built-in, and once through, the computation will give an answer tested to be within 0.2 mm for any point within the birdcage of orbiting GNSS satellites. No attempt is made to explain all the terms, but it can be inferred from Equations 7.37 and 7.38 that the solution is developed with respect to an auxiliary ellipsoid passing through the point at elevation $h'$.

$$b = a(1-f) \tag{7.33}$$

$$P^2 = X^2 + Y^2 \quad \text{and} \quad P = \sqrt{X^2 + Y^2} \tag{7.34}$$

$$r^2 = P^2 + Z^2 \quad \text{and} \quad r = \sqrt{P^2 + Z^2} \tag{7.35}$$

$$h' = r - a + \frac{(a-b)Z^2}{r^2} \tag{7.36}$$

$$a' = a + h' \tag{7.37}$$

$$b' = b + h' \tag{7.38}$$

$$\phi' = \arctan\left[\left(\frac{a'}{b'}\right)^2 \left(\frac{Z}{P}\right)\left(1 + \frac{e^4 h' a (Z^2 - P^2)}{4a'^4}\right)\right] \tag{7.39}$$

$$T = \frac{(P - h'\cos\phi')^2}{a^2} \tag{7.40}$$

$$U = \frac{(Z - h'\sin\phi')^2}{b^2} \tag{7.41}$$

$$h = h' + \frac{1}{2}\left[\frac{T + U - 1}{T/a + U/b}\right] \tag{7.42}$$

$$\phi = \arctan\left[\left(\frac{a}{b}\right)^2 \left(\frac{Z - e^2 h \sin\phi'}{P}\right)\right] \tag{7.43}$$

$$\lambda = \arctan\left(\frac{Y}{X}\right) \tag{7.44}$$

Notes about the BK1 and BK2 transformation include the following:

1. The BK1 and BK2 transformations are critical, and BK2 is probably the most difficult part of the GSDM. But, the rotation matrix in Equation 1.31 requires the latitude and the longitude of the standpoint. They are determined from the geocentric $X/Y/Z$ coordinates of the standpoint using the BK2 transformation.
2. The BK2 transformation is also described by others. See, for example, Soler and Hothem (1989), Hoffmann-Wellenhof et al. (2001), Ghilani (2010), You (2000), Hooijberg (1997), and Meyer (2010).
3. The integrity of any BK2 solution can be checked by using the computed latitude, longitude, and height in the BK1 equations. It should be possible to duplicate the $X/Y/Z$ values used in the BK2 transformation. There is no approximation in the BK1 equations.
4. The accuracy of results may also depend upon the significant figure capacity of the computer being used or by the way the software is written.

### EXAMPLE OF **BK1** TRANSFORMATION

Given: The GRS 80 ellipsoid and station REILLY, an A-order NAD 83 (1992) station located on the campus of New Mexico State University (NMSU), Las Cruces, New Mexico. Note that equations presume positive latitude north of the equator and negative latitude south of the equator. Longitude east of Greenwich is used as a positive value. If west longitude is used as a negative number, the results will be the same.

Ellipsoid: GRS 80, $a = 6,378,137.000$ m, $e^2 = 0.006694380023$
Latitude: $\phi = 32°\ 16'\ 55.''92906$ N
Longitude: $\lambda = 106°\ 45'\ 15.''16070$ W $= 253°\ 14'\ 44.''83930$ E
Ellipsoid height: $h = 1,167.57$ meters

Compute:

$$N = \frac{a}{\sqrt{1 - e^2 \sin^2 \phi}} = 6,384,235.5313 \text{ m}$$

$$P = (N + h)\cos\phi = 5,398,397.2940 \text{ m}$$

$$X = P\cos\lambda = -1,556,177.6148 \text{ m}$$

$$Y = P\sin\lambda = -5,169,235.3185 \text{ m}$$

$$Z = \left[N\left(1 - e^2\right) + h\right]\sin\phi = 3,387,551.7093 \text{ m}$$

## EXAMPLE OF **BK2** TRANSFORMATION—ITERATION

Ellipsoid: GRS 80, $a = 6{,}378{,}137.000$ m, and $e^2 = 0.006694380023$

Given: Station "K 785," an NAD 83 (1991) first-order GNSS station located on the campus of Oregon's Institute of Technology, Klamath Falls, Oregon.

$X = -2{,}490{,}977.042$ m
$Y = -4{,}019{,}738.192$ m
$Z = 4{,}267{,}460.404$ m

Compute:

$$\lambda = \arctan\left(\frac{Y}{X}\right);$$

$$\lambda = 238°\ 12'\ 50.''646096\,\mathrm{E}$$
$$= 121°\ 47'\ 09.''\,3904\,\mathrm{W}$$

$$P = \sqrt{X^2 + Y^2};\quad P = 4{,}728{,}981.0484\ \mathrm{m}$$

$$\phi_0 = \arctan\left(\frac{Z}{P(1-e^2)}\right);\quad \phi_0 = 42°\ 15'\ 17.''133952\,\mathrm{N}$$

$$N_0 = \frac{a}{\sqrt{1 - e^2 \sin^2 \phi_0}};\quad N_0 = 6{,}387{,}812.0556\ \mathrm{m}$$

$$\phi_1 = \arctan\left[\frac{Z}{P}\left(1 + \frac{e^2 N_0 \sin \phi_0}{Z}\right)\right];\quad \phi_1 = 42°\ 15'\ 17.''993923\,\mathrm{N}$$

Values for subsequent iterations are shown in Table 7.3.

Once the geodetic latitude is found, the ellipsoid height is computed as

$$h = \frac{P}{\cos \phi} - N;\quad h = 1{,}297.8797\ \mathrm{m}$$

---

## TABLE 7.3
## Summary of BK2 Iterations

| Iteration | Latitude | Difference | Normal | Difference |
|---|---|---|---|---|
| 0 | 42° 15′ 17.″133952 | — | 6,387,812.0556 m | |
| 1 | 42° 15′ 16.″993923 | −0.″140029 | 6,387,812.0412 m | −0.0144 m |
| 2 | 42° 15′ 16.″993408 | −0.″000515 | 6,387,812.0411 m | −0.0001 m |
| 3 | 42° 15′ 16.″993405 | −0.″000003 | 6,387,812.0411 m | −0.0000 m |

## EXAMPLE OF BK2 TRANSFORMATION—VINCENTY'S METHOD (SAME POINT)

Ellipsoid: GRS 80, $a = 6{,}378{,}137.000$ m, and $e^2 = 0.006694380023$

Given: Station "K 785," an NAD 83 (1991) first-order GNSS station located on the campus of Oregon's Institute of Technology, Klamath Falls, Oregon.

$X = -2{,}490{,}977.042$ m
$Y = -4{,}019{,}738.192$ m
$Z = 4{,}267{,}460.404$ m

Compute:

$$\lambda = \arctan\left(\frac{Y}{X}\right);$$

$$\lambda = 238°\ 12'\ 50.''646096\,\text{E}$$
$$= 121°\ 47'\ 09.''353904\,\text{W}$$

$$P^2 = X^2 + Y^2 \quad \text{and} \quad P = \sqrt{X^2 + Y^2}, \quad P = 4{,}728{,}981.0484 \text{ m}$$

$$r^2 = P^2 + Z^2 \quad \text{and} \quad r = \sqrt{P^2 + Z^2}, \quad r = 6{,}369{,}810.0486 \text{ m}$$

$$h' = r - a + \frac{(a-b)Z^2}{r^2}; \quad h' = 1{,}271.2291 \text{ m}$$

$$a' = a + h'; \quad a' = 6{,}379{,}408.2291 \text{ m}$$

$$b' = b + h'; \quad b' = 6{,}358{,}023.5433 \text{ m}$$

$$\phi' = \arctan\left[\left(\frac{a'}{b'}\right)^2 \left(\frac{Z}{P}\right)\left(1 + \frac{e^4 h' a (Z^2 - P^2)}{4a^4}\right)\right]; \quad \phi' = 42°\ 15'\ 17.''996289 \text{ N}$$

$$T = \frac{(P - h'\cos\phi')^2}{a^2}; \quad T = 0.54950876168$$

$$U = \frac{(Z - h'\sin\phi')^2}{b^2}; \quad U = 0.45049960785$$

$$h = h' + \frac{1}{2}\left( \frac{T + U - 1}{\frac{T}{a} + \frac{U}{b}} \right); \quad h = 1,297.8795 \text{ m}$$

$$\phi = \arctan\left[ \left( \frac{a}{b} \right)^2 \left( \frac{Z - e^2 h \sin \phi'}{P} \right) \right]; \quad \phi = 42° \ 15' \ 17.''993406$$

The advantage of Vincenty's method is that one needs to go through it only once. Although his method is an approximation, representative tests (even for very large values of $h$) confirm agreement within 0.2 mm. If that is not precise enough, an iteration solution is recommended.

## MERIDIAN ARC LENGTH

Throughout history, the size of the Earth has been determined by measuring a portion of a meridian arc and comparing that length to the corresponding angle subtended at the center of the Earth (this procedure is described as "degree measurement" in Chapter 6). Knowing any two of the three—arc length, central angle, or radius—the third is readily computed. Different values of arc-length-per-degree-of-latitude at various latitudes implied that the Earth is flattened at the poles. But, having selected an ellipsoid model for the Earth, differential geometry elements are integrated to compute meridian arc length or portions thereof; see Figure 7.5. Arc length of a

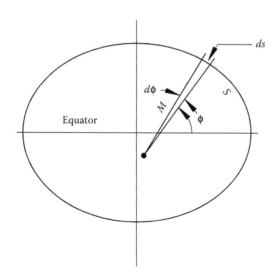

**FIGURE 7.5**   Meridian Arc Length

differential element, $ds$, equals the instantaneous radius of curvature, $M$, times the differential increment in geodetic latitude, $d\phi$.

$$ds = M d\phi \tag{7.45}$$

$S$, the meridian arc distance from one latitude to another, is obtained by integrating Equation 7.46 from one latitude to another. The meridian quadrant limits are $0°$ and $90°$ (in radian measure, $0$ and $\pi/2$).

$$S_{\phi_1 \to \phi_2} = \int_{\phi_1}^{\phi_2} M d\phi \tag{7.46}$$

The expression for $M$, Equation 7.18, is substituted into Equation 7.46 and the constant portion moved outside the integral to get

$$S_{\phi_1 \to \phi_2} = a\left(1 - e^2\right) \int_{\phi_1}^{\phi_2} \left(1 - e^2 \sin^2 \phi\right)^{-3/2} d\phi \tag{7.47}$$

Equation 7.47 contains an elliptical integral and cannot be integrated in closed form. That is, the expression inside the integral must be expressed in a series expansion containing ever-smaller terms that can be integrated individually. A solution is obtained by including all those terms that make a difference in the answer to the accuracy desired. Either a binomial series expansion or the MacLaurin series can be used with identical results (Rapp 1991). The result, Equation 7.48, can be used within any specified limits. Computational convention is to start at the southerly of two latitudes and end at the northerly limit. If the convention is switched, the result will be a negative arc distance. Latitude south of the equator should be used as a negative value.

The meridian arc distance (in meters given that $a$ is in meters) between latitude limits selected by the user is

$$
\begin{aligned}
S_{\phi_1 \to \phi_2} = a\left(1 - e^2\right) \Bigg[ & A\left(\phi_2 - \phi_1\right) - \frac{B}{2}\left(\sin 2\phi_2 - \sin 2\phi_1\right) \\
& + \frac{C}{4}\left(\sin 4\phi_2 - \sin 4\phi_1\right) - \frac{D}{6}\left(\sin 6\phi_2 - \sin 6\phi_1\right) \\
& + \frac{E}{8}\left(\sin 8\phi_2 - \sin 8\phi_1\right) - \frac{F}{10}\left(\sin 10\phi_2 - \sin 10\phi_1\right) \Bigg]
\end{aligned}
\tag{7.48}
$$

where the coefficients $A$ through $F$ are as given in Equations 7.49 and 7.50.

$$A = 1 + \frac{3}{4}e^2 + \frac{45}{64}e^4 + \frac{175}{256}e^6 + \frac{11,025}{16,384}e^8 + \frac{43,659}{65,536}e^{10} \tag{7.49}$$

$$B = \frac{3}{4}e^2 + \frac{15}{16}e^4 + \frac{525}{512}e^6 + \frac{2,205}{2,048}e^8 + \frac{72,765}{65,536}e^{10} \tag{7.50}$$

$$C = \frac{15}{64}e^4 + \frac{105}{256}e^6 + \frac{2,205}{4,096}e^8 + \frac{10,395}{16,384}e^{10} \tag{7.51}$$

$$D = \frac{35}{512}e^6 + \frac{315}{2,048}e^8 + \frac{31,185}{131,072}e^{10} \tag{7.52}$$

$$E = \frac{315}{16,384}e^8 + \frac{3,465}{65,536}e^{10} \tag{7.53}$$

$$F = \frac{693}{131,072}e^{10} \tag{7.54}$$

Notes for Equation 7.48 include the following:

1. The terms within the brackets are all unitless. That means, for example, that the latitude difference given by $\phi_2 - \phi_1$ must be in radians. The trigonometric ratios in the other terms are already unitless.
2. The coefficients need to be computed only once for each ellipsoid if they are stored (see Table 7.4). Depending upon the computational environment, it may be more efficient to compute and store them or to compute them as needed.
3. The meridian quadrant is computed by choosing limits of 0° and 90° (0 and $\pi/2$ in radians). In that case, Equation 7.48 reduces to Equation 7.55 because the trigonometric sines of the multiple angles are all zero. The meridian quadrant arc length for any Earth-model ellipsoid should be approximately 10,000,000 meters.

$$S_{0° \to 90°} = a\left(1 - e^2\right)\left[A\left(\pi/2\right)\right] \tag{7.55}$$

---

**TABLE 7.4**

**Meridian Coefficients and Quadrant Arc Length for Selected Ellipsoids**

|  | Clarke 1866 | GRS 80 | WGS 84 |
|---|---|---|---|
| $f$ | 0.00339007530393 | 0.0033528106812 | 0.0033528106647 |
| $e^2$ | 0.00676865799729 | 0.0066943800229 | 0.0066943799901 |
| $e^4$ | 0.00004581473108 | 0.0000448147239 | 0.0000448147234 |
| $e^6$ | 0.00000031010425 | 0.0000003000068 | 0.0000003000068 |
| $e^8$ | 0.00000000209899 | 0.0000000020084 | 0.0000000020084 |
| $e^{10}$ | 0.00000000001421 | 0.0000000000134 | 0.0000000000134 |
| $A$ | 1.00510892038799 | 1.0050525018131 | 1.0050525017882 |
| $B$ | 0.00511976506202 | 0.0050631086222 | 0.0050631085972 |
| $C$ | 0.00001086615776 | 0.0000106275903 | 0.0000106275902 |
| $D$ | 2.1524755323E−08 | 2.082037857E−08 | 2.082037826E−08 |
| $E$ | 4.1106495949E−11 | 3.932371371E−11 | 3.932371294E−11 |
| $F$ | 7.5116641593E−14 | 7.108453403E−14 | 7.108453229E−14 |
| Quadrant arc | 10,001,888.0430 m | 10,001,965.7292 m | 10,001,965.7293 m |

## LENGTH OF A PARALLEL

A parallel of latitude on the ellipsoid describes a circle whose plane is parallel to the equatorial plane and whose radius is $N \cos \phi$. If the Earth were spherical with radius r, the radius of a parallel of latitude would be $r \cos \phi$. Since each parallel is a circle, its circumference is simply $2\pi N \cos \phi$. Partial length of a parallel is computed as the proportionate part of the total circumference, or, as illustrated in Figure 7.6, arc length is computed directly using $L = R\theta$ with the longitude difference expressed in radians:

$$L_p = \left(\lambda_2 - \lambda_1\right) N \cos \phi = \frac{\Delta\lambda\, a \cos \phi}{\sqrt{1 - e^2 \sin^2 \phi}} \qquad (7.56)$$

## SURFACE AREA OF SPHERE

Surface area in a plane is length times width. Area on a curved surface can also be computed, but care must be taken because the distances are no longer "flat." Area of the uniformly curved surface of a sphere is computed using tools of differential geometry and integral calculus as shown in Figure 7.7. For illustration purposes, the sphere is cut into two equal pieces nominally called northern and southern hemispheres. The equator is the dividing plane and, using Equation 7.56, is a circle having a circumference of $2\pi R$ ($\Delta\lambda = 2\pi$, $N = R$, and $\cos 0° = 1$). The circumference of the sphere is taken to be the length of a plane rectangle. The width of the rectangle is an infinitesimally small differential element in the north/south direction. Referring to Figure 7.7, the differential north/south distance is $R\, d\phi$. If the small ring around the

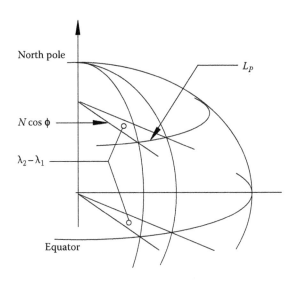

**FIGURE 7.6**   Length of a Parallel

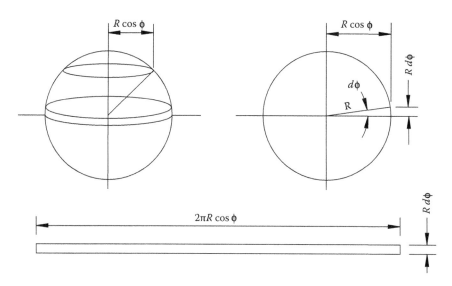

**FIGURE 7.7** Surface Area of a Sphere

sphere is then cut and rolled out flat, the differential area of that ring is computed as length $(2\pi R \cos \phi)$ times width $(R\, d\phi)$. The surface area of the entire sphere is found by adding up an infinite number of infinitely thin adjacent rings (integration). The surface area of the entire sphere is found by integrating Equation 7.58 from the South Pole to the North Pole ($\phi_1 = -90°$ and $\phi_2 = +90°$):

$$dA = (2\pi R \cos \phi) \times (R\, d\phi); \quad \text{surface area of thin ring} \qquad (7.57)$$

$$A = \int_{\phi_1}^{\phi_2} 2\pi R^2 \cos \phi\, d\phi = 2\pi R^2 \left[ \sin \phi_2 - \sin \phi_1 \right] \qquad (7.58)$$

$$\text{Surface area of entire sphere} = 4\pi R^2 \qquad (7.59)$$

## ELLIPSOID SURFACE AREA

Surface area of the ellipsoid is computed much the same way as surface area of a sphere. A differential surface area element on the ellipsoid is written, Equation 7.60, as the product of an elemental parallel distance and an elemental meridian distance as shown in Figure 7.8. A double integration is used to compute, first the area of a ring using longitude limits of 0 and $2\pi$ radians, then the ring areas between latitude limits selected by the user are computed and accumulated using Equation 7.62. Equation 7.63 can be used to compute the area of any rectangular block on the ellipsoid surface bounded by parallels and meridian as selected by the user. If limits of 0 to $2\pi$ for

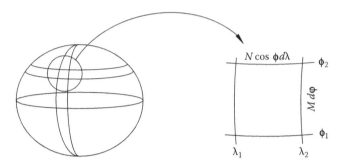

**FIGURE 7.8**   Ellipsoid Surface Area

longitude and limits of −90° to +90° for latitude are used in Equation 7.63, the result (omitting much algebraic manipulation) is Equation 7.64 which is used to compute total ellipsoid surface area:

$$dA = \left(N\cos\phi\,d\lambda\right)\left(M\,d\phi\right) = \frac{a^2\left(1-e^2\right)\cos\phi}{\left(1-e^2\sin^2\phi\right)}\,d\phi d\lambda \qquad (7.60)$$

$$A = a\left(1-e^2\right)\int_{\lambda_1}^{\lambda_2}\int_{\phi_1}^{\phi_2}\frac{\cos\phi}{\left(1-e^2\sin^2\phi\right)}\,d\phi d\lambda \qquad (7.61)$$

$$A = \left(\lambda_2-\lambda_1\right)a^2\left(1-e^2\right)\int_{\phi_1}^{\phi_2}\frac{\cos\phi}{\left(1-e^2\sin^2\phi\right)}\,d\phi \qquad (7.62)$$

$$A = \frac{\left(\lambda_2-\lambda_1\right)a^2\left(1-e^2\right)}{2}$$
$$\times\left[\frac{\sin\phi_2}{1-e^2\sin^2\phi_2}-\frac{\sin\phi_1}{1-e^2\sin^2\phi_1}+\frac{1}{2e}\ln\left(\frac{1+e\sin\phi_2}{1-e\sin\phi_2}\right)-\frac{1}{2e}\ln\left(\frac{1+e\sin\phi_1}{1-e\sin\phi_1}\right)\right] \qquad (7.63)$$

$$A_{ellipsoid} = 2\pi a^2\left(1-e^2\right)\left[\frac{1}{1-e^2}+\frac{1}{2e}\ln\left(\frac{1+e}{1-e}\right)\right] \qquad (7.64)$$

An interesting side note is that the equation for surface area of a sphere should be identical to surface area of an ellipsoid whose eccentricity is zero ($a = R$). If one attempts to insert $e = 0$ into Equation 7.64, an impasse is reached when working with the second term within the brackets. First, it is never permissible to divide by zero. As $e$ goes to 0, $1/2e$ becomes infinitely large. The next term involves taking the natural log of a term that goes to 1 as $e$ goes to 0. The second part of the second term goes to zero if one takes the natural log of 1. The interesting part is that L'Hopital's Rule can be used to compare the rates of each part of the second term as $e$ goes to zero. The ratio reduces to 1 over 1. Since the first term in the brackets goes to one, one plus one is two (for terms within the brackets). Two times the first part of Equation 7.64 gives an identical expression as Equation 7.59 for a sphere for $e = 0$.

## THE GEODETIC LINE

### DESCRIPTION

A geodetic line, also known the geodesic, is defined as the shortest distance between two points on the surface of the ellipsoid. The geodetic line on the ellipsoid is analogous to a great circle on a sphere. When drawn on a Mercator projection as shown in Figure 7.9, a geodetic line appears to be curved even though on the ellipsoid it bends neither to the right nor to the left.

If one were to start at any point on the equator and travel a geodetic line to a point on the opposite side of the world, the path would cross either the North Pole or the South Pole before ending up at the antipole (also on the equator). If one were to leave a point on the equator with a beginning azimuth of, say, 00° 01′ and travel a geodetic line, the path would miss the North Pole, and the final destination on the opposite side of the world would be between the antipole and the "lift-off point." Several comments about the geodetic line are as follows:

1. Beginning with an initial azimuth of 00° 01′ on the equator, the geodetic line will miss the North Pole less than 2 kilometers. Increase the azimuth at the equator to 1° and the geodetic line misses the pole something over 111 kilometers.
2. In all cases, there is some point along a geodetic line at which the distance to the pole is a minimum. At that point, the latitude is a maximum and the azimuth of the geodetic line is 90° (see Figure 7.9).

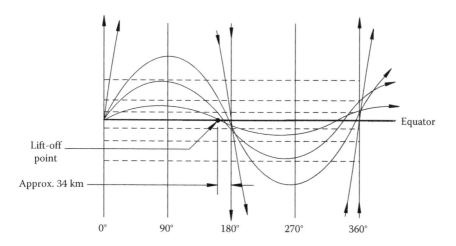

**FIGURE 7.9** Geodetic Lines around the Earth

3. The geodetic line has no lateral curvature but the underlying meridians are not parallel. Therefore, it should be apparent that the azimuth of the geodetic line changes continuously as it traverses the globe. One disadvantage of using a Mercator projection of the world to illustrate the behavior of the geodetic line is that meridians on a Mercator projection are parallel; meaning the geodetic line (which does not curve) must be shown as a curved line.
4. If one starts on the equator and travels east or west along the equator, the path of travel is a geodetic line only until reaching the lift-off point. Beyond that, the equator is not a geodetic line. No other parallel of latitude is a geodetic line.
5. Every point on the ellipsoid has an infinite number of geodetic lines going through it. The azimuth of a geodetic line through a point can have any value between 0° and 360°.

## CLAIRAUT'S CONSTANT

An important feature of a geodetic line is that each different line has its own unique number. The number, known as Clairaut's Constant, is defined as

$$K = N_1 \cos \phi_1 \sin \alpha_1 = N_2 \cos \phi_2 \sin \alpha_2 \qquad (7.65)$$

where:
$N$ is the ellipsoid normal.
$\phi$ is geodetic latitude of the point.
$\alpha$ is the azimuth of the geodetic line at the point.

Note that the value of Clairaut's Constant will be negative for geodetic line azimuths between 180° and 360°. Also note that while each geodetic line has its own unique constant, the value of Clairaut's constant does not change as one travels on a parallel of latitude. Hence, an unchanging value of Clairaut's Constant is not an exclusive property of a geodetic line.

But, a useful feature of Clairaut's Constant is that one can use it to determine the azimuth of a geodetic line at any latitude and, subsequently, the convergence between points. For example, if the azimuth of a line on the GRS 80 ellipsoid is 91° 16′ 28″ at Point C in Figure 7.10 (latitude = 38° 48′ 05.″3342 N), what is the azimuth of the same line at Point D (latitude = 38° 12′ 22.″5464 N)? Relating the problem to Figure 7.10 is important because the inverse sine function has two answers. The first answer (typically given by a calculator or computer) will be less than 90°. That would be a correct answer at Point A (at the same latitude as Point D) as shown in Figure 7.10, but the second correct answer is the supplement (180° − $x$) of the first.

Solution: First, compute Clairaut's Constant using Equation 7.65. Then rewrite Equation 7.65 to solve for $\sin \alpha_2$ as shown in Equation 7.66. The inverse sine function will provide two legitimate answers—each is correct at some point on the geodetic line. The user is expected to choose the correct answer of the two.

$$K = \frac{6,378,137.0 \cos\left(38° \ 48′ \ 05.″3342\right) \sin\left(91° \ 16′ \ 28″\right)}{\sqrt{1 - 0.006694380023 \sin^2\left(38° \ 48′ \ 05.″33423\right)}} = 4,975,935.539 \text{ m}$$

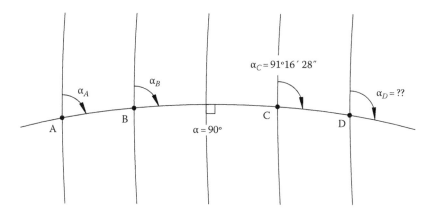

**FIGURE 7.10**  Geodetic Line Azimuth Using Clairaut's Constant

$$\sin \alpha_2 = \frac{K}{N_2 \cos \phi_2} = \frac{K\sqrt{1-e^2 \sin^2 \phi_2}}{a \cos \phi_2} \tag{7.66}$$

$$\sin \alpha_2 = \sin \alpha_A = \sin \alpha_D = \frac{4,969,560.178}{5,011,871.754} = 0.9915577296491$$

$$\alpha_A = 82° \ 32' \ 59'' \quad \text{and} \quad \alpha_D = 97° \ 27' \ 01''$$

Convergence of the meridians is defined as the difference in azimuth between two points on the same geodetic line.

$$Conv_{1 \to 2} \equiv \alpha_2 - \alpha_1 \tag{7.67}$$

In the current example, the convergence from C to D is

$$Conv_{C \to D} = \alpha_D - \alpha_C = 97° \ 27' \ 01'' - 91° \ 16' \ 28'' = 006° \ 10' \ 33''$$

Knowing that a geodetic line azimuth is 90° (sin 90° = 1) at the maximum latitude, it is also possible to use Equation 7.65 to determine the maximum latitude reached by a geodetic line. The solution involves considerable algebraic manipulation, but Equation 7.68 is derived by using 1.0 for $\sin \alpha_2$.

$$N_{max} \cos \phi_{max} \sin 90° = K$$

$$\cos \phi_{max} = \frac{K\sqrt{1-e^2}}{\sqrt{a^2 - K^2 e^2}} \tag{7.68}$$

Finally, the azimuth at which a geodetic line crosses the equator is found from Equation 7.65 by using $\phi_2 = 0°$, which gives

$$\sin \alpha_{eq} = \frac{K}{a} \qquad (7.69)$$

## GEODETIC AZIMUTHS

An azimuth is the angle that a line on the surface of the Earth makes with the meridian through the same point. The geodetic azimuth on the ellipsoid is the azimuth of the geodetic line but, when working with 3-D spatial data, it is often more convenient to work with the 3-D azimuth. The two are very nearly identical and, except for very precise applications, can be used interchangeably. Various azimuths are summarized here but a detailed analysis of the differences is given in Burkholder (1997):

1. If the angle is measured in the horizontal plane defined as perpendicular to the local plumb line, a Laplace correction is required to obtain an equivalent angle in the tangent plane to the ellipsoid at that point. As discussed in Chapter 9, if the deflection-of-the-vertical is zero or insignificant, the Laplace correction need not be applied.
2. If the angle is referenced to the physical spin axis of the Earth instead of the adopted mean geodetic position of the North Pole, a correction for polar motion is required. The polar motion correction is quite small and beyond the scope of this text. Additional information on polar motion can be found in texts such as Bomford (1971), Vaníček and Krakiwsky (1986), Leick (2004), or on an appropriate web site.
3. When looking through the telescope of a carefully leveled surveying instrument to a target at the other end of the line, there is a common line in 3-D space between Point A and Point B. A normal section is a line on the ellipsoid from point to point formed by the intersection of a plane containing the normal at the standpoint and the target at the forepoint. Interestingly enough, the normal section from Point A to Point B on the ellipsoid is not the same as the normal section from Point B to Point A because the directions of the ellipsoid normals at different latitudes are not parallel. The spatial vector between telescope and target (each way) is common to both planes. But, when the vertical plane at each end of the line is projected to the ellipsoid, the line (normal section) on the ellipsoid from Point A to Point B will be slightly different than the line from Point B to Point A. The difference between normal sections is exaggerated and shown in Figure 7.11.

    But, a geodetic line is defined as the shortest distance between two points on the ellipsoid surface. Several observations are as follows: First, the geodetic line has no lateral curvature but curves only in the plane containing the instantaneous normal section. Figure 7.11 gives the mistaken impression that a geodetic line has a double curvature. Second, the underlying meridians are not parallel on a globe but, on a Mercator map, the meridians

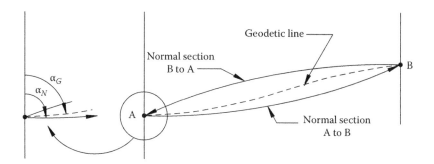

**FIGURE 7.11**   Normal Sections and the Geodetic Line

appear as parallel lines. This explanation for the wrong impression is that a three-dimensional phenomenon is portrayed on a two-dimensional diagram. The important point is that the geodetic line is one line between points on the ellipsoid surface and that the normal sections between points are slightly different depending on whether going from Point A to Point B or from Point B to Point A.

4. If the angle is measured to a target some distance above or below the ellipsoid, a target height correction may be required. As shown in Figure 7.12, the target height correction is required because the normal through the target (forepoint) is not parallel with the normal through the standpoint (instrument station). The elevation of the theodolite or total station above or below the ellipsoid is immaterial because the vertical axis of instrument is coincident with the vertex of the dihedral angle is being measured.

5. The 3-D azimuth (Burkholder 1997) lies in the tangent plane at the standpoint and is computed as the inverse tangent of $(\Delta e/\Delta n)$, the local geodetic horizon components of a 3-D vector defined by $\Delta X/\Delta Y/\Delta Z$.

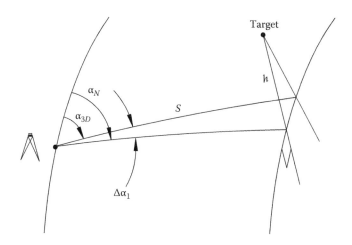

**FIGURE 7.12**   Target Height Correction

In summary, the following three azimuths are all very close to being identical and are often used interchangeably as a geodetic azimuth. The size of each correction can be used to decide whether the difference is significant or not.

- The geodetic line azimuth is the traditional standard as it is used in geodetic computations on the ellipsoid surface.
- The normal section azimuth is the azimuth of the line on the ellipsoid from standpoint to forepoint (on the ellipsoid) as observed from the standpoint.
- The 3-D azimuth is the spatial direction from the standpoint to the forepoint projected into the tangent plane at the standpoint. The 3-D azimuth is readily obtained from GNSS data and is the azimuth utilized in the GSDM.

There are two corrections that relate these three azimuths to each other. One is the target height correction. The other is called the geodesic correction from the normal section to the geodetic line. In many cases, the 3-D azimuth is the one routinely used. If and when a true geodetic line azimuth is required, the logical sequence would be to make the target height correction to obtain the normal section azimuth from the 3-D azimuth; then the geodesic correction is applied to get the geodetic line azimuth from the normal section azimuth.

## TARGET HEIGHT CORRECTION

The azimuth of the normal section is computed from the 3-D azimuth by adding the target height correction as shown in Figure 7.12:

$$\alpha_N = \alpha_{3D} + \Delta\alpha_1 \tag{7.70}$$

where:

$$\Delta\alpha_1 = \frac{\rho h e^2 \cos^2 \phi_1}{2N_1 (1 - e^2)} \left( \sin 2\alpha_{3D} - \frac{S}{N_1} \sin \alpha_{3D} \tan \phi_1 \right) \tag{7.71}$$

$\rho = 206{,}264.806247096355156$ seconds per radian.
$h$ = height of target in meters above or below the ellipsoid.
$e^2$ = eccentricity squared of the ellipsoid.
$\phi_1$ = the geodetic latitude of the standpoint.
$N_1$ = the length of the ellipsoid normal in meters at the standpoint.
$S$ = the distance from the standpoint to the forepoint in meters.
$\alpha_{3D}$ = the 3-D azimuth from the standpoint to the forepoint.

Notes related to using the target height correction include the following:

1. Rarely will the target height correction be greater than 0.5 arc seconds.
2. The correction is always added (subtraction is adding a negative number). Whether the correction is positive or negative is determined by $\sin(2\alpha_{3D})$.
3. If using Equation 7.70 backwards to compute the 3-D azimuth from the azimuth of the normal section, the normal section azimuth can be used in computing the correction instead of the 3-D azimuth.

## GEODESIC CORRECTION

Given the azimuth of a normal section, the azimuth of a geodetic line is found by adding a correction as shown in Equation 7.72:

$$\alpha_g = \alpha_N + \Delta\alpha_2 \tag{7.72}$$

where:

$$\Delta\alpha_2 = -\frac{\rho e^2 S^2}{12N_1^2} \cos^2 \phi_m \sin 2\alpha_N \tag{7.73}$$

$\rho = 206,264.806247096355156$ seconds per radian.
$e^2 =$ eccentricity squared of the ellipsoid.
$S =$ distance from standpoint to forepoint.
$N_1 =$ ellipsoid normal at standpoint.
$\phi_m =$ mean latitude between standpoint and forepoint.
$\alpha_N =$ azimuth of normal section standpoint to forepoint.

Notes regarding use of the geodesic correction include the following:

1. The magnitude of the geodesic correction is quite small and can be ignored in most cases. For example, regardless of standpoint location or azimuth of line, the geodesic correction will never exceed 0.0003 seconds of arc on a 10-kilometer line or 0.03 seconds of arc on a 100-kilometer line.
2. As written, the geodesic correction is a negative value. Depending upon the azimuth of the normal section, the correction may be positive or negative.
3. If Equation 7.72 is rewritten to find the azimuth of the normal section given the azimuth of the geodetic line, it is permissible to use the azimuth of the geodetic line in Equation 7.73 instead of the normal section azimuth.

## GEODETIC POSITION COMPUTATION—FORWARD AND INVERSE

Geometrical geodesy relationships are used extensively when computing geodetic traverses and inverses on the ellipsoid. A two-dimensional (2-D) geodetic traverse is typically called the geodetic forward (or direct) computation and computing the 2-D direction/distance between points is called the geodetic inverse. In order to avoid confusion with other similarly named computations (such as state plane coordinate forward and inverse transformations), the 2-D geodetic forward is henceforth referred to as a BK18 computation and the 2-D geodetic inverse is called a BK19 computation. Other BK designated computations are listed in Table 1.1 in the description of the GSDM.

## PUISSANT FORWARD (BK18)

The geodetic forward, BK18, computation computes the latitude and longitude of an unknown point based upon the known latitude and longitude of the beginning point and measured (or given) direction and distance from the known point to the unknown point. The direction is the geodetic line azimuth at the beginning point and the distance is the geodetic line distance along the ellipsoid surface. Due to ellipsoid flattening there is no closed-form equation that can be used without some approximation. Over the years, geodesists have devised numerous methods whereby the effect of required approximations is minimized and BK18 computations are performed according to specific procedures. One of the most popular procedures is the Puissant Method that is quite accurate for lines up to about 100 kilometers in length. The Puissant Method is illustrated in Figure 7.13 and summarized as

$$\phi_2 = \phi_1 + \Delta\phi \tag{7.74}$$

$$\Delta\phi'' = SB\cos\alpha_1 - S^2 C\sin^2\alpha_1 - D(\Delta\phi'')^2 - hS^2 E\sin^2\alpha_1 \tag{7.75}$$

where:

$$B = \frac{\rho}{M_1} \quad \text{seconds per meter}$$

$$h = SB\cos\alpha_1 \quad \text{seconds}$$

$$C = \frac{\rho\tan\phi_1}{2M_1 N_1} \quad \text{seconds per meter}^2$$

$$D = \frac{3e^2\sin\phi_1\cos\phi_1}{2\rho(1 - e^2\sin^2\phi_1)} \quad \text{per second}$$

$$E = \frac{(1 + 3\tan^2\phi_1)(1 - e^2\sin^2\phi_1)}{6a^2} \quad \text{per meter}$$

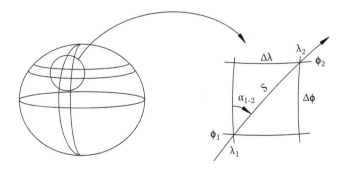

**FIGURE 7.13**   Geodetic Forward (BK18) and Inverse (BK19) Computations

The constants $h$, $B$, $C$, $D$, and $E$ are computed using $\rho = 206{,}264.8062470964$ seconds per radian; $S$ = the ellipsoidal distance; $\alpha_1$ = the geodetic line azimuth; $M$ = the radius of curvature in the meridian, Equation 7.18; $N$ = the ellipsoid normal, Equation 7.16; $a$ = the ellipsoid semimajor axis; and $e^2$ = the eccentricity squared. Note that $\Delta\phi''$ appears on both sides of Equation 7.75 requiring an iterative solution. Use $\Delta\phi'' = 0$ for the first iteration and pay close attention to the units of each term.

The longitude of point 2 is computed as

$$\lambda_2 = \lambda_1 + \Delta\lambda \quad \text{east longitude} \tag{7.76}$$

$$\Delta\lambda'' = \frac{\rho S \sin\alpha_1}{N_2 \cos\phi_2} = \frac{\rho S \sin\alpha_1 \sqrt{1 - e^2 \sin^2\phi_2}}{a \cos\phi_2} \tag{7.77}$$

Note that the latitude of point 2 must be computed before computing the longitude at point 2 because $\phi_2$ is used in Equation 7.77. The azimuth from point 2 to point 1 (the back azimuth) can be computed using Clairaut's Constant, but the Puissant Method uses the following solution based upon convergence, $\Delta\alpha$:

$$\alpha_{2\to1} = \alpha_1 + \Delta\alpha + 180° \tag{7.78}$$

$$\Delta\alpha'' = \frac{\Delta\lambda'' \sin\phi_m}{\cos\left(\dfrac{\Delta\phi}{2}\right)} + \frac{\left(\Delta\lambda''\right)^3 \sin\phi_m \cos^2\phi_m}{12\rho^2}; \quad \text{where } \phi_m = \frac{\phi_1 + \phi_2}{2} \tag{7.79}$$

## PUISSANT INVERSE (BK19)

The geodetic inverse, BK19, begins with the latitude and longitude of two points and computes the geodetic direction and distance between them. The Puissant Method for BK19 uses the same $B$, $C$, $D$, and $E$ coefficients as defined in BK18 and computes intermediate $x$ and $y$ values from which direction and distance are computed.

$$\Delta\phi = \phi_2 - \phi_1 \tag{7.80}$$

$$\Delta\lambda = \lambda_2 - \lambda_1 \quad \left(\text{east longitude}\right) \tag{7.81}$$

$$x = \frac{\Delta\lambda'' N_2 \cos\phi_2}{\rho} = S \sin\alpha_1 \tag{7.82}$$

$$y = \frac{1}{B}\left[\Delta\phi'' + Cx^2 + D\left(\Delta\phi''\right)^2 + E\left(\Delta\phi''\right)x^2\right] = S \cos\alpha_1 \tag{7.83}$$

$$\alpha_1 \left(\text{azimuth from north}\right) = \arctan\left(\frac{S \sin\alpha_1}{S \cos\alpha_1}\right) = \arctan\frac{x}{y} \tag{7.84}$$

$$S\left(\text{distance}\right) = \sqrt{x^2 + y^2}; \quad \text{in meters} \tag{7.85}$$

Advantages of the Puissant Method include the following: (1) it is quite accurate for distances up to about 100 kilometers (62 miles) and (2) it is a well-defined procedure that can be done by hand or by computer. Disadvantages of the Puissant Method are (1) the uncertainty of really knowing how accurate it is, (2) it is not obvious how various pieces fit in the solution, and (3) distance wise, it is limited in scope. Other traditional methods have similar advantages and disadvantages.

## NUMERICAL INTEGRATION

Jank and Kivioja (1980) published a numerical integration procedure for BK18 and BK19, which is summarized here and recommended for use as appropriate. The method is quite tedious if used for long-hand computations but, by user choice, the results can be as accurate as desired over any length of line. If the procedure is programmed, even on a small handheld computer, the tedious objection becomes moot. The Jank/Kivioja method utilizes differential geometry relationships for ellipsoidal triangles that are small enough to be treated as plane triangles. The key to preserving computational accuracy is maintaining the correct azimuth for each geodetic line element. That is done by using Clairaut's Constant.

## BK18: FORWARD

Figure 7.14 shows a diagram of the BK18 computation in which the computation begins at point 1 and ends at point 2. The overall geodetic distance is broken into as many increments as needed to preserve computational accuracy at the endpoint. Clairaut's Constant is used to update the geodetic line azimuth at the midpoint of each computational element. Conceptually, the process begins by getting on the geodetic line at point 1, moving up to the approximate midpoint of the element, using Clairaut's Constant to find an average azimuth for the entire element, computing latitude and longitude increments based upon midpoint values, adding the latitude and longitude increments to the beginning element values to get latitude and longitude at the end of the element, using Clairaut's Constant to update the azimuth at the element endpoint, then using the ending values of one element as beginning values for the next element, and repeating for the number of elements chosen by the user. Special considerations are required to compute across the meridian through the point of maximum latitude.

The following equations are written from the diagram in Figure 7.14 using the assumption that each element is small enough to be treated as a plane triangle:

$$M_m\left(\Delta\phi\right) = \Delta S \cos\alpha_m \Rightarrow \Delta\phi = \frac{\Delta S \cos\alpha_m}{M_m} \tag{7.86}$$

$$N_m \cos\phi_m\Delta\lambda = \Delta S \sin\phi_m \Rightarrow \Delta\lambda = \frac{\Delta S \sin\alpha_m}{N_m \cos\phi_m} \tag{7.87}$$

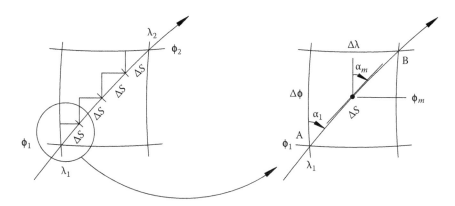

**FIGURE 7.14** Geodetic Line Numerical Integration

Steps for performing a geodetic line BK18 numerical integration are as follows:

1. $\Delta S = S$ divided by number of increments chosen by user.
2. Compute values of $M$ and $N$ at point 1 (Point A for first element), Equations 7.18 and 7.16.
3. Compute Clairaut's Constant for the line (see Equation 7.65).
4. Compute approximate change in latitude for the element using Equation 7.86 and the azimuth, radius of curvature $M$, and latitude at the *beginning* of the element.
5. Find latitude of element midpoint by adding half of approximate $\Delta\phi$ to latitude of beginning of element.
6. Compute $M$ and $N$ at element midpoint.
7. Use Clairaut's Constant to update geodetic line azimuth at midpoint.
8. Use Equations 7.86 and 7.87 to compute latitude and longitude increments.
9. Add latitude and longitude elements to latitude and longitude of beginning of element.
10. Compute $M$ and $N$ at endpoint of element.
11. Use Clairaut's Constant to update azimuth at endpoint.
12. Use endpoint values as beginning values for next element.
13. Repeat for number of elements chosen by user.

Jank and Kivioja (1980) claim that millimeter accuracy can be realized for 20,000 km lines (halfway around the world) if elements are limited to a length of 200 meters. For shorter lines, longer elements can be used while maintaining millimeter accuracy. Practically, increasing the number of increments and noting any change in the final computed position can be used to check the accuracy of a computed position for any line length. For example, geodetic latitude and longitude to five decimal places of seconds corresponds to about 0.0003 meter on the ellipsoid and is often used as a computational standard. If increasing (say doubling) the number of increments gives the same answer to five decimal places of seconds, enough elements were used the first time. The BK18 integration printout in Figure 7.15 shows a 2,000.000-meter line

BK18 geodetic forward computation    User: Earl F. Burkholder
GFORWARD (Ver B 11/91)    Date: August 3, 2000

Reference ellipsoid: GRS 80
    Semi-major axis of ellipsoid =    6,378,137.000 meters,
    **Reciprocal flattening of ellipsoid =**    298.2572221008800

Beginning point:

| | | | | | |
|---|---|---|---|---|---|
| Latitude | = | 32 16 55.929060 | CC | = | 3,816,545.2425 m |
| Longitude | = | 253 14 44.839300 | Dist | = | 2000.0000 m |
| Azi.(North) | = | 45 00 00.000000 | DS | = | 500.0000 m |

Section number 1

| | | | | | |
|---|---|---|---|---|---|
| Latitude (1) | = | 32 16 55.929060 | M1 | = | 6,353,629.8253 m |
| Longitude (1) | = | 253 14 44.839300 | N1 | = | 6,384,235.5313 m |
| Half Dphi | = | 5.738894 sec | MM | = | 6,353,631.4314 m |
| $\phi$ (Mid pt) | = | 32 17 01.667954 | NM | = | 6,384,236.0693 m |
| Azimuth (m) | = | 45 00 03.608278 | | | |
| Latitude (2) | = | 32 17 07.406645 | Dphi = | | 11.477585 sec |
| Longitude (2) | = | 253 14 58.350998 | Dlam = | | 13.511698 sec |
| Azimuth (2) | = | 45 00 07.216777 | | | |

Section number 2

| | | | | | |
|---|---|---|---|---|---|
| Latitude (2) | = | 32 17 07.406645 | M2 | = | 6,353,633.0376 m |
| Longitude (2) | = | 253 14 58.350998 | N2 | = | 6,384,236.6072 m |
| Half Dphi | = | 5.738691 sec | MM | = | 6,353,634.6437 m |
| $\phi$ (Mid pt) | = | 32 17 13.145336 | NM | = | 6,384,237.1452 m |
| Azimuth (m) | = | 45 00 10.825626 | | | |
| Latitude (3) | = | 32 17 18.883823 | Dphi = | | 11.477178 sec |
| Longitude (3) | = | 253 15 11.863641 | Dlam = | | 13.512643 sec |
| Azimuth (3) | = | 45 00 14.434695 | | | |

Section number 3

| | | | | | |
|---|---|---|---|---|---|
| Latitude (3) | = | 32 17 18.883823 | M3 | = | 6,353,636.2499 m |
| Longitude (3) | = | 253 15 11.863641 | N3 | = | 6,384,237.6831 m |
| Half Dphi | = | 5.738487 sec | MM | = | 6,353,637.8561 m |
| $\phi$ (Mid pt) | = | 32 17 24.622310 | NM | = | 6,384,238.2211 m |
| Azimuth (m) | = | 45 00 18.044114 | | | |
| Latitude (4) | = | 32 17 30.360593 | Dphi = | | 11.476770 sec |
| Longitude (4) | = | 253 15 25.377230 | Dlam = | | 13.513589 sec |
| Azimuth (2) | = | 45 00 21.653754 | | | |

Section number 4

| | | | | | |
|---|---|---|---|---|---|
| Latitude (4) | = | 32 17 30.360593 | M4 | = | 6,353,639.4623 m |
| Longitude (4) | = | 253 15 25.377230 | N4 | = | 6,384,238.7591 m |
| Half Dphi | = | 5.738283 sec | MM | = | 6,353,641.0685 m |
| $\phi$ (Mid pt) | = | 32 17 36.098876 | NM | = | 6,384,239.2971 m |
| Azimuth (m) | = | 45 00 25.263743 | | | |
| Latitude (5) | = | 32 17 41.836956 | Dphi = | | 11.476363 sec |
| Longitude (5) | = | 253 15 38.891765 | Dlam = | | 13.514535 sec |
| Azimuth (5) | = | 45 00 28.873954 | | | |

**Increments used for a better position = 10 (only the final values are printed.)**
    Latitude (10)    =    32 17 41.836956
    Longitude (10) =    253 15 38.891765
    Azimuth (10)   =    45 00 28.873954

**FIGURE 7.15**    BK18 Numerical Integration Printout

broken into four elements ($\Delta S = 500.000$ meters) with individual values printed out for each element. At the bottom of the printout, the results obtained using 10 elements (sections) shows the final results without the intermediate printouts. No change to six decimal places of seconds is noted. Conclusion: Results shown at the end of Section 4 in Figure 7.15 are accurate at least within 0.00003 meter (one magnitude better than the "standard").

## BK19: INVERSE

The BK19 numerical integration is really a misnomer in that numerical integration is only part of the procedure. Essentially, the BK19 computation utilizes a conventional geodetic inverse computation to obtain approximate answers (direction and distance). Those answers are then used in the BK18 computation to see how close the computed endpoint position is to the given values of latitude and longitude. Corrections to the approximate direction and distance are computed based upon the misclosure. With corrections applied, a second BK18 computation is performed and another misclosure is computed. The procedure (iteration) terminates when the user is satisfied that the computed position is acceptably close to the given latitude/longitude position. The direction and distance used in such a final BK18 computation is taken to be the BK19 solution.

Given two geometrical parameters for an ellipsoid and the latitude and longitude of two points, a listing of the steps given by Jank and Kivioja (1980) to find the direction and distance from one point to another (a BK19 solution) is as follows:

1. Find the azimuth of the normal section from point 1 to point 2. This is an approximation of the final geodetic line azimuth.

$$\cot A_n = \left( \frac{\tan \phi_2}{\left(1 + e'^2\right) \tan \phi_1} + \frac{e^2 V_2 \cos \phi_1}{V_1 \cos \phi_2} - \cos \Delta\lambda \right) \frac{\sin \phi_1}{\sin \Delta\lambda} \qquad (7.88)$$

where:

$$e'^2 = \frac{a^2 - b^2}{b^2} = \frac{a^2}{b^2} - 1 = \frac{e^2}{1 - e^2}$$

$$V_1 = \sqrt{1 + e'^2 \cos^2 \phi_1}$$
$$V_2 = \sqrt{1 + e'^2 \cos^2 \phi_2}$$

2. Compute Clairaut's Constant for the normal section azimuth:

$$K = \frac{a \cos \phi_1 \sin A_n}{\sqrt{1 - e^2 \sin^2 \phi_1}} \qquad (7.89)$$

3. Use Clairaut's Constant to compute the azimuth at the midlatitude between points 1 and 2:

$$A_m = \arcsin \frac{K\sqrt{1-e^2 \sin^2 \phi_m}}{a\cos\phi_m} \quad \text{where } \phi_m = \frac{\phi_1 + \phi_2}{2} \qquad (7.90)$$

4. Compute radii of curvature for prime vertical and meridian section at midlatitude:

$$M_m = \frac{a\left(1-e^2\right)}{\left(1-e^2 \sin^2 \phi_m\right)^{3/2}} \qquad (7.18)$$

$$N_m = \frac{a}{\sqrt{1-e^2 \sin^2 \phi_m}} \qquad (7.16)$$

5. Use Euler's formula, Equation 7.19, to compute the ellipsoid radius of curvature for the line (specific azimuth at given latitude):

$$R_{A_m} = \frac{M_m N_m}{M_m \sin^2 A_m + N_m \cos^2 A_m} \qquad (7.91)$$

6. Use the spherical law of cosines to compute the angle subtended at the center of the Earth by points 1 and 2 on the ellipsoid surface. Use this angle and radius from step 5 to compute the preliminary distance between points 1 and 2:

$$P_1 P_2 \; (\text{in radians}) = \arccos\left[\sin\phi_1 \sin\phi_2 + \cos\phi_1 \cos\phi_2 \cos(\Delta\lambda)\right] \qquad (7.92)$$

$$S_c = Dist\left(P_1 P_2\right) = R_{A_m}\left(P_1 P_2\right)_{rad} \qquad (7.93)$$

7. With the approximate distance known, compute the difference in azimuths of the geodetic line and the normal section. Apply the difference to the previously known normal section azimuth to get azimuth of geodetic line from point 1 to point 2. Update value of Clairaut's Constant to reflect traversing the geodetic line instead of the ellipsoid normal. Units in Equation 7.94 are seconds of arc.

$$\Delta A'' = A_n - A_g = \frac{\rho\left(V_1^4 - V_1^2\right)}{12} \times \frac{b^2}{a^4} \times S_c^2 \sin(2A_m) \qquad (7.94)$$

$$A_g = A_n - \Delta A'' \qquad (7.95)$$

$$K = \frac{a\cos\phi_1 \sin A_g}{\sqrt{1-e^2 \sin^2 \phi_1}} \qquad (7.96)$$

Although the difference is quite small, the value of $K$ in Equation 7.96 represents an improvement over the value in Equation 7.89 because the BK18 check computation follows the geodetic line, not the normal section.

8. Start from point 1 and use the preliminary distance along with the azimuth of the geodetic line in a BK18 computation to compute the latitude and longitude of point 2. The computed position of point 2 should be very close to given values as illustrated in Figure 7.17. The azimuth at the computed point is computed using Clairaut's Constant as

$$A_c = \arcsin\left(\frac{K}{N_2 \cos\phi_2}\right); \quad \text{Note approximation of } N_2 \text{ and } \phi_2 \text{ for } N_c \text{ and } \phi_c.$$

9. If the approximation of azimuth from step 7 and approximate distance from step 6 are good enough, the latitude/longitude position computed in step 8 will be close to the given position. In that case, the computation is done. Otherwise, corrections to the preliminary direction and distance need to be computed and the BK18 computation used again.

10. Specific steps for computing the corrections to direction and distance are given in Jank and Kivioja (1980) and only summarized here. With reference to Figure 7.16 the corrections are as follows:

$$\Delta\phi_c = \phi_2 - \phi_c; \quad \text{misclosure in north/south direction.} \tag{7.97}$$

$$\Delta\lambda_c = \lambda_2 - \lambda_1; \quad \text{misclosure in east/west direction, positive east } \lambda. \tag{7.98}$$

$$P_c P_2 = \sqrt{\left(N_2 \cos\phi_2 \, \Delta\lambda_c\right)^2 + \left(M_2 \Delta\phi_c\right)^2}; \quad \text{distance from } P_c \text{ to } P_2. \tag{7.99}$$

$$\tan\beta = \frac{M_2 \Delta\phi_c}{N_2 \Delta\lambda_c \cos\phi_2}; \quad \text{note that } \Delta\phi_c \text{ and } \Delta\lambda_c \text{ are in radians.} \tag{7.100}$$

$$\gamma = A_c - \beta; \quad \text{this equation is correct for the case shown; others exist.} \tag{7.101}$$

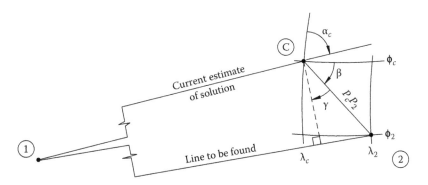

FIGURE 7.16  Misclosures in the Trail BK18 Computation

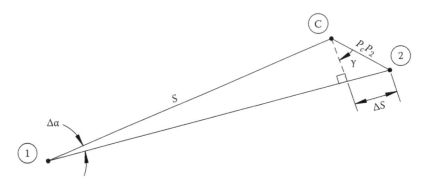

**FIGURE 7.17**   Corrections to Previous Direction and Distance

From Figure 7.17 it is readily seen that

$$\Delta s = P_c P_2 \sin \gamma; \quad \text{this is a correction to the distance.} \qquad (7.102)$$

$$\Delta A'' = \frac{\rho \, P_2 P_c \cos \gamma}{S_c}; \quad \text{this is a correction to geodetic line azimuth.} \quad (7.103)$$

11. With the corrections determined above, the BK18 computation is used again with new (better) values for distance and azimuth. Figure 7.17 shows only one of many possible combinations for applying the computed corrections. The user will need to assure that correct signs are used in applying the corrections. The misclosure should be much smaller at end of the second BK18 computation. If the agreement of the computed position with the given position is acceptable, the following direction and distance are the inverse (BK19) solution. If not, steps 7, 8, 9, 10, and 11 are repeated.

$$A_2 = A_g \pm \Delta A \qquad (7.104)$$

$$S_2 = S_c \pm \Delta s \qquad (7.105)$$

Comments on the Jank/Kivioja BK19 procedure include the following:

1. The Jank/Kivioja numerical integration procedure is tedious to perform but it does a good job and puts computational control in the hands of the user.
2. Ultimate accuracy of the Jank/Kivioja method is limited only by the significant figure capacity of the computer being used. But computational procedures and equations are arranged such that only modest significant figure capacity is required to achieve impressive results for both BK18 and BK19 computations.
3. The Jank/Kivioja procedure is included to show how "classical" geodesy approaches the BK19 computation. The 2-D computation is performed strictly on the ellipsoid surface.
4. The GSDM can be used in place of Equations 7.88 through 7.96. If needed, Equations 7.97 through 7.105 can be used to improve the GSDM solution. See the subsequent section in this chapter on GSDM 3-D geodetic computations.

## GEODETIC POSITION COMPUTATIONS USING STATE PLANE COORDINATES

State plane coordinates can also be used to perform geodetic direct and geodetic inverse computations. The use of state plane coordinates is described in Chapter 11. In the context of a map projection, the geodetic forward and inverse computations are performed using simple 2-D COGO relationships. The complex part is converting latitude and longitude to plane coordinates and plane coordinates to latitude and longitude (BK10 and BK11 computations). Computers have been programmed to perform those transformations and users are largely oblivious to the complexity involved. The advantage of using state plane coordinates for geodetic computations is the ease with which traverse and inverse computations are performed with plane coordinates. The disadvantage to using state plane coordinates to perform a geodetic direct computation is that local directions and distances must first be converted to grid azimuths and grid distances before performing the simple 2-D COGO computations. Similarly, the disadvantage to using state plane coordinates for the geodetic inverse computation is that the answer comes out in grid azimuth and grid distance. Many users desire local tangent plane distances and true bearings. That means grid azimuth and grid distance obtained from a state plane coordinate inverse must be converted to local tangent plane direction and distance or to ellipsoid direction and distance.

Procedures for using state plane coordinates and for computing a state plane traverse are included in Chapter 11. Steps for performing a single course geodetic traverse using state plane coordinates include the following:

1. Start with latitude/longitude of point 1 (state and zone must be named).
2. Convert latitude/longitude to state plane coordinates (use reliable software).
3. Reduce measured slope distance to grid distance using the following steps:
   a. Slope to horizontal (include curvature and refraction if needed)
   b. Horizontal to sea level (NAD 27) or to ellipsoid (NAD 83)
   c. Ellipsoid distance to grid distance (use Simpson's 1/6 rule for long lines)
   d. (Steps b and c are often done together using a combination factor)
4. Compute the grid azimuth of line from point 1 to point 2:
   a. Use field measured angle from known grid azimuth of reference line.
   b. Use solar/Polaris observation at point 1 and, as appropriate, apply
      i. Laplace correction to convert astronomic azimuth to geodetic azimuth
      ii. Convergence at point 1 to convert geodetic azimuth to grid azimuth
5. Compute state plane grid coordinates of point 2 using grid distance and grid azimuth (BK16):

$$E_2 = E_1 + (\text{Grid distance})\sin(\text{Grid azimuth}) \qquad (7.106)$$

$$N_2 = N_1 + (\text{Grid distance})\cos(\text{Grid azimuth}) \qquad (7.107)$$

6. Convert E/N state plane coordinates of point 2 to latitude/longitude (BK11).

Steps for performing a geodetic inverse using state plane coordinates include the following:

1. Start with latitude/longitude of point 1 and point 2 (state and zone must be named).
2. Convert latitude/longitude of each point to state plane coordinates.
3. Compute grid distance and grid azimuth between points (BK16):
   a. $Dist = \sqrt{\left(E_2 - E_1\right)^2 + \left(N_2 - N_1\right)^2}$ (grid distance)
   b. $\tan\alpha = \dfrac{E_2 - E_1}{N_2 - N_1} = \dfrac{\Delta e}{\Delta n}$ (grid azimuth—see Equations 4.11 through 4.13)
4. Convert (see Chapter 11)
   a. Grid distance to ellipsoid (NAD 83) or sea level (NAD 27) distance
   b. Ellipsoid (or sea level) distance to horizontal ground distance
5. Convert grid azimuth to geodetic azimuth using convergence:
   a. At point 1 to find geodetic azimuth point 1 to point 2
   b. At point 2 to find geodetic azimuth point 2 to point 1

*Note*: The geodetic azimuth point 1 to point 2 will not be the same (±180°) as the geodetic azimuth point 2 to point 1 because meridians on the Earth are not parallel.

## GSDM 3-D GEODETIC POSITION COMPUTATIONS

With the exception of state plane coordinate computations, the geodetic position computations described so far are part of classical geometrical geodesy computations and are conducted only on the ellipsoid surface, for example, 2-D. The argument could be made that since a map projection is strictly a 2-D model; the state plane coordinate exception is moot. The GSDM provides an alternative to classical geodesy methods and can be used to obtain results to any accuracy desired. The first solution is generally sufficient but, if needed, refinement of the solution uses the same misclosure and correction methods as the Jank/Kivioja method—Equations 7.97 through 7.105.

Geodetic position computations using the GSDM require extensive use of BK1 and BK2 computations. BK1 is very straight forward and uses Equations 7.22 through 7.25. BK2 is more complex and uses either an iteration procedure, Equations 7.26 through 7.32 or a more lengthy "recipe" procedure given by Equations 7.33 through 7.44. Given the procedures and equations described, the GSDM geodetic forward and inverse computations are as follows.

## FORWARD—BK3

The steps for performing a GSDM geodetic forward computation are as follows:

1. Given: Latitude and longitude for point 1. Ellipsoid height may also be given, but for a strict comparison with other examples, point 1 should be on the ellipsoid, $h = 0.00$ m.
2. Use BK1 to convert latitude/longitude/height coordinates of point 1 to geocentric $X/Y/Z$ ECEF coordinates.

3. The position of point 2 is computed as

$$X_2 = X_1 + \Delta X \tag{7.108}$$

$$Y_2 = Y_1 + \Delta Y \tag{7.109}$$

$$Z_2 = Z_1 + \Delta Z \tag{7.110}$$

where:
$\Delta X/\Delta Y/\Delta Z$ are components from a GNSS vector.
$\Delta e/\Delta n/\Delta u$ are components from total station observations rotated from the local geodetic horizon perspective to the geocentric perspective using the BK9 rotation matrix (Equation 1.22).
4. The geodetic latitude and longitude of point 2 are computed by transforming the X/Y/Z coordinates of point 2 to latitude/longitude/height using a BK2 transformation. Of course, the ellipsoid height, whether zero or not, is also a product of the BK2 computation.

A recent article (Burkholder 2015) describes the BK3 computation in terms of a third model for geodetic position computations. The same geodetic position is obtained by a comparison using geodesy equations, state plane computations, and geocentric (GSDM) computations. The concept being that, with no loss of geometrical integrity, the GSDM provides the alternative of performing the computations in 3-D space rather than using geodesy equations on the ellipsoid or cartographic 2-D mapping equations.

## INVERSE—BK4

Given the latitude and longitude (ellipsoid height must be zero) of Point A and Point B, the objective is to find the geodetic azimuth and ellipsoid distance between them using the GSDM. Steps in the GSDM geodetic inverse computation are as follows:

1. First, the latitude/longitude/height of each point are used to compute the ECEF geocentric coordinates for Point A and Point B.
2. Then, the geocentric inverse is computed as

$$\Delta X = X_2 - X_1 \tag{7.111}$$

$$\Delta Y = Y_2 - Y_1 \tag{7.112}$$

$$\Delta Z = Z_2 - Z_1 \tag{7.113}$$

3. The local $\Delta e/\Delta n/\Delta u$ components are computed using the BK8 rotation matrix as described in Equation 1.21.
4. Various azimuths and distances are computed depending upon choice of user.
   a.  The 3-D azimuth is computed as

$$\alpha_{3D} = \arctan\left(\Delta e/\Delta n\right). \qquad (7.114)$$

   If warranted, the geodetic azimuth is computed from the 3-D azimuth by applying the correction given in Equation 7.73. The target height correction is zero because $h = 0.0$ at the forepoint.
   b.  Given $h = 0.0$ m at both ends of the line, the mark-to-mark chord distance is either

$$D_{M-M} = \sqrt{\Delta X^2 + \Delta Y^2 + \Delta Z^2} \quad \text{or} \quad D_{M-M} = \sqrt{\Delta e^2 + \Delta n^2 + \Delta u^2} \qquad (7.115)$$

   c.  The local tangent plane horizontal distance from Point A to Point B is not really part of the geodetic inverse, but it can be easily computed as

$$HD(1) = \sqrt{\Delta e^2 + \Delta n^2} \qquad (7.116)$$

   d.  Using the concept of $L = R\theta$ ($\theta$ in radians), the geodetic arc distance is obtained from the chord distance between Point A and Point B and the "best" radius of the ellipsoid between the two points. Use Equation 7.115 for the chord distance and Equation 7.19 for the radius. If the line is not very long, Equation 7.20 may suffice for the radius.

$$D_{geod.} = R_\alpha \times 2\left(\arcsin\frac{D_{M-M}}{2R_\alpha}\right) \qquad (7.117)$$

   where:
   $R_\alpha$ = radius from Equation 7.19 or 7.20.
   $D_{M-M}$ = chord distance from Equation 7.115.

Note that, for long lines, the midpoint latitude geodetic line azimuth should be computed using Clairaut's Constant. This midpoint azimuth will give a better value of $R_\alpha$.

## GSDM INVERSE EXAMPLE: NEW ORLEANS TO CHICAGO

The great circle arc distance between New Orleans and Chicago was given as in Chapter 3 with promise of a similar ellipsoid computation in this chapter. Using the GRS 80 ellipsoid and the latitude/longitude positions (remember $h = 0.0$ m) listed in Chapter 3, the geocentric $X/Y/Z$ coordinates of the two points are computed using BK1:

GRS 80: $a = 6,378,137.000$ m and $e^2 = 0.006694380023$

| New Orleans (1) | Chicago (2) |
|---|---|
| $\phi = 30° \ 02' \ 17"$ N | $\phi = 42° \ 07' \ 39"$ N |
| $\lambda = 90° \ 09' \ 56"$ W | $\lambda = 87° \ 55' \ 12"$ W |
| $269° \ 50' \ 04"$ E | $272° \ 04' \ 48"$ E |
| $h = 0.0$ m | $h = 0.0$ m |
| $N = 6,383,493.2334$ m | $N = 6,387,764.6511$ m |
| $X = -15,967.7193$ m | $X = 171,947.3712$ m |
| $Y = -5,526,123.0752$ m | $Y = -4,734,389.6050$ m |
| $Z = 3,174,027.4177$ m | $Z = 4,256,117.7017$ m |

Using Equations 7.111 through 7.113, the geocentric components are

$$\Delta X = X_2 - X_1 = 187,915.0905 \text{ m}$$

$$\Delta Y = Y_2 - Y_1 = 791,733.4702 \text{ m}$$

$$\Delta Z = Z_2 - Z_1 = 1,082,091.2840 \text{ m}$$

Using the rotation matrix in Equation 1.21 at New Orleans, the local perspective components are

$$\Delta e = 185,627.6036 \text{ m}$$

$$\Delta n = 1,333,351.1820 \text{ m}$$

$$\Delta u = -144,197.4535 \text{ m}$$

The mark-to-mark chord distance (both points are on the ellipsoid) is computed using either geocentric components or local (New Orleans) perspective components as

$$D_{M-M} = \sqrt{\Delta X^2 + \Delta Y^2 + \Delta Z^2} = \sqrt{\Delta e^2 + \Delta n^2 + \Delta u^2} = 1,353,911.192 \text{ m} \qquad (7.115)$$

$$\alpha_{3D} = \tan^{-1}\left(\frac{\Delta e}{\Delta n}\right) = 7° \ 55' \ 32."4001 \qquad (7.114)$$

Equations 7.72 and 7.73 are used to compute the azimuth of the geodetic line at the standpoint (New Orleans):

$$\alpha_g = \alpha_{3D} - \frac{\rho \, e^2 D_{M-M}^2}{12 N_1^2} \cos^2 \phi_m \sin(2\alpha_{3D})(\phi_m = 36° \ 04' \ 58")$$

$$\alpha_g = 7° \ 55' \ 32."40016 - 0."92344 = 7° \ 55' \ 31."47666 \qquad (7.118)$$

In order to convert the chord distance to the equivalent ellipsoid arc distance using Equation 7.117, we need the azimuth of the line and the $R_\alpha$ radius of curvature at the midlatitude. Use Clairaut's Constant (computed at New Orleans) to get the midpoint azimuth and Equations 7.16, 7.18, and 7.19 to get the ellipsoid distance:

$$K = N_1 \cos\phi_1 \sin\alpha_1$$
$$= 6,383,493.232 \ \text{m}\cos\left(30°\ 02'\ 17''\right)\sin\left(7°\ 55'\ 31.''4767\right)$$
$$= 761,967.1120 \ \text{m}$$

$$M_m = \frac{a\left(1-e^2\right)}{\left(1-e^2\sin^2\phi_m\right)^{3/2}} = 6,357,570.4002 \ \text{m}$$

$$N_m = \frac{a}{\sqrt{1-e^2\sin^2\phi_m}} = 6,385,555.1108 \ \text{m}$$

The geodetic line azimuth at the midpoint latitude is found using Equation 7.66 as

$$\sin\alpha_m = \frac{K}{N_m\cos\phi_m} = \frac{761,966.113\ \text{m}}{5,160,594.603\ \text{m}} = 0.1476508371 \quad \alpha_m = 8°\ 29'\ 27.''92885$$

The "best" radius of curvature at the midpoint of the line is found using Equation 7.19 as

$$R_\alpha = \frac{M_m N_m}{M_m \sin^2\alpha_m + N_m \cos^2\alpha_m} = 6,358,177.8727 \ \text{m}$$

$$D_{ellipsoid} = R_\alpha \times 2\left(\sin^{-1}\frac{D_{M-M}}{2R_\alpha}\right)\frac{\pi}{180} = 1,356,482.2905 \ \text{m} \qquad (7.117)$$

As geodetic inverse computations go, New Orleans to Chicago is a long line. For shorter lines, the direction in Equation 7.114 and the distance in Equation 7.117 may be good enough but whether the line is long or short, the results should be checked using the BK18 forward computation. Remember the Earth is ellipsoidal, not spherical as was assumed in Equation 7.117. To check how good the answers are, we need to perform a numerical integration forward (BK18) computation using Equations 7.86 and 7.87. The BK18 printout shown in Figure 7.18 uses the latitude/longitude at New Orleans, the geodetic azimuth at New Orleans, and the ellipsoid distance from New Orleans to Chicago. Note that the line was broken into 20 pieces, each 67,824.1145 meters long. Those elements are not short enough for a good answer—but it puts us in the ballpark as shown at the end of element 20. In the same computation, the total distance was broken into 10,000 pieces (each 135.64829 meters long) but the intermediate results were not printed. A comparison of the results at the end of element 20 and the "final" solution shows very good agreement. But, the computed position in Chicago falls

BK18 geodetic forward computation                    User: Earl F. Burkholder
By: E. Burkholder 11/1991 (Ver B)                    Date: August 4, 2007

Reference ellipsoid: GRS 80
    Semimajor axis of ellipsoid =                  6,378,137.000 meters,
    **Reciprocal flattening of ellipsoid** =       298.2572221008827

Beginning point:
| | | | | | |
|---|---|---|---|---|---|
| Latitude | = | 30 02 17.000000 | CC | = | 761,966.1105 m |
| Longitude | = | 269 50 04.000000 | Dist | = | 1,356,482.2905 m |
| Azimuth (N) | = | 7 55 31.476600 | DS | = | 67,824.1145 m |

Section number 1
| | | | | | |
|---|---|---|---|---|---|
| Latitude | = | 30 02 17.000000 | M1 | = | 6,351,413.8649 m |
| Longitude | = | 269 50 04.000000 | N1 | = | 6,383,493.2334 m |
| Half Dphi | = | 1,090.788917 sec | MX | = | 6,351,707.5680 m |
| Midpoint | = | 30 20 27.788917 | NX | = | 6,383,591.6274 m |
| Azimuth | = | 7 56 59.514816 | | | |
| Latitude | = | 30 38 38.347137 | Dphi | = | 2,181.347137 sec |
| Longitude | = | 269 55 55.203701 | Dlam | = | 351.203701 sec |
| Azimuth | = | 7 58 28.891436 | | | |

Section number 2
| | | | | | |
|---|---|---|---|---|---|
| Latitude | = | 30 38 38.347137 | M1 | = | 6,352,002.9810 m |
| Longitude | = | 269 55 55.203701 | N1 | = | 6,383,690.5911 m |
| Half Dphi | = | 1,090.556747 sec | MX | = | 6,352,300.1327 m |
| Midpoint | = | 30 56 48.903885 | NX | = | 6,383,790.1343 m |
| Azimuth | = | 7 59 59.642822 | | | |
| Latitude | = | 31 14 59.223962 | Dphi | = | 2,180.876825 sec |
| Longitude | = | 270 01 50.813333 | Dlam | = | 355.609632 sec |
| Azimuth | = | 8 01 31.767361 | | | |

Sections 3 through 19 were deleted to save space.

Section number 20
| | | | | | |
|---|---|---|---|---|---|
| Latitude | = | 41 31 29.042976 | M1 | = | 6,363,502.1192 m |
| Longitude | = | 271 56 57.351362 | N1 | = | 6,387,540.4366 m |
| Half Dphi | = | 1,085.173096 sec | MX | = | 6,363,837.0364 m |
| Midpoint | = | 41 49 34.216072 | NX | = | 6,387,652.4954 m |
| Azimuth | = | 9 12 41.626309 | | | |
| Latitude | = | 42 07 39.009923 | Dphi | = | 2,169.966947 sec |
| Longitude | = | 272 04 47.840849 | Dlam | = | 470.489486 sec |
| Azimuth | = | 9 15 19.694725 | | | |

Increments used for following computed position = 10,000 (Only the final values are printed.)
| | | |
|---|---|---|
| Latitude | = | 42 07 38.997194 |
| Longitude | = | 272 04 48.034168 |
| Azimuth | = | 9 15 19.692856 |

**FIGURE 7.18**   Trial Geodetic Line New Orleans to Chicago—BK18

slightly south of the given latitude of 42° 07′ 39.″000 N. Also note that the computed longitude is slightly east of the given value of 272° 04′ 48.″000 E. Yes, the solution is close, but a better inverse solution is available if we use that information to compute corrections to both the approximate azimuth and distance. With the corrected values, the numerical integration BK18 computation will be performed again with better results.

At this point the 3-D inverse has done as good as it can do. From here on, the method of computing misclosures and corrections is taken from the inverse algorithm of Jank and Kivioja (1980).

The misclosures and corrections are computed using Equations 7.97 through 7.105 as follows:

$$\Delta\phi_c = \phi_2 - \phi_c = 42°\ 07'\ 39.''00000 - 42°\ 07'\ 38.''997194 = 0.''002806 \quad (7.97)$$

$$\Delta\lambda_c = \lambda_2 - \lambda_c = 272°\ 04'\ 48.''00000 - 272°\ 04'\ 48.''034168 = -0.''034168 \quad (7.98)$$

$$M_2 = \frac{a\left(1-e^2\right)}{\left(1-e^2\sin^2\phi_2\right)^{3/2}} = 6,364,172.255\ \text{m} \quad (7.18)$$

$$N_2 = \frac{a}{\sqrt{1-e^2\sin^2\phi_2}} = 6,387,764.651\ \text{m} \quad (7.16)$$

$$P_cP_2 = \sqrt{\left(N_2\cos\phi_2\Delta\lambda_c\right)^2 + \left(M_2\Delta\phi_c\right)^2} = 0.7895\ \text{m}\left(\Delta\phi\ \text{and}\ \Delta\lambda\ \text{in radians}\right) \quad (7.99)$$

$$\tan\beta = \frac{M_2\Delta\phi_c}{N_2\Delta\lambda_c\cos\phi_2} = \frac{0.08658\ \text{m}}{0.7848\ \text{m}} \Rightarrow \beta = 6°\ 17'\ 44'' \quad (7.100)$$

From printout at the end of Figure 7.18, we get $\alpha_c = 9°\ 15'\ 20''$.

$$\gamma = A_c - \beta = 9°\ 15'\ 20'' - 6°\ 17'\ 44'' = 002°\ 57'\ 36'' \quad (7.101)$$

$$\text{Correction to distance: } \Delta S = P_cP_2\sin\gamma = 0.0408\ \text{m} \quad (7.102)$$

$$\text{Correction to azimuth: } \Delta\alpha = \frac{P_cP_2\cos\gamma}{S_c} = 0.''120 \quad (7.103)$$

Using these corrections, the corrected direction and distance are

$$\text{Trial 2: Azimuth} = 7°\ 55'\ 31.''4766 - 0.''120 = 7°\ 55'\ 31.''3567 \quad (7.104)$$

$$\text{Distance} = 1,356,482.2905 - 0.0408\,\text{m} = 1,356,482.2497\,\text{m} \quad (7.105)$$

These corrected values were used in the (BK18) iteration software again and the final answer came within 1 cm of the latitude/longitude position given in Chapter 4. Another iteration could have been used to make the agreement even better. This method is tedious, but it puts the user in control and provides tools that can be used to obtain the best answer possible.

Notes about the GSDM geodetic inverse include the following:

1. The 3-D azimuth as computed by Equation 7.114 contains no approximation. But, especially for long line, the geodesic correction should be used
   to find the azimuth of the geodetic line. The BK18 computation provides a
   way to check the solution.
2. Since the Earth is ellipsoidal and not spherical, the geodetic distance computed with Equation 7.117 is an approximation—albeit a very good one.
   The integrity of the GSDM geodetic inverse can be checked the same way
   as the Jank/Kivioja iteration method. Once an approximate direction and distance [Equations 7.114 and 7.117] are obtained, the geodetic forward (BK18)
   is used to compute the latitude and longitude of point 2. If the misclosure is
   significant, corrections are computed and applied as outlined in Equations 7.97
   through 7.105. As in the iteration inverse, subsequent corrections are computed and applied as often as required to obtain a sufficiently precise answer.
3. The GSDM geodetic inverse is valid in true 3-D space and not limited to
   the 2-D example given above. Although the GSDM inverse can be used to
   duplicate a 2-D inverse on the ellipsoid, the GSDM inverse is more versatile
   and can be used to find any of the several horizontal distances described by
   Burkholder (1991).

Another 3-D inverse method is to compute $X/Y/Z$ values on the ellipsoid and then
compute the HD(3) chord between ellipsoid normals and the 3-D azimuth between
points. The geodesic arc distance is close to the HD(3) chord distance and is often
approximated by the circular arc between the ellipsoid normals. A comparison of
results (geodesic, chord, circular arc, and 3-D azimuth) is described in Burkholder
(2016). Integrity of the 3-D azimuth is as discussed in Burkholder (1997).

## REFERENCES

Bomford, G. 1971. *Geodesy*, 3rd ed. Oxford, U.K.: Oxford University Press.
Burkholder, E.F. 1991. Computation of horizontal/level distances. *Journal Surveying Engineering* 117 (3): 104–116.
Burkholder, E.F. 1997. The 3-D azimuth of a GPS vector. *Journal of Surveying Engineering* 123 (4): 139–146.
Burkholder, E.F. 2015. Comparison of geodetic, state plane, and geocentric computational methods. *Surveying and Land Information Science* 74 (2): 53–59.
Burkholder, E.F. 2016. 3-D geodetic inverse. *Surveying and Land Information Science* 75 (1): 17–21.
Ghilani, C.D. 2010. *Adjustment Computations: Statistics and Least Squares in Surveying and GIS*, 5th ed. Hoboken, NJ: John Wiley & Sons.
Hoffman-Wellenhof, B., H. Lichtenegger, and J. Collins. 2001. *GPS: Theory and Practice*, 5th ed. New York: Springer-Verlag Wien.
Hooijberg, M. 1997. *Practical Geodesy Using Computers* Berlin, Germany: Springer-Verlag.
Jank, W. and Kivioja, L.A. 1980. Solution of the direct and inverse problems on reference ellip-soids by point-by-point integration using programmable pocket calculators. *Surveying and Mapping* 40 (3): 325–337.
Leick, A. 2004. *GPS Satellite Surveying*, 3rd ed. New York: John Wiley & Sons.

Meyer, T. 2010. *Introduction to Geometrical and Physical Geodesy: Foundations of Geomatics.* Redlands, CA: ESRI Press.

Rapp, R. 1991. *Geometrical Geodesy—Part I.* Columbus, OH: Ohio State University, Department of Civil and Environmental Engineering and Geodetic Science.

Soler, T. and Hothem, L. 1989. Coordinate systems used in geodesy: Basic definitions and concepts. *Journal of Surveying Engineering* 114 (2): 84–97.

Vaníček, P. and Krakiwsky, E. 1986. *Geodesy: The Concepts.* New York: North-Holland.

Vincenty, T. 1980. On the transformation from three-dimensional to ellipsoidal coordinates (in German). *Zeitschrift fuer Vermessungswexen* 11 (1980): 519–521 (Stuttgart, Germany).

You, R. 2000. Transformation of Cartesian to geodetic coordinates without iteration. *Journal of Surveying Engineering* 126 (1): 1–7.

# 8 Geodetic Datums

## INTRODUCTION

As stated in the Preface, the reader is encouraged to consult the NGS web site (https://www.ngs.noaa.gov/ (accessed May 4, 2017)) to obtain information about current implementation of datums as related to spatial data applications and to plans for replacing both the NAD 83 and the NAVD 88 datums. The revisions to material on geodetic datums in this second edition are intended to reflect the impact of current practice on applications to the GSDM. However, due to the fluid nature of evolving practice and policy, it is impossible to make an unequivocal reading of the "crystal ball." Nonetheless, the anticipation is that the GSDM will be compatible with implementation of the 2022 geodetic datums.

A datum is a reference to which other values are related. In surveying, a vertical datum could be as simple as an arbitrary bench mark assigned a height of 100.000 meters (or feet, etc.). A horizontal datum could be defined by a stake pounded in the ground for a Point of Beginning (P.O.B.) and assigned arbitrary coordinates such as east = 5,000.000 meters, north = 10,000.000 meters. If both horizontal and vertical values are assigned to the same point, the result could be called a 3-D datum. The definition of units of measurement, orientation of azimuth, and coordinate system are all implicit in such a definition. Another important implicit assumption is that horizontal is perpendicular to the local plumb line or, as is the case with the GSDM, horizontal* is taken to be perpendicular to the ellipsoid normal through the standpoint or through the P.O.B. as defined in Chapter 1. Presumably, rectangular Cartesian coordinates, either 2-D or 3-D, are used to describe the location of all points with respect to the datum origin and with respect to each other. The permanence and value of such an assumed datum depend upon the stability of monumented points, the quality of coordinates on those points, and the extent to which assumptions associated with establishment of the origin are documented, followed, and made available to others. When using the GSDM, the user selects the P.O.B. and works with local flat-Earth components in a well-documented system.

In the larger view, a datum must accommodate more than a flat-Earth perspective. It has been known since before the birth of Christ that the Earth is not flat. Eratosthenes determined the size of the Earth several hundred years BC and the value he obtained (Carta 1962) was within about 16 percent of today's accepted value. Locally, it still makes sense to reference the horizontal location of an object with plane rectangular coordinates but, when describing the location of a point on the globe, it is more convenient to use the latitude/longitude graticule. Latitude is reckoned in angular (sexagesimal) units north or south from the equator and longitude

---

* It is acknowledged that various options for the definition of horizontal distance exist. A generic simple definition is that horizontal distance is the right-triangle component—HD(1) in Burkholder (1991)—of a slope distance referenced to the ellipsoid normal.

is similarly reckoned east and west from the arbitrary meridian through Greenwich, England. And, mean sea level has been used as a reference for vertical datums for many years. An observation here is that the use of mixed units—angular sexagesimal units for horizontal and length units for vertical—introduces a level of computational complexity for 3-D data that many users would like to avoid.

On one hand, the goal is to keep the datum definition as simple as possible. A simple 2-D or 3-D flat-Earth coordinate system is used successfully in many local applications to describe the location of points and objects. On the other hand, the definition of a datum should be sufficiently comprehensive to accommodate the entire world with geometrical integrity. According to Taylor (2004), Eratosthenes devised a rudimentary grid of latitude/longitude about 250 BC and several hundred years later, Claudius Ptolemy's *Geographica* "included a catalog of some eight thousand place-names, rivers, mountains, and peninsulas, each of them with its position defined by degrees of latitude and longitude." Presumably, the relative location of those positions was more valid than an absolute location due, in part, to uncertainties associated with a formal definition of a datum. Modern geodetic practice includes very detailed datum definitions, and the location of points anywhere within the birdcage of orbiting GNSS satellites can be determined within very small tolerances—either relative or absolute.

A datum and the GSDM are similar in that each is used to define a computational environment for handling geospatial data. But, there is also an important distinction—the GSDM provides a set of rules, equations, and relationships that can be used with various 3-D datums. It is also specifically noted that the GSDM should only be used on one datum at a time. The GSDM is not a tool for combining data from two separate datums. It is the users' responsibility to know at all times what underlying datum is being used as the basis for the data being manipulated. If a user is faced with combining data from several different 3-D datums, a separate transformation (such as a 7-parameter transformation described later in this chapter) should be employed.

Sometimes the exception can also be instructive. While it is true that the GSDM should not be used as a tool for combining datums, the stochastic model portion of the GSDM allows the user great flexibility with regard to working with spatial data. If the datum differences are at the 1-meter level and the data being manipulated has a larger standard deviation (say 5 meters), then the datum difference becomes insignificant by comparison and both data sets can be used along with their standard deviations. But the user must take responsibility for how the tool is used. In this case, lumping known systematic error with larger random errors may be expedient in a given circumstance, but such a decision should be documented carefully to avoid unintended consequences.

## HORIZONTAL DATUMS

### BRIEF HISTORY

Reviewing some of the history associated with horizontal and vertical datums provides additional perspective and promotes a better understanding of the GSDM.

A concise history of horizontal datums in the United States is given by Dracup (2006) as follows:

> In 1879 the first national datum was established and identified as the New England Datum. Station PRINCIPIO in Maryland, about midway between Maine and Georgia, the extent of the contiguous triangulation was selected as the initial point with its position and azimuth to TURKEY POINT determined from all available astronomical data, i.e. 56 determinations of latitude, 7 of longitude, and 72 for azimuth.
>
> Later its position was transferred to station MEADES RANCH in Kansas and the azimuth to WALDO by computation through the triangulation. The Clarke Spheroid of 1866 was selected as the computational surface for the datum in 1880, replacing the Bessel spheroid of 1841 used after 1843. Prior to 1843, there is some evidence that the Walbeck 1819 spheroid was employed.
>
> The datum was renamed the U.S. Standard Datum in 1901 and in 1913 the North American Datum (NAD) as Canada and Mexico adopted the system. In 1927 an adjustment of the first-order triangulation of the U.S., Canada and Mexico was began and completed about 1931. The end result was the North American Datum of 1927 (NAD 27).

More recent information is provided by Schwarz (1989) in connection with the readjustment of the horizontal network in the United States. He describes development of the North American Datum of 1983 (NAD 83) as being based upon the Geodetic Reference System 1980 (GRS 80) ellipsoid and includes other significant historical detail as well.

The Geodetic Glossary (NGS 1986) and the Glossary of the Mapping Sciences (ASCE et al. 1994) each contain descriptions and definitions for various datums used in the United States. For example, a number of new datums were established in various parts of Alaska when it was not convenient or possible to tie a new project to previously existing survey control. In each case, the goal was to define a mathematical model appropriate for that portion of the Earth being surveyed. A conflicting goal was to avoid making "unnecessary" changes. Consider that the Clarke Spheroid 1866 was adopted by the U.S. Coast & Geodetic Survey (USC&GS) in 1880 and used for projects throughout the United States. The International Ellipsoid was derived in 1909 by John Hayford of the USC&GS and adopted in 1924 by the International Association of Geodesy which recommended it for use by all member countries. But, the USC&GS continued using the Clarke Spheroid 1866 for the NAD 27 adjustment because the coordinates of many stations were already based upon the Clarke Spheroid 1866, because computational tables for the Clarke Spheroid 1866 were already published, and because the newer ellipsoid differed only slightly from the older one (Schwarz, 1989, ch. 4).

A brief history of horizontal datums in the United States would be incomplete without considering the transition from separate horizontal and vertical geodetic datums to an integrated 3-D datum that includes both horizontal and vertical. As described later in the section on "3-D Datums," the NGS has embarked upon an ambitious effort to replace both the current NAD 83 horizontal datum and the NAVD 88 vertical datum by the year 2022.

## NORTH AMERICAN DATUM OF 1927 (NAD 27)

Prior to the satellite era, it was impossible to establish accurate intercontinental ties. Consequently, effective application and extension of a datum was limited to a specific region or to one of the continental land masses. Datums lacking global extent are called regional geodetic datums. With the advent of the Space Age, the tools of satellite geodesy made it possible to survey the world as a whole and regional datums were no longer able to accommodate "big picture" observations adequately. The solution was to develop a best-fitting mathematical model for the whole Earth with the origin located at the Earth's center of mass. Such a model is called a global geodetic datum. By contrast, a regional geodetic datum has its origin located at some point on or near the surface of the Earth. The NAD 83 is a global geodetic datum and the NAD 27 is a regional geodetic datum with the origin located at triangulation station MEADES RANCH in Kansas. Parameters that define the NAD 27 are (NGS 1986)

$a$ = semimajor axis, Clarke Spheroid 1866 = 6,378,206.4 m.
$b$ = semiminor axis, Clarke Spheroid 1866 = 6,356,583.8 m.
$\phi$ = geodetic latitude of station = 39° 13′ 26."686 N.
$\lambda$ = geodetic longitude of station = 98° 32′ 30."506 W.
$\alpha$ = geodetic south azimuth to station Waldo = 75° 28′ 09."64.
$N$ = geoid height = 0.00 meter.

An implied condition is that the minor ellipsoid axis is parallel with the spin axis of the Earth. It was also intended that deflection-of-the-vertical be zero at Station MEADES RANCH but subsequent refinements in the geoid model indicate residual deflection components at the initial point. The NAD 27 is a 2-D datum.

## NORTH AMERICAN DATUM OF 1983 (NAD 83)

Adjustment of the national horizontal network and publication of the NAD 27 was an enormous accomplishment and that network served the control needs of a growing nation for more than half a century. But, the NAD 27 was not without its problems. Some problems were associated with the sparseness of the data, some problems were associated with the accumulation of newer high-quality data that were expected to "fit" the 1927 adjustment, and some problems were associated with the lack of computing "horsepower." Localized readjustments were used during the intervening decades to fix problems in various parts of the network. Such stop-gap measures accumulated to a point where the decision was made to readjust the entire North American network and, in the process, to define a new datum. The "North American Datum" (Committee on the North American Datum 1971) is a report prepared for the U.S. Congress by the National Academy of Sciences that justifies allocating resources for performing the readjustment. The North American Datum of 1983 (NAD 83) project began on July 1, 1974, and was completed on July 31, 1986, at a cost of approximately $37 million.

The NAD 83 is a global geodetic datum defined as follows:

1. The datum origin was located at the center of mass of the Earth as best determined at the time. As discussed later in this chapter, subsequent observations have yielded slightly different results.
2. The Z-axis is in the direction of the Conventional Terrestrial Pole as defined by the International Earth Rotation Service (IERS).
3. The X-axis coincides with zero degrees longitude defined by the Prime Meridian near the Greenwich Observatory—see Malys et al. (2016).
4. A reference ellipsoid is defined by the following four physical geodesy parameters:
   a. $a$ = semimajor axis of Geodetic Reference System 1980.
   b. $GM$ = the Geocentric Gravitational Constant.
   c. $J_2$ = zonal spherical harmonic coefficient of second degree.
   d. $\omega$ = rotational velocity of the Earth.

A mathematical ellipse is defined by two geometrical parameters. In the case of an ellipsoid for the Earth, the two parameters are typically $a$ and $b$, $a$ and $1/f$, or $a$ and $e^2$ as noted in Chapter 7. The geometrical parameters for the Geodetic Reference System 1980 (GRS 80) are derived from four physical geodesy parameters as adopted by the International Union of Geodesy and Geophysists meeting in Canberra, Australia in December 1979 and reported by Moritz (1980). The four parameters are used in an iterative algorithm to compute a value of $e^2$, which, in turn, is used to compute the value of $1/f$ for the GRS 80. The following derived value of $1/f$ was computed to 16 significant figures (Burkholder 1984):

$a$ = semimajor axis (exact) = 6,378,137 meters.
$1/f$ = reciprocal flattening (derived) = 298.2572221008827.

Although angle and distance observations were reduced to the geoid for the NAD 27 adjustment, angle and distance observations used in the NAD 83 adjustment were reduced to the GRS 80 ellipsoid. Scale and orientation for the overall NAD 83 network were provided by a combination of a precise transcontinental traverse (TCT), Doppler data, lunar laser ranging (LLR), Very Long Baseline Interferometry (VLBI), satellite laser ranging (SLR), and astronomical azimuths. Gravity data were used to compute geoid heights and deflections-of-the-vertical at all occupied control points. The end product of the NAD 83 adjustment was the two-dimensional latitude and longitude coordinates for each station. But, the NAD 83 is called a 3-D datum because it is ultimately based upon the underlying ECEF coordinate system defined by the DOD (see Chapter 1). Ellipsoid height is the third dimension.

## WORLD GEODETIC SYSTEM 1984

The U.S. DOD has been engaged in navigation and mapping activities all over the world for many years. They were among the first to develop geodetic models for the entire world and have continued to refine early efforts. The initial World Geodetic

System was dated 1960 (WGS 60). Since then the DOD has variously used WGS 66, WGS 72, and WGS 84 and practice has evolved to where the World Geodetic System 1984 (WGS 84) name is used both for an ellipsoid and for a datum. Similarity between the GRS 80 ellipsoid and the WGS 84 ellipsoid was mentioned in Chapter 7. While GRS 80 refers only to an ellipsoid, WGS 84 refers to both an ellipsoid and a datum. This distinction is essential in avoiding confusion about their respective uses. Given the high level of DOD implementation, the spatial data user should also understand that there are differences between a datum definition, a reference frame, and the realization of the datum in the form of coordinates. The WGS 84 as a datum also incorporates gravity, equipotential surfaces, and other physical geodesy concepts. However, the summary here is limited to describing how the WGS 84 relates to the collection and manipulation of spatial data primarily via GNSS equipment and observations. Materials for additional study include a comprehensive report by NIMA (1997) and a 2014 update by NGA (2014). Standard geodesy texts and a web search can also be very productive.

The geometrical parameters for the WGS 84 ellipsoid were originally intended to be the same as for the GRS 80 ellipsoid, but at one point in the computational process the DOD truncated an intermediate value prematurely and the resulting value of $1/f$ is noticeably different. Although the numbers in the $1/f$ value for WGS 84 are different, the impact of that difference is very small. For example, in Table 7.1, the computed value of c, the polar radius of curvature, is a large number and differs by only 0.0001 meter between the GRS 80 and WGS 84 ellipsoids. The geometrical parameters for the WGS 84 ellipsoid are as follows:

$a$ = semimajor axis (exact) = 6,378,137 meters.
$1/f$ = reciprocal flattening (derived) = 298.257223563.

There is no practical difference between GRS 80 and WGS 84 so far as ellipsoids are concerned. But, NAD 83 coordinates and WGS 84 coordinates may differ by a meter or more because the datum origins are at different locations and/or the coordinates were derived from disparate survey operations (Schwarz 1989, ch. 22).

### INTERNATIONAL TERRESTRIAL REFERENCE FRAME

The International Earth Rotation Service (IERS) provides the International Terrestrial Reference Service (ITRS) as one of several services and defines a reference frame for scientific uses the world over. A quote from the IERS web site (http://www.iers.org (accessed May 4, 2017)) is, "The ITRS is realized by estimates of the coordinates and velocities of a set of stations observed by VLBI, GPS, SRL, and DORIS. Its name is the International Terrestrial Reference Frame (ITRF)." The ITRF uses the meter as the unit of length and shares the center of mass of the Earth as its origin with other 3-D datums. Initially the orientation of the ITRF was the same as WGS 84 but the "time evolution of orientation" for the ITRF is chosen such that there is a net zero rotation with regard to horizontal tectonic plate movements over the entire Earth. The ITRF uses the same ECEF rectangular coordinate system as other 3-D datums and, although ECEF coordinates are the defining values, the GRS 80 ellipsoid should be used when expressing ITRF positions in latitude/longitude/height.

The point is that NAD 83, WGS 84, and the ITRF are all 3-D datums having their origin at the Earth's center of mass. They all use the meter as the unit of length and all employ the ECEF rectangular coordinate system. With those characteristics, the GSDM can be used equally well with any of them—one at a time. The difference between the datums lies in the location of the center of mass, the orientation of the coordinate system, and the realization of coordinates within the respective systems.

Miscellaneous comments are as follows:

1. The GNSS satellites orbit the physical center of mass of the Earth. The NAD 83 origin was positioned quite well with publication of the NAD 83 adjustment but, since then, better data for the position of the Earth's center of mass is consistent at the centimeter level and different from the NAD 83 origin by about 2 meters. The NAD 83 is realized by the coordinates of "fixed" points on the North American tectonic plate. Earthquake zones being an exception, that means the NAD 83 coordinates for monumented points throughout North America change very little, if at all, over time. New NAD 83 coordinates are best determined by adding precise *relative* $\Delta X/\Delta Y/\Delta Z$ baseline components to higher-accuracy control points.

2. The WGS 84 datum is updated periodically so that GNSS satellite orbits are computed with respect to the current approximation of the center of mass of the Earth. A WGS 84 datum update consists of computing and using revised coordinates for the DOD GPS tracking network. The updates are identified by WGS 84 (G730), WGS 84 (G873), and WGS 84 (G1150) where "G" means the update is based upon GPS observations and the number following is the GPS week since January 5, 1980. In general, *absolute* WGS 84 coordinates are obtained from code-phase GNSS observations and referenced to the broadcast ephemeris of the satellites. See Chapter 10 for more details.

3. The ITRF is developed by collaborators in the international community who compute yearly updates to the coordinates of a larger network of GNSS tracking stations. The ITRF solution is quite close to the WGS 84 realization and daily comparisons are made between the ephemerides of WGS 84 and ITRF. Although the differences between WGS 84 and ITRF do exist, they are small and statistically insignificant. A general statement is that ITRF utilizes both absolute and relative positioning data where NAD 83 is based primarily on relative GNSS data and WGS 84 primarily uses absolute GNSS data.

4. NAD 83, WGS 84, and ITRF are similar (they are all 3-D datums), but they are also quite different. Each exists for specific reasons and applications vary accordingly. Points to be remembered are as follows:
   a. The NAD 83 is "fixed" to the North American plate and is very stable. Typically, if newer coordinates for a control monument as published by the NGS are different than previous one, the changes really reflect

greater consistency in the network. The changes are most likely due to improved observations and readjustments by the NGS. The possibility of tectonic movement exists—especially in earthquake-prone areas. A postscript to this section is that NGS plans to replace the NAD 83 in 2022—see later section on "3-D Datums."

b. The WGS 84 is "native" to the NAVSTAR satellite system and is maintained/updated by the U.S. DOD. The quality is very high but modifications are strictly the prerogative of the U.S. DOD. Absolute positioning (as opposed to relative positioning) is the primary goal.

c. The ITRF is a collaborative product of the international community and serves a greater user base than either the NAD 83 or the WGS 84. The ITRF uses the "best" technology available irrespective of the source and is updated on a yearly basis to reflect the best current solution for coordinates of tracking stations and computation of precise orbits. Like the internet, it lacks the commitment of a sovereign to guarantee permanence. It could be argued that commitment of the user community is a better guarantee.

5. Relative geocentric differences are determined by carrier phase GNSS observations and define a vector from one point to another. For short lines (say less than 20 km), these $\Delta X/\Delta Y/\Delta Z$ components are very nearly identical in each of the three datums: ITRF, WGS 84, and NAD 83. A careful evaluation would need to be made for very precise applications and/or long lines.

6. In the past, standard geodetic surveying practice included building a network of precise vectors and attaching such a network to fixed points of greater accuracy. With the development of RTK and PPP surveying procedures, nontraditional control configurations are becoming more popular. Issues of redundancy notwithstanding, the GSDM provides various options for the spatial data analyst and accommodates well-formulated network adjustments, virtual network solutions, and side shot data collected in the radial mode. Admittedly, the GSDM permits reckless use (such as mixing datums or using unchecked data). But the GSDM supports many beneficial uses and the intent is that the GSDM will be used responsibly and competently. If a project is based upon appropriate control information and if observed spatial data components are entered with their appropriate standard deviations (and covariances), subsequently derived answers will reflect legitimate statistical properties of those data. Issues of local accuracy and network accuracy also need to be examined carefully (Burkholder 1999, Pearson 2004).

## HIGH ACCURACY REFERENCE NETWORK—HARN

During the 12 years that the NAD 83 was being computed, GNSS became a viable positioning tool embraced by the user community. One justification for readjusting the NAD 27 was that, over a period of decades, people in the user community gained access to better equipment and were able to survey more precisely than the network to which they were expected to attach their results. Ironically, GNSS positioning technology was not used in the NAD 83 readjustment and, by the late 1980s, that justification repeated itself. Immediately following publication of the NAD 83, the NGS came under pressure to support the control needs of GNSS-equipped persons and organizations in the user community. To forestall creation of local or proprietary networks, the NGS adopted a policy of upgrading portions of the network—state by state—based upon GNSS control surveys. These upgrades were known as High Precision Geodetic Networks (HPGN). Those GNSS surveys were conducted primarily to improve the horizontal latitude/longitude position of the GNSS control points and to establish new control points in places more accessible to the user public (Bodnar 1990). But, this time there was a big difference—the datum did not change; only the coordinates for the reobserved control points were improved. Although many parts of the national network have been readjusted and HPGN values began being published in 1990, it is still the NAD 83. It was not a new datum. The counter argument is, "if the coordinates for the same point are different, it is a new datum." Education helps spatial data users understand those differences.

Subsequently, the HPGN acronym gave way to the High Accuracy Reference Network (HARN), which is essentially the same as a HPGN. The following references, D'Onofrio (1991), Strange and Love (1991), and Doyle (1992), use the HARN acronym and describe the technical, logistical, and political considerations for upgrading the NAD 83 to support modern GNSS positioning capabilities.

Much could be said (Doyle 1994) about the evolving character and quality of the National Geodetic Reference System (NGRS)—also called the National Spatial Reference System (NSRS). An oversimplification is that the NAD 83 datum evolved from a 2-D datum to a 3-D datum (hence the name NGRS) and that the name was revised to NSRS in an attempt to "address the changing requirements for positional information." The original NAD 83 coordinates were simply called NAD 83 coordinates (as opposed to NAD 27 coordinates). The newly adjusted GNSS-derived HARN coordinates in a state were referred to as NAD 83 (19xx) where xx is the year in which the GNSS-derived values were published. In time, the original NAD 83 coordinates came to be known as NAD 83 (1986) values. Two considerations for using NAD 83 coordinates are (1) to be clear as to the (epoch) version of the coordinates being used and (2) to avoid mixing coordinates from two different epochs. Of course, there are times when coordinates from different epochs can be used without detrimental consequences but a user should not attempt to start a survey on a NAD 83 (19xx) control point, close on a NAD 83 (1986) point, and expect reliable results—it is like mixing apples and oranges or using different datums.

Another issue to be addressed was the boundary of adjustments between states. Although the newer adjustment results were typically within 0.5 meter of the previously published values, that level of difference needed to be "feathered" into the positions of adjoining states so that users could continue to enjoy consistent results even if starting a control survey in one state and closing it on control points in an adjacent state. An example of performing a GNSS survey across a state boundary (and different HARN adjustments) is given in Chapter 12. Such issues were rendered moot by a subsequent 2007 national readjustment of the HARN networks and will be avoided entirely with replacement of the NAD 83 in 2022.

## CONTINUOUSLY OPERATING REFERENCE STATION—CORS

A GNSS receiver connected to a permanently mounted antenna and collecting data continuously is called a continuously operating reference station (CORS). Depending upon the capabilities of the receiver, the design of the antenna, and associated software, a CORS installation may be a sophisticated high-end automated installation that collects meteorological data as well as satellite signals and posts collected data to the internet in real time. Some CORS installations (in the United States) are owned/operated by the National Geodetic Survey (NGS), some are owned/operated by the U.S. military, and others are owned and/or operated by federal/state/local agencies, corporations, businesses, or individuals. The NGS maintains a clearing house of CORS data that meet standards established for operation and maintenance of a national CORS network. CORS data archived by NGS are available via the internet from the NGS web site—http://www.ngs.noaa.gov (accessed May 4, 2017). The precise position of the fundamental CORS network stations are computed daily with respect to the ITRF and modifications are made to the published positions as warranted. The point here is that CORS positions published by NGS are of very high quality and are the "best" available to users in the United States. Using appropriate datum transformation software, the same ITRF CORS positions, standard deviations, and velocities may also be published on other datums such as the NAD 83 or WGS 84.

At the other end of the spectrum, a GNSS receiver running continuously and broadcasting raw data to remote receivers operating within local radio range of the "base" station is also referred to as a CORS or base station. Such an installation typically serves a small area and maybe even only one user. The reference position of the base station may have been established with respect to NAD 83, WGS 84, ITRF, or a user-defined datum at the prerogative of the owner. The position of the remote receiver(s) is determined during a survey with respect to the CORS base station within the guidelines (proven or unproven) chosen by the users/owner. Proving the quality or integrity of such results is the responsibility of the owner/user. It is not uncommon for local RTK surveying activities to be served by a local CORS or, more recently, a network of CORS operating in concert and known as a real-time GNSS CORS network (RTN). The value of such RTNs is enhanced to the extent the integrity and quality of the results are proven consistent and compatible with the National Spatial Reference System (NSRS) maintained by the NGS.

In the late 1990s, the NGS embarked upon a program known as "height modernization" in which state-by-state projects were implemented with the idea of improving the height component of the NSRS. Height modernization observations are typically referenced to a network of continuously operating reference stations (CORS) whose positions are determined rigorously by the NGS with respect to the ITRF datum. Those ITRF coordinates are converted to NAD 83 (CORS) coordinates using a 14-parameter transformation as documented in HTDP—see section "Datum Transformations" below. Data from many CORS networks are available via the internet.

As mentioned earlier, the NGS completed a project to readjust the entire national network of HARN networks to the CORS. The project was similar to the massive readjustment performed for the NAD 83 but was accomplished with many fewer people over a much shorter time frame. The datum did not change but the readjusted values are referred to as NAD 83 (NSRS2007). Completion of the readjustment was announced on February 10, 2007, to coincide with the 200th anniversary of the establishment of the "Survey of the Coast," predecessor of the National Geodetic Survey.

## NA2011

The following is a quote from the NGS web site, http://www.ngs.noaa.gov/web/surveys/NA2011/ (accessed May 4, 2017):

> As part of continuing efforts to improve the NSRS, on June 30, 2012, NGS completed the National Adjustment of 2011 Project. This project was a nationwide adjustment on NGS 'passive' control (physical marks that can be occupied with survey equipment, such as brass disk bench marks) positioned using Global Navigation Satellite System (GNSS) technology. The adjustment was constrained to current North American Datum of 1983 (NAD 83) latitude, longitude, and ellipsoid heights of NGS Continuously Operating Reference Stations (CORS). The CORS network is an 'active' control system consisting of permanently mounted GNSS antennas, and it is the geometric foundation of the NSRS. Constraining the adjustment to the CORS optimally aligned the GNSS passive control with the active control, providing a unified reference frame to serve the nation's geometric positioning needs.

## VERTICAL DATUMS

As stated at the beginning of this chapter, a simple vertical datum can be defined by assigning an arbitrary elevation to a specified bench mark and referencing other elevations to that assumed value. That practice may be legitimate for local relative elevation differences but it is not pursued here. More formally, a vertical datum is an equipotential surface used as a reference for elevation. In this sense, elevation is an absolute term associated with the third dimension. Other terms associated with elevation include words such as "altitude" and "height." The goal here is to retain the rigor of the formal definitions while acknowledging and building on the intuitive understanding of the reader.

Mean Sea Level (MSL) is widely understood as being a reference for elevation. Most humans stand erect and have some concept of oceans and sea level. "Up" goes higher and sea level seems to be a good place from which to start. But sea level moves up and down according to the tides, changing barometric pressure, ocean currents, and other factors. As discussed more in Chapter 9, the geoid is the formal reference surface for elevation and is approximated by the idealized mean sea level at rest. Although finding the precise geoid is quite difficult, the geoid (or MSL) has been used worldwide as a reference for elevation.

### Sea Level Datum of 1929 (Now NGVD 29)

The Sea Level Datum of 1929 was established in the United States on the basis of extensive readings of 26 tide gages scattered along the North American coast—21 of them in the United States and 5 of them in Canada. Zero elevation as determined by the mean readings at each tide gage was held in the adjustment, and elevations were computed for thousands of bench marks on the differential level loops that had been run throughout the United States.

In the decades following publication of Sea Level Datum of 1929 elevations, it became apparent, based upon precise level loops throughout North America, that apples were being compared with oranges. Elevation as determined by a tide gage at one location is not necessarily the same as the elevation as determined by precise leveling from a separate tide gage. The implication is that a zero elevation is not exactly the same as mean sea level. As a consequence, the name of the datum was changed from the Sea Level Datum of 1929 to the *National Geodetic Vertical Datum of 1929 (NGVD 29)* in May 1973. No published elevations were changed, only the name of the datum was changed (Berry 1976).

## INTERNATIONAL GREAT LAKES DATUM

When attempting to compute accurate hydraulic head for water stored in the Great Lakes system, scientists in the United States and Canada realized that a system of geopotential numbers and dynamic heights would provide a better model for their computations than does the concept of elevation. As described in more detail in Chapter 9, equipotential surfaces are separated by units of work (force times distance) where gravity is the force variable and elevation difference is the distance variable. The geopotential number for an undisturbed water surface is a constant and is computed as the infinite summation of the product of a force times distance accumulated along the path of the maximum gradient. A dynamic height is computed as the geopotential number divided by normal gravity. A dynamic height consists of numbers that look like an elevation but dynamic height is not the distance from the geoid. Dynamic heights are everywhere the same for the surface of a body of water at rest (e.g., one of the Great Lakes) and can be used for accurate hydraulic head computations.

Understanding the need for dynamic heights and recognizing that the Hudson Bay region is still rebounding from the glacial burden of 10,000 years ago, the International Great Lakes Datum (IGLD 55) was established jointly by United States and Canadian scientists in 1955 for the Great Lakes region of North America. It was also anticipated that the IGLD would need to be readjusted every 30 years or so due to continuing crustal rebound, hence the current version is IGLD 85. An update to the IGLD will be part of the effort to replace the NAVD 88 datum in 2022.

## NORTH AMERICAN VERTICAL DATUM OF 1988—NAVD 88

In addition to readjustment of the NAD 83, the NGS also performed a readjustment of the vertical network (Zilkoski et al. 1992) and published the results as the North American Vertical Datum of 1988 (NAVD 88). Several significant differences between the NGVD 29 and the NAVD 88 are (1) the internal consistency of the loops covering the nation is better and (2) while the NGVD 29 adjustment was constrained to the tide gage readings around the coast of North America, the NAVD 88 elevations are computed with respect to a single tide gage elevation at Father Point/Rimouski, Quebec, Canada.

Several items are as follows:

1. The single bench mark at Father Point had already been selected as the initial bench mark for the IGLD 85 by the Coordinating Committee on Great Lakes Basic Hydraulic and Hydrologic Data.
2. Using the Father Point bench mark also met the requirement that the datum shift from NGVD 29 to NAVD 88 would, to the extent possible, minimize recompilation of the National Mapping products by the U.S. Geological Survey.
3. The IGLD 85 and the NAVD 88 datums both have the same geopotential numbers on each published bench mark. From those geopotential numbers, the IGLD 85 dynamic heights are obtained by dividing the geopotential number by normal gravity at latitude 45° (980.6199 gals) and the NAVD 88 Helmert orthometric heights are obtained from the same geopotential numbers by dividing by the value of gravity at the station. Even so, there are slight differences between a dynamic height and an IGLD 85 height known as hydraulic correctors. For more information, see http://www.ngs. noaa.gov/TOOLS/IGLD85/igld85.shtml (accessed May 4, 2017).

## DATUM TRANSFORMATIONS

The GSDM defines an efficient environment for working with spatial data on a given datum but the GSDM is not a tool for transforming data from one datum to another. The topic of datum conversions is very important, and the following summary of specific datum transformations is provided for the convenience of the reader:

1. NAD 27 to NAD 83—2-D problem, use CORPSCON.
2. NAD 83 to HPGN—2-D/3-D problem, use CORPSCON.
3. NGVD 29 to NAVD 88—1-D problem, use CORPSCON.
4. NAD 83 (yyyy) to ITRF 3-D—3-D problem, use HTDP program from NGS.
5. NAD 83 and numerous other datums to WGS 84—3-D problem; see NGA (2014) and NIMA (1997, Appendices B, C).
6. From one epoch to another—use HTDP from NGS for 3-D data.
7. Write your own 3-D datum conversion—use 7- or 14-parameter transformation.

The following transformation procedures are all bidirectional, meaning there is a unique set of range and domain values in each case. Any transformation procedure should be checked by performing the reverse computation to confirm the integrity of the software and the process. The user should also be aware that most datum transformation procedures are approximations (although most are very good) and should not be used for critical or very precise applications unless done so under the responsible supervision of a knowledgeable professional.

## NAD 27 TO NAD 83 (1986)

This 2-D datum transformation can be performed using the NADCON program available from the NGS. The NADCON program was incorporated into the overall CORPSCON Windows-based program developed by the U.S. Army Corps of Engineers and is available free of charge.

## NAD 83 (1986) TO HPGN

This is a separate process. To complete the transformation of NAD 27 values to the NAD 83 (xxxx) in each state, the NGS provided an additional set of transformation parameters for each state and region as required to transform between the original NAD 83 (1986) adjustment and the GNSS state-wide upgrades. The CORPSCON combines the two-step process into a single operation. However, the user should confirm that the CORPSCON version being used contains the current HPGN files for the specific region needed.

## NAD 83 (xxxx) TO NAD 83 (yyyy)

Various NAD 83 adjustments have created a need for additional transformation programs. NGS has developed a 3-D coordinate transformation program, GEOCON 2.0, for use between various adjustments including NAD 83 (1986), NAD 83 (HARN), NAD 83 (NSRS2007) NAD 83 (2011), and others.

## NGVD 29 TO NAVD 88

This 1-D datum transformation can be performed using the VERTCON program available from the NGS. The VERTCON program was also incorporated into the overall CORPSCON Windows-based program developed by the Corps of Engineers and is available free of charge.

## HTDP

The horizontal time dependent positioning (HTDP) program is available gratis from the NGS web site as either a download or to use interactively. Based upon observed tectonic movements, the program permits the user to interpolate positions for any recent epoch anywhere within the United States and its territories. The HTDP software can also be used to convert between NAD 83 and ITRF.

### SOFTWARE SOURCES

The following web sites can be used to find free software:

CORPSCON: http://www.agc.army.mil/Missions/Corpscon.aspx (accessed May 4, 2017)
NADCON: http://www.ngs.noaa.gov (follow link to software) (accessed May 4, 2017)

VERTCON: http://www.ngs.noaa.gov (follow link to software) (accessed
   May 4, 2017)
HTDP: http://www.ngs.noaa.gov (follow link to software) (accessed May 4, 2017)
GEOCON: http://www.ngs.noaa.gov (follow link to software) (accessed
   May 4, 2017)

## 7-(14-) Parameter Transformation

A 7-paramter transformation is often used when converting 3-D coordinates from
one reference frame to another. There are 3 translation parameters, 3 rotation parameters, and one scale parameter. A generalized form of the 7-parameter transformation
in matrix form is

$$X_2 = K + S\,RX_1 \qquad\qquad (8.1)$$

where:
   $K$ = translation vector.
   $S$ = scalar—often close to 1.0.
   $R$ = rotation matrix frame 1 to frame 2.
   $X_1$ = vector of frame 1 coordinates.
   $X_2$ = vector of frame 2 coordinates.

The 7-paramter transformation algorithm as stated contains no approximation. In a
perfect world, the equations are exact and data in one reference frame can be converted
perfectly to another. But in the real world, coordinates are obtained from measurements
and measurements contain errors. Therefore, the stochastic properties of the data must
be evaluated to determine what is good enough and what isn't in a transformation. The
stochastic features of the GSDM accommodate those issues as well—see Chapter 14.
Note that a 14-parameter transformation accommodates velocities on the 7 parameters.

## 3-D DATUMS

An important distinction must be made with regard to transition to a 3-D geodetic
datum: geospatial data are referenced to a geodetic datum while spatial data may
be referenced to a datum that is not geodetic. For example, the geometric dimensioning and tolerancing (GD&T) community makes extensive use of 3-D datums in
numerous applications in which curvature of the Earth is not an issue. The American
Society of Mechanical Engineers has developed extensive standards for those applications (ASME 2009). With due consideration and use of a local P.O.B., the GSDM
accommodates both spatial and geospatial data and provides an efficient bridge
between a flat-Earth 3-D datum and a 3-D geodetic datum.
   A formal definition of a 3-D datum will include physical geodesy concepts and
is left to the scientists. But, for purposes of spatial data and using the GSDM, a 3-D
datum consists of the following:

- An Earth-centered Earth-fixed (ECEF) right-handed rectangular coordinate
   system.

- An origin located at the Earth's center of mass.
- The $Z$-axis directed to the Conventional Terrestrial Pole as defined by the International Earth Rotation Service (IERS).
- The $X/Y$ plane is the plane of the Equator and the $X$-axis is directed to zero longitude. The $Y$-axis completes the right-handed system.
- The meter (or metre) is the unit of length.

Subject to choices of the user with respect to datums, derived quantities include

- Latitude, longitude, and ellipsoid height based upon some chosen ellipsoid
- Orthometric height based upon ellipsoid height and geoid height
- State plane, UTM, or map projection coordinates based upon choice of user
- Direction, distance, and rectangular components between points in
  1. Local tangent plane through standpoint
  2. Tangent plane through P.O.B. selected by user

Currently, the NAD 83 replacement is called a "geometric reference frame" and the NAVD 88 replacement is called the "geopotential reference frame." Information on the status and progress of that effort can be obtained from the NGS web site (http://www.geodesy.noaa.gov (accessed May 4, 2017)).

## REFERENCES

American Society of Civil Engineers, American Congress on Surveying & Mapping, American Society of Photogrammetry & Remote Sensing. 1994. *Glossary of the Mapping Sciences.* New York: American Society of Civil Engineers, American Congress on Surveying & Mapping, and American Society of Photogrammetry & Remote Sensing.

American Society of Mechanical Engineers (ASME). 2009. *Geometric Dimensioning and Tolerancing Handbook: Applications, Analysis, & Measurement.* New York: American Society of Mechanical Engineers. https://www.asme.org/products/books/gdthdbk-geometric-dimensioning-tolerancing (accessed May 4, 2017).

Berry, R.M. 1976. History of geodetic leveling in the United States. *Surveying and Mapping* 36 (2): 137–153.

Bodnar, A.N. 1990. National Geodetic Reference System statewide upgrade policy. *Proceedings of the ACSM GIS/LIS 1990 Fall Convention*, Anaheim, CA, November 7–10, 1990.

Burkholder, E.F. 1984. Geometrical parameters of the geodetic reference system 1980. *Surveying and Mapping* 44 (4): 339–340.

Burkholder, E.F. 1991. Computation of horizontal/level distances. *Journal of Surveying Engineering* 117 (3): 104–116.

Burkholder, E.F. 1999. Spatial data accuracy as defined by the GSDM. *Surveying and Land Information Systems* 59 (1): 26–30.

Carta, M., trans. 1962. *Jordan's Handbook of Geodesy*, vol. 3. Washington, DC: U.S. Army Corps of Engineers, Army Map Service.

Committee on the North American Datum. 1971. *North American Datum.* Washington, DC: National Academy of Sciences.

D'onofrio, J.D. 1991. High precision geodetic network for California. Paper presented at *ASCE Specialty Conference*, Sacramento, CA, September 18–21, 1991.

Doyle, D.R. 1992. High accuracy reference networks: Development, adjustment, and coordinate transformation. Paper presented at *ACSM Annual Convention*, New Orleans, LA, February 15–18, 1992.

Doyle, D.R. 1994. Development of the National Spatial Reference System. https://www.ngs.noaa.gov/PUBS_LIB/develop_NSRS.html (accessed May 4, 2017).

Dracup, J.F. 2006. Geodetic surveys in the United States: The beginning and the next one hundred years; U.S. horizontal datums. https://www.ngs.noaa.gov/PUBS_LIB/geodetic_survey_1807.html (accessed May 4, 2017).

Malys, S., J. Seago, N.K. Pavlis, P.K. Seidelmann, and G.H. Kaplan. January 13, 2016. Prime Meridian on the move. *GPS World* (January): 44–45. http://gpsworld.com/prime-meridian-on-the-move/ (accessed May 4, 2017).

Moritz, H. 1980. Geodetic Reference System 1980. *Bulletin Geodesique* 54 (3): 395–405. http://www.gfy.ku.dk/~iag/handbook/geodeti.htm (accessed May 4, 2017).

National Geodetic Survey (NGS). 1986. *Geodetic Glossary*. Rockville, MD: National Geodetic Survey, National Ocean Service, National Oceanic and Atmospheric Administration.

National Geospatial-Intelligence Agency (NGA). 2014. Department of Defense World Geodetic System 1984: Its definition and relationships with local geodetic systems. NGA.STND.0036_1.0.0_WGS84. Arnold, MO: National Geospatial-Intelligence Agency.

National Imagery and Mapping Agency (NIMA). 1997. Department of Defense World Geodetic System 1984: Its definition and relationships with local geodetic systems. Technical Report 8350.2, 3rd ed. Bethesda, MD: National Imagery and Mapping Agency.

Pearson, C. 2004. National spatial reference systems readjustment of NAD83. https://www.ngs.noaa.gov/NationalReadjustment/Items/CFP.The%20national%20readjustment-1.html (accessed May 4, 2017).

Schwarz, C.R. (ed.). 1989. *North American Datum of 1983*. Rockville, MD: National Geodetic Survey, National Ocean Service, National Oceanic and Atmospheric Administration, U.S. Department of Commerce.

Strange, W.E. and J.D. Love. 1991. High accuracy reference networks: A national perspective. Paper presented at *ASCE Specialty Conference*, Sacramento, CA, September 18–21, 1991.

Taylor, A. 2004. *The World of Gerard Mercator: The Mapmaker Who Revolutionized Geography*. New York: Walker.

Zilkoski, D., Richards, J., and Young, G. 1992. Results of the general adjustment of the North American Vertical Datum of 1988. *Surveying & Land Information Systems* 52 (3): 133–149.

# 9 Physical Geodesy

## INTRODUCTION

Physical geodesy is the branch of science that relates the internal distribution of mass within the Earth to its corresponding gravity field. Under ideal circumstances, there would be no hills, valleys, mountains, or oceans on the Earth and the distribution of mass within the Earth would be uniform. Given those conditions, the plumb line (the vertical) would be coincident with the ellipsoid normal and there would be no difference between the ellipsoid and the geoid. But, the geoid is an equipotential surface that is always perpendicular to the local plumb line and the direction of the plumb line is dictated by the vector sum of forces acting on the plumb bob; for example, gravity is the sum of gravitational attraction and centrifugal force, as shown in Figure 9.1. The centrifugal force component is very predictable and can be computed. But, due to topography of the Earth and due to variations of density within the Earth, gravitational attraction varies from point to point and, although the difference is generally quite small, the resulting vertical is rarely coincident with the ellipsoid normal. That being the case, the geoid is not parallel to the ellipsoid and separation between the two surfaces varies with location. Ongoing geodetic research continues to improve knowledge of the relationship between the ellipsoid and the geoid. Evidence of continuing progress and improvement in geoid modeling in the United States is seen in publications of geoid models dated 1990, 1993, 1996, 1999, 2003, 2006 (Alaska), 2009, 2012A, 2012B, and so on.

Physical geodesy is also of concern to the geophysicist who studies gravity anomalies in search of patterns that indicate the presence of substrata oil deposits and to the geodesist who computes the trajectory of missiles and satellites flying above the Earth. Inferring the characteristics of underground strata and being able to predict how objects move are both important concepts, but beyond the scope of this book. The focus of this book is documenting where things are in terms of geometrical geodesy and the GSDM. Historically, physical geodesy and geometrical geodesy have shared the challenge of finding the elusive geoid and using it as a reference for elevation. The approach in this book is different in that elevation is viewed as a derived quantity computed from GNSS ellipsoid height and the ever-improving knowledge of geoid height.* Therefore, while a fundamental description of physical geodesy is viewed as useful in understanding the broader context of geospatial data and how they are used, the reader is encouraged to consult other sources for additional information on physical geodesy.

---

* Of course, the time-honored practice of starting on a bench mark of known elevation and taking simple backsight and foresight readings is still a legitimate method for determining the elevation of an unknown point.

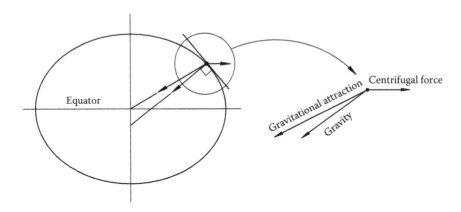

**FIGURE 9.1**   Vector Components of Gravity

## GRAVITY

Gravity is a vector quantity composed of the sum of gravitational attraction and centrifugal force due to the rotation of the Earth. Centrifugal force is always parallel to, and is greatest at, the equator. Centrifugal force is zero at the poles and gravity at the poles is the same as gravitational attraction. But, at the equator, centrifugal force is colinear with gravitational attraction and, because it acts in the opposite direction, the force of gravity is smaller at the equator than is gravitational attraction. Therefore, on the same equipotential surface, the force of gravity at the pole is greater than the force of gravity at the equator. The global implication is that level surfaces are not parallel. See Figure 9.2.

As described by Newton, gravitational attraction is the mutual attractive force between each and every particle in the universe:

$$F = \frac{kM_1M_2}{D^2} \tag{9.1}$$

where:
   $k$ = universal gravitational constant.
   $M_1$ and $M_2$ = masses of the particles.
   $D$ = the distance between particles.

The magnitude of attraction between respective paired centers of mass decreases by the square of the increasing distance between them. Particle pairs with a large separation react minimally and attractions at very large distances tend to be ignored. But, taken as a large collection of particles (a large body such as the Earth or sun can be treated as a point mass located at its center), the gravitational attractions interact to keep the planets in orbit about the sun in addition to keeping the moon and satellites in orbit about the Earth. The conventional practice on Earth is to express gravity as the force per unit mass with respect to the mass of the Earth concentrated at its center.

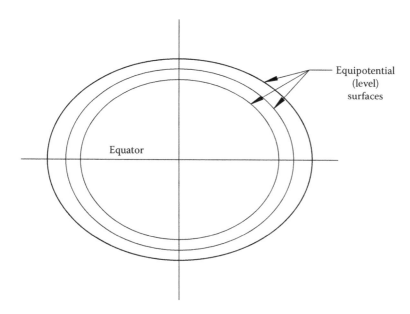

**FIGURE 9.2**   Level Surfaces Are Not Parallel

Two important consequences of the gravity vector are as follows:

1. As postulated by Newton and shown in Figure 9.1, the Earth is flattened at the poles because (except at the equator and at the poles) the gravity vector does not point directly to the center of mass of the Earth and because the geoid is always perpendicular to the plumb line.
2. On a global scale, equipotential surfaces are not parallel and, as illustrated in Figure 9.2, the distance between level surfaces is not constant. This means the definitions of elevation, level surfaces, orthometric heights, and other terms need to be very specific.

The intuitive equipotential surface most commonly understood is sea level. Sea level is physical, readily visible, and observed worldwide. Due to the influence of tides and other factors, the term "mean sea level" has long been associated with the geoid and served as the reference surface for the Sea Level Datum of 1929 in the United States. Recognizing that zero elevation is not necessarily the same as mean sea level, the name of the datum was changed in May 1973, to the National Geodetic Vertical Datum of 1929 (Berry, 1976). None of the published elevations were changed, only the name of the datum was changed.

## DEFINITIONS

The topic of units deserves particular attention when discussing concepts of physical geodesy. Spatial data users are primarily concerned with distance units (meters, feet, etc.) as related to location and elevation. However, when building on fundamental physical concepts, the physical separation between two equipotential surfaces is defined in terms of work, (i.e., force times distance). The following definitions presume an understanding that gravity is the force part of "work" and that elevation difference is the distance part. The following definitions are intended to be consistent with common usage, recent publications, and standard references, such as the National Geodetic Survey (NGS 1986); American Society of Civil Engineers, American Congress on Surveying and Mapping, and American Society of Photogrammetry and Remote Sensing (1994); and Meyer et al. (2004).

## ELEVATION (GENERIC)

Elevation is the distance above or below a reference surface. The geoid is an equipotential reference surface that has been widely used and is closely approximated by mean sea level. Unless specifically stated otherwise, a mean sea level elevation should probably be viewed as a generic elevation.

## EQUIPOTENTIAL SURFACE

An equipotential surface is a continuous surface defined in terms of work units with regard to its physical environment. Although not perfect, mean sea level is often given as an example. Two objects at rest having the same mass and located on the same equipotential surface store the same amount of potential energy. Work is required to move any object to a higher elevation. If the strength of gravity at Point A is greater than at Point B, then, for identical objects and for expenditure of the same work, the distance moved by the object at Point B will be greater than at Point A. The implication is that equipotential surfaces are parallel if and only if gravity is the same at both points on each respective surface. Although defined differently, a level surface and an equipotential surface are very nearly identical and, for most purposes, can be used interchangeably.

## LEVEL SURFACE

A level surface is a continuous surface that is always perpendicular to the local plumb line. A level surface can be at any elevation. Due to curvature of the Earth and variations of density within the Earth, the direction of the plumb line changes as one moves from point to point on or near the surface of the Earth. Consequently, a level surface (which is always perpendicular to the plumb line) is said to be "lumpy" due to these random changes in the direction of the plumb line. But, in most cases, changes in the direction of the plumb line are gradual and "lumps" in the geoid are gradual as well.

## GEOID

The geoid is an equipotential surface most closely represented by mean sea level in equilibrium all over the world (i.e., constant barometric pressure at the surface, no winds, no currents, uniform density layers of water, etc.). The "ideal" conditions do not exist and locating the geoid precisely on a global scale is an enormous challenge.

## GEOPOTENTIAL NUMBER

A geopotential number is a relative value computed as the infinite summation of the product of force times distance accumulated along a path of maximum gradient. A geopotential number has units of work and is rarely used in surveying and mapping applications. Dynamic heights are often used instead—see next subsection.

## DYNAMIC HEIGHT

Dynamic height is the geopotential number at a point divided by a constant reference gravity. Often, normal gravity at latitude 45° is used. Dynamic heights and geopotential numbers are useful when working with precise hydraulic grade lines over a large area. Standard geodesy texts contain additional information on these topics and their applications. Web searches can also be productive.

## ORTHOMETRIC HEIGHT

Orthometric height is the curved distance along the plumb line from the geoid to a point or surface in question. Few users make the distinction between the curved-line distance and the straight-line distance between the plumb line endpoints. In the past, orthometric height has been computed as the geopotential number of the equipotential surface divided by gravity at the point (Zilkoski et al. 1992). More recently (Meyer et al. 2004), orthometric height is more specifically called a Helmert orthometric height and is computed using ellipsoid heights (from GNSS) and geoid modeling procedures. The accuracy of such a derived height is dependent upon both the quality of the GNSS data and the integrity of the geoid modeling. Although elevation and orthometric height can often be used interchangeably, elevation is considered generic while orthometric height is specific.

## ELLIPSOID HEIGHT

Ellipsoid height is the distance as measured along the ellipsoid normal above or below the mathematical ellipsoid.

## GEOID HEIGHT

Discounting curvature of the plumb line, geoid height is taken to be the distance along the ellipsoid normal between the ellipsoid and the geoid. Geoid height is computed as the ellipsoid height minus the orthometric height. Within the conterminous United States, the orthometric height is always greater than the ellipsoid height, which means the geoid height is a negative number. But, on a worldwide basis, the simple definition is

$$h = H + N \qquad (9.2)$$

where:
  $h$ = ellipsoid height.
  $H$ = orthometric height.
  $N$ = geoid height.

## GRAVITY AND THE SHAPE OF THE GEOID

Everyone knows that the Earth is flat. At least that is our experience until we become convinced otherwise. That happens as we look at the bigger picture and watch ships sailing beyond the horizon or view the curved shadow of the Earth on the moon during an eclipse. Learning that the Earth is round and rotating on its axis also helps us understand the sunrise and sunset each day and the traverse of stars across the night sky. Furthermore, coming to understand that the Earth also revolves yearly around the sun helps explain why the sun and the stars appear to traverse the heavens at different rates. In considering observable physical phenomenon and studying the definitions in the previous section, everyone should agree that a flat-Earth view of the world is misleading.

On a global scale, Newton rationalized that the Earth must be flattened at the poles because a level surface is always perpendicular to the plumb line and, due to addition of force vectors, the plumb line does not point directly to the center of the Earth— unless one is on the equator or at one of the poles. But, that level of analysis does not differentiate between the various land masses of the continents or include the fact that some parts of the Earth are more dense than others. For example, the land mass and ice loading at the South Pole gives rise to an identifiable bulge in the geoid in the southern hemisphere. Therefore, the geoid is referred to as pear shaped.

Understanding that gravity is a vector having both direction and magnitude and realizing that the gravity vector is the sum of all external forces acting on a given mass at a given location, it should be understandable that the scope of physical geodesy encompasses much more material than is presented here. The magnitude of gravity affects the spacing of equipotential surfaces and variations in the direction of gravity affect the shape of an equipotential surface. That means that the geoid is not a regularly curved surface. Since the resultant geoid is related to both magnitude and direction of gravity, investigation of cause-effect relationships must include considerations of both. Stated differently, it is said that the magnitude of gravity affects where the geoid is located and that the direction of gravity determines the shape of the geoid. On a global scale, a stronger value of gravity will tend to pull the geoid in closer to center of the Earth but the shape of the geoid, both globally and locally, is always perpendicular to the direction of gravity.

Figure 9.3 illustrates an apparent paradox. While it is true that a stronger value of gravity tends to pull the geoid in closer to center of the Earth, a local mass concentration lying next to a mass deficiency (mountains and oceans) will deflect the plumb

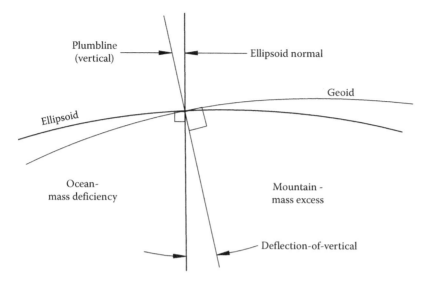

**FIGURE 9.3**   Deflection-of-the-Vertical

bob toward the mass excess and, contrary to statements in the previous paragraph, the geoid will rise under the mountains while dipping over the ocean. It remains a challenge to determine the location of the geoid precisely because the actual location of the geoid is the physical realization of many factors.

## LAPLACE CORRECTION

The Laplace correction is used to relate a geodetic azimuth to an astronomical azimuth. Such a correction is made at a point called a Laplace Station where the deflection-of-the-vertical components are known. In a broader sense, the Laplace equations are used to connect the physical world with a mathematical representation. The mathematical ellipsoid normal is a computational standard and is perpendicular to the tangent to the ellipsoid at a point. In the physical world, the direction of the plumb line is the result of physical forces acting on an object at a point. Although the difference is not significant for many applications, the two are different and must be considered in geodetic applications. The direction of the normal at a point is well defined and computable in the ECEF environment but, given the nonuniform distribution of mass within the Earth, the precise direction of the plumb line is less predictable. The irony of human experience is that we, standing erect, view the plumb line as being vertical. That is, we reference our view of the world to a changing feature that appears constant. In our view, up is always up. But, when we speak of the plumb bob being deflected by the mass of a mountain range, we are, in fact, referencing our perspective to the ellipsoid normal.

The difference between the ellipsoid normal and the vertical plumb line is called deflection-of-the-vertical and is realized several ways. First, the ellipsoid surface is not necessarily parallel with the level surface—the geoid. Given the two surfaces are not coincident, except for where they cross, there must be a physical distance between them. That difference is called geoid height and is studied under the name of geoid modeling. Second, the angular amount by which the two surfaces are not parallel is given by the deflection-of-the-vertical and expressed in terms of a north-south component and an east-west component. Any measurement or observation made with an instrument having a level bubble on it (whether a carpenter's level, a surveyor's total station, a differential level, a telescope in an observatory, or an inertial measuring unit) is physically referenced to the local vertical. On the other hand, vectors obtained from GNSS observations are referenced to the ellipsoid normal.

Specifically, astronomical observations of directions to stars are gravity based and yield astronomical positions, while GNSS-derived positions are normal based and provide geodetic positions. The two systems are related by the Laplace equations based upon the deflection-of-the-vertical components at a point that, using Greek letters, are called

$$\text{North-south}: \text{xi} = \xi \quad \text{and} \quad \text{East-west}: \text{eta} = \eta$$

The relationships between astronomical latitude, longitude, and azimuth and geodetic latitude, longitude, and azimuth are

$$\phi = \Phi - \xi \quad \text{latitude} \tag{9.3}$$

$$\lambda = \Lambda - \frac{\eta}{\cos\phi} \quad \text{longitude} \tag{9.4}$$

$$\alpha = A - \eta \tan\phi \quad \text{azimuth} \tag{9.5}$$

where:
$\phi$ = geodetic latitude.
$\lambda$ = geodetic longitude.
$\alpha$ = geodetic azimuth.
$\Phi$ = astronomic latitude.
$\Lambda$ = astronomic longitude.
$A$ = astronomic azimuth.

Equations 9.3 through 9.5 can be used two ways—if the geodetic latitude and longitude and the astronomic latitude and longitude are both known, then deflections-of-the-vertical can be computed at that point. If the deflection-of-the-vertical components are known, then geodetic latitude, longitude, and azimuth can be computed from astronomic latitude, longitude, and azimuth and vice versa. Probably Equation 9.5 is the most commonly used where the astronomic azimuth (observed from Sun shot or Polaris observation) is converted to the equivalent geodetic azimuth.

In the United States, deflection-of-the-vertical components are readily obtained from the NGS web site (https://www.ngs.noaa.gov/TOOLS/, accessed May 4, 2017) using the "deflect" program.

## MEASUREMENTS AND COMPUTATIONS

One way to understand physical geodesy and the geoid better is to look at some of the physical quantities that can be measured and to describe the quantities that might be derived from those observations. Given recent advances in technology, there are changes in the combinations of what is measured and what can be reliably computed from those measurements. The goal is that, as technology and procedures evolve, the geoid can be located with greater certainty than in the past and that the computational burden for spatial data users will be reduced.

### Interpolation and Extrapolation

The following discussion should also be viewed in terms of estimating and the difference between interpolation and extrapolation. Ideally, a reliable observation is available at every point where such a measurement is needed. That is, however, rarely the case. Representative measurements are made and other values are estimated—either by interpolation (often acceptable) or by extrapolation (often used as a last resort). The validity of any estimation hinges on factors such as the following:

1. What is the accuracy (uncertainty) of the available measurements?
2. Is the number and spacing (density) of the measurements consistent with observed rates of change? Would a greater density of measurements merely confirm what can be obtained by interpolation or are more measurements needed to provide a better picture of the quantity being estimated?
3. What is the rate of change of the quantity being measured? Is the observed rate of change reasonable and/or consistent with anticipated changes based upon current theories and the models being applied?
4. How sensitive are the indirect values being computed to changes in the quantity being measured?

Geoid height cannot be measured directly but must be inferred or computed from other quantities. With regard to locating the geoid or determining geoid heights, the list of measurable physical quantities includes

- Gravity
  1. Magnitude
     a. Absolute
     b. Relative
  2. Direction (deflection-of-the-vertical)
- Tide levels (location of mean sea level)
- Differential levels (changes in orthometric height)
- Ellipsoid heights (distance from ellipsoid)
- Time—related to gravity and location of geoid (Kleppner 2006)

There is a direct correlation between deflection-of-the-vertical and changes in geoid height. Deflection-of-the-vertical values express the "slope" of the geoid at a given point with respect to the ellipsoid normal. In the past, geoid heights were computed from an accumulation of deflection-of-the-vertical determinations at stations in the national triangulation network where the observed astronomical position was compared to the computed geodetic position. Such dual-value stations are called Laplace stations.

## Gravity

If a precise value of the magnitude of gravity was known at all points, the location and shape of the geoid could be computed without ambiguity. But gravity values are not known everywhere and, because the value of gravity changes from place to place, obtaining sufficient high-quality gravity measurements for precise geoid computations is not feasible. Absolute gravity measurements at the one-part-per-billion level (of gravitational acceleration at the Earth's surface) are possible (NGS 2005), but they require expensive equipment and rigorous observing procedures. Relative gravity measurements are more popular because they are easier to make and less expensive to obtain. A network of relative gravity measurements needs to be attached to an absolute gravity network in order to obtain information that is useful for geodetic computations.

In addition to ground-based measurements, gravity measurements are made at sea (both on the surface and underwater) and in the air (balloons, airplanes, and space craft). High-quality, land-based gravity measurements are typically more accurate because, while the measurement is being made, the gravity meter is motionless with respect to the Earth. Gravity measurements in a moving environment must be corrected for the location and motion of the platform—ship, airplane, etc. There is a trade-off between more costly high-quality, land-based gravity measurements (access to private property may also be an issue) and mobile gravity measurements obtained from a moving vehicle or platform. See https://www.ngs.noaa.gov/GRAV-D/, accessed May 4, 2017 for information on the NGS Grav-D observation program.

Measuring the direction of gravity is distinctly different than measuring the strength (magnitude) of gravity. A simple plumb bob is probably the best example of determining the direction of gravity—no matter where you go, the plumb bob always points down! Other devices for measuring the direction of gravity include the level vial as used on a carpenter's level, the bull's eye bubble as used on many surveying instruments, striding levels as used on precision theodolites and, for high accuracy applications, the undisturbed surface of a pool of mercury in a vacuum (precisely perpendicular to the local plumb line). Sensors in modern inertial measuring systems also detect the direction of gravity.

With regard to the direction of gravity, a challenge is to answer the question, "with respect to what?" Ironically, the direction of gravity is always vertical! After all, that is the definition. But, due to land masses and the nonuniform distribution of density within the Earth, the direction of gravity changes from point to point and ultimately the question is, "What is the direction of gravity at a point with respect to the ellipsoid normal at the same point?" That difference is called the deflection-of-the-vertical and is not measured directly but is inferred from other measurements and computations. During the era of nationwide triangulation, it was commonplace to compute the difference between the astronomic position observed at a point and the computed geodetic position for the same point based upon some datum and initial point. Deflection-of-the-vertical components were derived from those differences and defined the slope of the geoid at that point. With a sufficient number of dual-value Laplace stations and associated deflections, the shape of the geoid in that area could be inferred. Given that the shape of the geoid is attached to some initial point where the geoid height is zero, then absolute geoid heights can be determined in other locations throughout the country.

The U.S. Army Map Service (1967) published a 3-sheet map of "Geoid Contours in North America" based upon astrogeodetic deflections. The map shows 1 meter contours throughout North America on the North American Datum of 1927. The geoid height was held to be zero ($N = 0.0$ meter) at the Initial Point, triangulation station MEADES RANCH near Osborne, Kansas. Although the map is based upon astrogeodetic deflections, gravity data can be used to verify and/or augment the deflection-of-the-vertical computations.

## TIDE READINGS

Tide gage data have been collected at numerous locations by various organizations for many years. The First General Adjustment of geodetic leveling in the United States was published in 1900 and based upon mean sea level as determined by tide gages in five locations. Subsequent leveling network adjustments (1903, 1907, and 1912) connected to more tide gages and the 1929 General Adjustment (which became known as the Sea Level Datum of 1929) was based upon mean sea level as determined at 26 tide gage locations—21 in the United States and 5 in Canada (Berry 1976).

## DIFFERENTIAL LEVELS

As described by Zilkoski et al. (1992), loops of precise levels (orthometric height differences) within the United States were observed in support of the readjustment and definition of the North American Vertical Datum of 1988 (NAVD 88). The loop results are very good but, because the tide gages do not define the same geoid surface, the relative loop differences could not be attached to absolute tide gage elevations without distorting the observed relative differences. Therefore, one station, Romouski at Father Point, Quebec, was selected as the datum point and all NAVD 88 orthometric heights are stated with respect to that one arbitrary point. That being the case, it can be said that all NAVD 88 orthometric heights are relative and that there is no basis for absolute vertical accuracy statements with respect to the geoid. Such a statement is consistent with conventional leveling standards (Bossler 1984) that are given in terms of relative differences. Admittedly, NAVD 88 orthometric heights, being consistent with the definition of "absolute" in Chapter 3 and being on the same datum, are often used as absolute values.

## ELLIPSOID HEIGHTS

With the advent of GNSS and through the combined efforts of many people, the location* of the center of mass of the Earth is known within 1 cm (Schwarz 2005). Holding the center of mass as the origin and using the rectangular ECEF coordinate system as defined by the U.S. DOD, all spatial data computations within the birdcage of orbiting GNSS satellites follow the time-honored rules and equations of solid geometry. In that computational environment, ellipsoid heights are a derived quantity based upon the geocentric $X/Y/Z$ position of the point and the ellipsoid parameters chosen by the user. Several relevant issues are that spatial data components (coordinate differences) are derived from a variety of sensors and those relativistic effects should be modeled in the process of determining those spatial data components. Of course, the user must take responsibility for choosing and working on an appropriate datum such as NAD 83, WGS 84, or ITRF.

A related issue here is that GNSS-derived ellipsoid heights are typically quoted as less precise than the horizontal position. One reason is that those results are based only on signals from satellites visible from one side of the world. More data are available. Taken as a whole, GNSS is, among other things, a huge interpolation system for spatial data. Given that signals from orbiting satellites are transmitted to the Earth from all sides, it is conceivable, if data from all satellites are used simultaneously and if the worldwide GNSS network is treated as a deformable solid, that the observed vertical (radial) dimension could turn out to be the strongest component of a GNSS-derived position. That being the case, ellipsoid height can be determined very precisely for many points around the world (Burkholder 2003). The hypothesis is that absolute ellipsoid height can be determined routinely within a centimeter or less with respect to the center of mass of the Earth. Of course, with Earth tides, a monumented point (or CORS position) on the surface of the Earth will rise and fall twice a day as our moon goes around the Earth. A mean value of ellipsoid height for a point will be adopted and used for most purposes but, much like polar motion components, the daily differences should be available to those needing them. Many local spatial data applications will continue using relative ellipsoid height differences attached to a network of precisely determined absolute ellipsoid height points. Tectonic uplift and subsidence are nonregular movements monitored by those having responsibility for maintaining and updating the National Spatial Reference System.

Comparing ellipsoid height with orthometric height at a point is one of the best ways to find the geoid height. The problem is that, although obtaining a reliable ellipsoid height at a point is fairly routine, rarely are there enough reliable high-order bench marks available for making the comparison. Consider two extremes: If high-quality gravity values were known at a sufficient number of points in an area, it would be possible to compute geoid heights using gravity data. On the other hand, if sufficient high-quality orthometric heights were available in an area, it would be

---

* By definition, the location of the center of mass of the Earth is fixed and does not move—land masses and points on the surface of the Earth move with respect to the center of mass. The computed relative location of points on the surface of the Earth and orbit of satellites are consistent within 1 cm.

possible to compute geoid heights using Equation 9.2 rewritten as $N = h - H$. The challenge is to find and use the best (most efficient, reasonable, practical) combination of observations and procedures to determine acceptable geoid height values. Once primary reliable values for the geoid height are determined in a given area, other geoid heights in the same area can be interpolated using standard modeling techniques.

## TIME

Although this author will leave it to others, a recent hypothesis is that the absolute location of the geoid can be inferred from precise time measurement. In discussing "the great geoid search," Kleppner (2006, p. 11) states

> In the not-too-distant future, our ability to compare atomic frequency standards and clocks at different laboratories will be limited by our knowledge of the geoid.
> The obvious way to deal with the geoid problem is to reverse the argument and employ the gravitational redshift to explore the geoid. If, for instance, one had a portable atomic frequency standard accurate to 1 part in $10^{18}$ and if it could be compared to a primary standard with the same accuracy, the position of the geoid could be independently and relatively quickly determined to 1 cm.

The point of discussing the measurements and computations related to the geoid height is to focus on the quantities that can be determined with the greatest certainty. From a geometrical geodesy perspective, it appears that the combination of precise ellipsoid heights and high-order orthometric heights is the best way to determine geoid heights. Gravity measurements are used as collaborating data to verify results. But, from a physical geodesy perspective, a worldwide gravimetrically derived geoid is used as the basis for developing geoid models such as GEOID12B. Many more details are available from the NGS (2016) web site. And, if using atomic clocks to locate the geoid independently ever comes to fruition, current geoid modeling procedures may be significantly modified.

## USE OF ELLIPSOID HEIGHTS IN PLACE OF ORTHOMETRIC HEIGHTS

Given the elusive nature of the geoid (Kleppner 2006) and given that ellipsoid heights are readily available in the user community, it has been proposed (Burkholder 2002, 2006, Kumar 2005) that ellipsoid heights be used in place of orthometric heights for routine 3-D spatial data applications. Selling that change will be difficult because most people are comfortable with using sea level as a vertical reference. Sea level is a physical reference surface that people can see and understand. But, starting in 1973 with the change in the name of the vertical datum, the association of mean sea level with the zero elevation was broken, and with publication of the NAVD 88, the zero elevation is no longer "connected" to the tide gages. Technically, the change to using ellipsoid heights for orthometric heights can be managed effectively in the same way as any other datum upgrade. But, although selling that change may be difficult, there are several precedent setting examples to be considered—the definition of the North Pole and the definition of time.

The difference between the physical position of the North Pole (it moves) and the mathematically adopted conventional terrestrial pole (CTP) is an important concept but used only by a relatively small number of people. Polar motion is routinely factored into those computations needed to reconcile the physical position of the spin axis with the CTP and GNSS users the world over confidently use GNSS-derived positions without needing to model polar wandering (Leick 2004). The fact that few people actually use polar motion coordinates does not diminish the scientific importance of determining those quantities with exactitude.

Time is another analogy. In years past, noon was defined as the instant the sun crossed one's local meridian and the beginning of each day was defined as midnight, the instant the sun crosses ones meridian on the other side of the world. This definition is simple, physical, easily understood, and relies on the apparent motion of the sun across the sky. Two problems: the sun–meridian definition of noon is location dependent and, throughout the year, the time interval from noon to noon varies. Using the noon definition of time, each railroad train station in the 1800s had its own version of correct time and developing reliable train schedules was a real problem. The problem was solved (Howse 1980) by adopting a system of standard time zones in the United States in 1883. The worldwide system of time zones was adopted by the *International Meridian Conference* in Washington, DC, in 1884. Yes, selling standard times zones to the public was an enormous task but, in hindsight, apparently well worth the effort.

The second time problem is measuring the uniformity with which time progresses. The equation-of-time is the difference between time as determined by the motion of the Earth and the uniform progression of atomic time. The ancient Greeks recognized the existence of the equation-of-time but it was not until the late 1600s that John Flamsteed, the first Royal Astronomer of the Greenwich Observatory, quantified the equation-of-time. Since then, measurement of the uniform progression of time has evolved from a pendulum clock to a quartz crystal oscillator in an atomic clock capable of accuracies in the range of one part in $10^{14}$ (Jones 2000). Now most people take time zones for granted and the fact that the sun crosses the local meridian before or after 12:00:00 noon is of little consequence. But, for scientific and other purposes, the equation-of-time and other time scale differences are known, documented, and used by those for whom the difference matters.

Similarly, the location of the center of mass of the Earth is easier to locate, is more stable than the geoid, and offers obvious advantages when used as a single reference for terrestrial 3-D spatial data. Making such a change would in no way diminish the importance of geoid modeling research but it would relieve the spatial data user community of a significant computational burden imposed by continued use of separate origins for horizontal and vertical data. Issues associated with such a change include the following:

1. Water flows downhill. In most cases, ellipsoid height differences are sufficiently accurate to establish grades for highway alignments and gravity sewer flows. As an example, typical storm sewer manholes are spaced about

100 meters apart. If one assumes an acceptable as-built tolerance of 0.005 m for the invert orthometric height at each manhole ($0.005 \times \sqrt{2} = 0.007$ m), the tolerance for slope is arctan (0.007 m per 100 m) or about 15 seconds of arc (deflection-of-the-vertical is less than 15 seconds of arc in most places). Given that engineering involves finding, documenting, and using acceptable approximations, a study needs to be conducted and published to identify the severity of slope for other conditions and to identify acceptable toler-ances for other assumptions. Of course, more critical applications need to model and accommodate differences between gravity-based measurements (e.g., differential leveling) and normal-based measurements (e.g., GNSS). Examples include hydraulic grades lines (dynamic heights) over large areas, tunneling through a mountain from two ends, and establishing a true plane for atomic particle colliders. In such cases, geoid modeling should still be used by those for whom the difference matters.

2. In the conterminous United States, the geoid lies below the ellipsoid. If ellipsoid heights are used in place of orthometric heights, there are many places along the coast where one is clearly standing on dry ground but the ellipsoid height for the point is a negative number. We have learned to accept negative orthometric heights in places like Death Valley but negative elevations on dry land along the coast is a rather dramatic reminder that zero elevation is not the same as mean sea level.

## THE NEED FOR GEOID MODELING

Before getting into the details of geoid modeling, it may be beneficial to look at sev-eral assumptions associated with using or choosing to not use geoid height:

1. For the first approximation, the difference between the ellipsoid and the geoid can be assumed to be zero—that is, ignoring geoid heights altogether. On a global basis, the error of such an assumption could be as much as 100 meters. In the United States, the error of ignoring the geoid height is limited to about 50 meters. If the orthometric heights being used have a standard deviation of 100 meters or more, the noise of the data is greater than the feature being modeled and it makes little or no difference whether one uses ellipsoid height or orthometric height.

2. Another assumption is that the world is flat (for a small area) and that the ellipsoid and the geoid are two parallel planes. Under this assumption, if the geoid height at one place is known, the geoid height at nearby loca-tions is the same. For some applications, this assumption may be appro-priate but there is a real danger that limits of the assumption may not be recognized. The user should be aware of unintended, possibly even severe, consequences.

3. A better assumption (but still assuming a flat Earth) is that the ellipsoid and geoid are nonparallel plane surfaces—that is, one is tilted with respect to the other. If geoid heights are known at three points (required to define a plane) in a specified area, other geoid heights in the same general area can

be obtained by linear interpolation techniques. Extrapolation may provide reasonable answers in the same general area but uncontrolled extrapolation is to be avoided. If geoid heights are known for more than three points scattered throughout the project area, a local "best fit" geoid model may be the appropriate choice.

4. The best assumption underlying geoid modeling is that the two surfaces are both curved and not concentric. Curvature of the ellipsoid is mathematically well defined and can be computed. But, even though the geoid is a continuous surface, the geoid curves in an irregular manner and writing a mathematical function for the separation between the two curved surfaces involves complex modeling. The reader is referred to the NGS (2016) web site for additional information on geoid modeling and the latest geoid model available to the user community.

In spite of the advantages of using ellipsoid heights for orthometric heights, that change may never occur. We need to recognize that current practices have evolved to be what they are for specific reasons (we are where we are because of where we came from) and the prudent user needs to be familiar with current geoid modeling practices.

In spatial data applications, geoid modeling is used primarily to relate ellipsoid height to orthometric height. As noted in the previous section, "Use of Ellipsoid Heights in Place of Orthometric Heights," that application may be significantly reduced if ellipsoid heights are adopted and used in place of orthometric heights. A secondary application using geoid heights is reducing a horizontal distance at some elevation to either the geoid or to the ellipsoid. Whether using the ellipsoid or the geoid as the computational surface, demand for geoid modeling is driven by the desire for physical measurements to be modeled at a level of significance that preserves their geometrical integrity. The next section, "Geoid Modeling and the GSDM," considers the impact of geoid height on reducing horizontal distance to the ellipsoid.

Typically, a slope distance is measured between a standpoint and a forepoint, reduced to an equivalent horizontal distance, and further reduced to the ellipsoid for use in geodetic computations (Burkholder 1991). One could say that geoid modeling is not required if it makes no significant difference whether orthometric height or ellipsoid height is used for the elevation reduction computation. But, we also need to acknowledge practices of the past as we make decisions about current standards, specifications, and procedures. For example, triangles and quadrilaterals in the national horizontal network defining the North American Datum of 1927 (NAD 27) were computed on the geoid as if it were the Clarke Spheroid 1866. Of course, both the shape of the Clarke Spheroid 1866 and the location of the datum origin at triangulation station MEADES RANCH were selected so that, in the United States, the separation between the geoid and the Clarke Spheroid 1866 was minimized. That proximity mitigated the consequences of performing the computations on the geoid instead of the Clarke Spheroid 1866.

However, the North American Datum of 1983 (NAD 83) was computed on the ellipsoid, not on the geoid. The shape and location of the GRS 80 ellipsoid were chosen for a global best fit rather than a continental best fit, as was done for the NAD 27. The origin of the NAD 83 was located at the Earth's center of mass (as determined at the time) and the shape of the GRS 80 ellipsoid approximates the global shape of the geoid. An unintended consequence of a global best fit for the geoid is that differences between the ellipsoid and the geoid in North America are greater on NAD 83 than they are on NAD 27. Separately, the intended internal consistency of the NAD 83 was to be at least one magnitude better than for NAD 27. These factors, both singularly and collectively, mean that geoid height is an important consideration in distance reduction and should not be ignored in geodetic computation of positions on the NAD 83.

In an attempt to quantify the severity of using an orthometric height instead of ellipsoid height in the distance reduction, Burkholder (2004) examines the accuracy of the elevation reduction factor. Equation 9.6 is taken from that article and can be used to compute the elevation reduction factor standard deviation for any combination of values and is given as

$$\sigma_{ElevationFactor} = \sqrt{\left[\frac{h}{(r+h)^2}\right]^2 \sigma_r^2 + \left[\frac{-r}{(r+h)^2}\right]^2 \sigma_h^2} \qquad (9.6)$$

where:
  $h$ = ellipsoid height used in reduction.
  $r$ = radius of the Earth.
  $\sigma_h$ = uncertainty of ellipsoid height.
  $\sigma_r$ = uncertainty of radius of the Earth.

The four examples in Table 9.1 were computed using Equation 9.6 and are intended to show that

1. The uncertainty in ellipsoid height is a prominent factor in the reduction of a horizontal distance to the ellipsoid
2. Ellipsoid height and the uncertainty in Earth radius both contribute to the uncertainty in the elevation reduction factor, but not much

TABLE 9.1

**Four Examples of Elevation Reduction Factor Uncertainty**

|  | Uncertainty of Height | Height of Standpoint | Uncertainty in Earth Radius | Resulting Uncertainty Elevation Factor |
|---|---|---|---|---|
| 1. | 50 meters | 500 meters | 1,000 meters | 1:127,460 |
| 2. | 50 meters | 2,000 meters | 5,000 meters | 1:127,457 |
| 3. | 10 meters | 500 meters | 1,000 meters | 1:637,267 |
| 4. | 10 meters | 2,000 meters | 5,000 meters | 1:629,882 |

3. Using ellipsoid height instead of orthometric height or vice versa (ignoring a geoid height of 50 meters) may be acceptable at the 1:100,000 level
4. Ignoring a geoid height of 10 meters will affect results at the 1 ppm level

But, the real question to be addressed is whether geoid modeling is needed at all. Because the presumed answer is "yes," the question deserves careful consideration. When looking at computation and use of spatial data components, the GSDM stores geocentric $X/Y/Z$ coordinates and covariance values as the primary record of a point. A new point is established from an existing point using the 3-D forward (BK3) computation—either on the basis of GNSS-derived $\Delta X/\Delta Y/\Delta Z$ or on the basis of local normal-based $\Delta e/\Delta n/\Delta u$ components rotated to $\Delta X/\Delta Y/\Delta Z$. A more recent positioning technique is precise point positioning (PPP) that is gaining popularity. In this case the new position is determined independently based on timing and satellite orbits. A new position does not rely on other Earth-based control monuments. For all cases, the computation of a new point takes place in 3-D space, not on the geoid and not on the ellipsoid. No geoid modeling is required. An exception could be if the slope of the geoid with respect to the ellipsoid normal is needed to correct vertical-based measurements to normal-based measurements. In that case, deflection-of-the-vertical data are needed, not the precise geoid height. If the geodetic line distance on the ellipsoid between the standpoint and the forepoint is desired, it can be computed from a geodetic inverse (BK19)—geoid height is not an issue.

As described in Chapter 3, rectangular spatial data components are computed as coordinate differences in the geocentric rectangular coordinate system and rotated (without distortion or loss of integrity) to the local perspective for use as 3-D rectangular flat-Earth components. In that environment, horizontal distance from the standpoint and to the forepoint lies in the local tangent plane through the standpoint. The 3-D azimuth (Burkholder 1997) from standpoint to forepoint is computed simply as arctan $(\Delta e/\Delta n)$ and ellipsoid height is derived directly from the $X/Y/Z$ coordinates (see Equation 1.7). Geoid modeling is not an issue unless one needs to relate the orthometric heights (elevations) to corresponding ellipsoid heights. Two points are as follows:

1. Yes, orthometric heights are still used and geoid modeling is important as a way to find orthometric height from ellipsoid height. But for many flat-Earth applications (e.g., RTK construction layout staking over limited distances), the $\Delta u$ component of a local vector can be used as an orthometric height difference. If a curved Earth height difference is needed, the ellipsoid height difference ($\Delta h$) of the vector will suffice unless the slope of the geoid in that area is quite severe and/or very high precision is required for vertical. In that case, geoid modeling will be needed.
2. Reducing a horizontal distance to the ellipsoid for geodetic computation is no longer needed. Instead, a geodetic 3-D forward computation (BK3) based upon slope distance, zenith direction, and azimuth of the line is used to compute $X/Y/Z$ coordinates of the forepoint. Latitude/longitude/height of the forepoint are computed from the forepoint $X/Y/Z$ values. See the "Comparison of 3-D Computational Models" example in Chapter 15.

## GEOID MODELING AND THE GSDM

An important concept is that the simplest model that supports the integrity of the data is the most appropriate model to use. The 1-D flat-Earth model for leveling is probably the simplest spatial data model and is used extensively. The 2-D flat-Earth model of rectangular plane coordinates is used all over the world for simple local applications. A local 3-D rectangular model is also used for applications such as describing condominium space, local area topographic maps, and construction stakeout. But, geographers quickly find themselves beyond the range of an acceptable flat-Earth model and they utilize spherical latitude/longitude coordinates to express horizontal location. Orthometric height (or altitude) is used to describe vertical location. At an even higher level of complexity, geodesists, geophysicists, and other scientists need to use the flattened ellipsoid model in order to preserve geometrical integrity of measurements made all over the Earth. As noted earlier, the spherical and ellipsoidal Earth models involve the use of mixed units, that is, angular units for horizontal and length units for vertical. Map projections (see Chapter 11) were invented as a way to represent a portion of the curved Earth on a flat map. That essentially solves the mixed unit problem but map projections are severely limited, among others, by the fact that a map projection is strictly two dimensional. Spatial data are 3-D and modern practice needs to combine horizontal and vertical into the same database.

Whether using the NAD 83, WGS 84, or ITRF datum, the GSDM defines any point within the birdcage of orbiting GNSS satellites by a triplet of geocentric rectangular $X/Y/Z$ coordinates. Furthermore, each stored point has a covariance matrix associated with it and interdependencies between points are stored in a point-pair correlation submatrix. From these values stored in a BURKORD™ database, the user can select and compute derived quantities. In some cases, the assumptions are contained within the definition of the GSDM, while in other cases, the user must specify additional assumptions about the derived quantities. And, in all cases, the computations are bidirectional, meaning data can be transformed either way without loss of geometrical integrity.

Examples include the following:

1. Quantities computed directly from the $X/Y/Z$ coordinates and covariance matrix for a point are latitude, longitude, ellipsoid height, and standard deviations for the point in either the geocentric reference frame or the local reference frame. The local "up" standard deviation is the standard deviation of the ellipsoid height. The implicit assumptions are that the user has selected the datum and the ellipsoid parameters.
2. Other quantities that can be computed from the latitude and longitude of a point include UTM coordinates, state plane coordinates, or user defined projection coordinates. The implicit assumptions are that the user is responsible for using appropriate transformation equations, using the correct units, and staying on the same datum. The local reference frame ($e/n/u$) standard deviations are not changed by such a transformation.
3. Orthometric height can be computed from the ellipsoid height using geoid modeling and a rearrangement of Equation 9.2 as $H = h - N$. The geoid

modeling program, GEOID12B (and other versions) is described in the next section. The standard deviation of the computed orthometric height is based upon the standard deviation of the ellipsoid height and the standard deviation of the geoid height. It can be done either way, but this is one place where the user must exercise caution—the standard deviation of an absolute geoid height at a point is typically inferior to the standard deviation of the geoid height *difference* between points, as described later.

4. Other derived quantities are listed here for the sake of completeness but are discussed in more detail later. Given that a user selects a pair of points, the derived quantities include the mark-to-mark distance, the 3-D azimuth from standpoint to forepoint, the zenith (or vertical) direction from standpoint to forepoint, the local tangent plane horizontal distance from standpoint to forepoint, and the standard deviation of each quantity. Furthermore, if correlation submatrices have been stored, both the network and local accuracies can be computed for these derived quantities. The explicit condition here is that all quantities are with respect to the ellipsoid normal at the standpoint.

5. Finally, if the user selects a P.O.B. datum point (see Chapter 1), then any and all other points in the database can be viewed from that perspective and those local coordinate *differences* can be used as local flat-Earth plane coordinates with respect to the origin selected by the user. Standard deviations of all derived quantities (direction, distance, area, volume, etc.) are available and routinely reported. The implicit assumption is that all horizontal distances are within the tangent plane through the P.O.B. and that all directions are grid directions with respect to the true meridian through the P.O.B. A further assumption is that, unless modified by the user (see Chapter 11), the "up" component is the perpendicular distance from the forepoint to the tangent plane through the P.O.B. As described in Chapter 11, these derived values will see extensive use throughout the spatial data user community in a wide variety of applications.

## USING A GEOID MODEL

The following procedures for determining orthometric heights via GNSS and geoid modeling are similar, but not identical, to those procedures given by NGS in the publication "Guidelines for Establishing GPS-Derived Orthometric Heights" (NGS 2008). Given one is to fulfill the requirements of the NGS specification, those criteria supersede the suggestions offered here and should be followed. For purposes of local practical application, the following procedures have been proven to provide excellent results. A detailed example of using GNSS to establish a reliable NADV 88 elevation on a HARN station REILLY is discussed in Chapter 15.

As implemented in GEOID12B and other geoid modeling programs, the printout typically gives the geoid height to 3 decimal places of meters. The geoid height at that point is not necessarily within 1 millimeter as implied by the printout, but to compute the *difference* in geoid heights between two neighboring points does need that many decimal places. Stated differently, the relative accuracy is better than the

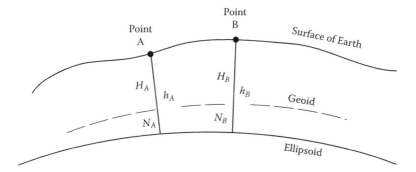

**FIGURE 9.4**  Determining Orthometric Height Using GPS and Geoid Modeling

absolute accuracy. That means the shape of the geoid is known better than the location of the geoid. The following procedure is recommended to take advantage of that characteristic of geoid modeling.

Given:

Known orthometric height at Point A = $H_A$
GNSS-based ellipsoid heights at Points A and B, $h_A$ and $h_B$
GEOID12B geoid heights at Points A and B, $N_A$ and $N_B$
Fundamental relationships: $h = H + N$, $H = h - N$, and $\Delta H = \Delta h - \Delta N$

Problem: As illustrated in Figure 9.4, determine the orthometric height of Point B using the orthometric height at Point A, GEOID12B (or other geoid model), and the GNSS vector from Point A to Point B.

Solution:

1. Determine the latitude, longitude, and ellipsoid height of Point A and Point B from the geocentric $X/Y/Z$ coordinates of the points. If the covariance matrix of each point is also stored, a program such as BURKORD™ will also provide, among others, the standard deviation of the height component.
2. Use GEOID12B and the latitude/longitude of each point to compute the geoid height for each point, $N_A$ and $N_B$. Note that in the conterminous United States, the geoid lies below the ellipsoid and values of $N$, geoid height, are negative. The sign conventions for ellipsoid height, geoid height, and orthometric height are pretty much standard the world over and the equations used herein are consistent with the Figure 9.4, but the negative sign of $N$ must be handled properly in the algebraic sense.
3. Combine the various components as follows:

$$\Delta h = h_B - h_A \quad \left(\text{from GNSS and/or geocentric } X/Y/Z\right)$$

$$N = N_B - N_A \quad \left(\text{from GEOID12B or similar model}\right)$$

$$\Delta H = \Delta h - \Delta N$$

$$H_B = H_A + \Delta H = H_A + \left( h_B - h_A \right) - \left( N_B - N_A \right) \tag{9.7}$$

Computing the standard deviation of the orthometric height at Point B is a natural extension of this discussion.

GEOID12B or update is available gratis from the National Geodetic Survey (NGS) web site—http://www.ngs.noaa.gov.

## REFERENCES

American Society of Civil Engineers, American Congress on Surveying & Mapping, and American Society of Photogrammetry & Remote Sensing. 1994. *Glossary of the Mapping Sciences*. New York: American Society of Civil Engineers, American Congress on Surveying & Mapping, American Society of Photogrammetry & Remote Sensing.

Berry, R.M. 1976. History of geodetic leveling in the United States. *Surveying & Mapping* 36 (2): 137–153.

Bossler, J. 1984. *Standards and Specifications for Geodetic Control Networks*. Rockville, MD: Federal Geodetic Control Committee.

Burkholder, E.F. 1991. Computation of level/horizontal distance. *Journal of Surveying Engineering* 117 (3): 104–116.

Burkholder, E.F. 1997. Three-dimensional azimuth of GPS vector. *Journal of Surveying Engineering* 123 (4): 139–146.

Burkholder, E.F. 2002. Elevations and the global spatial data model (GSDM). *Proceedings of the 58th Annual Meeting of the Institute of Navigation*, Albuquerque, NM, June 24–26, 2002.

Burkholder, E.F. 2003. The digital revolution begets the global spatial data model (GSDM). *EOS, Transactions, American Geophysical Union* 84 (15): 140–141.

Burkholder, E.F. 2004. Accuracy of elevation reduction factor. *Journal of Surveying Engineering* 130 (3): 134–137.

Burkholder, E.F. 2006. The digital revolution: Whither now? *GIM International* 20 (9): 25–27.

Howse, D. 1980. *Greenwich Time and the Discovery of Longitude*. Oxford, U.K.: Oxford University Press.

Jones, T. 2000. *Splitting the Second*. Boston, MA: Institute of Physics Publishing.

Kleppner, D. March 10–11, 2006. Time too good to be true. *Physics Today* 10–11.

Kumar, M. 2005. When ellipsoidal heights will do the job why look elsewhere! Paper presented at the *American Congress on Surveying & Mapping Annual Meeting*, Las Vegas, NV, March 2005. http://citeseerx.ist.psu.edu/viewdoc/download?doi=10.1.1.139.8498&rep=rep1&type=pdf.

Leick, A. 2004. *GPS Satellite Surveying*. New York: John Wiley & Sons.

Meyer, T.H., Roman, D.R., and Zilkoski, D.B. 2004. What does height really mean? Part I: Introduction. *Surveying and Land Information Science* 64 (4): 223–233.

National Geodetic Survey (NGS). 1986. *Geodetic Glossary*. Rockville, MD: National Geodetic Survey, U.S. Department of Commerce.

National Geodetic Survey (NGS). 2005. Gravimetry: Absolute gravity. https://www.ngs.noaa.gov/GRD/GRAVITY/ABSG.html, accessed May 4, 2017.

National Geodetic Survey (NGS). 2008. Guidelines for establishing GNSS-derived orthometric heights (Standards: 2 cm and 5 cm). https://www.ngs.noaa.gov/PUBS_LIB/NGS592008069FINAL2.pdf, accessed May 4, 2017.

National Geodetic Survey (NGS). 2016. GEOID12. https://www.ngs.noaa.gov/PC_PROD/ GEOID12/, accessed May 4, 2017.

Schwarz, K.-P. 2005. Personal communication. University of Calgary, Calgary, Alberta, Canada.

U.S. Army Map Service. 1967. *Geoid Contours in North America [Map]*. Washington, DC: U.S. Army Corps of Engineers.

Zilkoski, D.B., Richards, J.H., and Young, G.M. 1992. Results of the general adjustment of the North American Vertical Datum of 1988. *Surveying & Land Information Systems* 52 (3): 133–149.

# 10 Satellite Geodesy and Global Navigation Satellite Systems

## INTRODUCTION

Goals of satellite geodesy include using Earth-orbiting satellites to determine the size and shape of the Earth, to obtain a greater understanding of the Earth's gravity field, and to define the position or movement of points on or near the surface of the Earth and in near space. The first two goals are primarily scientific in nature, but the third goal is more applications oriented and of interest to many more people. The utility of position fostered by pervasive use of satellite systems in various applications means that many spatial data users are interested in learning more about fundamental concepts of satellite positioning. Many principles and systems are involved in meeting those goals and, because there are so many details, this chapter should be considered as an overview.

Satellite positioning technology evolved during the Cold War being triggered, in part, by the launch of Sputnik I by the Russians in October 1957. Since then, various satellite systems have been used for positioning—the TRANSIT satellite system, the NAVSTAR constellation of global positioning system (GPS) satellites, the Russian GLONASS system, the GALILEO satellite positioning system being developed by the Europeans, and other systems such as those being developed by India and China. As a point of clarification, the original NAVSTAR system developed by the United States is commonly referred to as GPS. Since then, others have also developed similar systems and the phrase global navigation satellite system (GNSS) is used generically to describe all satellite positioning systems as opposed to reference to a particular brand. The abbreviation GNSS is used in this book because the GSDM provides an appropriate bridge between the exacting rigor of the methods used by the "rocket scientists" and the much simpler flat-Earth methods typically employed in the spatial data user community.

Spatial data were defined in Chapter 3 as the distance between the endpoints of a line in 3-D space. As such, the definition is rather meaningless until such data are expressed with respect to a coordinate system—local, geodetic, or geocentric. The GSDM includes each of those coordinate systems and defines a common geometrical computational environment for all geospatial data users, including those who develop satellite positioning systems, those who collect and process physical measurements to obtain coordinates, those who use spatial data for an endless array of applications, and those who manage the collection, storage, manipulation, and use of geospatial data.

As stated in Chapter 2, "The Global Positioning System: Charting the Future" (NAPA 1995) is a report prepared by the National Academy of Public Administration that documents the development, applications, and future of GPS. In the Executive Summary, the report predicted worldwide revenue for GPS-related products and services to exceed $30 billion by 2005. A more recent estimate of economic impact as reported by GPS World (Leveson 2015) gives a range of $37.1 billion to $74.5 billion in 2013. Many highly skilled persons are involved in building the tools and supporting the systems used to generate quality 3-D spatial data.

"Geographic Information for the 21st Century" (NAPA 1998) is a report also prepared by the National Academy of Public Administration—in this case for the U.S. Bureau of Land Management, U.S. Forest Service, U.S. Geological Survey, and National Ocean Service—that is a formal study of civilian federal surveying and mapping activities. This arena includes many talented persons who use the measurement systems and related technology to generate and manipulate 3-D spatial data. But, perhaps more significantly, this arena also includes countless applications of spatial data within the context of a modern geographic information system (GIS).

Even though one could say the focus of those two publications is very different, there is an enormous overlap in that both rely on a shared foundation of 3-D spatial data. Given the impact of the digital revolution (Burkholder 2003), spatial data are now characterized as digital and three-dimensional (3-D) and modeled by the GSDM. Built on the premise of a single origin for 3-D data, the GSDM simultaneously accommodates the rigorous methods of the GNSS arena and the flat-Earth practicality of GIS applications. Establishing and tracking spatial data accuracy efficiently is an added bonus for all users.

Entire books are written on satellite geodesy and the processes used to convert physical measurements into spatial data components. The interested reader is referred to comprehensive texts such as Leick (1990, 1995, 2004), Leick et al. (2015), Seeber (1993), or Hofmann-Wellenhof et al. (2001) for more details. Current magazines such as *GNSS World*, *Inside GNSS*, and *Geospatial Solutions* provide additional information and web searches on GNSS will reveal other sources—especially for information on the newer satellite positioning systems. While not exhaustive, material in this chapter is intended to provide context for spatial data users who wish to gain a better understanding of GNSS satellite data and how these data are collected and used. Once physical observations are reliably processed to the point of being included in the GSDM as geocentric $X/Y/Z$ coordinates on a specific datum, then others should be able to use those spatial data with confidence.

Although the Russian GLONASS system and the European GALILEO system both deserve consideration, GPS is the primary satellite system discussed here. Coverage of GLONASS and GALILEO are included by extension of GPS concepts. This is done because the focus of this book is on the fundamental geometrical model for spatial data and the objective of this chapter is to describe how spatial elements and components are obtained from satellite data. As related to satellite geodesy and GNSS positioning, the following points are important:

*   As stated in Chapter 8 on datums, spatial data users encounter both absolute positions as represented by $X/Y/Z$ coordinates and networks of relative positions built from observed GNSS baselines. Absolute and relative concepts are both considered in this chapter.
*   Relative spatial data: Traditional ground-based measurements of angles and distances are used to determine relative positions. Spatial data components represented as vectors are derived from satellite data and computed indirectly from measurements of physical quantities such as voltage, current, time, and phase shifts. Direct measurements of angles and distances are rarely made in satellite geodesy or by GNSS surveying. Instead, by modeling the physical and geometrical circumstance of each measurement, vectors are computed from the raw data. Those vectors are analyzed for quality, adjusted as needed for geometrical consistency, and attached to previously "fixed" points. These new points are either used immediately or stored as coordinates in a database. Information from the database is then used to derive other quantities such as the distance and direction between points, areas, volumes, velocities, etc.
*   Absolute spatial data: Geocentric $X/Y/Z$ coordinates of a point are said to be absolute positions. If the coordinates of a point were computed on the basis of other known points and connecting vectors (from relative positioning), there may be correlation between the absolute positions of adjacent points. If, however, the coordinates were determined autonomously with a single instrument, the point position is independent of other adjacent points. In the past, such autonomous points were determined using a C/A code instrument measuring elapsed time of signal transmission from orbiting satellites. The nominal accuracy of such points was limited to approximately 5 meters. However, a newer technique called precise pointing positioning (PPP) uses precise timing and other sophisticated signal processing techniques to obtain an autonomous (independent) position rivaling the accuracy of vector-based points. Currently, the time of data collection for PPP is still much longer than that required for vector-based points.
*   When discussing layout for construction projects, the traditional procedure is to lay out angles and distances relative to control points previously established on-site. GNSS may have been used to establish those control points and construction drawings may show new features located with respect to such points. Traditional layout procedures may be used for years to come, but with the newer technology of GNSS base station networks and real-time kinematic (RTK) surveying procedures, the control monuments are effectively taken to be the satellites in the sky and, using satellite orbits as the control points, stakes are established according to construction drawings. Carried further, the construction stake is eliminated entirely because the design feature is defined in an electronic file, the file

is loaded into a computer carried on board the construction scrapper or bulldozer, a GNSS unit receives signals from the satellites, and the current position of the cutting blade is relayed to the onboard computer, which computes the difference between "actual" and "design" locations. The equipment operator, guided by an electronic display, cuts or fills according to the instructions shown on the display—all in the comfort of the heated or air-conditioned cab.

- Each spatial data component, whether measured directly or indirectly, has a standard deviation associated with it. And, if correlation exists between components, the appropriate covariance matrix should be part of the record for that point. Given that covariance information for each point is stored in a 3-D database, the standard deviation of any derived quantity is readily available via the stochastic model component of the GSDM. Correlation between points is also a consideration.

- The quality of spatial data is not established just because it came from GNSS or another measuring system. Metadata, statistical verification, quality control, or personal testimony may be needed to determine the quality of spatial data. However, once the quality of spatial data is determined, reliable information management procedures are essential for preserving the value of that information. The GSDM provides for efficient numerical storage of stochastic information for each point in the database.

- The GSDM provides a standard interface between two categories of activities; those efforts devoted to generating reliable spatial data and those efforts associated with using spatial data. Although some people undoubtedly operate in both categories, the goal in developing the GSDM was for the designers and builders of measurement systems to know specifically how far the processing needs to go until the spatial data can be stored and/or turned over to the user. By the same token, spatial data users deserve to know with assurance that data from a given measurement system and/or database conform to an underlying standard and that reliable information on spatial data accuracy is readily available.

## BRIEF HISTORY OF SATELLITE POSITIONING

It is generally agreed that the Space Age dawned in October 1957 with the launch of the Russian Sputnik satellite—the first to achieve Earth orbit. Scientists at John Hopkins University were able to track the signal broadcast by the satellite and, noting the Doppler Effect, were able to reconstruct the orbit of the satellite. The process was then inverted, making it possible to compute the position of a receiver on the ground by knowing the satellite orbit and observing the Doppler shift of the signals received from the satellites passing overhead. This was the basis for development of the U.S. TRANSIT satellite system, first funded in December 1958, operational in 1964, and released for commercial use in July 1967—all within a 10-year time frame.

Since the launch of Sputnik I in 1957, many spacecraft have been launched for various purposes; research, experimentation, communication, navigation, mapping, exploration, and, yes, even sending humans to the moon. An early tabulation of objects in orbit and decayed objects gives dates and other details of launch/decay for several hundred satellites between 1957 and 1963 (Mueller 1964, section 2.64).

With regard to geodesy, passive satellites and active satellites have both been used beneficially. The Echo 1 and Echo 2 satellites were large metallic balloons that could be seen from the Earth on clear nights in the early 1960s. Multiple precisely timed photographic exposures generated images (small dots) of the satellite moving across a background of stars. By synchronizing exposures recorded on glass plates at widely separated locations, it was possible to work out the geometrical separation of the observing stations. This ability to compute a geometrical connection from one continent to another represented significant progress for geodesy. Such use of satellites for positioning was sufficiently successful to justify the launch of a dedicated geodetic balloon satellite in 1966 called PAGEOS (*PA*ssive *GEO*detic *S*atellite) that was used to establish a geometrical worldwide network of 45 stations observed with BC4 cameras over a period of about 6 years.

Geodetic research has both contributed to and benefited from various satellite programs. But geometrical geodesy has probably benefited most from two satellite systems that were designed and built for navigation purposes—the TRANSIT system (NAVSAT) and the NAVSTAR global positioning system (GPS). In the early 1960s, the U.S. Polaris submarine fleet relied heavily upon inertial navigation instruments that needed periodic position updates, typically based upon astronomical observations. Such an update could not be done if the vessel was submerged or if clouds prevented the needed celestial observations. After the TRANSIT navigational system became operational in 1964, a position update could be obtained 24 hours a day, rain or shine, and, in the case of a submarine, only required exposure of a radio antenna instead of the vessel.

Decommissioned in 1996, the TRANSIT system was used extensively worldwide for both sea and land navigation until replaced by GPS. Recognizing that high-quality geodetic positions could be obtained from Doppler measurements, a number of companies developed instruments designed for land-based surveying applications. In addition to private sector use of Doppler positioning, the U.S. Bureau of Land Management (BLM) made extensive use of the Magnavox MX 1502 Doppler receiver during the 1970s and 1980s and, from 1973 to 1978, the National Geodetic Survey (Schwarz, 1989) collected Doppler data at approximately 600 stations in all 50 states in support of the realization of the North American Datum of 1983 (NAD 83). Additional information on development and use of the TRANSIT system can be found in Hoar (1982) and Stansell (1978).

The DOD first started development on the *NAV*igation *S*atellite *T*iming *A*nd *R*anging (NAVSTAR) satellite system in 1973 and called it the global positioning system (GPS). GPS was developed for the purpose of providing timely reliable positioning and navigation support for military activities all over the world. The first GPS satellite was launched in 1978 and, after completing a constellation of satellites that provided global coverage 24 hours per day, Initial Operational Capability (IOC) for GPS was declared on December 8, 1993. Full Operational Capability (FOC) was declared on July 17, 1995.

GPS consists of three segments—the space segment, the control segment, and the user segment. The space segment consists minimally of 24 satellites in six different planes; each plane inclines 55° to the equator. Each satellite orbits the Earth twice a day at an altitude of 20,183 km broadcasting information on two frequencies, L1 and L2. The L1 carrier frequency is 1575.42 MHz and has a wavelength of 19 cm, while the L2 carrier frequency is 1227.60 MHz and has a wavelength of 24 cm. Information about the satellite clock performance, the health/status of each satellite, and a prediction of (ephemeris for) each satellite orbit is modulated onto the carrier frequencies broadcast by each satellite. Credible information and operational details on GNSS can be found at Wikipedia and other web sites. See https://en.wikipedia.org/wiki/GNSS_applications, accessed May 4, 2017.

Although now containing 11 tracking stations, the GPS control segment was initially wholly owned, controlled, and operated by the U.S. DOD and included six tracking stations located around the world—Cape Canaveral, Florida; Hawaii in the Pacific Ocean; Ascension Island in the South Atlantic Ocean; Diego Garcia in the Indian Ocean; Kwajalein in the North Pacific Ocean; and Colorado Springs, Colorado. Data from all tracking stations are transmitted to the Master Control Station in Colorado Springs where the data are processed and the predicted orbit (ephemeris) for each satellite is computed. Ephemerides, clock corrections, and other messages are then transmitted back to each GPS satellite once a day (optionally more often) from one of the three upload stations located at Ascension, Diego Garcia, and Kwajalein. The WGS 84 ECEF coordinates of the electrical center of each antenna for the six tracking stations were determined for the epoch G1150 and were included in the addendum to a report published by the National Imagery and Mapping Agency (1997). As described in a follow-up "standardization document" by the National Geospatial-Intelligence Agency, the control segment now includes a total of 17 tracking stations and the coordinate position of each refers to the Antenna Reference Point (ARP) instead of the antenna electrical center as previous. The (G1762) Epoch 2005.0 ECEF coordinates of each station are listed in Table 10.1 (NGA 2014).

The user segment consists of all those persons and organizations that collect the GNSS signal from the satellites and use it for an increasing variety of applications. From the user perspective, GNSS is a passive system in that the end user only receives signals from the satellite. However, with the evolution of technology, it could be said that some GNSS receivers are active because the signal received at a given location is immediately rebroadcast to other GNSS units operating in the same local area for purposes of performing RTK surveys. For some continuously operating reference station (CORS) networks, the GNSS signal is collected and retransmitted via radio, cell phone, or the internet. Even so, GNSS is still considered a passive navigation, timing, and positioning system. Except for the possibility under consideration for newer systems, the end user does not transmit signals to the GNSS satellites.

During the 1980s, while the GPS constellation was still being developed, manufacturers began building receivers for civilian use. Two modes of operation were developed; one based upon the coarse acquisition (C/A) code modulated onto the L1 frequency and the other based upon observing the phase shift of the fundamental L1 carrier frequency as received at two separate locations. Using information broadcast by the GPS satellites, the autonomous location of a C/A code GPS unit

### TABLE 10.1
### ECEF Coordinates of DOD Global Control Stations

| Station Location | X (meters) | Y (meters) | Z (meters) |
|---|---|---|---|
| *Air Force Stations* | | | |
| Colorado Springs | −1,248,599.695 | −4,819,441.002 | 3,976,490.117 |
| Ascension | 6,118,523.866 | −1,572,350.772 | −876,463.909 |
| Diego García | 1,916,196.855 | 6,029,998.797 | −801,737.183 |
| Kwajalein | −6,160,884.028 | 1,339,852.169 | 960,843.154 |
| Hawaii | −5,511,980.264 | −2,200,246.752 | 2,329,481.004 |
| Cape Canaveral | 918,988.062 | −5,534,552.894 | 3,023,721.362 |
| *NGA Stations* | | | |
| Australia | −3,939,182.512 | 3,467,072.917 | −3,613,217.139 |
| Argentina | 2,745,499.034 | −4,483,636.563 | −3,599,054.496 |
| England | 4,011,440.890 | −63,375.739 | 4,941,877.084 |
| Bahrain | 3,633,910.105 | 4,425,277.147 | 2,799,862.517 |
| Ecuador | 1,272,867.304 | −6,252,772.044 | −23,801.759 |
| US Naval Observatory | 1,112,158.852 | −4,842,855.557 | 3,985,497.029 |
| Alaska | −2,296,304.083 | −1,484,805.898 | 5,743,078.376 |
| New Zealand | −4,749,991.001 | 520,984.518 | −4,210,604.147 |
| South Africa | 5,066,232.068 | 2,719,227.028 | −2,754,392.632 |
| South Korea | −3,067,863.250 | 4,067,640.938 | 3,824,295.770 |
| Tahiti | −5,246,403.943 | −3,077,285.338 | −1,913,839.292 |

could be determined within about 100 meters worldwide. In the early years of GPS, the DOD purposefully degraded the C/A code signal to deny ultimate accuracy to nonmilitary users. This policy of Selective Availability (SA) was discontinued May 1, 2000, and since then, the autonomous accuracy of a C/A code receiver worldwide is within about 10 meters. With enhancements such as augmentation or using differential corrections, submeter accuracy can be achieved using C/A code receivers.

On the other hand, it might seem that the carrier phase mode of GPS was developed specifically for surveying and mapping because, by using GPS, it is now possible to determine the location of an unknown point with respect to a known point very precisely (within mm or cm) even though the stations may be separated by 20, 50, or even 100 kilometers or more. Since 1985, the use of GNSS has completely revolutionized the surveying and mapping professions. Using GNSS, the surveyor no longer needs intervisibility between ground points and no longer needs to travel to a remote mountain top to use a control station listed in the NSRS. In fact, if working in an area covered by a GNSS CORS network (see Chapter 12), the modern surveyor takes a portable pole-mounted unit to any point having sky visibility and determines the position of a point within centimeters (or less) in real time. The overall point of this discussion is that by the time GPS was declared operational in 1993, civilian uses for the GPS signal had effectively outstripped military applications and now the utility of GNSS is taken for granted by many users. Depending upon the application,

GNSS receivers (code-phase, carrier-phase, or both) are now relatively inexpensive and are being used beneficially worldwide by novice and expert alike.

The Russians (former Soviet Union) have also built a satellite positioning system called Globalnaya Navigazionnaya Sputnikovaya Sistema (GLONASS) that is very similar to GPS but with specific differences. For example, the GLONASS satellites are closer to the Earth and their orbit period is slightly shorter than the GPS orbit period. GPS satellites all broadcast on the same two frequencies and each satellite is identified by a unique code. The GLONASS satellites broadcast on different frequencies. The first GLONASS satellite was launched on October 1982 and the system was declared operational on September 24, 1993. Various manufacturers either build or plan to build equipment capable of using signals from GPS, GLONASS, and/or GALILEO satellites.

In 2002, the European Space Agency committed to building, launching, and operating a satellite positioning system called GALILEO that will both complement and provide competition for the two existing systems. The first GALILEO satellite was launched in December 2005 and planned to be fully operational in 2014. Subjected to unanticipated logistical and budgeting problems GALILEO is not expected to be in service until 2020. GALILEO is being developed as a subscription-based system whereas GPS and GLONASS are supported respectively by the U.S. and Russian governments with no charges made for using the signals.

## MODES OF POSITIONING

Many variables need to be considered in using satellites for positioning. Some of the most obvious variables are the instantaneous location of each satellite, the time each signal is broadcast or received, the number of satellites broadcasting signals, the frequencies of signals being broadcast, the location of each receiver, the number of receivers collecting data, and the atmosphere through which the signals travel. Even with all those variables, GNSS positioning is ultimately based upon some combination of three physical concepts: elapsed time, Doppler shift, and interferometry.

### ELAPSED TIME

Distance is the product of rate and time. The time interval for the transit of a signal from a satellite to the receiver is measured very precisely. The speed of light (radio signal) is modeled for the intervening atmosphere and the distance from satellite to receiver is computed as the product of time interval and rate. Distances from a minimum of three satellites are needed to compute an intersection in three-dimensional space on or near the Earth. But, since a small correction to the receiver clock is also required, a minimum of four satellites must be observed—three for position and one for a clock correction. If signals from additional satellites are available, a position may be determined more quickly and with greater accuracy.

### DOPPLER SHIFT

As described by Christian Doppler (1803–1853) in the 1840s, the frequency received at a given place depends upon the frequency transmitted and whether the source and

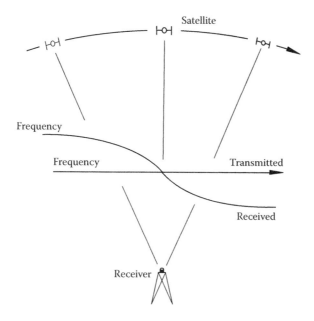

**FIGURE 10.1**    Example of Doppler Shift for Signal Received

receiver are stationary with respect to each other, moving closer together, or moving farther apart. Standing at a train station and listening to the whistle of a train going by is often used as an example. To a person riding on the train while the whistle sounds, the pitch heard is uniform and continuous (no movement between the source and the receiver). This is the trivial case. The sound heard by a person standing on the platform is of more interest. If the train is approaching the station, the frequency heard is higher than the frequency broadcast because the distance between the source and receiver is decreasing while the sound is traveling. On the other hand, if the train is going away from the station, the frequency heard is lower than the frequency broadcast. That is because the distance is growing larger while the sound is traveling from the source to the receiver.

GNSS satellites broadcast a steady precise frequency. As illustrated in Figure 10.1, the frequency received on the ground is higher or lower than the transmitted frequency depending upon the location of the observer and whether the distance between the satellite and observer is decreasing or increasing. Not surprisingly, the instant at which the two frequencies match—minimum distance between satellite and receiver—can be identified very precisely. This instant figures prominently in the calculations of the observer's position using Doppler data. But, it takes many passes of a TRANSIT Doppler satellite to determine an unknown position precisely. It is far easier to use Doppler data to verify the integrity of a computed solution than it is to compute an unknown position solely using Doppler data. Elapsed time and Doppler data are often used in combination to determine an autonomous position.

## INTERFEROMETRY

Using the concept that light (or radio signals) can be represented by a sine wave, interferometry is the term used to describe both the constructive and destructive interaction of two waves arriving simultaneously at a single location. When the signals are in phase, a double "high" is recorded. On the other extreme, when the phase of one signal is shifted 180° with respect to the other, the high of one signal cancels out the low of the other. The pattern of the changing phase shift, double-high-to-zero-and-back, is driven by a *difference* in the distance traveled by the signals and is itself a sine wave. Figure 10.2 illustrates the interference of light waves after passing through two slots in a barrier.

Interferometry concepts are used in processing GPS carrier frequency measurements. In general, the L1 signal is broadcast from a satellite and collected by a GPS receiver. The distance from a GPS satellite to the receiver is a huge number of 19-centimeter wavelengths plus a fractional part of a wavelength. The large number of integer wavelength is not known when the observations start and is called the integer ambiguity. However, as time progresses, the receiver keeps track of (counts) the wavelengths received and, at specified intervals (1 second, 5 seconds, 15 seconds, etc.), records the phase shift (fractional wavelength) of the incoming signal. If the signal is interrupted, it is said the receiver "loses lock" and the count of full wavelengths must start over. Such an interruption is called a cycle slip in the data. Uninterrupted signals collected simultaneously at two GPS carrier phase receivers over a period of time (e.g., 5–60 minutes) from four or more satellites are used to compute a three-dimensional space vector between receivers in terms of $\Delta X/\Delta Y/\Delta Z$ components in the ECEF geocentric coordinate system. And, because each 19-centimeter wavelength can be resolved into 100 fractional parts, it is said that the ultimate resolution of a carrier phase GPS measurement is about 1.9 mm.

As described in the next section, a *C/A code* (elapsed time) observation typically provides an economical autonomous *absolute* position with respect to a specific datum using one receiver. On the other hand, interferometric *carrier phase* processing requires data from two receivers and the resulting vector between points is a *relative* measurement of one point with respect to another. Although no longer necessarily true, carrier phase GNSS instruments have often been referred to as survey grade instruments and typically cost more than C/A code receivers.

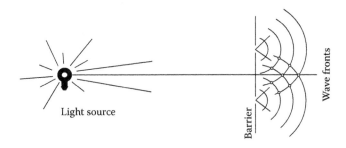

**FIGURE 10.2**   Illustration of Interferometry

## SATELLITE SIGNALS

Development of GPS grew out of a concept known as Very Long Baseline Interferometry (VLBI) which consists of collecting random pattern radio signals from distant quasars. Quasars are located extremely far from our solar system and emit radio waves that are received here on Earth. Two characteristics of interest are the pattern of signals received and the direction to the quasar. In order to determine the precise direction to a quasar, the same pattern must be received at widely separated locations and correlated according to their respective arrival times (as determined by very precise atomic clocks). Once the direction to a quasar is known, the process can be inverted to determine the 3-D space vector between receivers (Earth-based radio telescopes). Seeber (1993) reports a standard deviation of 1.7 cm on a 5,998 km baseline between the United States and Germany on data collected between 1984 and 1991.

GNSS is different from VLBI in that

- The source of the GNSS signal is not from a fixed direction in deep space but from one of many satellites orbiting the Earth
- The GNSS signal is not a random radio noise pattern but consists of very stable frequencies that are coded with important data related to the orbit, performance, and health of each satellite

Accurate timing is the heart of GNSS positioning. Jones (2000) describes development of atomic clocks: hydrogen maser (stable to $1:10^{15}$), cesium (stable to $1:10^{14}$), and rubidium (stable to $1:10^{13}$). Although hydrogen maser clocks demonstrate superior performance, cesium clocks are preferred over hydrogen maser clocks for their long-term stability, and rubidium atomic clocks are preferred for GNSS satellites due to their adequate performance, compact size, and comparatively lower cost.

GPS also provides an enormous benefit to everyone by providing continuous access to accurate atomic time all over the world. GPS time began at $0^h$ Coordinated Universal Time (UTC) on January 6, 1980, and progresses at the same rate as the International Atomic Time (TAI) scale. UTC is the time scale used all over the world by the general population. The rotation of the Earth on its axis is quite regular and was used as a time standard before the era of atomic clocks. Now however, because the rate of atomic clocks is more uniform than the rotation of the Earth, UTC is modified by a leap second from time to time to keep midnight at midnight. Therefore, GPS time differs from UTC by an integer number of seconds—17 seconds as of June 1, 2015, http://tycho.usno.navy.mil/leapsec.html, accessed May 4, 2017. Additional information on atomic time and time scales is also available from Leick (1990, 1995, 2004), Leick et al. (2015), Kleppner (2006), or a web search.

Details of the electromagnetic spectrum are listed in Table 10.2 and include gamma rays, X-rays, ultraviolet light, visible light, infrared rays, microwaves, and radio waves. The GNSS signal is located in the radio wave portion of the spectrum. During World War II, the radar band portion of the spectrum was assigned capital letter designators for various wavelengths. The range of frequencies from 1,000 to

**TABLE 10.2**

**Electromagnetic Spectrum**

| | Gamma Rays | X-Rays | Ultraviolet Rays | Visible Light | Infrared Rays | Radio Waves | | |
|---|---|---|---|---|---|---|---|---|
| | | | | | | (micro) | GPS L1/L2 | |
| Hertz | $10^{21}$ | $10^{18}$ | $10^{15}$ | | $10^{12}$ | $10^{9}$ | $10^{6}$ | $10^{3}$ |
| Wavelength | $10^{-14}$ m | $10^{-11}$ m | $10^{-8}$ m | | $10^{-5}$ m | $10^{-2}$ m | $10^{1}$ m | $10^{4}$ m |

2,000 MHz was called the L-band and includes the GNSS frequencies. The GPS signal structure is based upon a fundamental oscillator frequency of 10.23 MHz. The L1 frequency of 1,575.42 MHz (wavelength = 19.0 cm) is obtained as 154 times the fundamental frequency and the L2 frequency of 1,227.60 MHz (wavelength = 24.0 cm) is 120 times the fundamental frequency. Given the underlying carrier frequencies, the coarse acquisition (C/A) code is modulated onto the L1 frequency at 1.023 MHz and the precision (P) code is modulated onto both the L1 and the L2 carrier frequencies without alteration (i.e., at the original 10.23 MHz). The navigation message containing the broadcast ephemeris for each satellite—GPS time and other system parameters (health, etc.)—is modulated onto both L1 and L2 frequencies at a rate of 50 bits-per-second (bps).

The original intent was for all users to have access to the standard positioning service (SPS) based upon the C/A code on L1. The P-code was modulated onto both L1 and L2 frequencies to support what is known as the precise positioning service (PPS). But, access to the P-code was reserved for military users. In addition to civilian users not having access to the P-code, the L1 signal was purposefully degraded so that C/A code users could not count on autonomous stand-alone positioning accuracy any better than about 100 meters. This policy of selective availability (SA) was intended to provide the U.S. military access to the full capability of the GNSS system while denying ultimate functionality to others—especially users hostile to interests of the United States. As it turns out, this intent of SA was thwarted by invention and adoption of differential positioning techniques that obviated the impact of SA. Therefore, SA was formally discontinued on May 1, 2000, at the direction of the President of the United States.

A second security measure is known as anti-spoofing (A-S) that "guards against fake transmissions of satellite data by encrypting the P-code to form the Y-code." A-S was exercised intermittently through 1993 and implemented on January 31, 1994. Additional information on implementation of A-S is found at https://en.wikipedia.org/wiki/Selective_availability_anti-spoofing_module, accessed May 4, 2014.

Enhancements have been made to both the hardware and the signal structure since the first GPS satellites were built and launched. For example, Leick (2004), Leick et al. (2015) describes the P(Y) code as related to military uses and notes that manufacturers have developed proprietary methods that make the P(Y) code a nonissue for civilian uses. Some changes have to do with policy issues (e.g., S/A) and others are related to system performance. Changes of note for civilian users also include the addition of a new civil code on L2 (known as L2C) and a third civil frequency known as L5 that enhances robustness by mitigating effects of interference and supporting increased availability of precision navigation. Not surprisingly, some improvements have come about in response to competition from the two newer systems, GLONASS and GALILEO. A wealth of current information and status of all three GNSS systems is available via a web search.

## C/A Code

The observable for a C/A code phase receiver is the time delay of the signal to travel from the satellite to the receiver. The C/A code is time-tagged as it leaves the satellite and matched to a duplicate of the same code in the receiver by shifting the unique pattern on the receiver clock time scale. As illustrated in Figure 10.3, the shift required for the best match is a measure of the time interval of the signal from satellite to receiver. The distance from the satellite to the receiver is the product of the speed of light (corrected for delays) and the observed time interval. Theoretically, three such distances can be used to solve for the 3-D location of the unknown receiver. Regretfully, the signal does not travel in a vacuum and the clock in the receiver is not as precise as the atomic clocks in the satellites. Relativity is also a consideration. The time delay of the signal through the atmosphere is modeled to a close approximation (such details are beyond the scope of this book), but the receiver clock error must be observed and treated as an unknown. Therefore, four satellites must be observed to solve for four unknowns—the geocentric *X/Y/Z* coordinates of the antenna and a correction to the receiver clock. Note in Equations 10.2 through 10.5 that the receiver clock correction shows up as a correction to the distance to each satellite and is the same regardless of the satellite from which the signal originated.

The C/A code solution can be described as a 3-D application of the Pythagorean equation that states

$$Dist = \sqrt{\Delta X^2 + \Delta Y^2 + \Delta Z^2} \tag{10.1}$$

Using that form and writing an equation for the distance from the single receiver to each of four satellites gives

$$\text{To satellite A:} \quad D_A + \Delta t \times c = \sqrt{\left(X_A - X_R\right)^2 + \left(Y_A - Y_R\right)^2 + \left(Z_A - Z_R\right)^2} \tag{10.2}$$

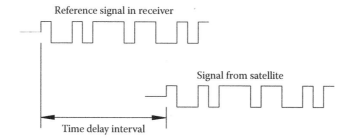

**FIGURE 10.3** Matching Signals from Receiver and Satellite

$$\text{To satellite B}: \quad D_B + \Delta t \times c = \sqrt{\left(X_B - X_R\right)^2 + \left(Y_B - Y_R\right)^2 + \left(Z_B - Z_R\right)^2} \quad (10.3)$$

$$\text{To satellite C}: \quad D_C + \Delta t \times c = \sqrt{\left(X_C - X_R\right)^2 + \left(Y_C - Y_R\right)^2 + \left(Z_C - Z_R\right)^2} \quad (10.4)$$

$$\text{To satellite D}: \quad D_D + \Delta t \times c = \sqrt{\left(X_D - X_R\right)^2 + \left(Y_D - Y_R\right)^2 + \left(Z_D - Z_R\right)^2} \quad (10.5)$$

where:
$D_A, D_B, D_C, D_D$ = observed distance to satellite A, B, C, D.
$X_A, Y_A, Z_A$ = known coordinates of satellite A.
$X_B, Y_B, Z_B$ = known coordinates of satellite B.
$X_C, Y_C, Z_C$ = known coordinates of satellite C.
$X_D, Y_D, Z_D$ = known coordinates of satellite D.
$X_R, Y_R, Z_R$ = unknown geocentric coordinates of receiver.
$\Delta t$ = correction to receiver clock.
$c$ = speed of light in a vacuum.
$\Delta t \times c$ = distance correction to each satellite (small).

With the observed distance to each satellite and the known $X/Y/Z$ coordinates of each satellite, there are only four unknowns in the four equations—$\Delta t$, $X_R$, $Y_R$, and $Z_R$. Equations 10.2 through 10.5 look innocent enough but using them to solve for the position as displayed by a GNSS receiver is no easy task. The GNSS receiver initially solves for geocentric $X/Y/Z$ coordinates, and then converts those values to latitude/longitude/height using the BK2 transformation equations.

Finding the autonomous position of a point with a small handheld receiver (or a chip in your cell phone) to the nearest 10 meters with respect to the Equator and the Greenwich Meridian is an incredible feat. But scientists and manufacturers are continually working to find ways to improve that performance. Differential corrections and augmentation are discussed in a subsequent section.

## CARRIER PHASE

The observable for a survey grade carrier phase GPS receiver is the fractional part of the 19-cm wavelength received at a specific time. The fractional parts (or phase shifts) recorded at two separate receivers from four or more satellites over time are used to compute a precise vector from one antenna to the other. The material discussed here is conceptual and the actual algorithms used by the various vendors involve many more details. Also, be aware that GNSS equipment designers do not restrict themselves to using just one principle or concept, but draw upon various techniques to build equipment that customers will purchase. For example, a vector (as used in surveying) from one point to another is determined using carrier phase observations but that is not to say that interferometry is the only concept used in the processing algorithm. Likewise, neither is it correct to say that carrier phase observations are never used in determining the autonomous position of a handheld receiver.

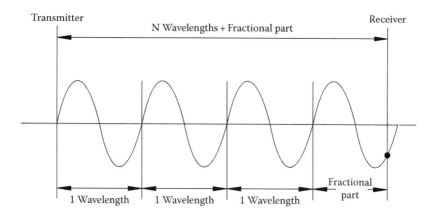

**FIGURE 10.4** Fundamental Wave Measurement

The fundamental concept exploited in carrier-frequency processing is that the distance between two points is represented by a sine wave. No matter the distance or the wavelength, there is always an integer number of waves and a fractional portion of the same wave as shown in Figure 10.4. When a GNSS measurement starts (a receiver "locks" onto a satellite), the observation is the fractional part of the wavelength. The integer number of wavelengths is unknown. But the receiver keeps track of (counts) the number of wavelengths as long as it maintains lock on the satellite and records the phase shift at specified time instants.

So long as a receiver maintains lock on a given satellite, the *change* in distance between the satellite and receiver is directly observed. But, if the signal is interrupted, a cycle slip is said to have occurred and the integer count for that satellite must begin again. When observing numerous satellites, a cycle slip on a given satellite may be inconsequential and may have little impact on the solution. But, in the early years of baseline processing, especially when attempting to get a good solution using a limited number of satellites, the process of fixing cycle slips typically involved user intervention and could be tedious and time-consuming. In some cases, fixing cycle slips was not possible and no solution could be obtained. In that case, the baseline had to be reobserved or eliminated from the solution. With the current full constellation of satellites and more robust automated baseline processing software, cycle slips are not the nuisance they once were. In fact, the modern user is often unaware of their occurrence.

## DIFFERENCING

The comment has been made that electrical engineers rule the world because nearly every aspect of modern life, in one way or another, relies upon electronic signal processing. That appears to be especially true with GNSS data. A common procedure when processing GNSS signals is to subtract one (sine wave) signal from the other. Various advantages are realized depending upon how the differences are formed and used. While details can be found in references such as Leick (2004),

Leick et al. (2015), Seeber (1993), Hofmann-Wellenhof et al. (2001), and CORPS (1990), a summary of differencing options includes the following.

## SINGLE DIFFERENCING

Three kinds of single differences are possible. First, and probably the most commonly used, the signal from one satellite is collected at a common time (epoch) by two different receivers. When these data are merged and differenced, the satellite clock error, orbit errors, and atmospheric delays cancel out because they are common to the signals received at both locations. Second, a single difference can be formed from the data collected at one receiver from two separate satellites. This difference combination effectively eliminates the impact of the receiver clock error. Thirdly, a single difference can be formed between epochs by a given satellite/receiver pair. The Doppler effect shows up in these data and is modeled accordingly.

## DOUBLE DIFFERENCING

A double difference is obtained as the difference between the single differences of a given type. Other double differences can be used, but when processing GNSS carrier phase data, the most common double-differencing procedure involves finding the differences between two receivers observing two satellites at the same epoch. This procedure effectively eliminates both the receiver clock errors and the satellite clock errors from the baseline solution.

A characteristic of such a double-difference solution is that the resulting wavelength ambiguity is an integer. That is, a baseline solution may be found quicker if, during the solution, the integer variable is allowed to take on a real number (known as a float solution). However, the unknown number of wavelengths in physical three-dimensional space is an integer and a stronger solution (known as a fixed solution) is obtained by constraining the solution to integer values for the number of wavelengths.

## TRIPLE DIFFERENCING

Triple differencing involves a difference of the double differences over time and is useful because the integer ambiguity cancels out of the observation equations. But, because of the time interval involved, the triple difference solution must also accommodate the Doppler effect. In the early days of GPS vector processing, triple-difference solutions were sometimes used, especially for longer lines where fixing the integers was problematic. However, with development of more robust processing algorithms for fixing integers and with using data from more than four satellites, triple difference solutions are seldom the final solution. But triple differencing remains a valuable tool for preliminary (or intermediate) solutions.

# RINEX

Of many things that could be said about the GNSS signals, this section includes a short description of the Receiver INdependent EXchange (RINEX) format.

Given more than one way to manipulate the signals received from the GNSS (and other) satellites, it is not surprising that data from one brand receiver is not necessarily compatible with data collected by other brands. Computing a baseline vector from data collected by Brand X receiver on one point and Brand Y receiver on another point is impossible without having the data in a common format. Therefore, the RINEX format was proposed as a convenience for both the manufacturers and GNSS users so that data collected by any of various brands of equipment could be combined in computing baselines or networks.

At a minimum, the RINEX format includes definitions for (there are more)

- Observation file
- Meteorological file
- Navigation file

A RINEX observation file uses ASCII characters and includes the basic observables, the phase measurement on both L1 and L2 (in cycles) and the range (in meters) to each satellite referenced to the receiver clock according to GPS (not UTC) time. The observation file header also contains information on the antenna type and the field-measured antenna height as determined by the user.

A web search on "RINEX" provides several hundred thousand hits but not all sites involve the use of satellite data. The site ftp://igs.org/pub/data/format/rinex302.pdf is a site containing excellent information on RINEX details, format, and history.

## PROCESSING GNSS DATA

Processing GNSS data involves a number of "flavors." In the early years of GPS, two primary modes of processing were code phase (autonomous) processing and carrier phase (vector) processing. Given the evolution of technology and practice, each "flavor" begins to look more like the other and exclusive descriptions become meaningless. The purpose here is to describe the two basic procedures and acknowledge the evolution of practices. Current literature (theoretical, technical, and promotional) is rich with detail. The point, as described below, is that the GSDM supports all geospatial data processing, including GNSS-derived data.

In years past, processing GNSS data involved a lot more user interaction than it does now. The computational load for the spatial data user has been significantly reduced and, in some cases, eliminated entirely. In the not-too-distant future, if not already being done, the collected data will be processed according to options and tolerances input by the user and the "final" answer will be available and used in real time. This is already happening with aircraft landing approaches, with GNSS-enabled Earth moving equipment, wide-area-augmentation for navigation, and, to some extent, with RTK surveying practices and procedures. Irrespective of the "flavor" and with regard to answers, two issues are involved: (1) getting an instant answer in units and datum expected and (2) being assured the answer meets the required spatial data accuracy. In terms of answers, the user must be knowledgeable of the options, must know specifically what kind of spatial data are expected, and must know what constitutes a finished task, job, project. In the case of an aircraft landing, the job is complete once the plane has landed and the tolerance for unacceptable results is very small. There are many other applications for real-time positioning—each with its own spatial data accuracy criterion. In each case, the GSDM provides for comparison of "actual" and "desired" differences and provides an efficient procedure for assessing the accuracy of such comparisons (Burkholder 1999).

## SPATIAL DATA TYPES

Spatial data types are listed in Chapter 3 and most of them relate to geospatial data as obtained from GNSS. Several important points of which each user should be aware are as follows:

1. The data types are independent of whether the observations were collected with GPS, GLONASS, or GALILEO.
2. The underlying spatial data types are also independent of the mode (code phase or carrier phase) by which the data were collected and processed. But, the covariance matrix associated with each point is greatly influenced by the mode of collection and processing. Generally, the carrier phase–based data will have smaller standard deviations than the code-phase-derived data.
3. The GSDM handles all spatial data the same—independent of the mode of collection (code phase or carrier phase) or whether the data were collected and processed on the NAD 83, WGS 84, or ITRF.
4. The GSDM does not move data from one datum to another. But neither does the GSDM discriminate. For example, a user could input Point A as defined on the NAD 83 and Point B as being on the WGS 84. Clearly that is not appropriate. But, if the standard deviation of either (or both points) is significantly larger than the known datum difference, then the 3-D inverse between Point A and Point B is legitimate *at the level of uncertainty derived from the 3-D inverse*. Used that way, the GSDM can be a dangerous tool.

It is presumed that position coordinates are associated with a named datum and represent the GNSS antenna location. Of course, a known offset may also be involved. Such an offset may be from the antenna to the cutting edge of a bulldozer or earthmover or, in the case of surveying, the offset is to the mark on the ground determined via the antenna height measurement. Offset measurements are the user's responsibility.

The seven spatial data types from Chapter 3 are as follows:

1. Absolute geocentric *X/Y/Z* coordinates—primary values stored in a 3-D database.
2. Absolute geodetic coordinates of latitude/longitude/height—derived from the stored *X/Y/Z* primary values.
3. Relative geocentric coordinate differences—this is the primary result of GNSS baseline vector processing. Relative geocentric coordinates are also obtained from local geodetic horizon coordinate differences that have been rotated into the geocentric reference frame.

4. Relative geodetic coordinate and ellipsoid height differences—obtained as the difference of compatible (common datum) geodetic coordinates.
5. Relative local coordinate differences—these are the local components of a GNSS vector rotated to the local *e/n/u* perspective. These local components are also the product of surveying total station observations and can often be used as flat-Earth components in the local tangent plane with an origin as selected by the user.
6. Absolute local coordinates—*e/n/u* are distances from some origin whose definition may be mathematically sufficient in three dimensions, two dimensions, or one dimension. Examples (see comments in Chapter 3) are
   • Point-of-Beginning (P.O.B.) datum coordinates as defined by the GSDM
   • Map projection (state plane) coordinates
   • Elevations on some named datum
7. Arbitrary local coordinates may be 1-D (assumed elevations), 2-D (assumed plane coordinates), or 3-D (spatial objects). Although useful in some applications, arbitrary local coordinates are generally not compatible with geospatial data and have limited value in the broader context of georeferencing.

Often, the type of spatial data expected will be dictated by the context. For example, a graphical display of local coordinate differences may be used to navigate to a point. It is one thing to return to camp following a hike in the woods (based upon code-phase observations) and something completely different to navigate to a point to be staked for construction or to find a fire hydrant buried under a snow bank (based upon RTK surveying procedures). Of course, the user may select a different type of data to be displayed (e.g., latitude/longitude/height, plane coordinates, etc.). In any case, the spatial data user is responsible for knowing specifically what to expect and verifying that the data being obtained are those expected.

Back to those automated aircraft landings—a web search on "GPS aircraft landings" returns over 600,000 hits. Innovative applications and research opportunities involving spatial data abound.

Admittedly, the following descriptions of "autonomous" and "vector" processing are over simplified but they can be useful for learning more about underlying concepts of how GNSS data are processed and used.

## Autonomous Processing

Autonomous processing of GNSS data has progressed remarkably during the past 10 years—primarily out of sight of the consumer. There is an untold number of GNSS-enabled devices that keep track of, and even report, the location of the sensor. While many concepts of "smart device" behind-the-scenes applications overlap with intentional use, this section was originally written to provide information about how GPS signals are transformed into a location of immediate interest to the user. In that context, the autonomous processing is performed within a C/A code-phase receiver and used for a variety of applications—recreation, navigation, tracking, and GIS. Once turned "on" and receiving data from a minimum number of satellites, an autonomous position is obtained from a single GNSS receiver with little or no input by the user. The results are displayed in the coordinate system, datum, and units as specified by the user. Internally, the GNSS signals are processed to find the geocentric $X/Y/Z$ coordinates of the antenna (spatial data type 1). But, since those rectangular $X/Y/Z$ values are difficult to visualize, the results are converted into the coordinate system as selected by the user (typically spatial data type 2—latitude/longitude/ellipsoid height). Other options may be available such as specifying a datum, a state plane coordinate zone, UTM coordinates, a national grid designation, or other user defined system. User-selectable options for a handheld C/A receiver typically include

- Datum
    - World Geodetic System 1984 (WGS 84)
    - North American Datum of 1983 (NAD 83)
    - International Terrestrial Reference Frame (ITRF), epoch XX
    - Other
- Units
    - Meters (standard)
    - International Foot
    - U.S. Survey Foot
    - Statute Mile
    - Nautical Mile
    - Other
- Display
    - Geocentric Earth-centered Earth-fixed (ECEF) $X/Y/Z$
    - Degrees, minutes, seconds (decimal)
    - Degrees and minutes (decimal)
    - Decimal degrees
- Time
    - Coordinated Universal Time (UTC)
    - Time zone offset
    - 24-hour mode or 12-hour mode with AM and PM

With SA enabled, routine accuracy expected from a C/A code receiver was originally plus/minus about 100 meters. But, SA was discontinued in May 2000 and the expected autonomous accuracy dropped to about 10 meters. That is impressive, but, using advanced processing techniques, vendors routinely claim submeter accuracy even for nonsurvey grade GNSS receivers.

## VECTOR PROCESSING

When processing carrier phase GNSS data, two separate receivers are needed to collect data from common satellites. The simplest form of vector processing uses GNSS signals recorded simultaneously in two receivers located at two different points. The data file and station message file from each receiver and an ephemeris file from one of the receivers are transferred to a computer having the appropriate processing software. Using those data, the baseline between stations is computed and reported in terms of $\Delta X/\Delta Y/\Delta Z$ between stations (spatial data type #3). The covariance matrix for the baseline is also computed and reported. *The ECEF coordinate differences and their covariances are the primary answers obtained from baseline processing.* Other pieces of information such as local component differences $(\Delta e/\Delta n/\Delta u)$, slope distance, ellipsoid height difference, forward geodetic azimuth, back geodetic azimuth, and standard deviations of the various answers can be derived from those primary answers.

Depending upon which of several options are used, various results can be obtained when computing GNSS baselines. Typically, a preliminary triple difference solution is computed first, even for shorter baselines, because the integer ambiguity value cancels out of the observation equation. If the baseline is quite long, cycle slips may be more of a problem and a triple difference solution may be an important intermediate step in determining a preliminary solution. However, once cycle slips are identified and fixed, a double difference solution is typically better than a triple difference solution. But, there are two levels of double difference solutions. The first double difference solution solves for the integer ambiguities as a real number. This is called a float solution because the values found for the integer wavelength numbers are allowed to "float" and to be something other than an integer. There are cases where the float solution may be the strongest solution available for a given set of data. But, the preferred solution is obtained when the integer ambiguities are "fixed" and forced to be integers—because, physically, the path of the signal contains an integer number of wavelengths plus a remainder. Modern baseline processing software is quite robust and, unlike the early days of GPS baseline processing, the user rarely needs to interject decisions as the processing proceeds. However, once baseline processing is complete, networks are formed, adjustments are performed, and statistics of the results are developed, the judgment of the user is still important in deciding upon the ultimate acceptability of the results. If the results are not acceptable, corrective measures may include actions like changing the elevation mask, eliminating signals from (turning off) a particular satellite or reobserving a given baseline. Vendor manuals and other GNSS users offer many suggestions for various corrective actions.

## MULTIPLE VECTORS

Added complexity in carrier phase processing arises if data are collected simultaneously with three or more receivers. Although the underlying algorithm for computing $\Delta X/\Delta Y/\Delta Z$ components for each baseline may be the same, when processing multiple baselines, two additional considerations are (1) avoiding the use of trivial vectors and (2) handling the correlation between baselines sharing a common station. A nonexclusive definition of a trivial baseline is any vector computed using two data sets that have already been used in computing another baseline. There are different ways of determining a trivial baseline but perhaps the easiest way is to look specifically at nontrivial vectors. A nontrivial vector is a baseline computed using, at least in part, data not used previously. If data collected at both ends of a line have been used previously, the result may be a trivial vector. If data at one end has been used

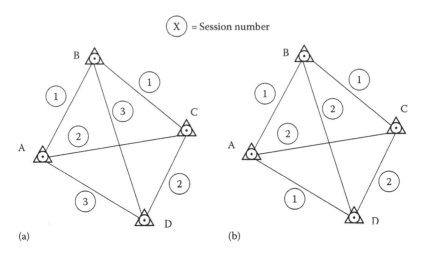

**FIGURE 10.5** GNSS Receivers and Observing Sessions for Nontrivial Vectors. (a) Using Three Receivers and (b) Using Four Receivers

but data at the other end have not been used, it is a nontrivial vector. Many baseline processing packages permit the user to choose the nontrivial baselines, but default choices built into most processing software are also popular.

Figure 10.5 shows two views of a network of four points to be surveyed using GNSS. The points are labeled A, B, C, and D and GNSS baselines between all points are represented by the six connecting lines. Scenario 1 is not illustrated but view (a) illustrates using three GNSS receivers and three observing sessions for the six nontrivial vectors while view (b) illustrates using four receivers and two observing sessions for the same six vectors.

*Scenario 1*: Two GNSS receivers are used to occupy each baseline in six separate sessions. It is a tedious and time-consuming method of GNSS data collection, but each observed baseline is a nontrivial vector.

*Scenario 2*: Three GNSS receivers are used simultaneously and two nontrivial vectors are obtained in each of three sessions. For example, in session 1, receivers occupy stations A, B, and C. Nontrivial vectors are AB (two new data sets) and BC (one new data set at C). For session 2, receivers occupy stations A, C, and D. Nontrivial vectors are CA and CD. In session 3, receivers occupy stations D, B, and A and the nontrivial vectors are DB and DA. Other combinations could also be used so long as each new vector includes station data not used previously.

*Scenario 3*: Four GNSS receivers are used simultaneously and three nontrivial vectors are obtained in each of two sessions. For example, in session 1, nontrivial vectors could be AB (two new data sets), BC (one new data set at C), and AD (one new data set at D). For session 2, the nontrivial vectors are CD (two new data sets), CA (one new data set at A), and DB (one new data set at B).

## TRADITIONAL NETWORKS

The simplest network scenario is to build a GNSS network of independent vectors. That would be the case if a network was built using independent baseline data collected from just two receivers. When combining vectors and computing such a multisession network adjustment, it is standard practice to anchor the network to a single fixed 3-D ($X/Y/Z$) point and to compute a minimally constrained least squares adjustment. The purpose of computing a minimally constrained network is to confirm the absence of blunders. If blunders exist, the offending baseline is reprocessed sans the blunder, reobserved, or eliminated entirely. Once the observed baselines are verified blunder-free, the network is constrained to the $X/Y/Z$ values of two or more existing high (or higher) order stations and the network is recomputed. A successful least squares adjustment provides adjusted $X/Y/Z$ coordinates (and covariance matrix) for each new station established during the survey. Such a network is both practically and statistically legitimate. A well-documented least squares network adjustment of independent vectors is posted at http://www.globalcogo.com/nmsunet1.pdf, accessed May 4, 2017.

As illustrated in Figure 10.5 where three or more GNSS receivers collect data simultaneously, there is correlation between those nontrivial vectors sharing a common station. When developing the weight matrix for a multistation session, the correlation between baselines should also be included in the covariance matrix of the observations. The network adjustment then becomes a multistation multisession solution, which is described in more detail by Seeber (1993). Examples include adjustment of a statewide HARN network, the national readjustment of the NAD 83 datum network, or, ultimately, even a global network of GNSS stations.

Whether building a network of independent vectors or building a multistation multisession network, the GSDM provides a "natural" computational environment for performing those computations. The final coordinates (spatial data type 1) from a least squares adjustment are absolute positioning values based upon the fixed absolute control selected by the user (also spatial data type 1) and the observed relative baseline vectors (spatial data type 3). The least squares adjustment also provides the covariance matrix for each point and the correlation matrix between points. A BURKORD™ database stores the $X/Y/Z$ coordinates for each point, the covariance matrix for each point, and (optionally) the correlation between named point pairs.

## ADVANCED PROCESSING

The concepts presented here are advanced in that they move beyond the fundamental procedures described earlier. But they are not advanced in that GNSS professionals routinely work with many of these "advanced" concepts. First, there are several issues deserving consideration that are omitted. It is well known and understood that the quality of long baselines can be enhanced by using a precise ephemeris instead of the (default) broadcast ephemeris. Fine-tuning the baseline computational process is left to others. And, in some cases, GNSS data are comingled with inertial data. That discussion is also very relevant but beyond the scope of this book. However, a closing thought is that once inertial data are "corrected" for deflection-of-the-vertical (see Chapter 9), even gravity-based measurements (total stations, levels, or an inertial measuring unit) can be used along with appropriate standard deviations in the GSDM.

Issues of absolute and relative spatial data need to be considered in moving from traditional network adjustments to procedures such as differential positioning, augmentation, RTK positioning, and using the NGS Online User Positioning Service (OPUS). It would be nice if absolute and relative issues could be considered separately but as technology and operational procedures evolve, the simple categories are no longer applicable. Admittedly it doesn't matter for many because, in all honesty, many end users are really interested in obtaining a reliable position with appropriate statistics without needing to worry about how it was obtained.

Differential GPS positioning, called DGPS, involves observing an autonomous position at two locations—one known, the other unknown. The difference of "known" minus "observed" at the known location is called a correction. This correction is applied at the unknown location to give a differentially corrected position. The process of applying the correction can take any of various forms. An abridged list of possible scenarios (some give better results than the others) includes the following:

1. One receiver is used by one person at multiple locations. One of the locations visited is the known point. The correction is computed and applied manually. The assumption is that the correction does not vary over (short periods of) time or with location.

2. Two GPS receivers are used in the field. One receiver remains on the known base station location while a second receiver (rover) is taken to various points. The correction as determined at the base is applied (manually) for each point visited. This procedure documents whether the correction is "fixed" or whether the correction varies with time. This procedure is better than the first but not very efficient, especially if security considerations mandate that a person stay with the base receiver.

3. The location of a permanent (secure) base station is surveyed precisely and the base station receiver records data continuously. A second receiver (rover) is used by one person to collect data at various points. Back at the office, data from the base station is retrieved and the correction is applied to points collected with the roving receiver. Software for automated processing increases efficiency.

4. The correction is computed automatically at the base station and transmitted via radio or cell phone from the base to the remote receiver. The "corrected" position at the rover is determined in real time and used immediately or stored for use elsewhere or by others.

5. The correction at the base is determined, not as a location difference, but as a time delay modification to the signal received from each satellite. The correction sent to the remote is not a location difference but a time-delay correction from each observed satellite. The DGPS position observed at the remote station is better than that obtained with other methods and is achieved with greater computational efficiency.

General statements are that DGPS provides an improved absolute position obtained from a C/A code receiver. The quality of such an observed position is enhanced by observing more than four satellites, by using the strongest geometry afforded by the existing satellite constellation, by observing for a longer time period, by observing the same point on at least two different occasions, and by using a different receiver for subsequent observations.

Augmentation can be described as DGPS on a big scale. The Federal Aviation Administration (FAA) has established a system of Wide Area Augmentation System (WAAS) for determining differential corrections at ground-based locations and transmitting that information to geosynchronous communication satellites that retransmit the signals over a large area—most of the United States. WAAS provides absolute position in real time and was designed primarily to support civil aviation. An informative web site for additional information on WAAS is https://www.faa.gov/about/ office_org/headquarters_offices/ato/service_units/techops/navservices/gnss/waas/, accessed May 4, 2017.

A summary of approximate *absolute* positions based upon code-phase receivers is

1. Original C/A code position with Selective Availability imposed: 100 m
2. C/A code position without Selective Availability imposed: 10 m
3. Typical DGNSS: 3–5 m
4. Typical WAAS positional accuracy: <3 m

The reader should also be aware that some manufacturers claim submeter accuracy for their autonomous GNSS receivers. Such accuracy may be obtained by exploiting information on the carrier frequency in addition to the code-phase data. Those claims, equipment, and observational processes should be discussed with the manufacturer. Tests can be conducted to establish the veracity of such claims.

RTK surveying procedures are based upon processing carrier-phase data received at two separate locations and provide a *relative* baseline vector between the two points. Data must be collected simultaneously and combined to compute a vector. One scenario (kinematic surveying) is to collect data with two or more receivers (one receiver on a known point) and bring the data back to the office for processing. But, if the raw carrier-phase data at the base station are transmitted to the remote unit, then the processing can take place at the remote unit in real time. Given further that the base station occupies a precisely surveyed point, then the position of the remote unit can be computed within mm or cm accuracy in real time. Here too, the answer is an absolute value (coordinates), but those coordinates are based upon a *relative* vector tied to the known base station position. One drawback of RTK surveying is lack of redundancy. This can be overcome by occupying the unknown point a second time or by computing a baseline to the unknown station from a second base station. Many localities either have or are in the process of establishing GNSS base station networks to cover specific regional areas and building a real-time network (RTN) such that precise RTK surveying can be conducted with one unit and one person. RTNs are discussed later in this chapter.

OPUS is another processing option that is available via the internet from the NGS. At least 2 hours of dual-frequency data are collected on an unknown point. The data are submitted via the internet to NGS along with antenna type and measured antenna height. Shortly, often within a matter of minutes, an email is returned to the sender and contains the position of the point computed on the basis of the CORS network in the general region. The answer is reported in both geocentric $X/Y/Z$ and conventional latitude/longitude/ellipsoid height and in both NAD 83 and ITRF coordinates. An estimate of the accuracy is also provided.

The NGS also supports a service called OPUS-RS (rapid static) that will provide a solution with as little as 15 minutes of static GNSS data. Details are available at the NGS web site (https://www.ngs.noaa.gov/OPUS/about.jsp; accessed May 4, 2017).

The GSDM fundamentally supports all varieties of GNSS processing including both absolute and relative considerations, NAD 83 or ITRF (not at the same time), and any GNSS. Furthermore, the GSDM handles spatial data accuracy for all types of spatial data. See Chapter 14 for a discussion of spatial data accuracy.

## THE FUTURE OF SURVEY CONTROL NETWORKS—HAS IT ARRIVED?

One goal in writing this book was to focus on fundamental concepts and characteristics of digital spatial data rather than the status of current technology. Writing a section predicting the future can be foolhardy at best—even with disclaimers. The views expressed in the first edition are reprinted in the following paragraphs. Although much of that information remains relevant, it seems that the future has arrived even quicker than anticipated. An update section is included at the end of this chapter. In time, those details and predictions will also be judged with 20/20 hindsight. Admitting an element of self-fulfilling prophecy, the aspiration is that the value of the underlying GSDM as a foundation for the spatial data infrastructure will be realized worldwide.

There are many speculative predictions about GNSS technology, the use of spatial data, and the future of survey control networks. Some parts of the speculation are already possible though not necessarily commonly realized in practice. However, even modest extrapolating exposes exciting possibilities for many spatial data users.

At the risk of rubbing the crystal ball too hard, the future of monumented control points is bleak and limited for control surveys. Using existing technology, it is possible, and becoming ever more practical, to determine a position anywhere in the world based upon signals received from satellites in the sky. Under that scenario, it is said the satellite orbits become the permanent reference monuments—see items 1 and 2 following. Of course, old habits die hard and, to some extent, there will always be a demand for a reliable physical point of beginning (P.O.B). Even so, the following issues need to be considered when evaluating access to the NSRS:

1. Satellite visibility (or lack thereof) will obviate some of the following points.
2. Traditional relative positioning methods will serve as a back-up if and when GNSS technology is not available or appropriate.
3. However, GNSS technology and equipment are already being used to establish locations all over the world within impressive levels of tolerance. Yes, it takes more sophisticated equipment and more exacting observing procedures to position a point within 1 cm than it does to position a point within 1 meter. Are smaller tolerances feasible? Yes.

4. Realized (surveyed) positions may be absolute or relative. In some cases, they may need to be both.

   a. An absolute position is given by the coordinates of the surveyed location. Many applications are satisfied with absolute data. Examples include answering questions such as—where is the point source pollution, where did that accident occur, where is the defective transformer, and other inventory-related questions.

   b. A relative position is important in other applications where the user needs to know a location with respect to other points. How far is it to my destination? How far have I traveled today? How far is the back of the curb from the right-of-way line? What is the direction/distance from one property corner to the next? How far is it from the airplane to the runway (automated landings)?

5. The results of a survey may be consumed instantaneously or archived for future use. The value of an instantaneous position may be the comfort of knowing where I am—no one with a GNSS unit ever needs to admit to not knowing where they are. On the other hand, information archived for future use may involve storing the waypoint of the trail head so I can return to my vehicle at the end of a hike. I may need to go back to a specific location so I can capture accurate "before" and "after" photographs. Or, I may need to return to the site of a recently discovered petroglyph or an underwater wreck that has been there for many years.

6. In order for a coordinate position to be meaningful, it needs to be compared with a previous value. Whether a hiker, photographer, anthropologist, or maritime archeologist (as noted above) the comparison may be casual, time-delayed, or approximate. However, the stakes are higher when using GNSS to land an airplane, to fly an unmanned aerial vehicle (UAV), or to control the movement of a driverless car. The location (3-D representation) of the runway—or other local features—must be in the database (computer memory), and the instantaneous position of the aircraft must be compared with values in the database to determine the separation between the aircraft and the runway. Whether landing an aircraft, using a UAV, or controlling a driverless car, accuracy is critical and spatial separation changes rapidly. Three important considerations are as follows:

   a. The accuracy of data in the database must be of proven quality—see item 7 following.

   b. The accuracy of the observed instantaneous position must be realized time wise within an acceptable tolerance.

   c. The comparison needs to occur instantly and answers must be available in real time.

   Understandably, all three issues become moot once the aircraft is safely on the ground. But relative position and spatial accuracy remain critical on a continuous basis when using an UAV or while controlling a driverless car.

7. In surveying, and other related applications, real-time considerations may be less critical than when landing an airplane, but relative and absolute positions still need to be considered along with spatial data accuracy issues.

Of course, as technological improvements keep coming, the tolerances will become smaller, the comparisons will become more economical, and there will be many more applications.

In the case of driverless cars, the issues of location and spatial data accuracy are essential for the safety of all concerned. In one scenario, the absolute location is reliably determined in real time and compared to an equally reliable database of absolute positions. Relative positions and relative (local) accuracy are both computed from absolute values. However, in addition to determining absolute values, current sensor technology excels at making relative measurements to other (adjacent) objects. In the extreme case, no absolute positions or accuracy are determined and the relative location of the vehicle is based solely on relative measurements. In most cases, a combination of absolute and relative data is used in controlling the motion of a driverless vehicle.

8. Especially with regard to high-end applications and comparisons, data in the database must be compatible with the position derived from satellites orbiting the center of mass of the Earth. Scientists, engineers, and manufacturers build equipment that determines the location of the receiver/antenna. That is only half of the solution. The quality of a relative location depends heavily upon the quality of information in the database. Three possible kinds of information in the database include the following:

   a. Design locations (virtual) are stored in the database and represent errorless quantities.

   b. Staked positions (intended) are marked on the ground in accordance with design locations using equipment/procedures capable of providing answers within a given tolerance. The error (spatial data accuracy) of such location is determined by the equipment and the procedures used during the layout process.

   c. Surveyed location (actual) is a measurement of the position of a marked point. In this case, the quality of the information in the database reflects both the integrity with which the point (monument) was established and the quality with which it was re-observed. Ideally those two error sources should both be controlled and "small." However, it could be that either the staked location of the point or the surveyed location of the point as stored in the database contains a large error. In either case, regardless of how good the current instantaneous position might be, a subsequent comparison with the value stored in the database will be bogus.

A huge caveat to this entire discussion is the understanding that all data, regardless of their quality, must be on the same datum. If database positions are not expressed in the same datum as the currently observed satellite position, then computed relative positions may contain unacceptable error.

The overall point of this chapter is that GNSS computations expressing location, whether conducted on ITRF, WGS 84, or NAD 83, can and should be accomplished in the ECEF environment of the GSDM. Two compelling justifications are that high-level scientists and "flat-Earth" end users can both use the same rectangular solid geometry equations and that spatial data accuracy is easily established, tracked, and made available in terms of the GSDM.

## REFERENCES

Burkholder, E.F. 1999. Spatial data accuracy as defined by the GSDM. *Surveying and Land Information Systems* 59 (1): 26–30.

Burkholder, E.F. 2003. The digital revolution begets the global spatial data model (GSDM). *EOS, Transactions, American Geophysical Union* 84 (15): 140–141.

Hoar, G.J. 1982. *Satellite Surveying: Theory, Geodesy, Map Projections: Applications, Equipment, Operations.* Torrance, CA: Magnavox Advanced Products and Systems.

Hofmann-Wellenhof, B., H. Lichtenegger, and J. Collins. 2001. *GPS: Theory and Practice*, 5th ed. Vienna, Austria: Springer-Verlag.

Jones, T. 2000. *Splitting the Second.* Boston, MA: Institute of Physics Publishing.

Kleppner, D. 2006. Time too good to be true. *Physics Today* 59 (3): 10–11.

Leick, A. 1990. *GPS Satellite Surveying.* New York: John Wiley & Sons.

Leick, A. 1995. *GPS Satellite Surveying*, 2nd ed. New York: John Wiley & Sons.

Leick, A. 2004. *GPS Satellite Surveying*, 3rd ed. Hoboken, NJ: John Wiley & Sons.

Leick, A., L. Rapoport, and D. Tatarnikov. 2015. *GPS Satellite Surveying*, 4th ed. Hoboken, NJ: John Wiley & Sons.

Leveson, I. 2015. The economic benefits of GPS. *GPS World* 26 (9): 36–42. http://gpsworld.com/the-economic-benefits-of-gps/, accessed and validated May 01, 2017.

Mueller, I.I. 1964. *Introduction to Satellite Geodesy.* New York: Frederick Ungar.

National Academy of Public Administration (NAPA). 1995. The global positioning system: Charting the future. Report by a Panel of the National Academy of Public Administration and a Committee of the National Research Council for the U.S. Congress and the Department of Defense. Washington, DC: National Academy of Public Administration.

National Academy of Public Administration (NAPA). 1998. Geographic information for the 21st Century: Building a strategy for the nation. Report by a Panel of the National Academy of Public Administration for the Bureau of Land Management, Forest Service, U.S. Geological Survey, and National Ocean Service. Washington, DC: National Academy of Public Administration.

National Imagery and Mapping Agency (NIMA). 1997. Addendum to NIMA TR 8350.2: Implementation of the World Geodetic System 1984 (WGS 84) Reference Frame G1150. In Department of Defense World Geodetic System 1984: Its definition and relationships with local geodetic systems. Technical Report TR8350.2, 3rd ed. Washington, DC: National Imagery and Mapping Agency. http://earth-info.nga.mil/GandG/publications/tr8350.2/Addendum%20NIMA%20TR8350.2.pdf, accessed May 4, 2017.

Schwarz, C.R., ed. 1989. North American Datum of 1983. NOAA Professional Paper NOS 2. Rockville, MD: National Geodetic Survey, National Ocean Service, National Oceanic and Atmospheric Administration.

Seeber, G. 1993. *Satellite Geodesy: Foundations, Methods, and Applications.* Berlin, Germany: Walter de Gruyter.

Stansell, T.A. 1978. *The TRANSIT Navigation Satellite System: Status, Theory, Performance, Applications.* Torrance, CA: Magnavox Government and Industrial Electronics.

U.S. Army Corps of Engineers. 1990. *Engineering and Design—NAVSTAR Global Positioning System Surveying.* Washington, DC: Department of the Army, U.S. Army Corps of Engineers.

# 11 Map Projections and State Plane Coordinates

## INTRODUCTION: ROUND EARTH—FLAT MAP

A map projection is a 2-D model whereby the curved surface of the Earth is portrayed on a flat map. If one looks only at a small portion of the surface of the Earth, such as a city map, it appears that local features on the map correctly portray the same features as viewed by a person walking or driving in that area. However, when dealing with larger and larger portions of the surface of the Earth, distortion and the challenge of true map representation grow exponentially. An extreme example of distortion is recognized by many elementary school children who look at a comparison of Alaska and Brazil—first on the globe, then on a Mercator world map. On the globe, Alaska appears noticeably smaller than Brazil but on a Mercator map projection of the world, Alaska appears much larger than Brazil. The problem is that, on the globe, all meridians converge at the North Pole and South Pole. On the projection at the Equator, the spacing of the meridians is identical to the meridian spacing on the globe. But the meridians remain parallel on the projection and features nearer the poles appear grossly exaggerated in size. In fact, neither the North Pole nor the South Pole can be shown on a Mercator map, which touches the Earth at the equator as shown in Figure 11.1a.

A spherical Earth is shown in Figure 11.1a with rays originating at the center and piercing the globe at each 15° latitude before striking the cylindrical surface of the Mercator projection. Following such graphical projection, the cylinder is cut down the back and rolled out flat to give the appearance of the graticule shown in Figure 11.1b. Notice that the 15° blocks of latitude and longitude near the Equator are nearly square but that the same 15° blocks become elongated further from the Equator. An area lying near either pole is grossly exaggerated in size when shown on such a Mercator projection.

Cartography is the science of making maps and includes various graphical portrayals of spatial data. Using cartographic definitions, a graticule is the grid-like appearance of parallels and meridians covering the Earth and a map projection is defined as a systematic arrangement of the graticule on a flat surface. The challenge is going from a curved surface to a flat map without distorting any geometric element. It can't be done. Most people know that when you peel an orange (even children enjoy making those pieces as big as possible), the curved peel will not lay flat on a table unless one presses it flat. In so doing, the peel is distorted. Either the peel tears or other parts of the peel are artificially compressed in the process of being flattened. However, if one considers only a small portion of the orange peel, it appears to be smooth and flat—even though it originated from a spherical "whole." So it is with the Earth. Small portions of the surface can be represented very well using the

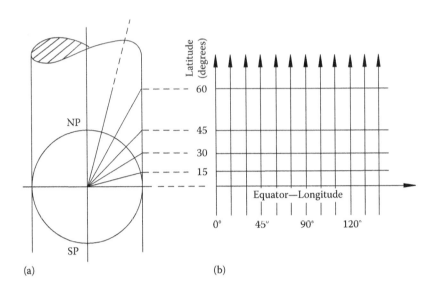

**FIGURE 11.1**   Globe and Mercator Grid. (a) Earth Wrapped in Cylinder, (b) Appearance of Graticle for Cylinder Rolled Out Flat

assumption that the Earth is flat. But, when dealing with larger and larger areas on the Earth, the inevitable distortions that occur between the curved surface and the flat map must be accommodated.

Maps range from simple to complex and serve many purposes. In some cases, a map communicates best by grossly distorting geometrical detail. In other cases, the geometrical detail of a map is the basis of its value to the user. Given that most spatial data are now digital and given further the proliferation of computerized data visualization tools, the opportunities for cartographers to develop creative and innovative representations of spatial data have become innumerable. The GSDM provides a concise set of rules equally applicable worldwide for generating, storing, manipulating, viewing, and otherwise using digital geospatial data. Although beneficial uses still exist, map projections have lost some of their utility because a map shows only 2-D relationships from a fixed perspective. Modern practice must accommodate 3-D digital spatial data and many users prefer the option of choosing a perspective. The GSDM allows flat-Earth relationships (including 2-D ones) to be computed, viewed, and used as local coordinate differences while the underlying ECEF coordinates retain their geometrical integrity and global uniqueness.

## PROJECTION CRITERIA

Use of the GSDM notwithstanding, concepts of map construction are still important and are summarized herein. It is impossible to generate a flat map that depicts the curved surface of the Earth accurately without distorting two or more of the following geometrical elements: angles, distances, or area. In the past, questions to be answered included "What elements will be distorted and by how much?" Another part of the same question is, "What element can be preserved in the projection without being distorted?" Various answers dictate a particular class of projection and maps are made accordingly. But, digital revolution has changed much of that because computers, equations, and talented cartographers can now manipulate digital spatial data in creative ways not previously feasible. Therefore, a basic understanding of map projections is still important.

Mathematical set theory includes the concept of domain and range. Bringing that analogy to map projections, the location of a point on the curved Earth is considered to be the domain while an equivalent expression for the same point on a flat map is the range. A map projection is the function (set of rules) whereby a discrete point in the domain is given equivalent expression in the range. And, the transformation rules must be bidirectional in order to preserve the unique point-to-point match between the domain and the range.

Given the impossibility of portraying a curved Earth on a flat map with true geometrical integrity, the "rules" used in the transformation function will distort some combination of angles, distances, and area. Of course, the reader should realize that if the area being mapped is relatively small, the distortion of a given geometrical element may be small enough to be of no consequence (the magnitude of permissible distortion will be discussed later). But, when considering larger and larger areas, a cartographer has the option of designing a map projection such that one of the elements—angles, distances, or area—can be preserved on the map. That choice provides the mathematical basis for three important classes of map projections.

1. *Conformal*: A conformal map projection is one in which a horizontal angle on the Earth is unchanged in its representation on the flat map. Distances and areas are distorted on a conformal projection but the distortion is controlled by limiting the area of the Earth being projected. Conformal map projections are used extensively in surveying and mapping applications and are discussed later in this chapter.

2. *Equidistant*: An equidistant map projection is one in which the distances on the Earth are faithfully represented on the map. Angles and areas are distorted.

3. *Equivalent*: An equivalent map projection is one in which the area of a given portion of the surface of the Earth is truthfully represented on the map. In geography, the equal-area map is used beneficially in many applications.

Another important issue is that map projections can be developed by graphical construction (often used for illustration purposes) or mathematical equations. Graphical projections are categorized by the origin of the imaginary light ray. A gnomonic projection is one where the light ray originates at the center of the Earth—see Figure 11.2a. A stereographic projection is one in which the light rays originate on the opposite side of the world—see Figure 11.2b. And, an orthographic projection is one in which the rays originate from a point source at an infinite distance. This means all rays arrive perpendicular to the projection surface—see Figure 11.2c.

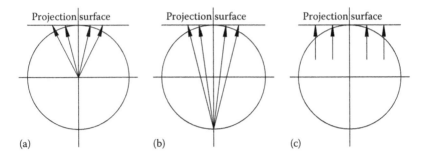

**FIGURE 11.2**   Location of Light Sources for Projection: (a) Gnomonic, (b) Stereographic, and (c) Orthographic

When using the P.O.B. option of the GSDM, as described in Chapter 1, the relative position of each point is plotted with respect to the P.O.B. according to their local latitudes and departures. The result is an orthographic projection of a point cloud to the tangent plane through the P.O.B. selected by the user. If scanned data are processed to obtain geocentric $X/Y/Z$ coordinates of each point in an image and if those points are stored pixel by pixel in a BURKORD™ database, then the P.O.B. plot of those pixels will result in a rectified map of those points. The impact of this feature will be significant for softcopy photogrammetry, use of scanned images, and photogrammetric mapping. Although the orthographic projection is not conformal over large areas, the distortion of angles and distances for small areas is inconsequential and each user has the option of selecting successive P.O.B.s such that the spacing between P.O.B.s will control distortion at an acceptable level. This feature of the GSDM needs additional study and further clarification.

The conformal map projections discussed in the remainder of this chapter are mathematical projections even though the graphical mode is used for illustration purposes.

## PROJECTION FIGURES

Map projections are also categorized according to whether the projection surface is a plane, a cone, or a cylinder, as illustrated in Figure 11.3. Conceptually, the basic difference between all three cases is the location of the apex of the cone. In one extreme, the apex of the cone is on the curved surface resulting in a tangent plane projection, Figure 11.3a. The cylinder illustrates the other extreme in which the apex of the cone is infinitely distant from the Earth, Figure 11.3c. Between those extremes, the cone contacts the Earth along a standard parallel of latitude as determined by the distance between the curved surface and the apex of the cone. As the distance to the apex becomes larger and larger, the standard parallel of latitude moves closer and closer to the Equator. Gerard Mercator (1512–1594) is credited with devising the cylindrical conformal Mercator projection while Johann Heinrich Lambert (1728–1777) is

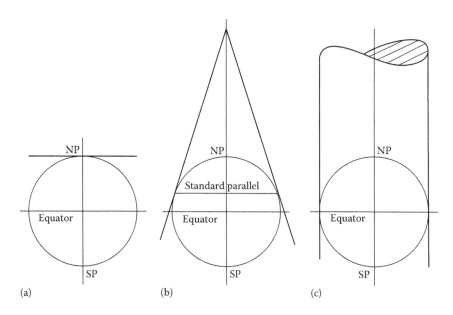

FIGURE 11.3 Map Projection Surfaces—Three Apex Locations: (a) Tangent Plane, (b) Cone, and (c) Cylinder

credited with developing the transverse Mercator projection and the conformal conic projection as an extension of Mercator's work.

The word "zone" is often used to describe a specific portion of the surface of the Earth that can be mapped with a single projection without exceeding some limit of distance distortion. For the state plane coordinate system (SPCS), the distortion limit in a zone is 1 part in 10,000 (exceptions exist), and for Universal Transverse Mercator (UTM) projections, the distortion limit in a zone is 1 part in 2,500. Of course, the goal is to include as much area in a zone as may be practical without exceeding the specified distortion limit. That means the zone must be as wide as possible and as long as practical. Two choices are typical when looking at zone length. The cartographic designer can choose either a Lambert conformal conic projection as shown in Figure 11.4a or a transverse Mercator projection as illustrated in Figure 11.4b. If a state to be covered has a long east-west extent (e.g., Tennessee), then a conic projection is appropriate. If the state to be covered has a long north-south extent, then a transverse Mercator projection is better (e.g., Illinois). In the case of the UTM projections, each zone is 6° wide and extends from latitude 80° S to 84° N.

When attempting to maximize the width of a zone without exceeding a given distortion limit, a further consideration is that the projection surface may be tangent to the Earth as shown in Figure 11.5a or secant as shown in Figure 11.5b. Distortion on a tangent projection stretches the distance from the curved surface to the mapping plane—the distortion is one sided. A wider zone is possible if a secant projection is used and the distortion is two sided—that is, if the distortion includes both compression and expansion. On a secant projection, the distortion near the center of

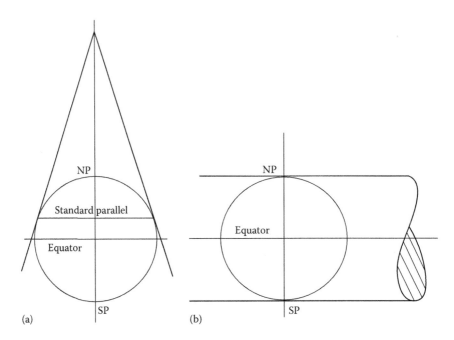

**FIGURE 11.4**    (a) Lambert Conformal Conic and (b) Transverse Mercator Projections

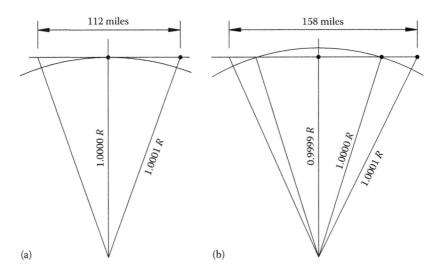

**FIGURE 11.5**    (a) Tangent and (b) Secant Projections

the projection compresses a distance element. Moving away from the center of the projection, the distortion diminishes, goes to zero where the two surfaces intersect, then increases without limit as one moves further and further away from the center of the zone. This means the nominal width of a zone is determined by an arbitrary choice of the designer regarding maximum allowable distortion.

## PERMISSIBLE DISTORTION AND AREA COVERED

The grid scale factor is used to describe distance distortion. The grid scale factor is represented by the letter $k$ and defined as the distance on the flat map grid divided by the curved distance on the ellipsoid:

$$k \equiv \frac{\text{Distance on the grid}}{\text{Distance on the ellipsoid}} \quad \text{example: } k = \frac{99.990 \text{ meters}}{100.00 \text{ meters}} = 0.99990 \quad (11.1)$$

Grid scale factor distortion can also be expressed as a ratio or parts-per-million (ppm). A distortion limit of 1 part in 10,000 is often used in the state plane coordinate system projections. The following are equivalent expressions of the same grid scale factor:

$$k = 0.99990 = 1 \text{ part in } 10,000 = 100 \text{ ppm}$$

When the 1 in 10,000 limit is applied to the state plane coordinate projections, the range of grid scale factors is as summarized in Table 11.1.

Using a radius of a spherical Earth as 6,372,000 meters (approx. 20,906,000 feet) and the grid scale factors above, the maximum zone widths in Figure 11.4 are computed as shown in Equations 11.2 and 11.3. To clarify, $R$ is in meters and contains four significant figures. Dividing by 1000 meters per kilometer makes the answer come out in kilometers. Converting to miles is left to the reader.

Tangent projection:

$$\text{Zone width} = \frac{\sqrt{(1.0001R)^2 - R^2}}{1000} \times 2 = 180 \text{ km} \approx 112 \text{ miles} \quad (11.2)$$

Secant projection:

$$\text{Zone width} = \frac{\sqrt{(1.0001R)^2 - (0.9999R)^2}}{1000} \times 2 = 255 \text{ km} \approx 158 \text{ miles} \quad (11.3)$$

---

### TABLE 11.1
### Comparison of Grid Scale Factors

**Tangent Projection:**

| | |
|---|---|
| Center of the zone | $k = 1.0000$ |
| Edge of the zone (imposed by 1/10,000 criterion) | $k = 1.0001$ |

**Secant Projection:**

| | |
|---|---|
| Center of zone | $k = 0.9999$ |
| Intersection of curved surface and mapping plane | $k = 1.0000$ |
| Edge of the zone (imposed by limit of 1 in 10,000) | $k = 1.0001$ |

---

When making decisions about what projection to use with the NAD 83, mapping professionals in several states chose to relax the 1:10,000 distance distortion criterion so that the entire state could be covered by a single zone. Minimum grid scale factors are computed using the constants in Appendix B and an equation to be presented later—Equation 11.35. On NAD 83, the state of South Carolina uses a single zone with a minimum grid scale factor of 0.999793656965 (1:4,846), the State of Montana uses 0.999392636277 (1:1,646), the State of Nebraska uses 0.999658595062 (1:2,938), and the State of Kentucky uses a single zone with a minimum grid scale factor of 0.999904942073 (1:10,520). All four are Lambert conformal conic projections based upon a single central parallel. The effective zone widths are

South Carolina:

$$\text{Zone width} = \frac{\sqrt{(1.000206343R)^2 - 0.9997936570R^2}}{1000} \times 2 = 366 \text{ km} \approx 227 \text{ miles}$$

Montana:

$$\text{Zone width} = \frac{\sqrt{(1.000607364R)^2 - 0.999392636R^2}}{1000} \times 2 = 628 \text{ km} \approx 390 \text{ miles}$$

Nebraska:

$$\text{Zone width} = \frac{\sqrt{(1.000341405R)^2 - 0.999658595R^2}}{1000} \times 2 = 471 \text{ km} \approx 293 \text{ miles}$$

Kentucky:

$$\text{Zone width} = \frac{\sqrt{(1.000095058R)^2 - 0.999904942R^2}}{1000} \times 2 = 249 \text{ km} \approx 154 \text{ miles}$$

## U.S. STATE PLANE COORDINATE SYSTEM (SPCS)

The SPCS zones in the United States were designed in the 1930s using a secant projection for use on the NAD 27. Although other projection options were considered for use on the NAD 83, the defining SPCS zone parameters were largely unchanged for implementation on the NAD 83. The SPCS on the NAD 83 consists of 54 transverse Mercator projections, 68 Lambert conformal conic projections, and 1 oblique Mercator projection. Some states are covered by a single zone but most states require more than one zone due to the limiting width of 158 miles and due to choosing SPCS zone boundaries to follow county boundaries. Other incidental changes were made during the transition from NAD 27 SPCS to NAD 83 SPCS and can be gleaned from two important publications. Claire (1968) is the "bible" for working with SPCS on the NAD 27 and Stem (1989) is the "bible" for working with SPCS on the NAD 83. Each booklet contains a description of the underlying map projections, a listing of the defining parameters for each zone, and a list of equations that can be used to perform bidirectional transformations between a latitude/longitude position and plane coordinates on the respective datum.

## HISTORY

The following quote is found in the section "SPCS—UTM and Oscar S. Adams" of the paper *Geodetic Surveys in the United States, the Beginning and the Next One Hundred Years, 1907–1940* by the late Joseph Dracup, a geodesist with the U.S. Coast & Geodetic Survey (USC&GS), now the National Geodetic Survey (NGS). The paper is available at https://www.ngs.noaa.gov/PUBS_LIB/geodetic_survey_1807.html (accessed May 4, 2017):

> In 1933-34, Oscar S. Adams ably assisted by Charles N. Claire developed the State Plane Coordinate System (SPCS) at the request of George F. Syme a North Carolina Highway engineer. Syme died shortly after the North Carolina system was developed being succeeded by O.B. Bestor to carry on the cause. Bestor was in charge of the State local control project established in 1933, later identified as the North Carolina Geodetic Survey. Most State and the few county projects involved in this program also were so named. Colonel C. H. Birdseye of the USGS, with a strong interest in Statewide coordinate grids also participated in the several conferences leading to the decision to honor Syme's request.
>
> The first tables for computing Lambert coordinates were developed for North Carolina and the first tables for the transverse Mercator grid were for New Jersey. Tables were prepared for all States early in 1934. For the first time all horizontal control stations previously defined only by latitudes and longitudes would be available in easy to use plane coordinates.

## FEATURES

The State Coordinate Systems (A Manual for Surveyors), otherwise known as Special Publication 235, is a booklet that describes details of the state plane coordinate systems (Mitchell and Simmons [1945] 1977). It is of both practical and significant historical value because it documents surveying policies and practices prior to the electronic revolution. Several important SPCS features described in Special Publication 235 include the following:

- The state plane coordinate system provides a method by which the latitude/longitude positions of the national triangulation network can be represented by plane coordinates. This meant local surveyors/engineers could continue using plane surveying procedures yet realize the benefits of basing their work on the national network of geodetic control points established by the federal agencies. This item is still valid in the 2-D arena (a subset of the 3-D arena). But, spatial data are 3-D and the GSDM does for 3-D data what the SPCS does for 2-D data.

- Normal land surveying measurements in the 1930s were made with a transit and steel tape. Expected accuracies were often in the range of 1:5,000 to 1:8,000 or better. Under those circumstances, a routine distance distortion of 1:10,000 could be tolerated without making a scale factor correction and without significant detrimental impact on the quality of the survey. With newer technology, this assumption is no longer valid because measurement accuracies today routinely exceed those of 80 years ago. Better accuracy is not a problem because high-quality computational results are obtained by applying the grid scale factor correction. With the corrections applied, the SPCS is fundamentally sound for 2-D applications. Elevation is typically used to handle the third dimension.

- There are two distance "corrections" to be made when working with the SPCS. The terminology encountered may vary depending on whether one is working with NAD 27 data or working with NAD 83 data (Burkholder 1993a). The "ellipsoid" and "elevation factor" are used when the SPCS is based on NAD 83 while "geoid" and "sea level factor" are used when the SPCS is based on NAD 27. Additionally, the phrase "grid scale factor" is currently used to avoid possible confusion with the term "scale factor" as historically used in topographic mapping. The two corrections are (1) the grid scale factor is used to correct for the distortion between the ellipsoid and the grid and (2) the elevation factor is needed to reduce a ground level horizontal distance to the ellipsoid. These two corrections (the grid scale factor and the elevation factor) are often multiplied together and called the "combination factor"—Stem (1986) calls it the "combined factor." The grid distance between the plumb lines through two points is the product of the horizontal ground distance and the combination factor. Special Publication 235 explains both factors quite well but, as discussed later, the grid/ground difference is the primary disadvantage of using the SPCS. Regretfully, when using the SPCS, a foot distance on the grid is not necessarily a foot distance on the ground. In many cases, such as centerline stationing on a highway project, the difference between grid and ground distances becomes intolerable (see Appendix III of Burkholder 1993b).

- Although the NGS has always performed and computed their geodetic surveys in metric units, the NAD 27 state plane coordinates were published in foot units—see sidebar discussion of the U.S. Survey Foot on page 278.

It is not true, as some have said, that the SPCS distorts distances by 1:10,000. It is true to say that, when compared to a distance on the map (projection surface), the equivalent distance on the ellipsoid *may* be distorted by up to 1:10,000. On a secant projection, the distortion is zero along the lines of exact scale where the ellipsoid and projection surfaces intersect and the distance on the map is the same as the distance on the ellipsoid. At the center of the zone, the distance is compressed by 1:10,000 or by whatever distortion value was selected by the zone designer. In some cases, a zone width of 158 miles was not quite sufficient to cover the area required and the distance distortion at the center of the zone is greater than 1:10,000 (i.e., the grid scale factor at the zone center is less than 0.9999)—see constants for California Zone 1, both Oregon zones, Zone 10 in Alaska, North Carolina, South Carolina, four of the five Texas zones, Utah Central Zone, and the offshore zone for Louisiana.

The grid scale factor is only part of the distortion. The elevation factor also contributes to the difference between a horizontal ground distance and the state plane grid distance. Modern practice looks more closely at the grid/ground distance difference (as a result of using the combination factor) and many resort to using surface coordinates or project datums as a way to avoid the mismatch between grid and ground distances. More recently, the use of a low-distortion projection (LDP) has been discussed as being a way to minimize the grid/ground distance distortion. Although the distance distortion issue is largely moot when using the GSDM, issues related to the use of a LDP are discussed later in this chapter.

## NAD 27 AND NAD 83

The NAD 27 was the only logical datum choice available when zones of the SPCS were chosen during the 1930s. The zones were selected by matching the projection type with the general geographic configuration of the state. Lambert conic projections were selected for states long in the east-west dimension while transverse Mercator projections were selected for states oriented primarily north-south. Some states have only one projection (zone); other states require more than one zone to cover the needed width; and some states have more than one projection type. For example, the State of Florida utilizes two transverse Mercator projections and one Lambert conic projection; New York employs three transverse Mercator projections and one Lambert conic projection; and the State of Alaska uses nine transverse Mercator projections, one Lambert conic projection, and one oblique Mercator projection.

In the 1930s, the USC&GS developed a "model law" that was promoted by the Council of State Governments for several decades. By 1971 the SPC model law was adopted in one form or another by 26 states (Mitchell and Simmons [1945] 1977). However, the Michigan Legislature adopted a different projection than that proposed by the USC&GS. Originally, Michigan was to be covered by three transverse Mercator projections but when the state plane coordinate law was written, professionals within the state opted instead for three Lambert projections based upon an elevated reference surface selected to minimize the need for the elevation reduction. The elevated system worked as intended and was deemed very beneficial but, because it was "non-standard," there was confusion, both in practice and in the published literature, about computing the correct combination factor for a line (Burkholder 1980). The Michigan state plane coordinate law for NAD 83 placed the reference surface on the GRS 80 ellipsoid.

## RELATIONSHIP BETWEEN THE METER, THE
## INTERNATIONAL FOOT, AND THE U.S. SURVEY FOOT

1. The length of the meter was established in the 1790s as 1/10,000,000 of the distance from the Equator to the North Pole as determined by a geodetic survey in France. Alder (2002) provides a fascinating account of that effort.

2. In the early 1800s, prototype meter bars were made and distributed to the nations of the world.

3. Although the meter has been used as the standard of length for geodetic surveys in the United States since establishment of the federal agency Survey of the Coast (predecessor to the NGS) in 1807, the meter length unit was declared legal for trade in the United States in 1866. The relationship between the foot and meter was stated in 1866 to be 39.37 feet = 12.00 meters exactly.

4. Leading up to and during World War II, Canada, the United States, and Great Britain each used a slightly different relationship between the foot and the meter.

    United States: 1.00 meter = 39.37 inches or 1 inch = 2.540005 cm
    England:                                1 inch = 2.539997 cm
    Canada:                                 1 inch = 2.540000 cm

5. Following World War II, machinists and aircraft mechanics, working under the auspices of NATO, discovered that parts of aircraft engines built according to the same blueprints were not interchangeable due to differences in unit definitions. A compromise was reached that adopted the Canadian relationship (1 inch = 2.54 centimeters exactly) as the International Foot (1 foot = 0.3048 meter exactly).

6. However, to avoid recomputing and republishing thousands of existing state plane coordinates, the United States retained use of 12 meters = 39.37 feet and gave that long-standing relationship a name—the U.S. Survey Foot. A 1959 Federal Register Notice (FRN 1959) stated that the U.S. Survey Foot should be used "until such time as it becomes desirable to readjust the basic geodetic networks in the United States, *after which the ratio of a yard, equal to 0.9144 meter, shall apply*" (emphasis added).

7. In 1960 the *11th General Conference on Weights and Measures* redefined the meter, but not the length. The redefinition made it possible to duplicate the 1.00 meter distance in terms of wavelengths of Krypton 86 gas instead of relying upon the distance between two marks on a prototype bar.

8. The definition of the length of the meter was changed again in 1983—this time in terms of the distance light would travel in vacuum in 1/299,792,458 second. The new definition is equivalent to saying that light travels 299,792,458 meters in one second.

Although the definition used for duplicating the length of the meter has evolved over the years, the fundamental unit of length has not changed. The relationship of 12.00 meters = 39.37 feet has existed in the United States for over 100 years. The name "U.S. Survey Foot" was developed in 1959 to describe the relationship already in existence. As a "state's rights" issue, the U.S. Survey Foot is designated for surveying and mapping activities on a state-by-state basis. "International Foot" is the name given to the relationship used before 1959 by Canada (1 foot = 0.3048 meter) and adopted for use around the world. Neither the U.S. Survey Foot nor the International Foot is part of the International System of Units (SI) adopted by the *11th General Conference on Weights and Measures* in 1960.

When the NAD 27 datum was readjusted and published as the NAD 83, the legislative intent was for the International Foot to be used as an alternate to meters. Recognizing that, a number of states included the International Foot in the state plane coordinate legislation written and adopted to accommodate the NAD 83. Other states objected and ultimately won. A notice published in the Federal Register Notice (1998), closes by saying, "The effect of this notice is to allow the U.S. Survey Foot to be used indefinitely for surveying and mapping in the United States. No other part of the 1959 notice is in any way affected by this notice." The NGS still uses meter units for all geodetic surveying operations.

The upshot is that NAD 83 state plane coordinates in the United States may be meters, U.S. Survey Feet, or International Feet. Although the GSDM is based exclusively on metric units, each user has the option of specifying different linear units when displaying or printing P.O.B. results. That is, provision is made for other derived units in the P.O.B. datum option. However, it is intended that the underlying ECEF coordinates will always be metric when using the GSDM.

## CURRENT STATUS—NAD 83 SPCS

Although developed for use on the NAD 27, design of the SPCS was revisited prior to publication of the NAD 83. Arguments were advanced and considered for taking advantage of the standardization offered by the UTM system and using 2° UTM zones on the NAD 83. After many discussions and consideration of various alternatives, the decision was to use the existing SPCS projections and zone parameters with the NAD 83. However, some exceptions are as follows:

- The reference surface for Michigan was placed on the GRS 80 ellipsoid instead of using an elevated reference surface.
- Zone 7 in California was eliminated. Zone 5 now covers that area.
- The states of Montana, Nebraska, and South Carolina elected to relax the arbitrary 1:10,000 criteria, covering each entire state with one zone. Kentucky has also adopted a single zone.

## Advantages

The advantages of using the SPCS today are largely the same as when the SPCS was first implemented. A map projection flattens a portion of the Earth and allows one to perform 2-D rectangular surveying computations within a defined zone using plane Euclidean geometry. Standardization and wide acceptance are two huge benefits. An incidental benefit of the SPCS is that the back azimuth of a line is the same as the forward azimuth plus 180°. This feature could also be called a disadvantage because it belies the fact that meridians are not parallel, but converge at the poles.

## Disadvantages

A disadvantage of the SPCS for the GIS community is the absence of uniqueness. For inventory, and other purposes, it is highly desirable for the description of any point location to be globally unique. State plane coordinates are unique within a zone but not globally. In addition to knowing the state plane coordinate values for a point, the spatial data user must also know what zone or map projection is associated with the point. Two points having the same (or nearly so) coordinate values may appear to be the same or very close together while they are, in fact, many kilometers apart. A triplet of ECEF rectangular $X/Y/Z$ metric coordinates used in the GSDM is unique within the "birdcage" of orbiting GPS satellites.

In the surveying, mapping, and engineering communities, the biggest disadvantage of using map projections and the SPCS is that they are strictly 2-D mathematical models and spatial data users work with 3-D data. The GSDM is a rigorous 3-D model. Specific drawbacks to using the SPCS are listed by Burkholder (1993a) as follows:

- Lack of accessibility: Control points are not easy to visit, permission, etc.
- Lack of proximity: Control points are too far away.
- Lack of quality: The published positions are not of sufficient high quality.
- Lack of understanding: Spatial data users need to learn more about the SPCS.
- Mapping distortion: Ground distance may differ too much from grid distance.

With the advent of GPS, continued densification of the control network, higher levels of support from NGS, and greater awareness within the spatial data user community, the first four listed disadvantages have been significantly mitigated. But, the grid/ground difference is more of a problem than ever because more and more people are using equipment and computational processes in which that systematic difference cannot be tolerated. An argument is that more education and better enforcement of minimum standards could overcome those disadvantages. Without discounting the benefits of more education, it is suggested that using the GSDM is another alternative in which spatial data users can fully exploit the 3-D characteristics of their data and in which 2-D applications are still supported as a subset of the 3-D model.

## PROCEDURES

Recognizing that full implementation of the GSDM will take some time, this section is included to help readers become more comfortable with the transition. Although a competent 3-D least squares adjustment of a network can more fully utilize the 3-D characteristics of GPS measurements, current practices involving 1-D and 2-D data will not be replaced instantaneously. Therefore, this section provides a summary of procedures commonly used when working with state plane coordinates. The overall point to remember is that computing a state plane coordinate traverse is the same as computing a regular plane surveying traverse with the following exception—one must use grid azimuths and grid distances. There are many sources of information and software available for instructions on how to compute and adjust a traverse. These points are only summarized here.

### Grid Azimuth

Conformal projections are used for the SPCS in the United States. That means that an angle measured on the ground is the same as the angle on the map and that a field measured angle added to or subtracted from a known grid azimuth will result in a grid azimuth. The implication is that one should always start with a grid azimuth. Two common methods for beginning with a grid azimuth are as follows:

1. Backsight another point having known state plane coordinates. Doing that, the grid azimuth from standpoint to the backsight is computed using the plane coordinate inverse, $\tan \alpha = \Delta e / \Delta n$, as used in Equations 5.11, 5.12, or 5.13, depending upon the quadrant.
2. Perform an astronomical observation using a star or the sun as the backsight to determine the astronomical azimuth to the foresight. Depending upon the quality of the grid azimuth required, two corrections are needed. A Laplace correction, Equation 9.5, is used to convert the observed/computed astronomical azimuth to a geodetic azimuth and the convergence (between geodetic north and grid north at the station) is used in Equation 11.4 to convert the geodetic azimuth to a grid azimuth. A generic diagram is shown in Figure 11.6 and shows that geodetic azimuth = grid azimuth + convergence.

$$\text{Grid azimuth} = \text{Geodetic azimuth} - \text{Convergence} \qquad (11.4)$$

Note that if the point lies west of the central meridian, the convergence is a negative quantity but Equation 11.4 remains valid. For lines over about 2 kilometers long, the $(t-T)$ correction may be needed to preserve high-quality results. See Stem (1989).

### Grid Distance

An important design feature of the SPCS is that the grid distance will approximate the reference surface distance within 1 part in 10,000. For SPCS based on NAD 83, the reference surface is the GRS 80 ellipsoid for SPCS based on NAD 27,

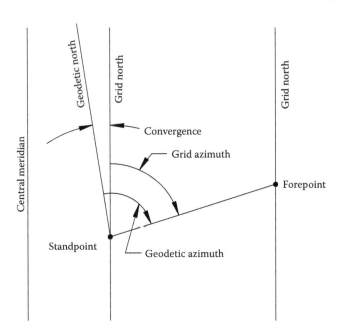

**FIGURE 11.6**   Convergence of Meridians

the reference surface is an approximation of the geoid (sea level). The grid scale factor is used to convert a reference surface distance to a grid distance and vice versa. Regretfully, most surveys are conducted at some elevation and not on the ellipsoid or geoid. Therefore an additional reduction is required to convert a ground level horizontal distance to a reference surface distance. Going back one step further, there are several options for computing a precise horizontal distance from observed slope distance and vertical (zenith) angles. For surveys of nominal accuracy, horizontal distance, HD(1), is computed as the right triangle component of slope distance and vertical angle. For surveys of higher accuracy and distances over about 2 km, horizontal distance is taken to be HD(2), the tangent plane distance between plumb lines, and involves computing a correction due to the plumb lines not being parallel. Such details are beyond the scope of this book but can be found in (Burkholder 1991).

A reliable horizontal distance is critical when using the SPCS because the horizontal distance must be reduced from horizontal to a reference surface distance and then to a grid distance. A high-quality grid distance relies upon the integrity of each part of the computational process. By contrast, the slope distance in 3-D space is used by the GSDM and the underlying model obviates reduction of slope distance to grid. Additional discussion of horizontal distance and the GSDM is included in Chapter 15.

Appropriate SPCS procedures should be followed when reducing a horizontal distance to a reference surface distance. To use the SPCS with NAD 83, the horizontal distance needs to be reduced to the ellipsoid distance as shown in Equation 11.6. To use the SPCS with NAD 27, the horizontal distance needs to be reduced to the sea-level distance as shown in Equation 11.7.

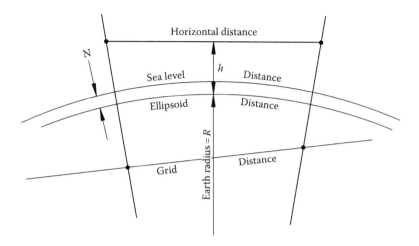

**FIGURE 11.7** Horizontal, Sea Level, Ellipsoid, and Grid Distance. *Note*: Within Conterminous United States, $N$ is a Negative Value

Finally, the reference surface distance must be reduced to the state plane grid. For a line less than 1 km long, one can use the grid scale factor computed at either end of the line or at the middle of the line. For a line longer than 1 km and less than about 4 km, it is acceptable to use the average grid scale factor. For a line longer than 4 km, the grid scale factor should be computed using the Simpson 1/6 Rule—see Stem (1989). Figure 11.7 shows a diagram illustrating the distance reductions.

As a summary, traditional state plane grid distances are computed as

1. Slope distance to horizontal distance:

$$HD(1) = (\text{Slope distance}) \times \sin(\text{Zenith direction}) \qquad (11.5)$$

$$HD(2) = \text{For a more precise option, see Burkholder (1991)}$$

2. Horizontal distance to reference surface distance: Use Equation 11.6 when the SPCS is based on NAD 83 and use Equation 11.7 when based on NAD 27.

a. $\text{Ellipsoid distance} = \text{Horizontal distance} \times \left( \dfrac{R}{R+h} \right) \qquad (11.6)$

b. $\text{Sea level distance} = \text{Horizontal distance} \times \left( \dfrac{R}{R+H} \right) \qquad (11.7)$

where:
   $R$ = Radius of Earth.
   $H$ = Elevation.
   $N$ = Geoid height.
   $h$ = Ellipsoid height = $H + N$.

Equations 11.6 and 11.7 both contain a ratio called "elevation factor" (Stem 1989, Burkholder 1993a). Those two ratios can be used interchangeably subject to conditions discussed by Burkholder (2004).

3. Reference surface distance to grid distance: Equation 11.8 is valid for all except a very long line. The difference is in computation of the appropriate grid scale factor.

$$\text{Grid distance} = \text{Reference surface distance} \times \text{Grid scale factor} \quad (11.8)$$

    a. For a "short" line, the grid scale factor for any part of the line can be used.

    b. For a line 2–4 km in length, the average grid scale factor for the line gives good results.

    c. For a line over 4 km long, use the Simpson's 1/6 Rule to compute the grid scale factor for a long line. See Stem (1989, p. 50).

Steps 2 and 3 above are often combined into a single step by using the combination factor for a line. The combination factor is the product of the grid scale factor and the elevation factor *at a point* and makes converting grid distance to ground distance (and vice versa) more efficient. But the temptation is to use a single combination factor for an entire project without investigating how it changes from point to point. It is true the same combination factor can be used for a given elevation over a specified area but each user should be aware of how the factor changes.

$$\text{Combination factor} = \text{Grid scale factor} \times \text{Elevation factor} \quad (11.9)$$

$$\text{Grid distance} = \text{Horizontal distance} \times \text{Combination factor} \quad (11.10)$$

$$\text{Horizontal distance} = \text{Grid distance/Combination factor} \quad (11.11)$$

The procedure for computing a combination factor as presented here is consistent with the use of state plane coordinates and can involve an approximation of elevation (or ellipsoid height). Depending on the length of the line, the computation may also include an approximation of the grid scale factor. A method for computing a closed-form combination factor for a line using the GSDM is presented in Burkholder (2016).

## TRAVERSES

The primary advantage of using a state plane coordinate traverse is that one can use simple plane surveying procedures to establish a "big picture" position on each traverse point—be it latitude/longitude or state plane coordinates. Two other huge benefits of using state plane coordinates are that the procedures are long adopted (standard) and that a state plane traverse can begin on one point and close on another, even distant, point. Otherwise, a traverse must return to the beginning point in order to determine the traverse misclosure as a check on possible blunders.

## Loop Traverse

A loop traverse is one that begins and ends at the same point—forming a closed loop. If one begins with state plane coordinates at the point of beginning and uses grid azimuths and grid distances, then it is a state plane coordinate traverse. The sum of the latitudes (north-south components of each course) and the sum of the departures (east-west components of each course) should each add up to zero. Any difference is the traverse misclosure and provides the basis of a traverse adjustment. Typically, a loop traverse is adjusted by the Compass Rule. Other methods exist, but the Compass Rule is quite simple to apply and, if used properly, delivers good results. Although a loop traverse may be quite useful, a point-to-point traverse is preferred because it provides better azimuth control and helps prevent possible scaling problems.

## Point-to-Point Traverse

A point-to-point traverse is a mathematically closed traverse that starts on one known point and ends on another. The traverse misclosure is determined as the difference of observed (computed) value minus the published (known) values and the traverse is typically adjusted by the Compass Rule. When using a point-to-point traverse, an angular misclosure and adjustment should be completed before the latitude and departure misclosures are computed. The purist will argue that a least squares adjustment is better than a Compass Rule adjustment and that may be true. But, by comparison, a Compass Rule adjustment is very easy to perform and achieves most of the benefit of a least squares adjustment.

## ALGORITHMS FOR TRADITIONAL MAP PROJECTIONS

The focus of this book is the 3-D GSDM, but the following map projection algorithms are included for the benefit of those needing them (Richardus and Alder 1972, Snyder 1987, Pearson 1990). The algorithms for the BK10 (forward) and BK11 (inverse) computations, described in Chapter 1, are given (as used in the northern hemisphere) for the Lambert conformal conic projection, the transverse Mercator projection, and the oblique Mercator projection. Specific rigorous equations for the BK14 and BK15 transformations (there are others) are described at the end of this chapter.

Although not technically prohibited from being used on the WGS 84 datum or the ITRF datum, the following algorithms are specifically intended to be used with the NAD 83 on the GRS 80 ellipsoid. Except where noted, the equations and symbols are intended to be consistent with those used in NOAA Manual NOS NGS 5, State Plane Coordinate System of 1983 (Stem 1989). Within a very small tolerance, the results obtained using these equations should be identical to those obtained using the equations and procedures given in said Manual 5. State plane coordinate zone parameters are listed in Appendix A of Stem (1989). The following algorithm is included in a paper presented by Burkholder (1985), which contains a description, the algorithm, flowchart, and FORTRAN listing of a computer program to compute zone constants and perform transformations (BK10 and BK11) on all three projections.

## LAMBERT CONFORMAL CONIC PROJECTION

The State Plane Coordinate System of 1983 is based upon the NAD 83, of which the GRS 80 is a part. The GRS 80 parameters are

$$a = \text{semimajor axis} = 6,378,137\,\text{m}\,(\text{exact}) \tag{11.12}$$

$$1/f = \text{reciprocal flattening} = 298.2572221008827 \tag{11.13}$$

Compute ellipsoid constants using:

$$e^2 = 2f - f^2 \quad \text{and} \quad e = \sqrt{e^2} \quad \text{eccentricity squared and eccentricity} \tag{11.14}$$

$$c_2 = \frac{e^2}{2} + \frac{5e^4}{24} + \frac{e^6}{12} + \frac{13e^8}{360} + \frac{3e^{10}}{160} \tag{11.15}$$

$$c_4 = \frac{7e^4}{48} + \frac{29e^6}{240} + \frac{811e^8}{11,520} + \frac{81e^{10}}{2,240} \tag{11.16}$$

$$c_6 = \frac{7e^6}{120} + \frac{81e^8}{1,120} + \frac{3,029e^{10}}{53,760} \tag{11.17}$$

$$c_8 = \frac{4,279e^8}{161,280} + \frac{883e^{10}}{20,160} \tag{11.18}$$

$$c_{10} = \frac{2,087e^{10}}{161,280} \tag{11.19}$$

Equations 11.15 through 11.19 are used to compute the following $F$ coefficients.

$$F_0 = 2(c_2 - 2c_4 + 3c_6 - 4c_8 + 5c_{10}) \tag{11.20}$$

$$F_2 = 8(c_4 - 4c_6 + 10c_8 - 20c_{10}) \tag{11.21}$$

$$F_4 = 32(c_6 - 6c_8 + 21c_{10}) \tag{11.22}$$

$$F_6 = 128(c_8 - 8c_{10}) \tag{11.23}$$

$$F_8 = 512c_{10} \tag{11.24}$$

The $F$ coefficients are in radian units and are used in the BK11 and BK15 transformations to compute geodetic latitude without iterating.

Input the defining parameters for the Lambert projection zone for the mapping or project area selected by the user:

$\phi_n$ = latitude of northern standard parallel.
$\phi_s$ = latitude of southern standard parallel.
$\phi_b$ = latitude of false origin (usually where northing = 0 meter).
$\lambda_0$ = longitude of central meridian, east longitude is + and west longitude is −.
$E_0$ = false easting on central meridian in meters.
$N_b$ = northing on false origin, usually 0 meter.

Compute coordinate system projection constants for the zone:

ln = natural logarithm.
exp($x$) = $\varepsilon^x$ where $\varepsilon$ = 2.71828... (base of natural logarithms).
$Q_i$ = isometric latitude for corresponding geodetic latitude.
$W_i$ = intermediate computational value at $\phi_n$ and $\phi_s$.

$$Q_n = \frac{1}{2}\left[\ln\left(\frac{1+\sin\phi_n}{1-\sin\phi_n}\right) - e\ln\left(\frac{1+e\sin\phi_n}{1-e\sin\phi_n}\right)\right] \tag{11.25}$$

$$Q_s = \frac{1}{2}\left[\ln\left(\frac{1+\sin\phi_s}{1-\sin\phi_s}\right) - e\ln\left(\frac{1+e\sin\phi_s}{1-e\sin\phi_s}\right)\right] \tag{11.26}$$

$$Q_b = \frac{1}{2}\left[\ln\left(\frac{1+\sin\phi_b}{1-\sin\phi_b}\right) - e\ln\left(\frac{1+e\sin\phi_b}{1-e\sin\phi_b}\right)\right] \tag{11.27}$$

$$W_n = \sqrt{1-e^2\sin^2\phi_n} \tag{11.28}$$

and

$$W_s = \sqrt{1-e^2\sin^2\phi_s} \tag{11.29}$$

$$\phi_0 = \sin^{-1}\left[\frac{\ln\left(W_n\cos\phi_s\right) - \ln\left(W_s\cos\phi_n\right)}{Q_n - Q_s}\right] \quad \text{latitude of central parallel}$$

$$\tag{11.30}$$

$$K = \frac{a\cos\phi_s\exp\left(Q_s\sin\phi_0\right)}{W_s\sin\phi_0} = \frac{a\cos\phi_n\exp\left(Q_n\sin\phi_0\right)}{W_n\sin\phi_0} \quad \text{mapping radius of equator}$$

$$\tag{11.31}$$

$$Q_0 = \frac{1}{2}\left[\ln\left(\frac{1+\sin\phi_0}{1-\sin\phi_0}\right) - e\ln\left(\frac{1+e\sin\phi_0}{1-e\sin\phi_0}\right)\right] \quad \text{isometric latitude of }\phi_0 \quad (11.32)$$

$$R_b = \frac{K}{\exp\left(Q_b \sin\phi_0\right)} \quad \text{mapping radius of latitude of origin} \quad (11.33)$$

$$R_0 = \frac{K}{\exp\left(Q_0 \sin\phi_0\right)} \quad \text{mapping radius of central parallel} \quad (11.34)$$

$$k_0 = \frac{R_0 \tan\phi_0 \sqrt{1 - e^2 \sin^2 \phi_0}}{a} \quad \text{grid scale factor at center of zone} \quad (11.35)$$

The preceding constants need to be computed only once for a given projection zone but they are used repeatedly in the following BK10 (forward) and BK11 (inverse) computations.

### BK10 Transformation for Lambert Conformal Conic Projection
Input:

$\phi$ = geodetic latitude (positive north).
$\lambda$ = geodetic longitude (positive east).

Compute:

$$Q_\phi = \frac{1}{2}\left[\ln\left(\frac{1+\sin\phi}{1-\sin\phi}\right) - e\ln\left(\frac{1+e\sin\phi}{1-e\sin\phi}\right)\right] \quad \text{isometric latitude of point}\,(\phi,\lambda) \quad (11.36)$$

$$R_\phi = \frac{K}{\exp\left(Q_\phi \sin\phi_0\right)} \quad \text{mapping radius of point}\,(\phi,\lambda) \quad (11.37)$$

$$\gamma = \left(\lambda - \lambda_0\right)\sin\phi_0 \quad \text{convergence at point}\,(\phi,\lambda) \quad (11.38)$$

$$k = \frac{R_\phi \sin\phi_0 \sqrt{1 - e^2 \sin^2 \phi}}{a \cos\phi} \quad \text{grid scale factor at point}\,(\phi,\lambda) \quad (11.39)$$

$$E = E_0 + R_\phi \sin\gamma \quad \text{easting for point}\,(\phi,\lambda) \quad (11.40)$$

$$N = R_b + N_b - R_\phi \cos\gamma \quad \text{northing for point}\,(\phi,\lambda) \quad (11.41)$$

### BK11 Transformation for Lambert Conformal Conic Projection
Input:

$E$ = easting of point within defined map projection.
$N$ = northing of point within defined map projection.

Compute:

$$R' = R_b - N + N_b \quad \text{intermediate value} \tag{11.42}$$

$$E' = E - E_0 \quad \text{intermediate value} \tag{11.43}$$

$$\gamma = \tan^{-1}\left(\frac{E'}{R'}\right) \quad \text{convergence at point}(E,N) \tag{11.44}$$

$$R_\phi = \sqrt{R'^2 + E'^2} \quad \text{mapping radius at point}(E,N) \tag{11.45}$$

$$Q_\phi = \frac{\ln K - \ln R_\phi}{\sin \phi_0} \quad \text{isometric latitude at point}(E,N) \tag{11.46}$$

$$\chi = 2\tan^{-1}\left(\frac{\exp(Q_\phi)-1}{\exp(Q_\phi)+1}\right) \quad \text{conformal latitude at point}(E,N) \tag{11.47}$$

$$\phi = \chi + \sin\chi\cos\chi\left(F_0 + \cos^2\chi\left(F_2 + \cos^2\chi\left(F_4 + \cos^2\chi\left(F_6 + F_8\cos^2\chi\right)\right)\right)\right)$$
$$\text{geodetic latitudeat point}(E,N) \tag{11.48}$$

$$\lambda = \lambda_0 + \frac{\gamma}{\sin\phi_0} \quad \text{geodetic longitude(east)at point}(E,N) \tag{11.49}$$

$$k = R_\phi \sin\phi_0 \frac{\sqrt{1-e^2\sin^2\phi}}{a\cos\phi} \quad \text{grid scale factor at point}(E,N) \tag{11.50}$$

The State of Oregon uses a Lambert projection for its state plane coordinate system. Figure 11.8 is a computer printout showing an example of Lambert conformal conic BK10 and BK11 transformations at Station MEDIAN 2 (PID NY0996) on the campus of the Oregon Institute of Technology.

## TRANSVERSE MERCATOR PROJECTION

The State Plane Coordinate System of 1983 is based upon the NAD 83 of which the GRS 80 ellipsoid is a part. The GRS 80 parameters are

$$a = \text{semimajor axis} = 6{,}378{,}137\,\text{m}\,(\text{exactly}) \tag{11.51}$$

$$1/f = \text{reciprocal flattening} = 298.2572221008827 \tag{11.52}$$

Compute ellipsoid constants using:

$$e^2 = 2f - f^2 \quad \text{and} \quad e = \sqrt{e^2} \quad \text{eccentricity squared and eccentricity} \tag{11.53}$$

```
                    PROGRAM: LOCALCOR
              COPYRIGHT 2001 BY GLOBAL COGO, INC.
                 LAS CRUCES, NEW MEXICO 88003
                    WWW.GLOBALCOGO.COM
```

USER: Earl F. Burkholder
DATE: 12 June 2007

LAMBERT CONIC CONFORMAL COORDINATE TRANSFORMATIONS
PROJECTION NAME: Oregon South Zone - 3602

REFERENCE ELLIPSOID: GEODETIC REFERENCE SYSTEM 1980
                     A = 6378137.0000 METERS
                   1/F = 298.2572221008827

ZONE PARAMETERS:
    NORTH STANDARD PARALLEL            44  0 0.000000
    SOUTH STANDARD PARALLEL            42 20 0.000000
    FALSE ORIGIN LATITUDE              41 40 0.000000
    CENTRAL MERIDIAN (W)              120 30 0.000000
    FALSE EASTING ON CM                  1500000.0000 METERS
    NORTHING AT FALSE ORIGIN                   0.0000 METER

ZONE CONSTANTS:
    CENTRAL PARALLEL                   43 10 6.919559
    SCALE FACTOR ON CENTRAL PARALLEL   0.999894607592090
    MAPPING RADIUS OF EQUATOR            12033772.69836 METERS
    MAPPING RADIUS OF FALSE ORIGIN       6976289.23822 METERS
    NORTHING OF CENTRAL PARALLEL ON CM    166836.95660 METERS
    CONFORMAL LATITUDE CONSTANTS:   F(0) = 0.006686920927
        F(2) = 0.000052014583      F(4) = 0.000000554458
        F(6) = 0.000000006718      F(8) = 0.000000000089

TRANSFORMATIONS:

NAME OF STATION: MEDIAN 2 (1998) - PID NY0996        FORWARD (BK10)

        LATITUDE:    42 15 15.611960   NORTHING        66102.3042 METERS
        LONGITUDE:  121 47 25.985950   EASTING      1393505.6444 METERS
        CONVERGENCE:       0-52 58.54  SCALE FACTOR:   1.000020826193

NAME OF STATION: MEDIAN 2 (1998) - PID NY0996        INVERSE (BK11)

        LATITUDE:    42 15 15.611959   NORTHING        66102.3042 METERS
        LONGITUDE:  121 47 25.985950   EASTING      1393505.6444 METERS
        CONVERGENCE:       0-52 58.54  SCALE FACTOR:   1.000020826194
```

**FIGURE 11.8**   Example BK10 and BK11 Transformations for Lambert Projection
```

$$n = \frac{f}{(2-f)} \quad \text{intermediate value} \tag{11.54}$$

$$r = a(1-n)(1-n^2)\left(1 + \frac{9n^2}{4} + \frac{225n_4}{64}\right) \quad \text{intermediate value} \tag{11.55}$$

$$u_2 = \frac{-3n}{2} + \frac{9n^3}{16} \tag{11.56}$$

$$u_4 = \frac{15n^2}{16} - \frac{15n^4}{32} \tag{11.57}$$

$$u_6 = \frac{-35n^3}{48} \tag{11.58}$$

$$u_8 = \frac{315n^4}{512} \tag{11.59}$$

The preceding intermediate values of $u$ are used only in equations that follow:

$$U_0 = 2(u_2 - 2u_4 + 3u_6 - 4u_8) \tag{11.60}$$

$$U_2 = 8(u_4 - 4u_6 + 10u_8) \tag{11.61}$$

$$U_4 = 32(u_6 - 6u_8) \tag{11.62}$$

$$U_6 = 128u_8 \tag{11.63}$$

The preceding values of $U$ are used to compute zone constants and in the BK10 transformation:

$$v_2 = \frac{3n}{2} - \frac{27n^3}{32} \tag{11.64}$$

$$v_4 = \frac{21n^2}{16} - \frac{55n^4}{32} \tag{11.65}$$

$$v_6 = \frac{151n^3}{96} \tag{11.66}$$

$$v_8 = \frac{1097n^4}{512} \tag{11.67}$$

The preceding values of $v$ are used only in the equations that follow:

$$V_0 = 2\left(v_2 - 2v_4 + 3v_6 - 4v_8\right) \tag{11.68}$$

$$V_2 = 8\left(v_4 - 4v_6 + 10v_8\right) \tag{11.69}$$

$$V_4 = 32\left(v_6 - 6v_8\right) \tag{11.70}$$

$$V_6 = 128v_8 \tag{11.71}$$

The preceding values of $V$ are used in the BK11 transformation.

Input the defining parameters for a transverse Mercator projection for the mapping or project selected by the user:

$\lambda_0$ = longitude of central meridian, east longitude is + and west longitude is −.
$E_0$ = false easting on central meridian in meters.
$k_0$ = grid scale factor on central meridian.
$\phi_0$ = latitude of false origin, usually where northing = 0 meter.
$N_0$ = false northing at false origin, usually 0 meter.

Compute coordinate system projection constants for the zone:

$$\omega_0 = \phi_0 + \sin\phi_0 \cos\phi_0 \left(U_0 + \cos^2\phi_0 \left(U_2 + \cos^2\phi_0 \left(U_4 + U_6\cos^2\phi_0\right)\right)\right)$$

$$\text{rectifying latitude of origin} \tag{11.72}$$

$$S_0 = rk_0\omega_0 \quad \text{distance on grid from equator to origin} \tag{11.73}$$

The preceding zone constants need to be computed only once for a given projection but they are used repeatedly in the following BK10 (forward) and BK11 (inverse) transformations. The nominal grid scale factor used on the central meridian for the state plane coordinate systems is 0.9999. Specific values are given in Appendix B. Other values are chosen when designing a custom system.

## BK10 Transformation for Transverse Mercator Projection

Input:

$\phi$ = geodetic latitude (positive north).
$\lambda$ = geodetic longitude (positive east).

Compute:

$$L = \left(\lambda - \lambda_0\right)\cos\phi \quad L \text{ in radians and positive east of central meridian} \tag{11.74}$$

$$t = \tan\phi \tag{11.75}$$

$$\eta^2 = \frac{e^2 \cos^2 \phi}{1-e^2} \tag{11.76}$$

$$\omega = \phi + \sin\phi\cos\phi\left(U_0 + \cos^2\phi\left(U_2 + \cos^2\phi\left(U_4 + U_6\cos^2\phi\right)\right)\right) \quad \text{rectifying latitude} \tag{11.77}$$

$$S = rk_0\omega \quad \text{arc distance on grid to parallel through point}(\phi,\lambda) \tag{11.78}$$

$$R = \frac{ak_0}{\sqrt{1-e^2\sin^2\phi}} \tag{11.79}$$

$$A_1 = -R \tag{11.80}$$

$$A_2 = \frac{1}{2}Rt \tag{11.81}$$

$$A_3 = \frac{1}{6}\left(1-t^2+\eta^2\right) \tag{11.82}$$

$$A_4 = \frac{1}{12}\left(5-t^2+\eta^2\left(9+4\eta^2\right)\right) \tag{11.83}$$

$$A_5 = \frac{1}{120}\left(5-18t^2+t^4+\eta^2\left(14-58t^2\right)\right) \tag{11.84}$$

$$A_6 = \frac{1}{360}\left(61-58t^2+t^4+\eta^2\left(270-330t^2\right)\right) \tag{11.85}$$

$$A_7 = \frac{1}{5,040}\left(61-479t^2+179t^4-t^6\right) \tag{11.86}$$

$$E = E_0 + A_1L\left(1+L^2\left(A_3+L^2\left(L_5+A_7L^2\right)\right)\right) \quad \text{easting of point}(\phi,\lambda) \tag{11.87}$$

$$N = N_0 + S - S_0 + A_2L^2\left(1+L^2\left(A_4+A_6L^2\right)\right) \quad \text{northing of point}(\phi,\lambda) \tag{11.88}$$

$$C_1 = -t \tag{11.89}$$

$$C_2 = \frac{1}{2}\left(1+\eta^2\right) \tag{11.90}$$

$$C_3 = \frac{1}{3}\left(1+3\eta^2+2\eta^4\right) \tag{11.91}$$

$$C_4 = \frac{1}{12}\left(5 - 4t^2 + \eta^2\left(9 - 24t^2\right)\right)$$  (11.92)

$$C_5 = \frac{1}{15}\left(2 - t^2\right)$$  (11.93)

$$\gamma = C_1 L\left(1 + L^2\left(C_3 + C_5 L^2\right)\right) \quad \text{convergence at point}\,(\phi, \lambda)$$  (11.94)

$$k = k_0\left(1 + C_2 L^2\left(1 + C_4 L^2\right)\right) \quad \text{grid scale factor at point}\,(\phi, \lambda)$$  (11.95)

### BK11 Transformation for Transverse Mercator Projection

Input:

$E$ = easting of point within defined map projection.
$N$ = northing of point within defined map projection.

Compute:

$$\omega = \frac{N - N_0 + S_0}{k_0 r}$$  (11.96)

$$\phi_f = \omega + \sin\omega\cos\omega\left(V_0 + \cos^2\omega\left(V_2 + \cos^2\omega\left(V_4 + V_6\cos^2\omega\right)\right)\right)$$  (11.97)

$$\eta_f^2 = \frac{e^2\cos^2\phi_f}{1 - e^2}$$  (11.98)

$$R_f = \frac{a k_0}{\sqrt{1 - e^2\sin^2\phi_f}}$$  (11.99)

$$Q = \frac{E - E_0}{R_f} \quad \text{radian units}$$  (11.100)

$$B_2 = \frac{-t_f\left(1 + \eta_f^2\right)}{2}$$  (11.101)

$$B_3 = \frac{-\left(1 + 2t_f^2 + \eta_f^2\right)}{6}$$  (11.102)

$$B_4 = \frac{-\left(5 + 3t_f^2 + \eta_f^2\left(1 - 9t_f^2\right) - 4\eta_f^4\right)}{12}$$  (11.103)

$$B_5 = \frac{\left(5 + 28t_f^2 + 24t_f^4 + \eta_f^2\left(6 + 8t_f^2\right)\right)}{120} \tag{11.104}$$

$$B_6 = \frac{\left(61 + 90t_f^2 + 45t_f^4 + \eta_f^2\left(46 - 252t_f^2 - 90t_f^4\right)\right)}{360} \tag{11.105}$$

$$B_7 = \frac{-\left(61 + 662t_f^2 + 1320t_f^4 + 720t_f^6\right)}{5040} \tag{11.106}$$

$$L = Q\left(1 + Q^2\left(B_3 + Q^2\left(B_5 + B_7Q^2\right)\right)\right) \tag{11.107}$$

$$\phi = \phi_f + B_2 Q^2\left(1 + Q^2\left(B_4 + B_6 Q^2\right)\right) \quad \text{geodetic latitude at point}(E, N) \tag{11.108}$$

$$\lambda = \lambda_0 + \frac{L}{\cos\phi_f} \quad \text{geodetic longitude(east) at point}(E, N) \tag{11.109}$$

$$D_1 = t_f = \tan\phi_f \tag{11.110}$$

$$D_2 = \frac{1 + \eta_f^2}{2} \quad G_2 \text{ in Manual NOS NGS5} \tag{11.111}$$

$$D_3 = \frac{-\left(1 + t_f^2 - \eta_f^2 - 2\eta_f^4\right)}{3} \tag{11.112}$$

$$D_4 = \frac{1 + 5\eta_f^2}{12} \quad G_4 \text{ in Manual NOS NGS5} \tag{11.113}$$

$$D_5 = \frac{2 + 5t_f^2 + 3t_f^4}{15} \tag{11.114}$$

$$\gamma = D_1 Q\left(1 + Q^2\left(D_3 + D_5 Q^2\right)\right) \quad \text{convergence at point}(N, E) \tag{11.115}$$

$$k = k_0\left(1 + D_2 Q^2\left(1 + D_4 Q^2\right)\right) \quad \text{grid scale factor at point}(N, E) \tag{11.116}$$

The State of New Mexico uses a transverse Mercator projection for its state plane coordinate system. Figure 11.9 is a computer printout showing example transverse Mercator BK10 and BK11 transformations at Station REILLY (PID AI5445) on the New Mexico State University campus. The transformations were computed using the equations in this section and numerical values match those shown on the NGS data sheet for the same station—NAD 83 (1992).

```
                         PROGRAM: LOCALCOR
                COPYRIGHT 2001 BY GLOBAL COGO, INC.
                   LAS CRUCES, NEW MEXICO 88003
                       WWW.GLOBALCOGO.COM
USER: Earl F. Burkholder
DATE: 12 June 2007


TRANSVERSE MERCATOR PROJECTION TRANSFORMATIONS
PROJECTION NAME: New Mexico Central Zone - 3002


REFERENCE ELLIPSOID: GEODETIC REFERENCE SYSTEM 1980
                     A = 6378137.0000 METERS
                     1/F = 298.2572221008827


ZONE PARAMETERS:


        CENTRAL MERIDIAN (W)                    106 15 0.000000
        LATITUDE OF FALSE ORIGIN                 31  0 0.000000
        FALSE NORTHING AT FALSE ORIGIN                0.0000 METER
        FALSE EASTING ON CENTRAL MERIDIAN        500000.0000 METERS
        SCALE FACTOR ON CENTRAL MERIDIAN         0.999900000000


ZONE CONSTANTS:


        RECTIFYING SPHERE RADIUS              6367449.1458 METERS
        RECTIFYING LATITUDE CONSTANTS:
            U(0) = -0.005048250776    V(0) =    0.005022893948
            U(2) =  0.000021259204    V(2) =    0.000029370625
            U(4) = -0.000000111423    V(4) =    0.000000235059
            U(6) =  0.000000000626    V(6) =    0.000000002181


        RECTIFYING LATITUDE OF FALSE ORIGIN     30 52 21.720626
        GRID MERIDIAN ARC TO FALSE ORIGIN    3430631.2260 METERS


TRANSFORMATIONS:


NAME OF STATION: REILLY (1992) - PID AI5445           FORWARD (BK10)


        LATITUDE:     32 16 55.929060   NORTHING      142268.7414 METERS
        LONGITUDE:   106 45 15.160700   EASTING       452506.4804 METERS
        CONVERGENCE:      0-16  9.48    SCALE FACTOR:   0.999927806946


NAME OF STATION: REILLY (1992) - PID AI5445           INVERSE (BK11)


        LATITUDE:     32 16 55.929059   NORTHING      142268.7414 METERS
        LONGITUDE:   106 45 15.160700   EASTING       452506.4804 METERS
        CONVERGENCE:      0-16  9.48    SCALE FACTOR:   0.999927806946
```

**FIGURE 11.9** Example BK10 and BK11 Transformations for Transverse Mercator Projection

## OBLIQUE MERCATOR PROJECTION

The State Plane Coordinate System of 1983 is based upon the NAD 83, of which the GRS 80 is a part. The GRS 80 parameters are

$$a = \text{semimajor axis} = 6,378,137\,\text{m}\,(\text{exact}) \qquad (11.117)$$

$$1/f = \text{reciprocal flattening} = 298.2572221008827 \qquad (11.118)$$

Compute ellipsoid constants:

$$e^2 = 2f - f^2 \quad \text{and} \quad e = \sqrt{e^2} \quad \text{eccentricity squared and eccentricity} \qquad (11.119)$$

$$e'^2 = \frac{e^2}{1-e^2} \quad \text{second eccentricity squared} \qquad (11.120)$$

$$c_2 = \frac{e^2}{2} + \frac{5e^4}{24} + \frac{e^6}{12} + \frac{13e^8}{360} + \frac{3e^{10}}{160} \qquad (11.121)$$

$$c_4 = \frac{7e^4}{48} + \frac{29e^6}{240} + \frac{811e^8}{11,520} + \frac{81e^{10}}{2,240} \qquad (11.122)$$

$$c_6 = \frac{7e^6}{120} + \frac{81e^8}{1,120} + \frac{3,029e^{10}}{53,760} \qquad (11.123)$$

$$c_8 = \frac{4,279e^8}{161,280} + \frac{883e^{10}}{20,160} \qquad (11.124)$$

$$c_{10} = \frac{2,087e^{10}}{161,280} \qquad (11.125)$$

Equations 11.119 and 11.121 through 11.125 are used once in computing the $F$ coefficients. These are the same $F$ coefficients as used in Equations 11.20 through 11.24.

$$F_0 = 2\left(c_2 - 2c_4 + 3c_6 - 4c_8 + 5c_{10}\right) \qquad (11.126)$$

$$F_2 = 8\left(c_4 - 4c_6 + 10c_8 - 20c_{10}\right) \qquad (11.127)$$

$$F_4 = 32\left(c_6 - 6c_8 + 21c_{10}\right) \qquad (11.128)$$

$$F_6 = 128\left(c_8 - 8c_{10}\right) \qquad (11.129)$$

$$F_8 = 512c_{10} \tag{11.130}$$

The $F$ coefficients are radian units and used in the BK11 and BK15 transformations to compute geodetic latitude without iterating.

Input the defining parameters for the oblique Mercator projection for the mapping or project area selected by the user:

$\phi_c$ = latitude of local origin.
$\lambda_c$ = longitude (east) of local origin.
$k_0$ = grid scale factor along projection axis.
$N_0$ = false northing at $(u, v)$ origin.
$E_0$ = false easting at $(u, v)$ origin.
$\alpha_c$ = positive skew axis ($u$-axis) azimuth at local origin.

Compute coordinate system projection constants for the zone:

$$B = \sqrt{1 + e'^2 \cos^4 \phi_c} \tag{11.131}$$

$$W_c = \sqrt{1 - e^2 \sin^2 \phi_c} \tag{11.132}$$

$$A = \frac{aB\sqrt{1 - e^2}}{W_c^2} \tag{11.133}$$

$$D = \frac{Ak_c}{B} \tag{11.134}$$

$$Q_c = \frac{1}{2}\left[ \ln\left(\frac{1 + \sin\phi_c}{1 - \sin\phi_c}\right) - e\ln\left(\frac{1 + e\sin\phi_c}{1 - e\sin\phi_c}\right) \right] \tag{11.135}$$

$$C = \cosh^{-1}\left(\frac{B\sqrt{1 - e^2}}{W_c \cos\phi_c}\right) - BQ_c \tag{11.136}$$

*Note:* $\cosh^{-1} x = \ln\left(x + \sqrt{x^2 - 1}\right)$

$$F = \sin\phi_0 = \frac{a\sin\alpha_c \cos\phi_c}{AW_c} \tag{11.137}$$

$$G = \cos\alpha_0 \tag{11.138}$$

$$I = \frac{Ak_c}{a} \tag{11.139}$$

$$\lambda_0 = \lambda_c + \frac{1}{B}\sin^{-1}\left[\frac{\sin\alpha_0 \sinh(BQ_c + C)}{\cos\alpha_0}\right] \tag{11.140}$$

*Note*: $\sinh x = \dfrac{e^x - e^{-x}}{2}$, $e$ = base of natural logarithms.

## BK10 Transformation for Oblique Mercator Projection

Input:

$\phi$ = geodetic latitude (positive north).
$\lambda$ = geodetic longitude (positive east).

Compute:

$$L = (\lambda_0 - \lambda)B \tag{11.141}$$

$$Q = \left[\ln\frac{1 + \sin\phi}{1 - \sin\phi} - e\ln\frac{1 + e\sin\phi}{1 - e\sin\phi}\right] \tag{11.142}$$

$$J = \sinh(BQ + C) \tag{11.143}$$

$$K = \cosh(BQ + C) \tag{11.144}$$

*Note*: $\cosh x = \dfrac{e^x + e^{-x}}{2}$, $e$ = base of natural logarithms

$$u = D\tan^{-1}\left[\frac{JG - F\sin L}{\cos L}\right] \tag{11.145}$$

$$v = \left(\frac{D}{2}\right)\ln\left[\frac{K - FJ - G\sin L}{K + FJ + G\sin L}\right] \tag{11.146}$$

$$E = E_0 + u\sin\alpha_c + v\cos\alpha_c \quad \text{easting of point}(\phi,\lambda) \tag{11.147}$$

$$N = N_0 + u\cos\alpha_c - v\sin\alpha_c \quad \text{northing of point}(\phi,\lambda) \tag{11.148}$$

$$\gamma = \tan^{-1}\left[\frac{F - JG\sin L}{KG\cos L}\right] - \alpha_c \quad \text{convergence at point}(\phi,\lambda) \tag{11.149}$$

$$k = \frac{I\sqrt{1 - e^2\sin^2\phi}\cos\left(\dfrac{u}{D}\right)}{\cos\phi\cos L} \quad \text{grid scale factor at point}(\phi,\lambda) \tag{11.150}$$

## BK11 Transformation for Oblique Mercator Projection
Input:

$E$ = easting of point within defined map projection.
$N$ = northing of point within defined map projection.

Compute:

$$u = (E - E_0)\sin\alpha_c + (N - N_0)\cos\alpha_c \qquad (11.151)$$

$$v = (E - E_0)\cos\alpha_c - (N - N_0)\sin\alpha_c \qquad (11.152)$$

$$R = \sinh\left(\frac{v}{D}\right) \qquad (11.153)$$

$$S = \cosh\left(\frac{v}{D}\right) \qquad (11.154)$$

$$T = \sin\left(\frac{u}{D}\right) \qquad (11.155)$$

$$Q = \frac{1}{B}\left[\frac{1}{2}\ln\frac{S - RF + GT}{S + RF - GT} - C\right] \qquad (11.156)$$

$$\chi = 2\tan^{-1}\left(\frac{e^Q - 1}{e^Q + 1}\right) \quad e = \text{base of natural logarithms} = 2.718281828\ldots \quad (11.157)$$

$$\phi = \chi + \sin\chi\cos\chi\left(F_0 + \cos^2\chi\left(F_2 + \cos^2\chi\left(F_4 + \cos^2\chi\left(F_6 + F_8\cos^2\chi\right)\right)\right)\right)$$
geodetic latitude at point $(E, N)$ $\qquad (11.158)$

$$\lambda = \lambda_0 + \left(\frac{1}{B}\right)\tan^{-1}\left[\frac{RG + TF}{\cos(u/D)}\right] \quad \text{geodetic longitude (east) at point}(E,N)$$
$$(11.159)$$

$$L = (\lambda - \lambda_0)B \qquad (11.160)$$

$$J = \sinh(BQ + C) \qquad (11.161)$$

$$K = \cosh(BQ + C) \qquad (11.162)$$

$$\gamma = \tan^{-1}\left[\frac{F - JG\sin L}{KG\cos L}\right] - \alpha_c \quad \text{convergence at point}(E,N) \qquad (11.163)$$

$$k = \frac{I\sqrt{1-e^2 \sin^2 \phi}\cos(u/D)}{\cos\phi\cos L} \quad \text{grid scale factor at point}\left(E,N\right) \quad (11.164)$$

Alaska Zone 1 is the only state plane coordinate system zone in the United States that uses the oblique Mercator projection. Figure 11.10 is a computer printout showing example oblique Mercator BK10 and BK11 transformations at Station JNU C

```
                       PROGRAM: LOCALCOR
               COPYRIGHT 2001 BY GLOBAL COGO, INC.
                   LAS CRUCES, NEW MEXICO 88003
                       WWW.GLOBALCOGO.COM
USER: Earl F. Burkholder
DATE: 12 June 2007


OBLIQUE MERCATOR COORDINATE TRANSFORMATIONS
PROJECTION ZONE: Alaska Zone 1 - 5001


REFERENCE ELLIPSOID: GEODETIC REFERENCE SYSTEM 1980
     A = 6378137.0000 METERS     1/F = 298.2572221008827


ZONE PARAMETERS:
     LATITUDE OF ZONE CENTER          57 00  0.000000
     LONGITUDE OF ZONE CENTER         133 40 0.000000
     SCALE FACTOR ALONG AXIS          0.999900000000
     FALSE EASTING AT (U,V) ORIGIN      5000000.0000 METERS
     FALSE NORTHING AT (U,V) ORIGIN    -5000000.0000 METERS
     AZIMUTH OF AXIS AT ORIGIN        323 7 48.368475


ELLIPSOID AND ZONE CONSTANTS
     FIRST ECCENTRICITY SQUARED        0.006694380022903416
     SECOND ECCENTRICITY SQUARED       0.006739496775481622
     CONFORMAL LATITUDE CONSTANTS:  F(0) =    0.006686920927
        F(2)  =     0.000052014583   F(4) =    0.000000554458
        F(6)  =     0.000000006718   F(8) =    0.000000000089
        A     =     6388718.8623050  F    = -0.32701295544998
        B     =     1.00029646140436 G    =  0.94501985532997
        C     =     0.00442683392641 I    =  1.00155891766182
        D     =     6386186.7325316


TRANSFORMATIONS:


NAME OF STATION: JNU C (1992) - PID AI4906          FORWARD (BK10)

     LATITUDE:       58 21 14.364490   NORTHING    726233.5912 METERS
     LONGITUDE:     134 34 26.891220   EASTING     765542.9612 METERS
     CONVERGENCE:        0-45 42.27    SCALE FACTOR:   0.999928422781


NAME OF STATION: JNU C (1992) - PID AI4906      INVERSE (BK11)

     LATITUDE:       58 21 14.364490   NORTHING    726233.5912 METERS
     LONGITUDE:     134 34 26.891220   EASTING     765542.9612 METERS
     CONVERGENCE:        0-45 42.27    SCALE FACTOR:   0.999928422781
```

FIGURE 11.10    Example BK10 and BK11 Transformations for Oblique Mercator Projection

(PID AI4906) near Juneau, Alaska. The transformations were computed using the equations in this section and numerical values match those shown on the NGS data sheet for the same station.

## LOW-DISTORTION PROJECTION

A LDP is a formal name assigned to a specific option used in Box 12 of Figure 1.4. The advantage of using a LDP is that measured ground distances will closely match the computed grid inverse distance between the same two points. Although LDPs are being successfully used in various places, three drawbacks to use of a LDP are (1) that a LDP covers a limited area, making multiple projection zones necessary to cover larger areas, (2) that there is a lack of standardization within the technical literature and professional practice in the use of LDPs, and (3) that LDPs are strictly 2-D while spatial data are 3-D. The first two drawbacks have been significantly mitigated in the recent past but the fact remains that LDPs are a 2-D model being applied to 3-D spatial data. All three drawbacks can be avoided by adopting and using the GSDM because the GSDM consists of one set of proven solid geometry equations equally applicable worldwide for handling 3-D spatial data. An added benefit to using the GSDM is that the GSDM also provides the framework, methods, and procedures for handling spatial data accuracy.

## REFERENCES

Alder, K. 2002. *The Measure of All Things: The Seven-Year Odyssey and Hidden Error That Transformed the World*. New York: Free Press.

Burkholder, E.F. 1980. Use of the Michigan scale factor. *Proceedings of the ACSM Annual Meeting*, St. Louis, MO, March 18–25, 1980.

Burkholder, E.F. 1985. State plane coordinates on the NAD 1983. Paper presented at the *ASCE Spring Convention*, Denver, CO, April 29–May 3, 1985.

Burkholder, E.F. 1991. Computation of horizontal/level distances. *Journal of Surveying Engineering* 117 (3): 104–116.

Burkholder, E.F. 1993a. Design of a local coordinate system for surveying, engineering, and LIS/GIS. *Surveying & Land Information Systems* 53 (1): 29–40.

Burkholder, E.F. 1993b. Using GPS results in true 3-D coordinate system. *Journal of Surveying Engineering* 119 (1): 1–21.

Burkholder, E.F. 2004. Accuracy of elevation factor. *Journal of Surveying Engineering* 130 (3): 134–137.

Burkholder, E.F. 2016. Using the global spatial data model (GSDM) to compute combined factors. *Journal of Surveying Engineering* 142 (4): 06016001.

Claire, C.N. 1968. State plane coordinates by automatic data processing. Publication 62-4. Rockville, MD: Coast & Geodetic Survey, National Information Branch, National Geodetic Survey, National Oceanic and Atmospheric Administration.

Federal Register Notice. 1959. *Federal Register* 24 (128): 5348.

Federal Register Notice. May 16, 1998. *Federal Register*.

Mitchell, H.C. and L. Simmons. [1945] 1977. The state coordinate systems (a manual for surveyors). Special Publication No. 235. Washington, DC: Coast & Geodetic Survey. U.S. Department of Commerce.

Pearson, F., II. 1990. *Map Projectionss: Theory and Applications*. Boca Raton, FL: CRC Press.

Richardus, P. and Adler, R.K. 1972. *Map Projections for Geodesists, Cartographers and Geographers*. Amsterdam, the Netherlands: North-Holland.

Snyder, J.P. 1987. Map projections—A working manual. U.S. Geological Survey Professional Paper 1395. Washington, DC: Government Printing Office.

Stem, J.E. 1989. State plane coordinate system of 1983. NOAA manual NOS NGS 5. Rockville, MD: National Geodetic Survey.

# 12 Spatial Data Accuracy

## INTRODUCTION

This book attempts to

1. Describe the mathematical and geometrical characteristics of geospatial data based on the assumption of a single origin for 3-D data
2. View the transition from the past to the future in terms of the digital revolution
3. Identify a common spatial data model equally useful to those who build, operate, or use measuring systems and those who use spatial data
4. Identify equations and procedures that can be used to organize spatial data computations—from measurements to end user applications—more efficiently
5. Highlight the importance of spatial data quality while providing well-defined procedures for establishing, tracking, storing, and using spatial data accuracy
6. Enhance opportunities for adopting and using the GSDM by including examples and applications

Following comments related to the first four objectives, this chapter* focuses on objective five, spatial data accuracy. Objective six is covered in Chapters 13 through 15.

## FORCES DRIVING CHANGE

The digital revolution is characterized by using the electronic computer and related devices to collect, store, analyze, and report information about the world and human activities. Using computers (and more recently "the cloud") to keep track of spatial data is only a small part of the digital revolution—but it is an important part. Scientists, mathematicians, cartographers, and others have been observing, recording, and describing our world for generations. Although many others also deserve credit, two persons are recognized for specific contributions: Gerard Mercator (1512–1594) is widely acclaimed as the mapmaker who revolutionized cartography, and René Descartes (1596–1650), perhaps better known as a philosopher, was a mathematician who systematized analytical geometry and gave us the Cartesian coordinate system. The profound impact of their combined legacy has permeated mapping and the use of spatial data for over 400 years. Without making light of their work, the digital revolution justifies this new look at fundamental assumptions associated with the collection, manipulation, and use of 3-D digital spatial data. As a consequence, the GSDM is seen as ideally suited for handling 3-D digital spatial data in various disciplines.

---

* This chapter makes extensive use of material from Burkholder (2004). Used with permission.

Although others can also be identified, forces driving the digital revolution include

- The transistor
- Miniaturization of circuits and physical devices
- Development of information technology/science/management
- Electronic signal processing
- Satellite positioning
- Enhanced spatial literacy

Other factors may be identified as consequences of the digital revolution:

- Reduction in privacy
- Better knowledge of where things are
- More efficient movement of people, products, and resources
- Greater access to information
- Nanotechnology

Recognizing the impossibility of identifying mutually exclusive cause-and-effect properties of the various factors, the focus of this book is on spatial data. With this qualifier, the assumption is that electronic measurements (GNSS and others), computer databases, information management, and the exponentially growing demand for reliable spatial data are the forces driving this reevaluation of spatial data models.

## TRANSITION

This book may not be an easy read for those with a low tolerance for detail. The attitude of some is "just tell me what I need to know and let's move on." An observation is that just-in-time learning is preferred to not learning. A new piece of equipment or software is purchased and users are trained how to use it. With practice, users become proficient and productive, making the enterprise profitable. Given the rapid pace of technological developments, such training is essential. However, whether learned in a formal classroom setting or in individual study, an understanding of fundamental concepts provides a foundation for such just-in-time learning. Building this foundation is the purpose of this book. The first three chapters identify details and concepts of the GSDM. Starting from Chapter 4, fundamental mathematical concepts are summarized, and a careful logical building process is included in subsequent chapters. While not all things to all users, the goal is to focus on developing an understanding of the concepts. With an ever-increasing number of persons using GNSS and with the GSDM providing an improved context for understanding spatial data, the collective passion for spatial literacy will be enhanced and a multitude of readers will be able to apply innovative 3-D concepts to an exponentially expanding array of spatial data applications.

It seems that society has been complacent in applying analytical geometry concepts to mapping and to geospatial data. The fundamental theorems of solid geometry and vector algebra were proved long ago and a map is a map is a map. Prior to the digital revolution and burgeoning use of 3-D data, there was little to get excited about. Oh yes, geodesists developed complex mathematical expressions for describing the size and shape of the Earth and cartographers developed an endless array of map projections. Such work is truly impressive. But, for many spatial data users, the most useful map projections are those that make it possible to perform "flat-Earth" computations using simple rules of plane 2-D Euclidean geometry. Because few really understand what it takes to make a good map, such beneficial features are taken for granted and, in part because humans walk erect, the Earth is normally viewed in terms of horizontal and vertical—separately. Again, without complaint, those concepts are learned as part of growing up and the spatial data user community largely accepts the traditional 1-D and 2-D views. Only recently have high-level policy professionals given serious consideration to using an integrated 3-D database. With implementation of the GSDM, spatial data disciplines all over the world will enjoy the luxury of working more efficiently when exchanging and using 3-D digital spatial data.

Of course, the digital revolution has already had an enormous impact on traditional mapping and the way spatial data are used. Existing processes were computerized and automated so that better work could be done more efficiently. Maps are now stored in a digital format in electronic files instead of flat files, and a new map can be plotted from the database at any time. Furthermore, the same database can support a wide range of map products having different themes, scales, or purposes. This added flexibility is very beneficial and has had a significant impact on the use of maps. This is very important, but is not the real issue.

The real issue is that digital spatial data have replaced the analog map as the primary storage medium. A recent web search of "digital Earth" returned over 420,000 hits (a similar search returned only 250,000 hits in 2007) and it seems that many organizations have a preference for a particular format for storing digital spatial data. The underlying characteristics of digital 3-D spatial data were examined carefully in the process of formulating the GSDM. Starting with the assumption of a single origin for 3-D data, rules of solid geometry and vector algebra governed the development of a simple consistent logical 3-D model. Of course, the historic value of maps as a record of the development of civilization remains enormously significant and is to be accommodated (Harvey 2000). Thankfully, existing 2-D applications are fully supported as subordinate uses of 3-D data. But, so far, it appears that the transition to using digital maps is evolving in a fragmented manner that begs careful evaluation. A concise summary of three evolutionary stages is as follows:

- In the past, spatial data were analog and separated into horizontal and vertical components. Maps or photographs were used as the primary storage medium.
- In the interim, spatial data are digital with separate horizontal and vertical components. Digitizing existing maps became an important professional/technical activity. Datums, projections, units, and coordinate systems all affect interoperability.
- In the future, the 3-D characteristics of spatial data measurements are preserved in the computational processes and spatial data are stored in a global integrated 3-D database. Yes, interoperability details are still

important, but metadata and covariance of all spatial data going into the database are the responsibility of the "owner" of the database. Then each user is able to retrieve spatial data from the database and, based upon a user-selected tolerance, is able to rely upon the proven quality of those data. From there each user has the freedom to manipulate the spatial data in accordance with application specifications. Existing 1-D and 2-D uses can continue to be employed as a subset of 3-D. On the other hand, 3-D interoperability is greatly enhanced because the GSDM provides an efficient bidirectional link between rigorous scientific uses and local "flat-Earth" applications.

A transition from the past to the future is both a challenge and an opportunity for spatial data users. Kuhn (1996) describes the processes involved in such a transition in "The Structure of Scientific Revolutions." Several quotes are as follows:

> ...the awareness of anomaly had lasted so long and penetrated so deep that one can appropriately describe the fields affected by it as being in a state of growing crisis (p. 67).
> ...a crisis may end with the emergence of a new candidate for paradigm and with the ensuing battle over its acceptance (p. 84).
> Probably the single most prevalent claim advanced by proponents of a new paradigm is that they can solve problems that have led the old one to a crisis (p. 153).
> Because scientists are reasonable men, one or another argument will ultimately persuade many of them. But there is no single argument that can or should persuade them all (p. 158).

This author is encouraged by the adaptability of the younger generation and their capacity for visualizing spatial relationships. Yes, adapting to policies consistent with efficient use of 3-D spatial data is more difficult for those having a 2-D plus 1-D mind-set, but, somehow, the younger generation seems to be saddled with fewer conceptual obstacles. In time, using the GSDM will be as comfortable for the spatial data user as is the automatic transmission for automobile drivers. Yes, there are still those who, for whatever reason, prefer to use the clutch and standard shift. Such 2-D and 1-D derivative uses of spatial data remain fully supported by the 3-D GSDM. Understandably, while 2-D and 1-D applications are fully supported by a 3-D database, building a 3-D database from incomplete 2-D and 1-D data set is not possible. The recommended procedure is to add only competent 3-D observations to an existing 3-D database.

## CONSEQUENCES

There are consequences associated with using the GSDM. Some have already occurred and are accepted as routine, some are a matter of recognizing the impact of using existing technologies, and other consequences will involve policies related to realization of benefits. Careful planning and implementation will go hand-in-hand.

It is impossible to identify all the consequences but several of the more obvious ones are as follows:

- All spatial data measurements going into the GSDM will need to be 3-D (or even time-stamped, making them 4-D).
- Spatial data are processed and stored under the assumption of a single origin for geospatial data. Elevation becomes a derived quantity—that is computed from existing *X/Y/Z* coordinates. Local observed differential elevation differences remain valid subject to deflection-of-the-vertical considerations.
- A World Vertical Datum yyyy will be adopted in which ellipsoid height is used as the third dimension. Arguments in favor of such adoption are given by Burkholder (2002, 2006), Kumar (2005, 2007), and Soler (2007). The need for geoid modeling will be enormously reduced for many flat-Earth applications but the importance of geoid modeling remains. Geoid heights will continue to be used by those for whom the difference matters, for example, those needing to relate current measurements to legacy data.
- Spatial data users all over the world will be able to work with local flat-Earth differences while enjoying specific, reliable connectivity to the world at large via the ECEF coordinates stored by the GSDM.
- Two groups of spatial data professionals are equally well served by information stored in a common 3-D database:
  1. Surveying, engineering, mapping, and navigation: The *relative* difference of one point with respect to another point is critical and relied upon. The P.O.B. feature in Box 10 of Figure 1.4 means that flat-Earth plane surveying components can be used to obtain ground-level tangent plane direction and distance between points. Consequently, the grid-ground distance difference issue becomes moot and the need for low-distortion projection coordinates goes away. Furthermore, the GSDM defines a mathematical process by which the local accuracy and network accuracy of computed quantities can be reliably determined.
  2. GIS, planning, inventory, and navigation: The *absolute* unique location of a point is of primary consideration and preserved via the ECEF geocentric coordinates. Of course, a point defined by ECEF coordinates can be equivalently expressed by 2-D map projections such as state plane, UTM, or other map projection coordinates.

The alert reader will note that "navigation" appears in both categories.

## ACCURACY

### INTRODUCTION

The stochastic model portion of the GSDM is described in Chapter 1 and addresses issues of spatial data accuracy. Additional information on stochastic models is given by Mikhail (1976), Ghilani (2010), and Burkholder (1999, 2004). Spatial data accuracy is an umbrella term that includes concepts such as uncertainty, standard deviation, positional tolerance, confidence intervals, and error ellipses. Of those, the GSDM uses standard deviation as the underlying concept for quantifying spatial data accuracy.

The stochastic model is used to answer the question, "accuracy with respect to what?" Two obvious possibilities are as follows:

- What is the absolute accuracy of a point with respect to the datum?
- What is the relative accuracy of a point with respect to another identified point?

Although both answers are readily available, they are somewhat different. Absolute datum accuracy is given in terms of standard deviations at the point—allowing a different standard deviation in each of three orthogonal directions—either in the ECEF reference frame or, easier for humans to visualize, in the local east/north/up reference frame. The bidirectional relationship between the ECEF perspective and the local perspective is given by matrix equations (1.32) and (1.33).

Relative accuracy is the standard deviation of the computed distance and/or the direction from the standpoint to the forepoint and can be either network or local. Depending upon choices made by the user and available covariance information, matrix equation (1.36) can be used to obtain either network accuracy (the two points are statistically independent) or local accuracy (reflecting the correlation given by the covariance information) between the two points.

To emphasize the difference between absolute accuracy and relative accuracy, absolute accuracy gives the uncertainty of a *point* with respect to the datum while relative accuracy gives the uncertainty of a distance *between* points—either network or local.

The stochastic model portion of the GSDM is optional. In some cases, the standard deviation information may be readily available, but not needed or used. In other cases, no standard deviations are known or given—meaning that the default value of zero is used for the standard deviation. If no standard deviations are available, the stored $X/Y/Z$ coordinates and associated derived spatial data elements are all used as an exact value. In all cases, functional model computations can still be performed whether standard deviations are available or not. The GSDM can accommodate whatever stochastic information the user provides but providing reliable stochastic data is ultimately the responsibility of each user. Like fire, competent use of the GSDM can provide enormous benefits, but, as with fire, misuse can also be very harmful. Responsible use is essential and consists of inputting and using the standard deviation of each observation or a justifiable estimate thereof.

Not restricted to any one discipline, the GSDM facilitates collection, storage, manipulation, exchange, and use of spatial data worldwide because the same 3-D model accommodates both those activities that generate spatial data and those activities that use spatial data, whether in high-level scientific research or in local "flat-Earth" applications. And, regardless of application, questions regarding spatial data accuracy can be handled with a common set of stochastic model equations. The spatial data accuracy discriminator—selected by the user—is the magnitude of the standard deviation, component by component and can be judged against FGDC (Federal Geographic Data Committee 1998, Table 2.1) standards for spatial data accuracy.

## DEFINITIONS

The intent is to use standard definitions and conventions. However, in case of ambiguity, the following definitions are used for purposes of this chapter.

*Spatial data uncertainty* is given by its standard deviation in each of three dimensions. One standard deviation (1 sigma) provides a 68 percent confidence level. Many spatial data users routinely use a 95 percent (2 sigma) confidence level as the basis for making comparisons and/or inferences.

As stated in Chapter 3, *spatial data* are defined as the distance between endpoints of a line in Euclidean space. Even though a line is the path of a moving point, a distance (not a point) is viewed as the spatial data primitive because the location of a point is meaningless unless and until described with coordinates (distances).

Physical geodesists use a definition of a geodetic datum that also includes the gravity field (Nation Imagery and Mapping Agency 1997). The 3-D GSDM is used in a subordinate manner to the geodetic datums as defined by the international scientific community. For purposes of describing location and spatial data accuracy—and consistent with those official datums—the 3-D GSDM datum is taken to be an ECEF right-handed rectangular $X/Y/Z$ coordinate system whose

1. Origin is at the center of mass of the Earth
2. $Z$-axis coincides (very nearly) with the mean spin axis of the Earth, which means $X/Y$ coordinates are in the plane of the equator
3. $X$-axis is coincident with $0°$ longitude (the Greenwich Meridian), which means the $Y$-axis lies at $90°$ east longitude
4. Distance unit is meters
5. Ellipsoid is defined by two parameters that permit computation of equivalent latitude, longitude, and ellipsoid height coordinates from geocentric $X/Y/Z$ coordinates

As discussed in Chapter 8, the GSDM and error propagation concepts described herein work equally well with the NAD 83, the WGS 84, or with the ITRF. The GSDM is also viewed as compatible with the new datums being developed by the NGS for publication in 2022. Of course, it is not appropriate to mix coordinate values from different datums (unless the datum differences are well below the tolerance threshold of spatial data accuracy selected by the user for a given application). Mixing values from different datums is the responsibility of the user. See section on "Datum Transformations" in Chapter 8.

## ABSOLUTE AND RELATIVE QUANTITIES

The origin of a well-defined coordinate system is the basis of *absolute quantities* (coordinates and elevations) within that system. A *relative quantity* is the difference between two absolute quantities expressed in the same system. Recognizing that the origin of any system is relative to some "larger" system it can be argued that such definitions are not mutually exclusive. For example, geocentric ECEF coordinates are used as absolute quantities, but the center of mass of the Earth is also relative to the center of mass of our solar system etc. As a general rule, the value assigned to a well-defined origin is an absolute quantity.

Comments on absolute and relative relationships:

- ECEF coordinates are absolute quantities. The standard deviations of those absolute quantities are referred to as *datum accuracy*.
- The difference in coordinates between two points (in any given system) is a relative quantity.
- An angle, being the difference between two directions, is relative.
- It is possible for an absolute quantity to be treated as a relative quantity. This could happen if the origin has units of zero. If zero is subtracted from an absolute quantity, the result can be considered a relative value because it represents the difference of two absolute quantities.
- Elevations and time are quantities. Each may look like an absolute quantity, but both are used as relative values due to the ambiguity of their physical origins.
  1. Traditional vertical datums are referenced to an arbitrary zero elevation surface or point that implies datum elevations are all relative.
  2. Time is either counted from the "big bang" (Hawking 1988), from the birth of Christ (BC and AD), from the vernal equinox (the instant of the sun's zero declination), from the daily transit of the sun over a stated meridian (a.m. or p.m.), or from some arbitrary zero computed from the readings of a group of atomic clocks. Whether in years, months, days, hours, or seconds, time is an interval between two specified events—a relative quantity.
- The geoid, often referred to as mean sea level, enjoys a simple physical definition as a "zero" equipotential surface and serves as the origin for orthometric heights. Globally, it is very difficult to locate that zero surface precisely. For the NAVD 88 datum, an existing (arbitrary) orthometric height was adopted for a single station, Father Point/Rimouski, Quebec, Canada. The NGS has committed significant resources to locating the geoid as precisely as possible for the gravimetric vertical datum to be published in 2022.
- Time differences and elevation differences can each be measured quite precisely and that information can be quite useful. But, accuracy statements regarding time and elevation should be limited to relative accuracy statements. In terms of absolute accuracy, there is nothing to be gained from adding a precise interval to an absolute quantity of dubious value.
- Ellipsoid height is a derived quantity with respect to the ellipsoid (ultimately with respect to the center of mass of the Earth). Because the origin is well defined and measurable, an ellipsoid height can be considered an absolute quantity. Ellipsoid height differences are relative quantities.

## SPATIAL DATA TYPES AND THEIR ACCURACY

Spatial data types were listed in Chapter 3 and have been subsequently used throughout this book. With regard to all spatial data components, both absolute and relative, each one can have a standard deviation associated with it. If the standard deviation of any component is zero, the quantity is either known very precisely or the value (e.g., a control point) is being used as a "fixed" quantity. Standard deviations of subsequently computed spatial data components are based upon propagation of the measurement error, and standard deviations of the computed points are determined through the network adjustment process. Given a successful network adjustment and computation of coordinates, the implied accuracy statement is "with respect to the points held fixed by the user." Maybe the beginning point was a hub pounded in the ground. Maybe it was a section corner of the U.S. Public Land Survey System. Maybe it was a HARN point or a CORS point published by the NGS. Or maybe it was the orbit parameters of the GNSS satellites. Understandably, the value of a completed project is greatly enhanced if explicit accuracy statements are made. But, making or not making an explicit statement is not the real issue.

### Accuracy Statements

The real issue is being able to make one of the following statements related to Table 2.1 of the FGDC (1998) standards and supported by appropriate statistics.

1. "The absolute datum (choose one—NAD 83, WGS 84, ITRF—and the appropriate epoch) accuracy of Point X in three dimensions is $\sigma_X = $ _____, $\sigma_Y = $ _____, and $\sigma_Z = $ _____." An equivalent statement, derived from the first, gives the standard deviations in the local reference frame as $\sigma_e = $ _____, $\sigma_n = $ _____, and $\sigma_u = $ _____. The absolute accuracy statement involves only one point and is with respect to the datum selected/named by the user. If the project was a 2-D survey (i.e., based on state plane coordinates), only two components would be named.
2. "The relative network accuracy of the direction and distance from Point 1 to Point 2 is $\sigma_{AZ} = $ _____ and $\sigma_{DIST} = $ _____." Relative accuracy applies to the difference between two independent points having individual absolute accuracy values in the same datum.
3. From Point 1 to Point 2, "the relative network accuracy of the height difference ($\Delta h$) or perpendicular distance from the local tangent plane ($\Delta u$) is $\sigma_{\Delta h} = $ _____ or $\sigma_{\Delta u} = $ _____."
4. "The relative local accuracy of Point 2 with respect to Point 1 is $\sigma_{AZ} = $ _____, $\sigma_{DIST} = $ _____, $\sigma_{\Delta h} = $ _____ or $\sigma_{\Delta u} = $ _____." Relative local accuracy exploits and is largely governed by the statistical correlation that exists between two directly connected points in the same datum. The procedure for computing each of the listed accuracies is given in Equation 1.36.

### But Everything Moves

Most spatial data activities involve using a database such as a GIS. The importance of the basic geodetic control in a GIS is well documented by the National Research

Council (NRC 1983) and others. Ideally, the geodetic control information upon which the database is built should be of such quality that it could be held "fixed," that is, having a zero standard deviation. Here again the question "with respect to what?" becomes relevant. A stable point that is well monumented in one system (e.g., NAD 83) may, in fact, be moving in another (WGS 84 or ITRF). With the advent of GNSS positioning, it is now possible to determine the location of control points much more accurately than before and the scientific community now has conclusive evidence that points once thought to be permanent are, in fact, moving—with respect to what? An oversimplified answer is that "everything moves."

A better answer is required. More specifically, the administrators and users of a database (whether local, regional, national, or global) deserve explicit information as to the stability and accuracy for the various categories of points in the database. And, if they arc moving, what is the velocity vector of the point? This chapter is primarily about 3-D uncertainties but, given that points move, time must be added as the fourth dimension and the epoch must enjoy equal standing with the coordinates. Software for converting $X/Y/Z$ coordinates from one epoch to another is called HTDP and is available gratis from the NGS at https://www.ngs.noaa.gov, accessed May 4, 2017. HTDP can also be used to convert $X/Y/Z$ coordinates from one 3-D datum to another (Snay 1999).

With respect to movement, a simple question must be asked, "Is the observer standing on the train watching the station go by or is the observer standing at the station watching the train go by?" The center of mass of the Earth is the location reference for the entire globe. Points on the surface of the Earth or anywhere within the Earth may move with respect to the center of mass. But, the reference is fixed by definition—the reference does not move. Admittedly, with respect to stable well-monumented points, statements are made that the center of mass of the Earth moves. The implied perspective is considered subordinate to the explicit statement, "the center of mass of the Earth does not move." Of course, with respect to the sun (and the solar system) the Earth rotates on its axis and the center of mass of the Earth revolves about the sun.

The ITRF is defined such that the net tectonic movement of all continental Earth plates is zero (Snay and Soler 1999, Part 3). But, points on the Earth's surface still move with respect to the Earth's center of mass and with respect to each other. Therefore the locations of the ITRF monuments are defined with both coordinates and velocities. Spatial data users in North America will be reassured to know that the NAD 83 datum is the one to use because, except for areas of tectonic activity, points on the NAD 83 remain "fixed" to the North American plate and move together. Such oversimplification is dangerous. The monumented NAD 83 control points on the ground may be stable, but the satellite orbits are and the NGS CORS coordinates are both published in the ITRF reference frame. (NGS also publishes NAD 83 coordinates for the CORS). The issue to be aware of is that the absolute coordinates (for points on the ground) may be in one reference frame and the relative coordinate differences (obtained from GNSS) may be in a different reference frame. Since the NAD 83 and ITRF local relative coordinate differences can be very similar, it can be permissible to attach ITRF relative coordinate differences to absolute NAD 83 datum coordinates. But mixing absolute datum coordinates in the same solution should be avoided—especially on long lines (>10 km).

Spatial data users should be aware of at least three competing 3-D geodetic datums—NAD 83, WGS 84, and ITRF. The reader should be aware that the NGS will be publishing a new horizontal (3-D datum) and a new vertical datum in 2022—see Chapter 8, Geodetic Datums. Each datum has a reason for existing and each has a role to fill. At a gross level of accuracy, it does not matter which datum is used because of their similarities. But, as the tolerance for uncertainty gets smaller and smaller, it does matter which datum is used. The GSDM can be used with each datum individually and provides a consistent method for identifying and tracking the uncertainties in a given datum—whatever they are. Comparing uncertainties (standard deviations) between datums is beyond the scope of this book. Those larger issues are addressed by others such as Han et al. (2008).

## OBSERVATIONS, MEASUREMENTS, AND ERROR PROPAGATION

In many ways, observations and measurements are very similar and the terms are often used interchangeably. But, a mathematical distinction is that observations are always independent quantities and measurements may be either independent or correlated. Stated differently, any observation may be called a measurement, but a measurement can be called an observation only if it is an independent quantity. As listed in Chapter 3, there is only a limited number of quantities that can be directly measured. But, whether the measurement is a length, time, voltage, temperature, etc., spatial data components are determined indirectly from those measurements using appropriate models and computations. The standard deviation of each component is determined by propagating the measurement uncertainty through the variance/covariance equation given by the following matrix formulation:

$$\Sigma_{YY} = J_{YX}\Sigma_{XX}J_{XY}^{t} \tag{12.1}$$

where:
   $\Sigma_{YY}$ = covariance matrix of computed result.
   $J_{YX}$ = Jacobian matrix of partial derivatives of the result with respect to the variables (measurements).
   $\Sigma_{XX}$ = covariance matrix of variables (measurements) used in the computations.

To reiterate, the variables in the measurement covariance matrix are independent and considered to be observations if and only if there is no correlation in the measurement covariance matrix.

### FINDING THE UNCERTAINTY OF SPATIAL DATA ELEMENTS

In the process of establishing the spatial data uncertainty of each point, the user must first decide which datum will be used. Mixing datum values is permissible only if the datum differences are smaller than the resolution of data added to the database. For example, if 10-meter data are being used and if the datum differences are at the

1-meter level, it makes no difference which datum is used. On the other hand, if 10-millimeter data are being used and datum differences are at the 1-meter level, the choice of datum does matter.

Second, each project should be based upon reliable control points having $X/Y/Z$ geocentric coordinates in the appropriate datum. One control point may be sufficient to put a new project on the chosen datum, but making a connection to two or more points is standard practice. If the basic control points are assigned a zero standard deviation, then that means subsequent accuracy statements should be made "with respect to the control points selected and held fixed by the user." Better statements regarding datum accuracy statement can be made if realistic standard deviations are assigned to the points used to control the project. The covariance matrix for each new point and the correlation between points in the network are a standard by-product of a least squares adjustment. When the network adjustment is done in terms of geocentric coordinates and coordinate differences, the resulting covariance matrix is in terms of the geocentric reference frame. The geocentric environment is more efficient for storage and computer operations, but, because of the human perspective, the local covariance matrix is preferred as being more intuitive—giving sigma east, sigma north, and sigma up as the square root of the diagonal elements.

The GSDM includes both the geocentric and local covariance matrices for each point, but, since one can be derived from the other, a BURKORD™ database stores only the geocentric covariance matrix. The local covariance matrix is computed upon demand. Both covariance matrices contain the same datum accuracy of each point component by component, but because of perspective, the numbers are different. Each of the two covariance matrices is a $3 \times 3$ symmetrical matrix containing the following elements:

Geocentric Covariance Matrix

$$\Sigma_{X/Y/Z} = \begin{bmatrix} \sigma_X^2 & \sigma_{XY} & \sigma_{XZ} \\ \sigma_{XY} & \sigma_Y^2 & \sigma_{YZ} \\ \sigma_{XZ} & \sigma_{YZ} & \sigma_Z^2 \end{bmatrix} \tag{12.2}$$

and
Local Covariance Matrix

$$\Sigma_{e/n/u} = \begin{bmatrix} \sigma_e^2 & \sigma_{en} & \sigma_{eu} \\ \sigma_{en} & \sigma_n^2 & \sigma_{nu} \\ \sigma_{eu} & \sigma_{nu} & \sigma_u^2 \end{bmatrix} \tag{12.3}$$

where:
   $\sigma_X^2, \sigma_Y^2, \sigma_Z^2$ = variances for geocentric coordinates for the point.
   $\sigma_{XY}, \sigma_{XZ}, \sigma_{YZ}$ = covariance elements for geocentric coordinates.
   $\sigma_e^2, \sigma_n^2, \sigma_u^2$ = local perspective variances for the point.
   $\sigma_{en}, \sigma_{eu}, \sigma_{nu}$ = local perspective covariance elements for the point.

The two covariance matrices are related by the following rotation matrix evaluated at the latitude/longitude of the standpoint (local origin):

$$\boldsymbol{R} = \begin{bmatrix} -\sin\lambda & \cos\lambda & 0 \\ -\sin\phi\cos\lambda & -\sin\phi\sin\lambda & \cos\phi \\ \cos\phi\cos\lambda & \cos\phi\sin\lambda & \sin\phi \end{bmatrix} \tag{12.4}$$

The matrix expression for the relationship between the two covariance matrices is

$$\Sigma_{e/n/u} = R\Sigma_{X/Y/Z}R^t \tag{12.5}$$

$$\Sigma_{X/Y/Z} = R^t\Sigma_{e/n/u}R \tag{12.6}$$

Points that are part of a network adjustment enjoy an interrelationship described by correlation. The correlation is especially significant for adjacent points that have been connected by a direct measurement. Correlation exists between points not directly connected but the influence drops as the number of courses between points increases (correlation is the reason crossties serve to strengthen a network). If the significant correlations between points are stored along with the covariance matrix for each point, the local accuracy of one point with respect to the other is readily computed along with the inverse direction and distance. If correlations are not stored (or if they are assumed to be zero), an inverse computation will readily provide the direction and distance between points and the two endpoint covariance matrices will provide the basis of the network accuracy associated with the relative differences.

## USING POINTS STORED IN A *X/Y/Z* DATABASE

Each stored *X/Y/Z* location is unique within the birdcage of orbiting GNSS satellites. Three application modes for using the stored *X/Y/Z* locations include the following:

- Single point (unique location for inventory tag).
- Point-pair (used to create lines, surfaces, and objects).
- "Cloud" (mapping). Even though stored as *X/Y/Z*, the location of any point can also be readily expressed in 2-D latitude and longitude, UTM, or state plane coordinates along with 1-D ellipsoid height.

The uncertainty of a single point is given by the datum accuracy as computed from the geocentric covariance matrix. These uncertainties (standard deviations, variances, and other covariance elements) can be viewed in either the geocentric reference frame or in the local reference frame. The geocentric reference frame is more efficient for data storage and computerized manipulation, but the local reference frame is more convenient for viewing because horizontal and vertical is the human perspective. If a point is a bench mark (vertical control point) it should have a small standard deviation on the vertical component. By contrast, horizontal control points should have small standard deviations on the east and or north components. A 3-D control point should have standard deviations on all three components.

The point-pair application provides the relative location of one point with respect to another. A map is generated by extensive successive use of the point-pair mode. The geometrical integrity of any such map is a function of the quality of data in the $X/Y/Z$ database and the relative spacing of the points from which the map is plotted. Local maps can be plotted from one P.O.B. but small scale maps will be best constructed if plotted from multiple P.O.B.s. Additional studies are needed to determine trade-off criteria between P.O.B. spacing and plotting accuracy.

Specifically, in the point-pair mode, point 1 is defined by $X_1/Y_1/Z_1$ and point 2 is defined by $X_2/Y_2/Z_2$. As given by Equations 1.34 and 1.35, the matrix formulation of the 3-D geocentric inverse from point 1 to point 2 is

$$\begin{bmatrix} \Delta X \\ \Delta Y \\ \Delta Z \end{bmatrix} = \begin{bmatrix} -1 & 0 & 0 & 1 & 0 & 0 \\ 0 & -1 & 0 & 0 & 1 & 0 \\ 0 & 0 & -1 & 0 & 0 & 1 \end{bmatrix} \begin{bmatrix} X_1 \\ Y_1 \\ Z_1 \\ X_2 \\ Y_2 \\ Z_2 \end{bmatrix} \tag{12.7}$$

The matrix of coefficients to the variables is called the Jacobian matrix and the general error propagation formulation in the form of Equation 12.1 is

$$\Sigma_\Delta = J \Sigma_{1 \to 2} J^t \tag{12.8}$$

Using the Jacobian matrix of 1's and 0's from Equation 12.7, having the geocentric covariance matrix of point 1 and point 2 both available, and using the correlation between point 1 and point 2, the covariance matrix of the inverse is computed using Equation 12.8 as

$$\Sigma_\Delta = \begin{bmatrix} -1 & 0 & 0 & 1 & 0 & 0 \\ 0 & -1 & 0 & 0 & 1 & 0 \\ 0 & 0 & -1 & 0 & 0 & 1 \end{bmatrix} \begin{bmatrix} \begin{bmatrix} \sigma_{X_1}^2 & \sigma_{X_1Y_1} & \sigma_{X_1Z_1} \\ \sigma_{X_1Y_1} & \sigma_{Y_1}^2 & \sigma_{Y_1Z_1} \\ \sigma_{X_1Z_1} & \sigma_{Y_1Z_1} & \sigma_{Z_1}^2 \end{bmatrix} & \begin{bmatrix} \sigma_{X_1X_2} & \sigma_{X_1Y_2} & \sigma_{X_1Z_2} \\ \sigma_{Y_1X_2} & \sigma_{Y_1Y_2} & \sigma_{Y_1Z_2} \\ \sigma_{Z_1X_2} & \sigma_{Z_1Y_2} & \sigma_{Z_1Z_2} \end{bmatrix} \\ \begin{bmatrix} \sigma_{X_1X_2} & \sigma_{Y_1X_2} & \sigma_{Z_1X_2} \\ \sigma_{X_1Y_2} & \sigma_{Y_1Y_2} & \sigma_{Z_1Y_2} \\ \sigma_{X_1Z_2} & \sigma_{Y_1Z_2} & \sigma_{Z_1Z_2} \end{bmatrix} & \begin{bmatrix} \sigma_{X_2}^2 & \sigma_{X_2Y_2} & \sigma_{X_2Z_2} \\ \sigma_{Y_2X_2} & \sigma_{Y_2}^2 & \sigma_{Y_2Z_2} \\ \sigma_{Z_2X_2} & \sigma_{Z_2Y_2} & \sigma_{Z_2}^2 \end{bmatrix} \end{bmatrix} \begin{bmatrix} -1 & 0 & 0 \\ 0 & -1 & 0 \\ 0 & 0 & -1 \\ 1 & 0 & 0 \\ 0 & 1 & 0 \\ 0 & 0 & 1 \end{bmatrix} \tag{12.9}$$

The off-diagonal submatrices reflect the correlation between point 1 and point 2.

*Datum accuracy* of point 1 and point 2 is included in Equation 12.9 as their respective covariance submatrices.

The following concise mathematical statements are the basis for the definitions of local accuracy and network accuracy given earlier in Chapter 1.

*Local accuracy* of the inverse between point 1 and point 2 is obtained by using the full covariance matrix in Equation 12.9. Correlation between point 1 and point 2 is included.

*Network accuracy* of the inverse between point 1 and point 2 is obtained if the correlation between point 1 and point 2 is either nonexistent or taken to be zero.

## EXAMPLE

The following example is a summary of a fully documented least squares network solution posted at http://www.globalcogo.com/nmsunet1.pdf, May 4, 2017. The GNSS network includes seven GNSS vectors and is based upon two A-order HARN points—station REILLY (PID AI5445) located in the central horseshoe of the New Mexico State University (NMSU) campus and station CRUCESAIR (PID CX1939) located at the Las Cruces airport some 16 kilometers west of the campus. The network consists of seven independent baselines connecting four additional points to the existing HARN stations as shown in Figure 12.1.

The GNSS baselines shown and used were collected on four different dates over a period of 5 years. These are not the only baselines on campus nor are they the only observations between the points in question. These baselines were selected because they show excellent consistency, are independent, and include often used points. The network is included here to illustrate use of the GSDM and computation of both network accuracy and local accuracy.

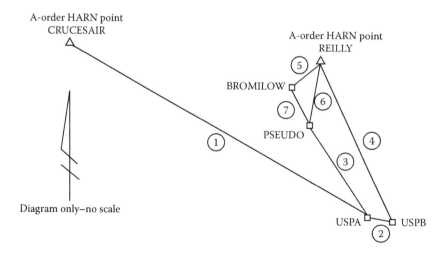

**FIGURE 12.1**   GNSS Survey Network on NMSU Campus

The original network adjustment documented in the link above and as listed in the first edition is based upon the 1992 NGS published values for stations REILLY and CRUCESAIR. The same vectors are used in a subsequent adjustment holding the 2011 NGS published values for the control stations. The results of the adjustment holding the 2011 values are reported herein. The adjustment results holding the 1992 NGS control station values are included in the first edition of this book.

## CONTROL VALUES AND OBSERVED VECTORS

The NAD 83 (2011) geocentric $X/Y/Z$ coordinates for A-order HARN stations REILLY and CRUCESAIR are as published by the NGS and were held fixed in this exercise. They are

| Station REILLY (2011) | Station CRUCESAIR (2011) |
|---|---|
| $X = -1,556,177.595$ m | $X = -1,571,430.649$ m |
| $Y = -5,169,235.284$ m | $Y = -5,164,782.254$ m |
| $Z = 3,387,551.720$ m | $Z = 3,387,603.202$ m |

Single-frequency Trimble GPS receivers were used to collect static data, 57 minutes being the shortest common observation time for any of the seven baselines. The baseline components and the covariance matrix for each observed baseline as determined by Trimble software using default processing parameters are

### Baseline 1—CRUCESAIR to USPA—Observed 3/28/02 (Use Subscript CA):

| | | Sxx | Syy | Szz |
|---|---|---|---|---|
| $\Delta X_{CA} = 15,752.080$ m | Sxx | 6.321492E–06 | | |
| $\Delta Y_{CA} = -5,179.102$ m | Syy | 1.545948E–05 | 4.739877E–05 | |
| $\Delta Z_{CA} = -903.089$ m | Szz | –1.061303E–05 | –3.184780E–05 | 2.388036E–05 |

### Baseline 2—USPA to USPB—Observed 11/12/03 (Use Subscript AB):

| | | Sxx | Syy | Szz |
|---|---|---|---|---|
| $\Delta X_{AB} = 14.964$ m | Sxx | 1.412453E–06 | | |
| $\Delta Y_{AB} = -15.365$ m | Syy | 1.285418E–06 | 4.653209E–06 | |
| $\Delta Z_{AB} = -16.664$ m | Szz | –5.669127E–07 | –1.658118E–06 | 1.872469E–06 |

### Baseline 3—USPA to PSEUDO—Observed 3/28/02 (Use Subscript AP):

| | | Sxx | Syy | Szz |
|---|---|---|---|---|
| $\Delta X_{AP} = -528.036$ m | Sxx | 9.505016E–08 | | |
| $\Delta Y_{AP} = 560.657$ m | Syy | 8.957064E–08 | 3.729339E–07 | |
| $\Delta Z_{AP} = 585.897$ m | Szz | –5.022282E–08 | –2.221975E–07 | 3.363763E–07 |

**Baseline 4—USPB to REILLY—Observed 3/28/02 (Use Subscript BR):**

|  |  | Sxx | Syy | Szz |
|---|---|---|---|---|
| $\Delta X_{BR} = -514.003$ m | Sxx | 3.650165E−07 |  |  |
| $\Delta Y_{BR} = 741.438$ m | Syy | 9.024127E−07 | 2.796189E−06 |  |
| $\Delta Z_{BR} = 868.293$ m | Szz | −6.189027E−07 | −1.881145E−06 | 1.410196E−06 |

**Baseline 5—BROMILOW to REILLY—Observed 12/10/98 (Use Subscript MR):**

|  |  | Sxx | Syy | Szz |
|---|---|---|---|---|
| $\Delta X_{MR} = 32.134$ m | Sxx | 2.762550E−07 |  |  |
| $\Delta Y_{MR} = 51.175$ m | Syy | 3.200312E−07 | 6.870545E−07 |  |
| $\Delta Z_{MR} = 94.198$ m | Szz | −2.008940E−07 | −4.006259E−07 | 4.661596E−07 |

**Baseline 6—PSEUDO to REILLY—Observed 1/23/02 (Use Subscript PR):**

|  |  | Sxx | Syy | Szz |
|---|---|---|---|---|
| $\Delta X_{PR} = 29.000$ m | Sxx | 1.325760E−07 |  |  |
| $\Delta Y_{PR} = 165.422$ m | Syy | 1.317165E−07 | 5.265054E−07 |  |
| $\Delta Z_{PR} = 265.719$ m | Szz | −7.253348E−08 | −3.020965E−07 | 5.006575E−07 |

**Baseline 7—BROMILOW to PSEUDO—Observed 1/23/02 (Use Subscript MP):**

|  |  | Sxx | Syy | Szz |
|---|---|---|---|---|
| $\Delta X_{MP} = 3.136$ m | Sxx | 3.367818E−07 |  |  |
| $\Delta Y_{MP} = -114.242$ m | Syy | 3.937476E−07 | 8.766570E−07 |  |
| $\Delta Z_{MP} = -171.527$ m | Szz | −5.186521E−07 | −8.977932E−07 | 1.446501E−06 |

## BLUNDER CHECKS

In order to verify the absence of blunders in the baselines, misclosures were computed for each component ($X/Y/Z$) as follows:

**Traverse Including Baselines 1, 2, and 4 (from CRUCESAIR to REILLY):**

|  | X | Y | Z |
|---|---|---|---|
| Station CRUCESAIR | −1,571,430.649 m | −5,164,782.254 m | 3,387,603.202 m |
| Baseline 1 | 15,752.080 m | −5,179.102 m | −903.089 m |
| Baseline 2 | 14.964 m | −15.365 m | −16.664 m |
| Baseline 4 | −514.003 m | 741.438 m | 868.293 m |
| Computed value | −1,556,177.618 m | −5,169,235.283 m | 3,387,551.7428 m |
| Station REILLY | −1,556,177.595 m | −5,169,235.284 m | 3,387,551.720 m |
| Misclosures | −0.023 m | 0.001 m | 0.022 m |

*(Continued)*

The loop including baselines 2-3-7-5-4 (being careful to preserve sign convention):

| Baseline 2 | −14.964 m | 15.365 m | 16.664 m |
| Baseline 3 | −528.036 m | 560.657 m | 585.897 m |
| Baseline 7 | −3.136 m | 114.242 m | 171.527 m |
| Baseline 5 | 32.134 m | 51.175 m | 94.198 m |
| Baseline 4 | 514.003 m | −741.438 m | −868.293 m |
| Misclosures | 0.001 m | 0.001 m | −0.007 m |

The loop including baselines 5-6-7 (being careful to preserve sign convention):

| Baseline 5 | 32.134 m | 51.175 m | 94.198 m |
| Baseline 6 | −29.000 m | −165.422 m | −265.719 m |
| Baseline 7 | −3.136 m | 114.242 m | 171.527 m |
| Misclosures | −0.002 m | −0.005 m | 0.006 m |

## LEAST SQUARES SOLUTION

All baselines have been included in the checks and all misclosures are acceptable. Therefore, it is legitimate to perform a least squares adjustment of the seven baselines to determine the "best" adjusted position for points USPA, USPB, PSEUDO, and BROMILOW. Any adjustment should also provide information on the quality of the answers (i.e., what is the standard deviation of the computed position?) in both the geocentric ($X/Y/Z$) reference frame and in the local (*east/north/up*) reference frame. The posted paper—http://www.globalcogo.com/nmsunet1.pdf, accessed May 4, 2017 NAD 83 (1992)—includes three different weighting schemes and shows a comparison of the various answers. The example here—NAD 83 (2011)—only shows the least squares results obtained using the full covariance matrix of each observed baseline in the "indirect observations" least squares model using one equation for each observation (3 observations per baseline and 7 baselines = 21 observations). The weight matrix was computed as the inverse of the covariance matrix of the observations with an *a priori* reference variance of 1.0:

$$\mathbf{v} + \mathbf{B}\Delta = f \qquad (12.10)$$

$$W = \left(1.0\right)\Sigma^{-1} \qquad (12.11)$$

The posted paper shows formulation of the matrices used in the solution for all three weighting possibilities. Those details are not included here, but the solution shown below was formulated as a linear problem and the matrix solution was obtained as

$$\Delta = \left(\mathbf{B}^{t}\, W\, \mathbf{B}\right)^{-1} \mathbf{B}^{t}\, W\, f \quad \text{or stated differently,} \quad \Delta = N^{-1}\, \mathbf{B}^{t}\, W\, f \qquad (12.12)$$

where:
   $\Delta$ = vector of parameters (answers).
   $N$ = matrix of normal equations ($B^t$ $W$ $B$). ($N^{-1}$ contains statistics for the answers.)
   $W$ = weight matrix obtained from baseline covariance matrices.
   $B$ = matrix of coefficients for the unknown parameters.
   $f$ = vector of constants computed from known values and observations.
   $v$ = vector of residuals.
   $N^{-1}$ = normal part of a least squares adjustment, shown in Figure 12.2.

The estimated (*a posteriori*) reference variance was computed as

$$\sigma_0^2 = \frac{v^t W v}{r} = 26.2668 \text{ m}^2 \tag{12.13}$$

where:
   $v$ = vector of residuals.
   $W$ = weight matrix.
   $r$ = the redundancy.

and the covariance matrix of the computed parameters is the product of the reference variance and the $N^{-1}$ matrix as shown in Figure 12.3.

## RESULTS

The geocentric $X/Y/Z$ coordinates of the four unknown points as shown in Figure 12.4 were computed directly in the least squares adjustment while the geodetic latitude, geodetic longitude, and ellipsoid height were computed from the $X/Y/Z$ values using the BK2 transformation. The covariance matrix of each new point is the 3 × 3 submatrix shown in Figure 12.3, and the standard deviations of the geocentric $X/Y/Z$ coordinates were computed as the square root of the variances as found in Figure 12.3. The standard deviations at each point in the local reference frame ($e/n/u$) were computed using Equation 12.5 and the computed latitude/longitude at each station.

## NETWORK ACCURACY AND LOCAL ACCURACY

Datum accuracy, network accuracy, and local accuracy are defined mathematically in Equation 12.9. Datum accuracy is a statement of how well the position of a single point is known with respect to the published datum. Network accuracy can be intuitively understood to be a statement of accuracy between points based upon how well the positions are known with respect to the control held by the user. It is presumed the points are independent—that is, there is no correlation of one with respect to the other as might be determined by a direct tie between them. Alternatively, local accuracy can be understood to be a statement of accuracy between points based upon a direct measurement between the points. The following paragraphs describe the results of computing both network accuracy and network accuracy from point USPA to point PSEUDO. The Excel spreadsheet shown in Appendix C (the file is called "3-D inverse with statistics.xlsx") was used to generate the values in Table 12.1 and can be obtained gratis from the author at http://www.globalcogo.com, accessed May 4, 2017.

|  | USPA | | | USPB | | | PSEUDO | | | BROMILOW | | |
|---|---|---|---|---|---|---|---|---|---|---|---|---|
| **USPA** | 1.7718E2-07 | 1.8375E-07 | -1.1720E-07 | 4.3066E-08 | 8.2451E-08 | 5.8009E-08 | 9.7813E-08 | 1.0060E-07 | -6.9907E-08 | 5.4180E-08 | 5.2480E-08 | -2.3281E-08 |
|  | 1.8374E-07 | 6.7135E-07 | -3.9266E-07 | 8.8343E-08 | 2.8447E-07 | -1.9218E-07 | 1.0135E-07 | 3.4076E-07 | -2.0173E-07 | 5.3394E-08 | 1.6539E-07 | -7.3629E-08 |
|  | -1.1720E-07 | -3.9266E-07 | 5.4022E-07 | -8.6569E-08 | -2.6605E-07 | 2.1867E-07 | -6.9275E-08 | -2.0122E-07 | 2.8992E-07 | -1.1552E-08 | -6.0629E-08 | 7.8919E-07 |
| **USPB** | 4.3066E-08 | 8.8343E-08 | -8.6569E-08 | 2.5318E-07 | 5.4453E-06 | -3.6494E-07 | 2.8437E-08 | 4.5314E-08 | -4.5972E-08 | 1.6720E-08 | 1.7811E-08 | -1.2879E-08 |
|  | 8.2451E-08 | 2.8447E-07 | -2.6605E-07 | 5.4453E-06 | 1.7148E-06 | -1.1089E-06 | 4.6961E-08 | 1.4918E-07 | -1.3958E-07 | 1.6056E-08 | 6.3792E-08 | -3.9397E-08 |
|  | -5.8009E-08 | -1.9218E-07 | 2.1867E-07 | -3.6494E-07 | -1.1089E-06 | 8.6421E-07 | -3.3394E-08 | -9.7816E-08 | 1.1975E-07 | -9.6270E-09 | -3.5276E-08 | 3.9427E-08 |
| **PSEUDO** | 9.7813E-08 | 1.0135E-07 | -6.9275E-08 | 2.8437E-08 | 4.6961E-08 | -3.3394E-08 | 1.0616E-07 | 1.0508E-07 | -7.2079E-08 | 5.8828E-08 | 5.5454E-08 | -2.4082E-08 |
|  | 1.0060E-07 | 3.4076E-07 | -2.0122E-07 | 4.5314E-08 | 1.4918E-07 | -9.7816E-08 | 1.0508E-07 | 3.6175E-07 | -2.1721E-07 | 5.4821E-08 | 1.7443E-07 | -7.8620E-08 |
|  | -6.9907E-08 | -2.0173E-07 | 2.8992E-07 | -4.5972E-08 | -1.3958E-07 | 1.1975E-07 | -7.2079E-08 | -2.1721E-07 | 3.3551E-07 | -7.3860E-09 | 5.9960E-08 | 8.8734E-08 |
| **BROMILOW** | 5.4180E-08 | 5.3394E-08 | -1.1552E-08 | 1.6720E-08 | 1.6056E-08 | -9.6270E-09 | 5.8828E-08 | 5.4821E-08 | -7.3850E-09 | 1.7206E-07 | 1.8485E-07 | -1.3826E-07 |
|  | 5.2480E-08 | 1.6539E-07 | -6.0629E-08 | 1.7811E-08 | 6.3792E-08 | -3.5276E-08 | 5.5454E-08 | 1.7443E-07 | -5.9960E-08 | 1.8485E-07 | 4.4358E-07 | -2.8219E-07 |
|  | -2.3281E-08 | -7.3629E-08 | 7.8919E-07 | -1.2879E-08 | -3.9397E-08 | 3.9427E-08 | -2.4082E-08 | -7.8626E-08 | 8.8734E-08 | -1.3826E-07 | -2.8219E-07 | 3.7445E-07 |

**FIGURE 12.2**　N Inverse Matrix From Least Squares Adjustment

| | USPA | | | USPB | | | PSEUDO | | | BROMILOW | | |
|---|---|---|---|---|---|---|---|---|---|---|---|---|
| USPA | 4.6540E-06 | 4.8265E-06 | -3.0783E-06 | 1.1312E-06 | 2.1657E-06 | 1.5237E-06 | 2.5692E-06 | 2.6423E-06 | -1.8362E-06 | 1.4231E-06 | 1.3785E-06 | -6.1152E-07 |
| | 4.8263E-06 | 1.7634E-05 | -1.0314E-05 | 2.3205E-06 | 7.4721E-06 | -.0480E-06 | 2.6621E-06 | 8.9506E-06 | -5.2988E-06 | 1.4025E-O6 | 4.3444E-06 | -1.9340E-06 |
| | -3.0783E-06 | -1.0314E-05 | 1.4190E-05 | -2.2739E-06 | -6.9882E-06 | 5.7437E-06 | -1.8196E-06 | -5.2854E-06 | 7.6153E-06 | -3.0343E-07 | -1.5925E-06 | 2.0729E-05 |
| USPB | 1.1312E-06 | 2.3205E-06 | -2.2739E-06 | 6.6501E-O6 | 1.4303E-04 | -9.5859E-06 | 7.4695E-07 | 1.1903E-06 | -1.2075E-06 | 4.3918E-07 | 4.6784E-07 | -3.3829E-07 |
| | 2.1657E-06 | 7.4721E-06 | -6.9882E-06 | 1.4303E-05 | 4.5041E-05 | -2.9127E-05 | 7.4695E-07 | 3.9185E-06 | -3.6664E-06 | 4.2174E-07 | 1.6756E-06 | -1.0348E-06 |
| | -1.5237E-06 | -5.0480E-06 | 5.7437E-06 | -9.5859E-06 | -2.9127E-05 | 2.2700E-05 | -8.7715E-07 | -2.5693E-06 | 3.1455E-06 | -2.5287E-07 | -9.2659E-07 | 1.0356E-06 |
| PSEUDO | 2.5692E-06 | 2.6621E-06 | -1.8196E-06 | 7.4676E-07 | 1.2335E-06 | -8.7715E-07 | 2.7884E-06 | 2.7601E-06 | -1.8933E-06 | 1.5452E-06 | 1.4566E-06 | -6.3256E-07 |
| | 2.6423E-06 | 8.9506E-06 | -5.2854E-06 | 1.1903E-06 | 3.9185E-06 | -2.5693E-06 | 2.7601E-06 | 9.5021E-06 | -5.7054E-06 | 1.4400E-06 | 4.5817E-06 | -2.0651E-06 |
| | -1.8362E-06 | -5.2988E-06 | 7.6153E-06 | -1.2075E-06 | -3.6664E-06 | 3.1455E-06 | -1.8933E-06 | -5.7054E-06 | 8.8154E-06 | -1.9401E-07 | 1.5750E-06 | 2.3308E-06 |
| BROMILOW | 1.4231E-06 | 1.4025E-06 | -3.0343E-07 | 4.3918E-07 | 4.2174E-07 | -2.5287E-07 | 1.5452E-06 | 1.4400E-06 | -1.9401E-07 | 4.5195E-06 | 4.8554E-06 | -3.6316E-06 |
| | 1.3785E-06 | 4.3444E-06 | -1.5925E-06 | 4.6784E-07 | 1.6756E-06 | -9.2659E-07 | 1.4566E-06 | 4.5817E-06 | 1.5750E-06 | 4.8554E-06 | 1.1651E-05 | -7.4121E-06 |
| | -6.1152E-07 | -1.9340E-06 | 2.0729E-05 | -3.3829E-07 | -1.0348E-06 | 1.0356E-06 | -6.3256E-07 | -2.0653E-06 | 2.3308E-06 | -3.6316E-06 | -7.4121E-06 | 9.8355E-06 |

**FIGURE 12.3**  Covariance Matrix of Computed Points in Network

| Geocentric and ECEF Sigma | Geodetic and Local Sigma |
|---|---|

Station USPA—NAD 83 (2011)

$X =$ −1,555,678.559 m +/− 0.0022 m $\quad$ $\phi =$ 32° 16′ 23.″00 123 N +/− 0.0024 m (N)
$Y =$ −5,169,961.360 m +/− 0.0042 m $\quad$ $\lambda =$ 106° 44′ 48.″90 785 W +/− 0.0018 m (E)
$Z =$ 3,386,700.102 m +/− 0.0038 m $\quad$ $h =$ 1,177.988 m $\qquad$ +/− 0.0052 m (U)

Station USPB—NAD 83 (2011)

$X =$ −1,555,663.594 m +/− 0.0026 m $\quad$ $\phi =$ 32° 16′22.″36 344 N +/− 0.0017 m (N)
$Y =$ −5,169,976.726 m +/− 0.0067 m $\quad$ $\lambda =$ 106° 44′ 48.″19 118 W +/− 0.0014 m (E)
$Z =$ 3,386,683.431 m +/− 0.0048 m $\quad$ $h =$ 1,177.882 m $\qquad$ +/− 0.0083 m (U)

Station PSEUDO—NAD 83 (2011)

$X =$ −1,556,206.595 m +/− 0.0017 m $\quad$ $\phi =$ 32° 16′45/74 754 N +/− 0.0020 m (N)
$Y =$ −5,169,400.704 m +/− 0.0031 m $\quad$ $\lambda =$ 106° 45′ 14.″39 942 W +/− 0.0014 m (E)
$Z =$ 3,387,285.999 m +/− 0.0030 m $\quad$ $h =$ 1,165.614 m $\qquad$ +/− 0.0039 m (U)

Station BROMILOW—NAD 83 (2011)

$X =$ −1,556,209.730 m +/− 0.0021 m $\quad$ $\phi =$ 32° 16′52.″33 508 N +/− 0.0019 m (N)
$Y =$ −5,169,286.460 m +/− 0.0034 m $\quad$ $\lambda =$ 106° 45′15/77239 W +/− 0.0016 m (E)
$Z =$ 3,387,457.523 m +/− 0.0031 m $\quad$ $h =$ 1,165.495 m $\qquad$ +/− 0.0045 m (U)

**FIGURE 12.4** Geocentric and Local Reference Frame Positions and Standard Deviations

When using the Excel spreadsheet, the user keys information into the spreadsheet and answers appear instantaneously. Input includes the names of the two stations, the geocentric $X/Y/Z$ coordinates of the two points, and the covariance information. When computing the inverse, the direction and distance will remain the same but the standard deviations will be different depending upon the covariance information input by the user. Choices for entering covariance information are as follows:

1. All standard deviations are entered as zeros. This means there is no standard deviation available and the $X/Y/Z$ coordinate data are used as being "fixed." The spreadsheet will still compute the local tangent plane direction and distance between points, but there will be no standard deviations associated with the inverse direction and distance.
2. The user can enter the standard deviations of the geocentric $X/Y/Z$ coordinates as variances (standard deviations squared). These covariance data are entered on the diagonal of the geocentric covariance matrix for each point. The spreadsheet computes the local reference frame covariance matrix (showing the local component $e/n/u$ standard deviations of each point), the inverse direction and distance standpoint to forepoint, and the standard deviation of the direction and the distance. Local and network accuracy will be identical because no correlation data were entered.
3. The user can enter the full covariance matrix for each point. This is the "best" inverse one can get without also providing correlation information as in step 4. This answer is "network" accuracy and presumes the coordinates of the two

## TABLE 12.1
## Comparison of Network and Local Accuracies NAD 83 (2011)

| USPA (Standpoint) | PSEUDO (Forepoint) |
|---|---|
| $X = -1,555,678.559$ m | $X = -1,556,206.595$ m |
| $Y = -5,169,961.360$ m | $Y = -5,169,400.704$ m |
| $Z = 3,386,700.102$ m | $Z = 3,387,285.999$ m |

Coordinate differences from standpoint to forepoint are

| Geocentric Differences | Local Differences |
|---|---|
| $\Delta X = -528.036$ m | $\Delta e = -667.190$ m |
| $\Delta Y = 560.656$ m | $\Delta n = 700.810$ m |
| $\Delta Z = 585.898$ m | $\Delta u = -12.448$ m |

Local inverse from standpoint to forepoint is

$$\text{Dist} = \sqrt{\Delta e^2 + \Delta n^2} \text{ and } \tan \alpha = \frac{\Delta e}{\Delta n} \text{ (same for each case following)}$$

| | | | Network Accuracy | Local Accuracy |
|---|---|---|---|---|
| 1. | No standard deviations | Distance = 967.615 m | ±0.0000 m | 0.0000 m |
| | | Direction = 316° 24′ 28.″1 | ±0.00 sec. | 0.00 sec. |
| 2. | Standard deviations of | Distance = 967.615 m | ±0.0044 m | 0.0044 m |
| | X/Y/Z values only | Direction = 316° 24′ 28.″1 | ±0.77 sec | 0.77 sec. |
| 3. | Full covariance matrix of | Distance = 967.615 m | ±0.0027 m | 0.0027 m |
| | each X/Y/Z point | Direction = 316° 24′ 28.″2 | ±0.59 sec. | 0.59 sec. |
| 4. | Full covariance matrix and | Distance = 967.615 m | ±0.0027 m | 0.0015 m |
| | correlation submatrix | Direction = 316° 24′ 28.″1 | ±0.59 sec. | 0.35 sec. |

points are statistically independent of one another. Local accuracy will compute as being identical to network accuracy because no correlation data are provided.

4. Or, the user may enter the full covariance matrix at each point as well as the correlation matrices between points. The correlation of the forepoint with respect to the standpoint is the transpose of the correlation of the standpoint with respect to the forepoint. It is redundant, but both correlation matrices need to be entered (the astute Excel user will quickly rekey the appropriate cells so that correlation data needs to be entered only once).

NGS geodesists (Soler and Smith 2010) took exception to the characterization and computation of local accuracy as described in this chapter. Differences of opinion were published in Discussion (Burkholder 2012) and Closure (Soler and Smith 2012), apparently without resolution. A follow-up rebuttal by Burkholder (2013) confirms that the derivation and the use of local accuracy as presented herein is correct, complete, and rigorous. Additional information about network accuracy and local accuracy for short, medium, and long lines is also included in the rebuttal. Details for those and other issues are discussed in Chapter 14.

## REFERENCES

Burkholder, E.F. 1999. Spatial data accuracy as defined by the GSDM. *Journal of Surveying and Land Information Systems* 59 (1): 26–30. http://www.globalcogo.com/accuracy.pdf, accessed May 4, 2017.

Burkholder, E.F. 2002. Elevations and the global spatial data model (GSDM). Paper presented at the *58th Annual Meeting of the Institute of Navigation*, Albuquerque, NM, June 24–26, 2002. http://www.zianet.com/globalcogo/elevgsdm.pdf, accessed May 4, 2017.

Burkholder, E.F. 2004. Fundamentals of spatial data accuracy and the global spatial data model (GSDM). Washington, DC: Filed with the U.S. Copyright Office. http://www.globalcogo.com/fsdagsdm.pdf, accessed May 4, 2017.

Burkholder, E.F. 2006. The digital revolution: Whither now? Global spatial data model. *GIM International* 20 (9): 25–27. http://www.globalcogo.com/Digital%20RevR.pdf, accessed May 4, 2017.

Burkholder, E.F. 2012. Discussion of "Rigorous Estimation of Local Accuracies". *Journal of Surveying Engineering* 138 (1): 46–48.

Burkholder, E.F. 2013. Standard deviation and network/local accuracy of geospatial data. Washington, DC: Filed with the U.S. Copyright Office. http://www.globalcogo.com/StdDevLocalNetwork.pdf, accessed May 4, 2017.

Federal Geographic Data Committee (FGDC). 1998. *Geospatial Positioning Accuracy Standards, Part 2: Standards for Geodetic Networks*. Reston, VA: Federal Geographic Data Committee, U.S. Geological Survey. http://www.fgdc.gov/standards/standards_publications, accessed May 4, 2017.

Ghilani, C. 2010. *Adjustment Computations*, 5th ed. Hoboken, NJ: John Wily & Sons.

Han, J.Y., van Gelder, B.M., Soler, T., and Snay, R.A. 2008. Geometric combination of multiple terrestrial network solutions. *Journal of Surveying Engineering* 134 (4): 126–131.

Harvey, M. 2000. *The Island of Lost Maps: A True Story of Cartographic Crime*. New York: Broadway.

Hawking, S.W. 1988. *A Brief History of Time*. New York: Bantam.

Kuhn, T.S. 1996. *The Structure of Scientific Revolutions*, 3rd ed. Chicago, IL: University of Chicago Press.

Kumar, M. 2005. When ellipsoid height will do the job, why look elsewhere! *Surveying and Land Information Science* 65 (2): 91–94.

Kumar, M. 2007. Ellipsoidal heights and engineering applications: Research with real data proves they will work! *Journal of Surveying Engineering* 133 (3): 96–97.

Mikhail, E. 1976. *Observations and Least Squares*. New York: Harper & Row.

National Imagery and Mapping Agency (NIMA). July 4, 1997. Department of Defense World Geodetic System 1984: Its definition and relationships with local geodetic systems, 3rd ed., NIMA TR8350.2. Bethesda, MD: National Imagery and Mapping Agency.

National Research Council (NRC). 1983. *Panel on a Multipurpose Cadastre: Procedures and Standards for a Multipurpose Cadaster*. Washington, DC: National Academy Press.

Snay, R.A. 1999. Using the HTDP software to transform spatial coordinates across time and between reference frames. *Surveying and Land Information Systems* 59 (1): 15–25.

Snay, R.A. and Soler, T. 1999. Modern terrestrial systems—Part 3: WGS 84 and ITRF. *Professional Surveyor* 20 (3): 24–28. http://www.ngs.noaa.gov/CORS/Articles/Reference-Systems-Part-3.pdf, accessed May 4, 2017.

Soler, T. 2007. Editorial: Practitioner's forum revisited. *Journal of Surveying Engineering* 133 (3): 95.

Soler, T. and Smith, D. 2010. Rigorous estimation of local accuracies. *Journal of Surveying Engineering* 136 (3): 120–125.

Soler, T. and Smith, D. 2012. Closure to "Rigorous estimation of local accuracies". *Journal of Surveying Engineering* 138 (1): 48–50.

# 13 Using the GSDM to Compute a Linear Least Squares GNSS Network

## INTRODUCTION

As stated in Chapter 4, least squares is an important tool when using the GSDM. This chapter describes formulation of a *linear* least squares solution for geocentric networks. Such an example is no substitute for a more generalized nonlinear least squares solution but it shows the power and efficiency of using geocentric coordinate differences in a least squares solution. A linear least squares solution has an advantage in that time-consuming iterations are avoided. This could be particularly important in time-critical circumstances such as real-time navigation or monitoring the motion of a drone or driverless vehicle. Yes, computer speeds are exceedingly fast, but achieving least squares results quickly in a real-time environment can benefit from both fast computers and efficient algorithms. Following a description of how a linear least squares solution is formulated, a detailed example network based upon RINEX data downloaded from nine stations of the Wisconsin Department of Transportation CORS (WISCORS) network is included. Other applications are described in subsequent chapters. Further clarification is provided in the last section of this chapter, "Notes Pertaining to Adjustment."

## PARAMETERS AND LINEARIZATION

Familiarity with terminology is a part of understanding use of a least squares adjustment. The topic of least squares is first introduced in Chapter 4 and described in generic terms. Greater specificity is given here. First, Chapter 4 notes that the goal of a least squares adjustment is to find the best answers to a problem. The word "parameters" is also parenthetically equated with the word "answers." More formally, parameters are elements of a problem that have no known numeric value prior to an adjustment. The objective of a least squares adjustment is to find the most probable numerical values for the parameters.

Chapter 4 also states that linearization is used to solve problems whose mathematical formulation includes nonlinear relationships. Since matrices are used extensively in computer solutions of least squares problems and since matrix manipulation only handles sets of linear equations, the "standard" least squares approach is to approximate nonlinear functions with the first two (linear) terms of a Taylor series expansion applied "near" the final solution. This method is very effective and is used extensively. A consequence of using matrices and least squares to solve nonlinear problems is that iterations are required to find a solution. The computational burden required for iteration can be significant, but with fast computers, the small price to pay (in added time) for a reliable solution is inconsequential in most applications. Nonetheless, applications that can be formulated as a linear problem can be solved more quickly than those requiring linearization and iteration.

## BASELINES AND VECTORS

The words "baseline" and "vector" are often used interchangeably when discussing spatial data. Depending on the context, sometimes one word is more appropriate than the other. Even so, a distinction between the two may depend on differing interpretation of the context. Generically, a vector is a mathematical entity that has both a magnitude and a direction. Similarly, a GNSS baseline is the direction and distance between two stations computed from simultaneous observations from two GNSS receivers. In that context, a baseline has the characteristics of a vector but a vector is not necessarily a baseline.

## OBSERVATIONS AND MEASUREMENTS

As noted in Chapter 12 the words "measurements" and "observations" are also often used interchangeably but, due to statistical independence, there is a difference that becomes important. Strictly speaking, an observation is statistically independent of any other observation, the associated weight matrix of a collection of observations is a diagonal matrix and there is no correlation between observations. When the observations are considered as measurements, as is done here, the associated weight matrix can contain information on correlation between the measurements. That correlation is found in the off-diagonal elements of the weight matrix. When observations are correlated, they are called measurements rather than observations. This occurs when using each component of a geocentric baseline as an observation—called a measurement in many cases. The example included herein uses baselines computed from independent GNSS observations. The three $\Delta X/\Delta Y/\Delta Z$ components of a baseline are correlated and the weight matrix for the overall adjustment consists of a diagonal arrangement of $3 \times 3$ submatrices—one block of $3 \times 3$ elements for each baseline. If there is correlation between baselines—such as happens if two baselines share common observations at a station—that correlation appears in appropriate $3 \times 3$ off-diagonal submatrices of the covariance matrix. Incidentally, this is where an attempt to include trivial vectors in an adjustment becomes futile because the covariance matrix accommodates only independent and correlated (not dependent) observations. Restating, the example discussed in this chapter includes independent vectors where each component is called a measurement (as opposed to an observation). Correlated vectors can be legitimate but are not included in this example. *Trivial vectors are not to be used in a least squares adjustment.*

## COVARIANCE MATRICES AND WEIGHT MATRICES

Mathematically, a weight matrix is the inverse of the associated covariance matrix. The weight matrix is used in a least squares adjustment and the covariance matrix reflects the stochastic properties of the observations or measurements. Variances are computed as the square of the standard deviations and appear as the diagonal elements of the covariance matrix. If correlation exists between the measurements, that information shows up in the off-diagonal elements of the covariance matrix. Any of several options may be encountered when building a covariance matrix:

1. If no correlation exists, the measurements are called independent observations, the off-diagonal elements are all zeros, and the standard deviation squared (the variance) of each observation appears in the diagonal elements of the covariance matrix. Elements of the weight matrix (used in a least squares adjustment) are simply computed as 1/variance of each observation and the weight matrix is a diagonal matrix. If the observations are independent, two diagonal covariance matrix possibilities are as follows:
   a. The standard deviations of the observations are all the same. This means the adjustment will be equally weighted and the weight matrix makes no contribution to the adjustment.
   b. The standard deviation squared (variance) of each independent observation is included as a diagonal element of the covariance matrix. The weight matrix is still the inverse of the covariance matrix but, because of the independence, each diagonal element of the weight matrix is more simply computed as 1 divided by the standard deviation squared.
2. If correlation exists in the covariance matrix, the observations are called measurements and the covariance matrix has elements on the off-diagonal. The weight matrix is still computed as the inverse of the covariance matrix.

Discussing option 2 further, the question to be addressed is, "How are the off-diagonal elements of the covariance determined?" Error propagation procedures are discussed in Chapter 4. Specifically, the truly independent observations (remember, the covariance matrix of independent observations is diagonal with elements containing the standard deviation squared of each observation) are used in the error propagation equation process as defined in Equation 4.23. The partial derivative of each measurement with respect to each observation forms the Jacobian matrix of partial derivatives. The resulting covariance matrix (left side of Equation 4.23) is computed from the Jacobian matrix of partial derivatives and the covariance matrix of the observations (right side of Equation 4.23). These procedures are derived in texts such as Mikhail (1976), Ghilani (2010), and others.

Vendor-specific baseline processing software provides the geocentric $\Delta X/\Delta Y/\Delta Z$ components for each vector when processing GNSS data. The same software package typically also provides the $3 \times 3$ covariance matrix for each baseline. Be aware that some vendors report elements of the covariance matrix as variances and covariances as discussed here. Another option for populating the covariance matrix is to report standard deviations (on the diagonal) and correlations on the off-diagonals. The two options are related by Equation 4.22.

## TWO EQUIVALENT ADJUSTMENT METHODS

There are two general approaches to solving a least squares problem—the method of indirect observations and the method of observations-only. In advanced least squares applications, the two approaches are combined under a broader umbrella called "a unified approach to least squares adjustment" (Mikhail 1976). The same two approaches are called a "conditional adjustment" and a "parametric adjustment" by Ghilani (2010) who describes combining the two procedures as a "general least squares method." The example in this chapter includes GNSS vectors and utilizes the method of indirect observations in which one equation is written for each observation (really a measurement) in terms of known constants, unknown parameters, and residuals. The GNSS network is built from independent vectors but the $\Delta X/\Delta Y/\Delta Z$ components are called measurements because they are correlated. If the components are not correlated, they can be called observations and the covariance matrix is diagonal. If, however, baseline components are correlated, they should be called measurements and the covariance matrix will have elements on the off-diagonals. In either case, the weight matrix used in the adjustment is the inverse of the covariance matrix.

The observations-only method of least squares adjustment is popular when solving interconnecting leveling loops because linear condition equations can be written in terms of loop closures. The least squares adjustment solves for the adjusted observations. The answers (elevations, unknowns, parameters) are then computed using the adjusted observations. A side note is that GNSS loops can also be solved using observations only. One difference is that a GNSS loop carries three different linear components simultaneously. The method of observations-only is discussed in surveying adjustment texts such as Mikhail (1976), Mikhail and Gracie (1981), and Ghilani (2010).

One key to successful solution of a least squares problem is building the appropriate matrices successfully. A wrong digit, or a digit in the wrong place, can derail an otherwise elegant solution. Once the measurements, parameters, fixed values, and weights of a given project are coded into the appropriate matrices, the least squares solution is reduced to manipulating those matrices according to proven methods given by others, for example, Ghilani (2010) and Mikhail (1976), and used in the section "Formulation of Matrices—Indirect Observations." Many software tools—including Excel spreadsheets—are readily available for matrix manipulation. Some of the points are as follows:

1. Although it can be tedious, the formulation of a least squares problem can be reduced to rote procedures of building matrices. Standard matrix manipulation tools can be used to obtain a solution. That puts a lot of power into the hands of the user.
2. Preliminary loop and/or point-to-point computations should be performed on the raw vector components to insure small misclosures prior to performing

an adjustment. If any baseline or measurement cannot be included in a misclosure computation, it will not be part of the least squares adjustment. If any misclosure is larger than acceptable, the reader must know that a least squares solution cannot be used to "fix" that problem. All input data should be investigated to identify any inconsistency that might be considered a blunder. Although a least squares adjustment cannot "fix" a blunder, it is also true that a preliminary least squares adjustment may be helpful in isolating blunders.

3. Interpreting the results of a least squares adjustment ranges from the obvious to the esoteric. On one hand, it is easy to look at the residuals found during an adjustment and see that none of the observations needs to be "corrected" by an unreasonable amount. If any residual is deemed to be too large, the reason for the deficiency should be investigated.

4. If, following an adjustment, any residual appears to be too large (assuming that the initial misclosures were acceptable), the investigation should include examining the weighting assumptions to make sure they are realistic. This too calls for more esoteric investigation.

5. With practice and experience, esoteric becomes routine.

## FORMULATIONS OF MATRICES—INDIRECT OBSERVATIONS

The method of indirect observations writes one equation for each measurement. The fundamental equation is given as

$$Y = f(x) + v \tag{13.1}$$

where:
  $Y$ = a numerical value either known or unknown.
  $f(x)$ = combination of parameters, fixed values, and one measurement.
  $v$ = the residual (a "small" correction to the measurement).

With one equation written for each measurement, the number of equations will be the same as the number of measurements. If the number of unknown parameters exceeds the number of equations, there is no redundancy and no least squares solution is possible. The number of measurements and the number of unknowns are important because they are subscripts to the matrices to be formed as shown below. If the number of observations equals (is the same as) the number of unknown parameters, there is a unique solution (also no redundancy) and no adjustment is possible. Redundancy is a prerequisite for a least squares adjustment and is the difference between number of measurements and number of unknowns ($r = n - u$). In statistics, redundancy is the same as degrees of freedom. Typically, more redundancy will give a better solution.

  $n$ = number of measurements.
  $u$ = number of unknowns (parameters).
  $1$ = indicates that the matrix is a vector, typically $(n, 1)$ or $(u, 1)$.

As stated in Chapter 4, the least squares solution in matrix form is formulated as

$$v + B\Delta = f \tag{13.2}$$

$$W = \sigma_0^2 \Sigma^{-1} = 1 \times \Sigma^{-1} \tag{13.3}$$

$$\Delta = \left( B^t W B \right)^{-1} B^t W f \quad \text{or, as often stated,} \quad \Delta = N^{-1} B^T W f \tag{13.4}$$

where:
$v_{n,1}$ = vector of residuals, one residual for each measurement.
$B_{n,u}$ = matrix of partial derivatives—in this case, 1s and 0s.
$\Delta_{u,1}$ = vector of unknown parameters—answers to be found.
$f_{n,1}$ = constants computed from known values and measurements.
$\Sigma_{n,n}$ = covariance matrix of measurements.
$W_{n,n}$ = weight matrix is the inverse of the covariance matrix.
$N_{u,u} = N = B^t W B$, $N$ is a square matrix of "normal" equations.
$\sigma_0^2$ = *a priori* reference variance. Can be anything but is often 1 (unity).

$$\hat{\sigma}_0^2 = a \text{ posteriori estimate of the reference variance} = \frac{v^t W v}{r}. \tag{13.5}$$

$$\Sigma_{\Delta\Delta}^2 = \hat{\sigma}_0^2 \times N^{-1} \text{ is the covariance matrix of the parameters.} \tag{13.6}$$

The reference variance is a single number. The *a priori* reference variance (before the adjustment) is often taken to be unity—that is 1. But, in reality, the *a priori* reference variance is a scalar and can be any number. A value other than 1 is sometimes used to keep the magnitude of the values encountered during an adjustment in a reasonable range. The magnitude of the *a posteriori* reference variance (after the adjustment) can be very informative. Simply, if the *a posteriori* reference variance is close to unity (1), that is an indication that the collection of variances associated with the measurements is reasonable. If the *a posteriori* reference variance is significantly less than 1, that is an indication that the variances associated with the measurements are probably smaller than those used in the adjustment—in other words, that the measurements are more precise than assumed before computing the adjustment. If the *a posteriori* reference variance is significantly greater than unity, that is an indication that there may a problem with the weighting—typically, that the variances associated with the measurements are larger than first assumed. This is especially true if the residuals were all within the expected range. Hypothesis testing can be employed to analyze the impact of unexpected *a posteriori* reference variances—either too large or too small. See Ghilani (2010) for more information on hypothesis testing.

As used in a least squares adjustment, the covariance matrix represents the stochastic properties of the measurements. There is one line in a covariance matrix for each measurement and the standard deviations squared (the variances) of the measurements are on the diagonal of the covariance matrix. If there are correlations between the measurements, they will appear as off-diagonal elements in the covariance matrix. Options for building the covariance matrix include the following:

1. If the measurements are independent (truly observations) and if all have the same standard deviation, then the adjustment will be "equally weighted" and the weight matrix makes no contribution to the adjustment. In that case, Equation 13.4 is written as

$$\Delta = \left( B^t B \right)^{-1} B^t f \text{ or as often stated, } \Delta = N^{-1} B^T f \qquad (13.7)$$

2. If the measurement has a standard deviation, the variance (standard deviation squared) appears on the diagonal of the covariance matrix. If the measurement qualifies as an observation, then it is not correlated with any other measurements.
3. If the measurement has a standard deviation and is correlated with other measurements, then the off-diagonal elements of the covariance matrix will show the impact of the correlation.

The GNSS network example shown in Chapter 12 includes the full covariance matrix of each baseline in the adjustment. All three weighting options (equal weights, variances of baseline components only, and full baseline covariance) for that project are included in a report of that project posted at http://www.globalcogo.com/nmsunet1.pdf, accessed May 4, 2017.

Equations 13.2 through 13.4 are all matrix equations that lend themselves to efficient manipulation. But, substantial effort is needed to get the answers in Equation 13.4. Matrices and computers make the effort worthwhile. A detailed example of building the matrices is as follows:

1. The first step is to use the measurements, one at a time, according to Equation 13.1. There will be "$n$" equations and each parameter needs a place in each equation. If a given parameter is not included in that equation, the coefficient is 0. In that case, 0 is a "place holder" in the subsequent $B$ coefficient matrix.
2. The observation equations (step 1) are rewritten in the format of Equation 13.2. From there, the residuals form a $(n, 1)$ vector, the $B$ matrix $(n, u)$ is formed from the coefficients of the parameters, the parameters form a $(u, 1)$ vector, and the constant $f$ vector $(n, 1)$ is a number computed from all the known elements in the equation.
3. The weight matrix is the inverse of the covariance matrix discussed previously.

With all matrices completed, the solution proceeds as dictated by the matrix manipulations listed in Equation 13.4. After the parameters are found, the residuals are computed using the found parameters and Equation 13.2. The *a posteriori* reference variance is found using the computed residuals in Equation 13.5. The variances (and standard deviations) of the computed parameters are computed using Equation 13.6.

## EXAMPLE GNSS NETWORK PROJECT IN WISCONSIN

The Wisconsin Department of Transportation (WisDOT) operates the WISCORS network that covers the entire state. RINEX data for nine network stations in southeastern Wisconsin were downloaded and processed using Trimble Business Center software to compute the vectors consisting of baseline components and associated covariance matrix for each baseline. Figure 13.1 is a map of the statewide network. Figure 13.2 shows the nine stations used with names and numbered vectors. Table 13.1 shows the NAD 83 (2011) values for control stations FOLA and KEHA

**FIGURE 13.1**   Statewide WISCORS Network (Courtesy of WisDOT, Madison, WI)

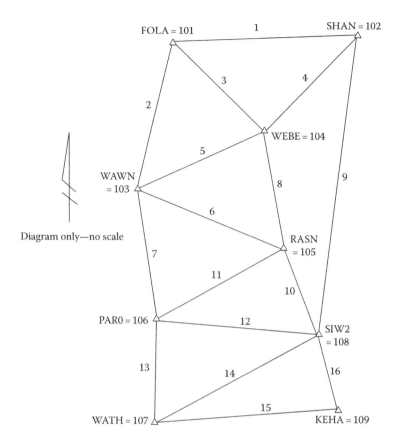

**FIGURE 13.2** GNSS Network in Wisconsin

as obtained from WisDOT and used in the adjustment. A listing of computed baseline components and associated covariance values is given in Table 13.2. Although only one adjustment is reported herein, several different adjustments were computed using tolerance values of 0.020 m, 0.100 m, 1.00 m, and 5.00 m on the two control stations to show that the local accuracy between CORS control stations remains fairly constant even though the standard deviations of the published values of those two control stations were allowed to be different from one adjustment to another. The purpose of the adjustment as described was not to duplicate the station values as obtained from WisDOT (although they were all quite close). The real purpose of the adjustment was to obtain the covariance matrix of the results and to use those results to compute both network accuracy and local accuracy as determined using Equation 12.9. A summary of the network/local accuracy project is discussed in Chapter 14, which shows that local (relative) accuracy between points is readily available and quite consistent even though network (absolute) accuracy suffers according to the user's choice for diminished quality of the control points. A partial printout for the 0.020 m case is included in Chapter 14.

**TABLE 13.1**

**Data for Control Stations—NAD 83 (2011)**

FOLA (Geocentric $X/Y/Z$ Values Obtained from WisDOT, Covariance Values Are Assumed[a])

| | | $S_{XX}$ | $S_{YY}$ | $S_{ZZ}$ |
|---|---|---|---|---|
| $X =$ 124,517.6445 m | $S_{XX}$ | 0.000004 | | |
| $Y = -4,609,735.8115$ m | $S_{YY}$ | 0.000000 | 0.000004 | |
| $Z =$ 4,391,808.8308 m | $S_{ZZ}$ | 0.000000 | 0.000000 | 0.000004 |

KEHA (Geocentric $X/Y/Z$ Values Obtained from WisDOT, Covariance Values Were Assumed[a])

| | | $S_{XX}$ | $S_{YY}$ | $S_{ZZ}$ |
|---|---|---|---|---|
| $X -$ 170,846.0946 m | $S_{XX}$ | 0.000004 | | |
| $Y = -4,702,276.0741$ m | $S_{YY}$ | 0.000000 | 0.000004 | |
| $Z =$ 4,291,680.3174 m | $S_{ZZ}$ | 0.000000 | 0.000000 | 0.000004 |

[a] Normal practice is to hold control points "fixed." The assumption here is that a standard deviation of 0.002 m for each $X/Y/Z$ component is very tight. The impact of assuming other tolerances for the control points is reported in Chapter 14 where network accuracy and local accuracy are discussed.

## RINEX DATA USED TO BUILD THE WISCONSIN NETWORK

RINEX data were downloaded from the WISCORS web site for nine stations such that the following baselines could be computed. Each baseline is independent because each baseline included data that were not used in another baseline. That is, the pair of station data sets used to compute each baseline is unique, thus preventing any opportunity for a trivial vector to affect the adjustment. The RINEX files were used and baselines were computed using Trimble software along with the NGS precise ephemeris. The $\Delta X/\Delta Y/\Delta Z$ baseline components and the (symmetric) covariance matrix obtained for each baseline are shown in Table 13.2.

## BLUNDER CHECKS

Misclosures are computed in this section to verify the absence of blunders in the data. Those misclosures could be computed one triangle at a time or the check could be completed more efficiently by including each vector in a series of baselines head-to-tail so long as the string of vectors ends where it started and so long as each vector is included in the string. With the given configuration of vectors, it is easier to use three multistation loops. For loop 1, the chosen sequence of vectors in this case is: 1, 9, 10, 11, 12, 16, 15, 13, 7, and 2. The second loop includes vectors 5, 3, 1, 4, 8, and 6. Using a vector a second time is no problem. The third loop includes vectors 12, 13, and 14. The sign of each component is important when computing the misclosures. While the loop check follows the "from-to" sequence in the data, the baseline components carry the same sign as observed. In those cases where the loop goes against the "from-to" sequence in the observation, the signs of each component of the baseline must be reversed. That reversal is indicated in Table 13.3 by

## TABLE 13.2
## Baselines Computed from WISCORS RINEX Files

**Baseline 1, FOLA to SHAN**

|  |  | $S_{XX}$ | $S_{YY}$ | $S_{ZZ}$ |
|---|---|---|---|---|
| $\Delta X_{FS} = -57,891.383$ m | $S_{XX}$ | 2.8502E–6 |  |  |
| $\Delta Y_{FS} = 1,679.130$ m | $S_{YY}$ | –1.1776E–6 | 2.86377E–5 |  |
| $\Delta Z_{FS} = 3,825.279$ m | $S_{ZZ}$ | 5.4830E–7 | –2.22834E–5 | 2.23555E–5 |

**Baseline 2, FOLA to WAWN**

|  |  | $S_{XX}$ | $S_{YY}$ | $S_{ZZ}$ |
|---|---|---|---|---|
| $\Delta X_{FW} = -21,811.232$ m | $S_{XX}$ | 1.38454E–5 |  |  |
| $\Delta Y_{FW} = -48,033.426$ m | $S_{YY}$ | 2.91732E–5 | 1.41957E–4 |  |
| $\Delta Z_{FW} = -50,037.232$ m | $S_{ZZ}$ | –2.08675E–6 | –9.07974E–5 | 6.88237E–5 |

**Baseline 3, FOLA to WEBE**

|  |  | $S_{XX}$ | $S_{YY}$ | $S_{ZZ}$ |
|---|---|---|---|---|
| $\Delta X_{FB} = 25,381.565$ m | $S_{XX}$ | 1.9546E–6 |  |  |
| $\Delta Y_{FB} = -27,977.353$ m | $S_{YY}$ | –1.8561E–6 | 3.22602E–5 |  |
| $\Delta Z_{FB} = -30,092.793$ m | $S_{ZZ}$ | 1.2268E–6 | –2.49196E–5 | 2.53440E–5 |

**Baseline 4, SHAN to WEBE**

|  |  | $S_{XX}$ | $S_{YY}$ | $S_{ZZ}$ |
|---|---|---|---|---|
| $\Delta X_{SB} = -32,510.268$ m | $S_{XX}$ | 2.7268E–6 |  |  |
| $\Delta Y_{SB} = -26,298.224$ m | $S_{YY}$ | –3.5647E–6 | 8.55221E–5 |  |
| $\Delta Z_{SB} = -26,267.509$ m | $S_{ZZ}$ | 1.5143E–6 | –5.28146E–5 | 4.08228E–5 |

**Baseline 5, WEBE to WAWN**

|  |  | $S_{XX}$ | $S_{YY}$ | $S_{ZZ}$ |
|---|---|---|---|---|
| $\Delta X_{BW} = -47,192.795$ m | $S_{XX}$ | 2.8605E–6 |  |  |
| $\Delta Y_{BW} = -20,056.051$ m | $S_{YY}$ | –2.1611E–6 | 3.64246E–5 |  |
| $\Delta Z_{BW} = -19,944.432$ m | $S_{ZZ}$ | 2.9093E–6 | –3.153363E–5 | 3.31404E–5 |

**Baseline 6, WAWN to RASN**

|  |  | $S_{XX}$ | $S_{YY}$ | $S_{ZZ}$ |
|---|---|---|---|---|
| $\Delta X_{WR} = 50,353.209$ m | $S_{XX}$ | 5.6081E–6 |  |  |
| $\Delta Y_{WR} = -9,089.568$ m | $S_{YY}$ | –2.9970E–6 | 6.93793E–5 |  |
| $\Delta Z_{WR} = -11,155.809$ m | $S_{ZZ}$ | 2.0288E–6 | –4.40001E–5 | 3.58558E–5 |

**Baseline 7, WAWN to PAR0**

|  |  | $S_{XX}$ | $S_{YY}$ | $S_{ZZ}$ |
|---|---|---|---|---|
| $\Delta X_{WP} = 12,264.525$ m | $S_{XX}$ | 2.4541E–6 |  |  |
| $\Delta Y_{WP} = -22,491.525$ m | $S_{YY}$ | 2.2576E–6 | 4.60535E–5 |  |
| $\Delta Z_{WP} = -24,372.823$ m | $S_{ZZ}$ | –2.5089E–6 | –3.26178E–5 | 2.99862E–5 |

*(Continued)*

**TABLE 13.2 (*Continued*)**
**Baselines Computed from WISCORS RINEX Files**

Baseline 8, WEBE to RASN

|  |  | $S_{XX}$ | $S_{YY}$ | $S_{ZZ}$ |
|---|---|---|---|---|
| $\Delta X_{BR} = 3{,}160.419$ m | $S_{XX}$ | 1.3630E−6 |  |  |
| $\Delta Y_{BR} = -29{,}145.645$ m | $S_{YY}$ | −2.029E−7 | 2.71011E−5 |  |
| $\Delta Z_{BR} = -31{,}100.235$ m | $S_{ZZ}$ | −1.928E−7 | −2.13529E−5 | 2.24759E−5 |

Baseline 9, SHAN to SIW2

|  |  | $S_{XX}$ | $S_{YY}$ | $S_{ZZ}$ |
|---|---|---|---|---|
| $\Delta X_{SS} = -17{,}613.230$ m | $S_{XX}$ | 1.6345E−6 |  |  |
| $\Delta Y_{SS} = -67{,}771.572$ m | $S_{YY}$ | −2.3304E−6 | 4.31998E−5 |  |
| $\Delta Z_{SS} = -71{,}093.815$ m | $S_{ZZ}$ | 2.8629E−7 | −2.86799E−5 | 3.03883E−5 |

Baseline 10, RASN to SIW2

|  |  | $S_{XX}$ | $S_{YY}$ | $S_{ZZ}$ |
|---|---|---|---|---|
| $\Delta X_{RS} = 11{,}736.622$ m | $S_{XX}$ | 2.1715E−6 |  |  |
| $\Delta Y_{RS} = -12{,}327.695$ m | $S_{YY}$ | −3.1324E−6 | 2.86377E−5 |  |
| $\Delta Z_{RS} = -13{,}726.078$ m | $S_{ZZ}$ | 3.3230E−7 | 3.69809E−5 | 3.84384E−5 |

Baseline 11, RASN to PAR0

|  |  | $S_{XX}$ | $S_{YY}$ | $S_{ZZ}$ |
|---|---|---|---|---|
| $\Delta X_{RP} = -38{,}088.692$ m | $S_{XX}$ | 1.3707E−6 |  |  |
| $\Delta Y_{RP} = -13{,}401.941$ m | $S_{YY}$ | −2.506E−7 | 2.79621E−5 |  |
| $\Delta Z_{RP} = -13{,}217.015$ m | $S_{ZZ}$ | −1.501E−7 | −2.17399E−5 | 2.24557E−5 |

Baseline 12, SIW2 to PAR0

|  |  | $S_{XX}$ | $S_{YY}$ | $S_{ZZ}$ |
|---|---|---|---|---|
| $\Delta X_{SP} = 49{,}825.313$ m | $S_{XX}$ | 1.3474E−6 |  |  |
| $\Delta Y_{SP} = 1{,}074.247$ m | $S_{YY}$ | −5.96E−8 | 2.61021E−5 |  |
| $\Delta Z_{SP} = -509.066$ m | $S_{ZZ}$ | −3.773E−7 | −2.01012E−5 | 2.05163E−5 |

Baseline 13, PAR0 to WATH

|  |  | $S_{XX}$ | $S_{YY}$ | $S_{ZZ}$ |
|---|---|---|---|---|
| $\Delta X_{PW} = 591.015$ m | $S_{XX}$ | 1.4192E−6 |  |  |
| $\Delta Y_{PW} = -26{,}179.546$ m | $S_{YY}$ | −6.85E−8 | 2.89509E−5 |  |
| $\Delta Z_{PW} = -28{,}308.041$ m | $S_{ZZ}$ | −3.123E−7 | −2.25179E−5 | 2.31770E−5 |

Baseline 14, SIW2 to WATH

|  |  | $S_{XX}$ | $S_{YY}$ | $S_{ZZ}$ |
|---|---|---|---|---|
| $\Delta X_{SW} = 49{,}234.298$ m | $S_{XX}$ | 1.8818E−6 |  |  |
| $\Delta Y_{SW} = 27{,}253.794$ m | $S_{YY}$ | 1.138E−7 | 2.64428E−5 |  |
| $\Delta Z_{SW} = 27{,}798.977$ m | $S_{ZZ}$ | 3.332E−7 | −1.98930E−5 | 2.09629E−5 |

*(Continued)*

---

**TABLE 13.2 (*Continued*)**

**Baselines Computed from WISCORS RINEX Files**

**Baseline 15, WATH to KEHA**

| | | $S_{XX}$ | $S_{YY}$ | $S_{ZZ}$ |
|---|---|---|---|---|
| $\Delta X_{WK} = 55{,}284.138$ m | $S_{XX}$ | 4.8227E–6 | | |
| $\Delta Y_{WK} = 4{,}164.224$ m | $S_{YY}$ | –9.042E–7 | 6.82233E–5 | |
| $\Delta Z_{WK} = 2{,}589.570$ m | $S_{ZZ}$ | 1.0047E–6 | –4.17516E–5 | 3.36817E–5 |

**Baseline 16, SIW2 to KEHA**

| | | $S_{XX}$ | $S_{YY}$ | $S_{ZZ}$ |
|---|---|---|---|---|
| $\Delta X_{SK} = -6{,}049.840$ m | $S_{XX}$ | 3.7414E–6˙ | | |
| $\Delta Y_{SK} = 23{,}089.552$ m | $S_{YY}$ | –5.3433E–6 | 8.19327E–5 | |
| $\Delta Z_{SK} = 25{,}209.414$ m | $S_{ZZ}$ | 1.9981E–6 | –5.11704E–5 | 3.92520E–5 |

All 16 vectors have been included in a misclosure check and the misclosures indicate an absence of blunders. Therefore, it appears that a least squares adjustment
of the network will be legitimate.

---

a minus (–) sign following each vector whose component signs are changed. Other
loop check configurations are legitimate so long as each vector is included in a loop
of baselines.

## BUILDING MATRICES FOR A LINEAR LEAST SQUARES SOLUTION

This section contains "how to" instructions for developing input files for a linear least
squares adjustment tailored to use with the GSDM. The theory and development of
matrix content is documented elsewhere—see standard least squares texts for example Ghilani (2010), Mikhail (1976), and Mikhail and Gracie (1981), or http://www.
globalcogo.com/nmsunet1.pdf, accessed May 4, 2017. for an actual example.

Matrices as used in least squares solutions have subscripts that reflect characteristics of the problem being solved. That is

　　1 = the second subscript of a matrix that is a vector.
　　$n$ = the number of equations—same as number of observations.
　　$u$ = the number of unknown parameters to be found in the solution.

Three matrices are used in this example, $f$, $B$, and $Q$.

The $f$ vector contains one observation for each equation. The $B$ matrix contains
only 1s and 0s as determined by whether a parameter is in the equation. The cofactor
$Q$ matrix contains the weighting information.

### $f$ VECTOR—$n$, 1

The $f$ vector contains the observations used by the program to find the adjusted
values (parameters). The $f$ vector contains only one observation in each equation.

## TABLE 13.3
## Misclosure Computations Showing Absence of Blunders

### Sequence of Vectors—Beginning at FOLA

|  |  | $\Delta X$ (m) | $\Delta Y$ (m) | $\Delta Z$ (m) |
|---|---|---|---|---|
| 1 |  | 57,891.838 | −1,679.130 | −3,825.279 |
| 9 |  | −17,613.230 | −67,771.572 | −71,093.815 |
| 10 | (−) | −11,736.622 | 12,327.695 | 13,726.078 |
| 11 |  | −38,088.692 | −13,401.941 | −13,217.015 |
| 12 | (−) | 49,825.313 | 1,074.247 | −509.066 |
| 16 |  | 6,049.840 | −23,089.552 | −25,209.414 |
| 15 | (−) | −55,284.138 | −4,164.224 | −2,589.570 |
| 13 | (−) | −591.015 | 26,179.546 | 28,308.041 |
| 7 | (−) | −12,264.525 | 22,491.525 | 24,372.823 |
| 2 | (−) | <u>21,811.232</u> | <u>48,033.426</u> | <u>50,037.232</u> |
| Total sum for loop 1 |  | 0.001 | 0.020 | 0.015 |

### Sequence of Vectors—Beginning at WAWN

|  |  | $\Delta X$ (m) | $\Delta Y$ (m) | $\Delta Z$ (m) |
|---|---|---|---|---|
| 5 | (−) | 47,192.795 | 20,056.051 | 19,944.432 |
| 3 | (−) | −25,381.565 | 27,977.353 | 30,092.793 |
| 1 |  | 57,891.838 | −1,679.130 | −3,825.279 |
| 4 |  | −32,510.268 | −26,298.224 | −26,267.509 |
| 8 |  | 3,160.419 | −29,145.645 | −31,100.235 |
| 6 | (−) | <u>−50,353.209</u> | <u>9,089.568</u> | <u>11,155.809</u> |
| Total sum for loop 2 |  | 0.010 | 0.027 | 0.011 |

### Sequence of Vectors—Beginning at SIW2

|  |  | $\Delta X$ (m) | $\Delta Y$ (m) | $\Delta Z$ (m) |
|---|---|---|---|---|
| 12 |  | −49,825.313 | −1,074.247 | 509.066 |
| 13 |  | 591.015 | −26,179.546 | −28,308.041 |
| 14 | (−) | <u>49,234.298</u> | <u>27,253.794</u> | <u>27,798.977</u> |
| Total sum for loop 3 |  | 0.000 | 0.001 | 0.002 |

Therefore, the dimensions of the vector are $n$, 1. For this linear least squares adjustment, there are two kinds of observations—the "fixed" stations and the observed baselines. In this case, the fixed stations are considered observations because the $X/Y/Z$ values of the station are assigned uncertainties. In this case, standard deviations of 0.002 m were assigned to each $X/Y/Z$ coordinate value. Other scenarios (for example, as discussed in Chapter 14) include an assumption of 0.020 m (10 times greater) for the uncertainty of the control points. In a generic case, the entire covariance matrix of the fixed point may be available from a previous adjustment. Or, if the control stations really are "fixed," the observation equations will include only baseline vectors. Such is the case in the least squares example discussed earlier and posted at http://www.globalcogo.com/nmsunet1.pdf, accessed May 4, 2017.

As derived in the theory and development of equations (as shown in link above), the sign of each observation is changed when included in the $f$ vector. In this case, the elements of the $f$ vector are listed in Table 13.4.

## *B* MATRIX—*n, u*

The order of the parameters appearing in each equation is arbitrary, but once established, that order must be followed exactly. Convention used here is that each triplet uses $X/Y/Z$ for the order of the parameters for a given station. The order of stations for this example is as shown in Table 13.5.

Elements in the $B$ matrix are partial derivatives of the result of each equation with respect to each parameter. Because it is a linear adjustment the partial derivative values are "1" for the "from" station and "−1" for the "to" station.

$B$ matrix is 54 rows (1 row for each observation equation) and 27 columns (9 stations with 3 unknowns each).

## *Q* MATRIX—*n, n*

The generic format for the $Q$ matrix is $3 \times 3$ block diagonal and symmetrical. Each of the different weight options discussed earlier is characterized by values included in each $3 \times 3$ block submatrix. Because the baseline *components* are correlated, the following example includes $3 \times 3$ block diagonal for each of sixteen baselines. Because the baselines are all uncorrelated, the off-diagonal submatrices are all zeros. The covariance matrix for each control station shows only the standard deviation squared (variances) assumed for the network adjustment. Other assumptions for the control points are discussed in Chapter 14. A generic covariance matrix format is shown below along with more specific elements for the station and baseline covariance matrices:

Generic

$$
\begin{bmatrix}
Variance_X & Covariance_{XY} & Covariance_{XZ} \\
Covariance_{XY} & Variance_Y & Covariance_{YZ} \\
Covariance_{XZ} & Covariance_{YZ} & Variance_Z
\end{bmatrix}
$$

Station       Baseline

$$
\begin{bmatrix}
\sigma_X^2 & \sigma_{XY} & \sigma_{XZ} \\
\sigma_{XY} & \sigma_Y^2 & \sigma_{YZ} \\
\sigma_{XZ} & \sigma_{YZ} & \sigma_Z^2
\end{bmatrix}
\ \text{or} \
\begin{bmatrix}
\sigma_{\Delta X}^2 & \sigma_{\Delta X \Delta Y} & \sigma_{\Delta X \Delta Z} \\
\sigma_{\Delta X \Delta Y} & \sigma_{\Delta Y}^2 & \sigma_{\Delta Y \Delta Z} \\
\sigma_{\Delta X \Delta Z} & \sigma_{\Delta Y \Delta Z} & \sigma_{\Delta Z}^2
\end{bmatrix}
$$

Specifically, the $Q$ matrix for all sixteen baselines and two control stations is listed in Table 13.6 and is built from the $3 \times 3$ submatrices for the various stations and baselines. Inconsequential off-diagonal $3 \times 3$ block submatrices (all zeros) are omitted to save space.

**TABLE 13.4**

**Elements of the *f* Vector**

$f_{(1, 1)} = -124,517.6445$
$f_{(2, 1)} = 4,609,735.8115$          Control station FOLA
$f_{(3, 1)} = -4,391,808.8307$

$f_{(4, 1)} = -57,891.838$
$f_{(5, 1)} = 1,679.130$               Vector 1
$f_{(6, 1)} = 3,825.279$

$f_{(7, 1)} = 21,811.232$
$f_{(8, 1)} = 48,033.426$              Vector 2
$f_{(9, 1)} = 50,037.232$

$f_{(10, 1)} = -25,381.565$
$f_{(11, 1)} = 57,977.353$             Vector 3
$f_{(12, 1)} = 30,092.793$

$f_{(13, 1)} = 32,510.268$
$f_{(14, 1)} = 26,298.224$             Vector 4
$f_{(15, 1)} = 26,267.509$

$f_{(16, 1)} = 47,192.795$
$f_{(17, 1)} = 20,056.051$             Vector 5
$f_{(18, 1)} = 19,944.432$

$f_{(19, 1)} = -50,353.209$
$f_{(20, 1)} = 9,089.568$              Vector 6
$f_{(21, 1)} = 11,155.809$

$f_{(22, 1)} = -12,264.525$
$f_{(23, 1)} = 22,491.525$             Vector 7
$f_{(24, 1)} = 24,372.823$

$f_{(25, 1)} = -3,160.419$
$f_{(26, 1)} = 29,145.645$             Vector 8
$f_{(27, 1)} = 31,100.235$

$f_{(28, 1)} = 17,613.230$
$f_{(29, 1)} = 67,771.572$             Vector 9
$f_{(30, 1)} = 71,093.815$

$f_{(31, 1)} = -11,736.622$
$f_{(32, 1)} = 12,327.695$             Vector 10
$f_{(33, 1)} = 13,726.078$

$f_{(34, 1)} = 38,088.692$
$f_{(35, 1)} = 13,401.941$             Vector 11
$f_{(36, 1)} = 13,217.015$

$f_{(37, 1)} = 49,825.313$
$f_{(38, 1)} = 1,074.247$              Vector 12
$f_{(39, 1)} = -509.066$

$f_{(40, 1)} = -591.015$
$f_{(41, 1)} = 26,179.546$             Vector 13
$f_{(42, 1)} = 28,308.041$

*(Continued)*

---

**TABLE 13.4 (Continued)**

**Elements of the f Vector**

| | |
|---|---|
| $f_{(43, 1)} = 49{,}234.298$ | |
| $f_{(44, 1)} = 27{,}253.794$ | Vector 14 |
| $f_{(45, 1)} = 27{,}798.977$ | |
| $f_{(46, 1)} = -55{,}284.138$ | |
| $f_{(47, 1)} = -4{,}164.224$ | Vector 15 |
| $f_{(48, 1)} = -2{,}589.570$ | |
| $f_{(49, 1)} = -6{,}049.840$ | |
| $f_{(50, 1)} = 23{,}089.552$ | Vector 16 |
| $f_{(51, 1)} = 25{,}209.414$ | |
| $f_{(52, 1)} = -170{,}846.0946$ | |
| $f_{(53, 1)} = 4{,}702{,}276.0741$ | Control station KEHA |
| $f_{(54, 1)} = -4{,}291{,}680.3174$ | |

---

With the $f$, $B$, and $Q$ matrices in hand, the following matrix operations were performed:

1. Transpose matrix $B$ to get $B^T$
2. Invert the $Q$ matrix to get $W$
3. Multiply $B^T \times W$ to get $B^T W$
4. Multiply again by $B$ to get $B^T WB$ (Rename to $N = B^T WB$)
5. Multiply $B^T W$ by $f$ to get $B^T Wf$
6. Invert $N$ (Normal Equations) to get $N$ *inverse* (*Ninv*)
7. Multiply *Ninv* inverse by $B^T Wf$ to get Delta (these are the answers)
8. Multiply $B$ by *Delta* to get $B$ *Delta*
9. Subtract $f$ minus $B$ *Delta* to get $v$ = residuals
10. Transpose vector $v$ to get $v^T$
11. Multiply $v^T$ by $W$ to get $v^T W$
12. Multiply $v^T W$ by $v$ to get $v^T Wv$
13. Divide $v^T Wv$ by $(n - u)$ to get *a posteriori* reference variance ($1 \times 1$ scalar)
14. Multiply $\dfrac{v^T W v}{n - u}$ by *Ninv* to get covariance matrix of parameters (answers)

Using the three matrices $f$, $B$, and $Q$, the following annotated computer printout shows the *Ninv* matrix, the *Delta* vector (answers), the residuals, the *a posteriori* reference variance, and the covariance matrix of the computed coordinates for all nine stations.

## COMPUTER PRINTOUTS

1. Of the least squares adjustment—Figure 13.3 shows a computer printout of the least squares adjustment of the network. The input matrices are listed in Tables 13.4 through 13.6. The information in Figure 13.3 includes the adjusted $X/Y/Z$ coordinate values for all stations in the network, the residuals needed to be added to the observations, the inverse of the normal equations ($N$ inverse), and the computed reference variance. The covariance

**TABLE 13.5**

**B Matrix of Partial Derivatives**

| From - to | FOLA | | | SHAN | | | WAWN | | | WEBE | | | RASN | | | PAR0 | | | SIW2 | | | WATH | | | KEHA | | |
|---|---|---|---|---|---|---|---|---|---|---|---|---|---|---|---|---|---|---|---|---|---|---|---|---|---|---|---|
| To FOLA (no "from" station) | -1 | 0 | 0 | 0 | 0 | 0 | 0 | 0 | 0 | 0 | 0 | 0 | 0 | 0 | 0 | 0 | 0 | 0 | 0 | 0 | 0 | 0 | 0 | 0 | 0 | 0 | 0 |
| | 0 | -1 | 0 | 0 | 0 | 0 | 0 | 0 | 0 | 0 | 0 | 0 | 0 | 0 | 0 | 0 | 0 | 0 | 0 | 0 | 0 | 0 | 0 | 0 | 0 | 0 | 0 |
| | 0 | 0 | -1 | 0 | 0 | 0 | 0 | 0 | 0 | 0 | 0 | 0 | 0 | 0 | 0 | 0 | 0 | 0 | 0 | 0 | 0 | 0 | 0 | 0 | 0 | 0 | 0 |
| Vector 1 FOLA to SHAN | 1 | 0 | 0 | -1 | 0 | 0 | 0 | 0 | 0 | 0 | 0 | 0 | 0 | 0 | 0 | 0 | 0 | 0 | 0 | 0 | 0 | 0 | 0 | 0 | 0 | 0 | 0 |
| | 0 | 1 | 0 | 0 | -1 | 0 | 0 | 0 | 0 | 0 | 0 | 0 | 0 | 0 | 0 | 0 | 0 | 0 | 0 | 0 | 0 | 0 | 0 | 0 | 0 | 0 | 0 |
| | 0 | 0 | 1 | 0 | 0 | -1 | 0 | 0 | 0 | 0 | 0 | 0 | 0 | 0 | 0 | 0 | 0 | 0 | 0 | 0 | 0 | 0 | 0 | 0 | 0 | 0 | 0 |
| Vector 2 FOLA to WAWN | 1 | 0 | 0 | 0 | 0 | 0 | -1 | 0 | 0 | 0 | 0 | 0 | 0 | 0 | 0 | 0 | 0 | 0 | 0 | 0 | 0 | 0 | 0 | 0 | 0 | 0 | 0 |
| | 0 | 1 | 0 | 0 | 0 | 0 | 0 | -1 | 0 | 0 | 0 | 0 | 0 | 0 | 0 | 0 | 0 | 0 | 0 | 0 | 0 | 0 | 0 | 0 | 0 | 0 | 0 |
| | 0 | 0 | 1 | 0 | 0 | 0 | 0 | 0 | -1 | 0 | 0 | 0 | 0 | 0 | 0 | 0 | 0 | 0 | 0 | 0 | 0 | 0 | 0 | 0 | 0 | 0 | 0 |
| Vector 3 FOLA to WEBE | 1 | 0 | 0 | 0 | 0 | 0 | 0 | 0 | 0 | -1 | 0 | 0 | 0 | 0 | 0 | 0 | 0 | 0 | 0 | 0 | 0 | 0 | 0 | 0 | 0 | 0 | 0 |
| | 0 | 1 | 0 | 0 | 0 | 0 | 0 | 0 | 0 | 0 | -1 | 0 | 0 | 0 | 0 | 0 | 0 | 0 | 0 | 0 | 0 | 0 | 0 | 0 | 0 | 0 | 0 |
| | 0 | 0 | 1 | 0 | 0 | 0 | 0 | 0 | 0 | 0 | 0 | -1 | 0 | 0 | 0 | 0 | 0 | 0 | 0 | 0 | 0 | 0 | 0 | 0 | 0 | 0 | 0 |
| Vector 4 SHAN to WEBE | 0 | 0 | 0 | 1 | 0 | 0 | 0 | 0 | 0 | -1 | 0 | 0 | 0 | 0 | 0 | 0 | 0 | 0 | 0 | 0 | 0 | 0 | 0 | 0 | 0 | 0 | 0 |
| | 0 | 0 | 0 | 0 | 1 | 0 | 0 | 0 | 0 | 0 | -1 | 0 | 0 | 0 | 0 | 0 | 0 | 0 | 0 | 0 | 0 | 0 | 0 | 0 | 0 | 0 | 0 |
| | 0 | 0 | 0 | 0 | 0 | 1 | 0 | 0 | 0 | 0 | 0 | -1 | 0 | 0 | 0 | 0 | 0 | 0 | 0 | 0 | 0 | 0 | 0 | 0 | 0 | 0 | 0 |
| Vector 5 WEBE to WAWN | 0 | 0 | 0 | 0 | 0 | 0 | -1 | 0 | 0 | 1 | 0 | 0 | 0 | 0 | 0 | 0 | 0 | 0 | 0 | 0 | 0 | 0 | 0 | 0 | 0 | 0 | 0 |
| | 0 | 0 | 0 | 0 | 0 | 0 | 0 | -1 | 0 | 0 | 1 | 0 | 0 | 0 | 0 | 0 | 0 | 0 | 0 | 0 | 0 | 0 | 0 | 0 | 0 | 0 | 0 |
| | 0 | 0 | 0 | 0 | 0 | 0 | 0 | 0 | -1 | 0 | 0 | 1 | 0 | 0 | 0 | 0 | 0 | 0 | 0 | 0 | 0 | 0 | 0 | 0 | 0 | 0 | 0 |
| Vector 6 WAWN to RASN | 0 | 0 | 0 | 0 | 0 | 0 | 1 | 0 | 0 | 0 | 0 | 0 | -1 | 0 | 0 | 0 | 0 | 0 | 0 | 0 | 0 | 0 | 0 | 0 | 0 | 0 | 0 |
| | 0 | 0 | 0 | 0 | 0 | 0 | 0 | 1 | 0 | 0 | 0 | 0 | 0 | -1 | 0 | 0 | 0 | 0 | 0 | 0 | 0 | 0 | 0 | 0 | 0 | 0 | 0 |
| | 0 | 0 | 0 | 0 | 0 | 0 | 0 | 0 | 1 | 0 | 0 | 0 | 0 | 0 | -1 | 0 | 0 | 0 | 0 | 0 | 0 | 0 | 0 | 0 | 0 | 0 | 0 |

*(Continued)*

**TABLE 13.5 (Continued)**

**B Matrix of Partial Derivatives**

| From - to | FOLA | | | SHAN | | | WAWN | | | WEBE | | | RASN | | | PARO | | | SIW2 | | | WATH | | | KEHA | | |
|---|---|---|---|---|---|---|---|---|---|---|---|---|---|---|---|---|---|---|---|---|---|---|---|---|---|---|---|
| Vector 7 WAWN to PARO | 0 | 0 | 0 | 0 | 0 | 0 | 1 | 0 | 0 | 0 | 0 | 0 | 0 | 0 | 0 | −1 | 0 | 0 | 0 | 0 | 0 | 0 | 0 | 0 | 0 | 0 | 0 |
| | 0 | 0 | 0 | 0 | 0 | 0 | 0 | 1 | 0 | 0 | 0 | 0 | 0 | 0 | 0 | 0 | −1 | 0 | 0 | 0 | 0 | 0 | 0 | 0 | 0 | 0 | 0 |
| | 0 | 0 | 0 | 0 | 0 | 0 | 0 | 0 | 1 | 0 | 0 | 0 | 0 | 0 | 0 | 0 | 0 | −1 | 0 | 0 | 0 | 0 | 0 | 0 | 0 | 0 | 0 |
| Vector 8 WEBE to RASN | 0 | 0 | 0 | 0 | 0 | 0 | 0 | 0 | 0 | 1 | 0 | 0 | −1 | 0 | 0 | 0 | 0 | 0 | 0 | 0 | 0 | 0 | 0 | 0 | 0 | 0 | 0 |
| | 0 | 0 | 0 | 0 | 0 | 0 | 0 | 0 | 0 | 0 | 1 | 0 | 0 | −1 | 0 | 0 | 0 | 0 | 0 | 0 | 0 | 0 | 0 | 0 | 0 | 0 | 0 |
| | 0 | 0 | 0 | 0 | 0 | 0 | 0 | 0 | 0 | 0 | 0 | 1 | 0 | 0 | −1 | 0 | 0 | 0 | 0 | 0 | 0 | 0 | 0 | 0 | 0 | 0 | 0 |
| Vector 9 SHAN to SIW2 | 0 | 0 | 0 | 1 | 0 | 0 | 0 | 0 | 0 | 0 | 0 | 0 | 0 | 0 | 0 | 0 | 0 | 0 | −1 | 0 | 0 | 0 | 0 | 0 | 0 | 0 | 0 |
| | 0 | 0 | 0 | 0 | 1 | 0 | 0 | 0 | 0 | 0 | 0 | 0 | 0 | 0 | 0 | 0 | 0 | 0 | 0 | −1 | 0 | 0 | 0 | 0 | 0 | 0 | 0 |
| | 0 | 0 | 0 | 0 | 0 | 1 | 0 | 0 | 0 | 0 | 0 | 0 | 0 | 0 | 0 | 0 | 0 | 0 | 0 | 0 | −1 | 0 | 0 | 0 | 0 | 0 | 0 |
| Vector 10 RASN to SIW2 | 0 | 0 | 0 | 0 | 0 | 0 | 0 | 0 | 0 | 0 | 0 | 0 | 1 | 0 | 0 | 0 | 0 | 0 | −1 | 0 | 0 | 0 | 0 | 0 | 0 | 0 | 0 |
| | 0 | 0 | 0 | 0 | 0 | 0 | 0 | 0 | 0 | 0 | 0 | 0 | 0 | 1 | 0 | 0 | 0 | 0 | 0 | −1 | 0 | 0 | 0 | 0 | 0 | 0 | 0 |
| | 0 | 0 | 0 | 0 | 0 | 0 | 0 | 0 | 0 | 0 | 0 | 0 | 0 | 0 | 1 | 0 | 0 | 0 | 0 | 0 | −1 | 0 | 0 | 0 | 0 | 0 | 0 |
| Vector 11 RASN to PARO | 0 | 0 | 0 | 0 | 0 | 0 | 0 | 0 | 0 | 0 | 0 | 0 | 1 | 0 | 0 | −1 | 0 | 0 | 0 | 0 | 0 | 0 | 0 | 0 | 0 | 0 | 0 |
| | 0 | 0 | 0 | 0 | 0 | 0 | 0 | 0 | 0 | 0 | 0 | 0 | 0 | 1 | 0 | 0 | −1 | 0 | 0 | 0 | 0 | 0 | 0 | 0 | 0 | 0 | 0 |
| | 0 | 0 | 0 | 0 | 0 | 0 | 0 | 0 | 0 | 0 | 0 | 0 | 0 | 0 | 1 | 0 | 0 | −1 | 0 | 0 | 0 | 0 | 0 | 0 | 0 | 0 | 0 |
| Vector 12 SIW2 to PARO | 0 | 0 | 0 | 0 | 0 | 0 | 0 | 0 | 0 | 0 | 0 | 0 | 0 | 0 | 0 | −1 | 0 | 0 | 1 | 0 | 0 | 0 | 0 | 0 | 0 | 0 | 0 |
| | 0 | 0 | 0 | 0 | 0 | 0 | 0 | 0 | 0 | 0 | 0 | 0 | 0 | 0 | 0 | 0 | −1 | 0 | 0 | 1 | 0 | 0 | 0 | 0 | 0 | 0 | 0 |
| | 0 | 0 | 0 | 0 | 0 | 0 | 0 | 0 | 0 | 0 | 0 | 0 | 0 | 0 | 0 | 0 | 0 | −1 | 0 | 0 | 1 | 0 | 0 | 0 | 0 | 0 | 0 |
| Vector 13 PARO to WATH | 0 | 0 | 0 | 0 | 0 | 0 | 0 | 0 | 0 | 0 | 0 | 0 | 0 | 0 | 0 | 1 | 0 | 0 | 0 | 0 | 0 | −1 | 0 | 0 | 0 | 0 | 0 |
| | 0 | 0 | 0 | 0 | 0 | 0 | 0 | 0 | 0 | 0 | 0 | 0 | 0 | 0 | 0 | 0 | 1 | 0 | 0 | 0 | 0 | 0 | −1 | 0 | 0 | 0 | 0 |
| | 0 | 0 | 0 | 0 | 0 | 0 | 0 | 0 | 0 | 0 | 0 | 0 | 0 | 0 | 0 | 0 | 0 | 1 | 0 | 0 | 0 | 0 | 0 | −1 | 0 | 0 | 0 |

*(Continued)*

**TABLE 13.5 (Continued)**
**B Matrix of Partial Derivatives**

| From - to | FOLA | | | SHAN | | | WAWN | | | WEBE | | | RASN | | | PAR0 | | | SIW2 | | | WATH | | | KEHA | | |
|---|---|---|---|---|---|---|---|---|---|---|---|---|---|---|---|---|---|---|---|---|---|---|---|---|---|---|---|
| Vector 14 SIW2 to WATH | 0 | 0 | 0 | 0 | 0 | 0 | 0 | 0 | 0 | 0 | 0 | 0 | 0 | 0 | 0 | 0 | 0 | 0 | 1 | 0 | 0 | -1 | 0 | 0 | 0 | 0 | 0 |
| | 0 | 0 | 0 | 0 | 0 | 0 | 0 | 0 | 0 | 0 | 0 | 0 | 0 | 0 | 0 | 0 | 0 | 0 | 0 | 1 | 0 | 0 | -1 | 0 | 0 | 0 | 0 |
| | 0 | 0 | 0 | 0 | 0 | 0 | 0 | 0 | 0 | 0 | 0 | 0 | 0 | 0 | 0 | 0 | 0 | 0 | 0 | 0 | 1 | 0 | 0 | -1 | 0 | 0 | 0 |
| Vector 15 WATH to KEHA | 0 | 0 | 0 | 0 | 0 | 0 | 0 | 0 | 0 | 0 | 0 | 0 | 0 | 0 | 0 | 0 | 0 | 0 | 0 | 0 | 0 | 1 | 0 | 0 | -1 | 0 | 0 |
| | 0 | 0 | 0 | 0 | 0 | 0 | 0 | 0 | 0 | 0 | 0 | 0 | 0 | 0 | 0 | 0 | 0 | 0 | 0 | 0 | 0 | 0 | 1 | 0 | 0 | -1 | 0 |
| | 0 | 0 | 0 | 0 | 0 | 0 | 0 | 0 | 0 | 0 | 0 | 0 | 0 | 0 | 0 | 0 | 0 | 0 | 0 | 0 | 0 | 0 | 0 | 1 | 0 | 0 | -1 |
| Vector 16 SIW2 to KEHA | 0 | 0 | 0 | 0 | 0 | 0 | 0 | 0 | 0 | 0 | 0 | 0 | 0 | 0 | 0 | 0 | 0 | 0 | 1 | 0 | 0 | 0 | 0 | 0 | -1 | 0 | 0 |
| | 0 | 0 | 0 | 0 | 0 | 0 | 0 | 0 | 0 | 0 | 0 | 0 | 0 | 0 | 0 | 0 | 0 | 0 | 0 | 1 | 0 | 0 | 0 | 0 | 0 | -1 | 0 |
| | 0 | 0 | 0 | 0 | 0 | 0 | 0 | 0 | 0 | 0 | 0 | 0 | 0 | 0 | 0 | 0 | 0 | 0 | 0 | 0 | 1 | 0 | 0 | 0 | 0 | 0 | -1 |
| Station KEHA (no "from" Station) | 0 | 0 | 0 | 0 | 0 | 0 | 0 | 0 | 0 | 0 | 0 | 0 | 0 | 0 | 0 | 0 | 0 | 0 | 0 | 0 | 0 | 0 | 0 | 0 | -1 | 0 | 0 |
| | 0 | 0 | 0 | 0 | 0 | 0 | 0 | 0 | 0 | 0 | 0 | 0 | 0 | 0 | 0 | 0 | 0 | 0 | 0 | 0 | 0 | 0 | 0 | 0 | 0 | -1 | 0 |
| | 0 | 0 | 0 | 0 | 0 | 0 | 0 | 0 | 0 | 0 | 0 | 0 | 0 | 0 | 0 | 0 | 0 | 0 | 0 | 0 | 0 | 0 | 0 | 0 | 0 | 0 | -1 |
| | FOLA | | | SHAN | | | WAWN | | | WEBE | | | RASN | | | PAR0 | | | SIW2 | | | WATH | | | KEHA | | |

## TABLE 13.6
## Q Matrix of Baseline 3 × 3 Covariance Submatrices

| | FOLA | | | BL 1 | | | BL 2 | | | ⋯ | BL 16 | | | KEHA | | |
|---|---|---|---|---|---|---|---|---|---|---|---|---|---|---|---|---|---|
| FOLA | 0.000004 | 0 | 0 | 0 | 0 | 0 | 0 | 0 | 0 | | 0 | 0 | 0 | 0 | 0 | 0 |
| | 0 | 0.000004 | 0 | 0 | 0 | 0 | 0 | 0 | 0 | | 0 | 0 | 0 | 0 | 0 | 0 |
| | 0 | 0 | 0.000004 | 0 | 0 | 0 | 0 | 0 | 0 | | 0 | 0 | 0 | 0 | 0 | 0 |
| BL.1 | 0 | 0 | 0 | 2.8502E−6 | −1.9173E−6 | 5.4830E−7 | 0 | 0 | 0 | | 0 | 0 | 0 | 0 | 0 | 0 |
| | 0 | 0 | 0 | −1.1776E−6 | 2.8638E−5 | −2.2283E−5 | 0 | 0 | 0 | | 0 | 0 | 0 | 0 | 0 | 0 |
| | 0 | 0 | 0 | 5.4830E−7 | −2.2283E−5 | 2.2356E−5 | 0 | 0 | 0 | | 0 | 0 | 0 | 0 | 0 | 0 |
| BL.2 | 0 | 0 | 0 | 0 | 0 | 0 | 1.3845E−5 | 2.9173E−5 | −2.0868E−6 | | 0 | 0 | 0 | 0 | 0 | 0 |
| | 0 | 0 | 0 | 0 | 0 | 0 | 2.9173E−5 | 1.4196E−4 | −9.0794E−5 | | 0 | 0 | 0 | 0 | 0 | 0 |
| | 0 | 0 | 0 | 0 | 0 | 0 | −2.0868E−6 | −9.0794E−5 | 6.8824E−5 | | 0 | 0 | 0 | 0 | 0 | 0 |
| BL.3 | 0 | 0 | 0 | 0 | 0 | 0 | ⋮ | | | | 1.9546E−6 | −1.8561E−6 | 1.2268E−6 | | | |
| | 0 | 0 | 0 | 0 | 0 | 0 | ⋮ | | | | −1.8561E−6 | 3.22602E−5 | −2.49196E−5 | | | |
| | 0 | 0 | 0 | 0 | 0 | 0 | ⋮ | | | | 1.2268E−6 | −2.49196E−5 | 2.53440E−5 | | | |
| BL.4 | 0 | 0 | 0 | 0 | 0 | 0 | ⋮ | | | | 2.27268E−6 | −3.5647E−6 | 1.5143E−6 | | | |
| | 0 | 0 | 0 | 0 | 0 | 0 | ⋮ | | | | −3.5647E−6 | 8.55221E−5 | −5.28146E−5 | | | |
| | 0 | 0 | 0 | 0 | 0 | 0 | ⋮ | | | | 1.5143E−6 | −5.28146E−5 | 4.08228E−5 | | | |
| BL.5 | 0 | 0 | 0 | 0 | 0 | 0 | ⋮ | | | | 2.8605E−6 | −2.1611E−6 | 2.9093E−6 | | | |
| | 0 | 0 | 0 | 0 | 0 | 0 | ⋮ | | | | −2.1611E−6 | 3.64246E−5 | −3.15363E−5 | | | |
| | 0 | 0 | 0 | 0 | 0 | 0 | ⋮ | | | | 2.9093E−6 | −3.15363E−5 | 3.31404E−5 | | | |
| BL.6 | 0 | 0 | 0 | 0 | 0 | 0 | ⋮ | | | | 5.6081E−6 | −2.9970E−6 | 2.0288E−6 | | | |
| | 0 | 0 | 0 | 0 | 0 | 0 | ⋮ | | | | −2.9970E−6 | 6.93793E−5 | −4.40001E−5 | | | |
| | 0 | 0 | 0 | 0 | 0 | 0 | ⋮ | | | | 2.0288E−6 | −4.40001E−5 | 3.58558E−5 | | | |
| BL.7 | 0 | 0 | 0 | 0 | 0 | 0 | ⋮ | | | | 2.4541E−6 | 2.2576E−6 | −2.5089E−6 | | | |
| | 0 | 0 | 0 | 0 | 0 | 0 | ⋮ | | | | 2.2576E−6 | 4.60533E−5 | −3.26178E−5 | | | |
| | 0 | 0 | 0 | 0 | 0 | 0 | ⋮ | | | | −2.5089E−6 | −3.26178E−5 | 2.99862E−5 | | | |
| BL.8 | 0 | 0 | 0 | 0 | 0 | 0 | ⋮ | | | | 1.3630E−6 | −2.029E−7 | −1.928E−7 | | | |
| | 0 | 0 | 0 | 0 | 0 | 0 | ⋮ | | | | −2.029E−7 | 2.71011E−5 | −2.13529E−5 | | | |
| | 0 | 0 | 0 | 0 | 0 | 0 | ⋮ | | | | −1.928E−7 | −2.13529E−5 | 2.24759E−5 | | | |

*(Continued)*

**TABLE 13.6 (Continued)**
## Q Matrix of Baseline 3 × 3 Covariance Submatrices

| | FOLA | BL 1 | BL 2 | ⋯ | BL 16 | KEHA |
|---|---|---|---|---|---|---|
| **BL 9** | 0 0 0<br>0 0 0<br>0 0 0 | 0 0 0<br>0 0 0<br>0 0 0 | ⋮<br>⋮<br>⋮ | 1.6345E−6  −2.3304E−6  2.8629E−6<br>−2.3304E−6  4.31998E−5  −2.86799E−5<br>2.8629E−6  −2.86799E−5  3.03883E−5 | | 0 0 0<br>0 0 0<br>0 0 0 |
| **BL 10** | 0 0 0<br>0 0 0<br>0 0 0 | 0 0 0<br>0 0 0<br>0 0 0 | ⋮<br>⋮<br>⋮ | 2.1715E−6  −3.1324E−6  3.3230E−6<br>−3.1324E−6  4.31998E−5  3.69809E−5<br>3.3230E−6  3.69809E−5  3.84384E−5 | | 0 0 0<br>0 0 0<br>0 0 0 |
| **BL 11** | 0 0 0<br>0 0 0<br>0 0 0 | 0 0 0<br>0 0 0<br>0 0 0 | ⋮<br>⋮<br>⋮ | 1.3707E−6  −2.506E−7  −1.501E−7<br>−2.506E−7  2.79621E−5  −2.17399E−5<br>−1.501E−7  −2.17399E−5  2.24557E−5 | | 0 0 0<br>0 0 0<br>0 0 0 |
| **BL 12** | 0 0 0<br>0 0 0<br>0 0 0 | 0 0 0<br>0 0 0<br>0 0 0 | ⋮<br>⋮<br>⋮ | 1.3474E−6  −5.96E−8  −3.773E−7<br>−5.96E−8  2.61021E−5  −2.01012E−5<br>−3.773E−7  −2.01012E−5  2.0516E−5 | | 0 0 0<br>0 0 0<br>0 0 0 |
| **BL 13** | 0 0 0<br>0 0 0<br>0 0 0 | 0 0 0<br>0 0 0<br>0 0 0 | ⋮<br>⋮<br>⋮ | 1.4192E−6  −6.85E−8  −3.123E−7<br>−6.85E−8  2.89509E−5  −2.25179E−5<br>−3.123E−7  −2.25179E−5  2.3170E−5 | | 0 0 0<br>0 0 0<br>0 0 0 |
| **BL 14** | 0 0 0<br>0 0 0<br>0 0 0 | 0 0 0<br>0 0 0<br>0 0 0 | ⋮<br>⋮<br>⋮ | 1.8818E−6  1.138E−7  3.332E−7<br>1.138E−7  2.64428E−5  −1.98930E−5<br>3.332E−7  −1.98930E−5  2.09629E−5 | | 0 0 0<br>0 0 0<br>0 0 0 |
| **BL 15** | 0 0 0<br>0 0 0<br>0 0 0 | 0 0 0<br>0 0 0<br>0 0 0 | | 4.28227E−6  −9.042E−7  1.0047E−6<br>−9.042E−7  6.82233E−5  −4.17516E−5<br>1.0047E−6  −4.17516E−5  3.36817E−5 | 0 0 0<br>0 0 0<br>0 0 0 | 0 0 0<br>0 0 0<br>0 0 0 |
| **BL 16** | 0 0 0<br>0 0 0<br>0 0 0 | 0 0 0<br>0 0 0<br>0 0 0 | ⋮<br>⋮<br>⋮ | | 3.7414E−6  −5.3433E−6  1.9981E−6<br>−5.3433E−6  8.1933E−5  −5.1170E−5<br>1.9981E−6  −5.1170E−5  3.9252E−5 | 0 0 0<br>0 0 0<br>0 0 0 |
| **KEHA** | 0 0 0<br>0 0 0<br>0 0 0 | 0 0 0<br>0 0 0<br>0 0 0 | | | 0 0 0<br>0 0 0<br>0 0 0 | 0.000004  0  0<br>0  0.000004  0<br>0  0  0.000004 |

Program name/version: LSAdjGPSvectors: Version 2014-A

The user is:                          Earl F. Burkholder

Program used on:                      August 4, 2016

The input file for program is:        Ch13GPS2011.dat

The name of this output file is:      Ch13GPS2011Ans.out                    File renamed Figure 13.3.docx

The client is: WISCORS-16 baselines and holding FOLA and KEHA at 0.002 m NAD 83(2011)

The project is: This file documents the network adjustment for Chapter 13.

Three input matrices are $f$, B, and Q as shown in Tables 13.4 through 13.6. The original computer printout contained a listing of all input data. They are not repeated here.

The weight matrix is the inverse of the Q matrix and could have been printed but was omitted. The inverse of the normal equations (N inverse) is:

*Note:* Output has been annotated for readability.

$Ninv'_{27, 27}$

| | | | | | | |
|---|---|---|---|---|---|---|
| 2.63676102E-6 | -5.48656843E-8 | -1.73752545E-8 | 2.27832949E-6 | -1.07866090E-7 | 9.01748207E-8 | 2.25082689E-6 |
| -2.55361577E-7 | 1.43925117E-7 | 2.33330639E-6 | -1.34603208E-7 | 6.75031951E-8 | 2.18174054E-6 | -2.68577345E-7 |
| 1.79038825E-7 | 2.10628549E-6 | -2.06452266E-7 | 1.08947787E-7 | 1.98521312E-6 | -1.49578782E-7 | 8.54083646E-8 |
| 2.06220315E-6 | -8.00321968E-8 | -4.50111475E-8 | 1.36323898E-6 | 5.48656843E-8 | 1.73752545E-8 | -7.28276329E-8 |
| -5.48656843E-8 | 3.52354163E-6 | -4.38046131E-7 | -3.69590972E-8 | 3.14952180E-6 | -3.68419523E-7 | 2.62941676E-6 |
| 2.86332553E-6 | -2.73668395E-7 | -4.77689970E-8 | 2.99134358E-6 | -2.84278309E-7 | -6.08022815E-8 | -2.54771406E-7 |
| -1.55994478E-7 | -5.39977630E-8 | 2.48846774E-6 | -2.47740287E-7 | -3.19868348E-8 | 2.21835276E-6 | -1.11976465E-8 |
| -5.06115621E-8 | 2.43994893E-6 | -3.73419489E-7 | 5.48656843E-6 | 4.76458374E-7 | 4.38046131E-7 | 2.42114160E-7 |
| -1.73752545E-8 | -4.38046131E-7 | 3.24014452E-6 | 2.21002617E-8 | -1.31060303E-7 | 2.64198632E-6 | 1.52692915E-6 |
| -5.63660689E-8 | 2.43885305E-6 | -2.38478679E-9 | -1.02299575E-7 | 2.60050998E-6 | 4.30809387E-9 | 2.56568212E-6 |
| 2.08342665E-6 | -1.05698511E-8 | 6.50465644E-8 | 1.97607092E-6 | -5.06829902E-9 | 2.25515225E-7 | -1.03102662E-7 |
| -4.15590258E-8 | -5.04741844E-8 | 1.85116012E-6 | 1.73752545E-8 | 4.38046131E-7 | 7.59855482E-7 | |
| 2.27832949E-6 | -3.69590972E-8 | 2.10026172E-8 | 3.26459410E-6 | -7.97622278E-7 | 5.16134788E-7 | |
| -2.04696456E-8 | -8.91994411E-8 | 2.64303916E-6 | -2.03717204E-7 | 1.23491295E-7 | 2.61667266E-6 | |

FIGURE 13.3    Computer Printout of Least Squares Adjustment

*(Continued)*

| | | | | | | |
|---|---|---|---|---|---|---|
| -3.65912865E-8 | 2.57596221E-6 | -1.65488995E-7 | -4.04933363E-8 | 2.47910436E-6 | -2.31261152E-7 | 2.41134976E-8 |
| 2.63770488E-6 | -2.48525168E-7 | -7.75339712E-8 | 1.72167051E-6 | 3.69590972E-8 | -2.21002617E-8 | -2.45069480E-8 |
| -1.07866090E-7 | 3.14952180E-6 | -1.31060303E-7 | -7.97622278E-7 | 1.98737030E-5 | -1.24747383E-5 | 8.82118075E-6 |
| 7.31158351E-6 | -3.56244498E-6 | -2.28849922E-7 | 8.13160691E-6 | -4.42367301E-6 | -1.17836212E-7 | -3.61195951E-6 |
| -4.91214086E-6 | -1.26246235E-7 | 8.06089011E-6 | -4.24313886E-6 | -1.36768220E-7 | 7.17426443E-6 | -1.24443128E-7 |
| -2.15470176E-7 | 8.72447156E-6 | -4.75884420E-6 | 1.07866090E-7 | 8.50478202E-7 | 1.31060303E-7 | -4.69116582E-6 |
| 9.01748207E-8 | -3.68419523E-7 | 2.64198632E-6 | 5.16134788E-7 | -1.24747383E-5 | 1.41495616E-5 | 4.73253912E-6 |
| -3.54004288E-6 | 5.46755265E-6 | 1.31937467E-7 | -4.46461824E-6 | 6.74700589E-6 | -3.31637603E-8 | 3.61611379E-6 |
| 6.51043604E-6 | -6.71050236E-8 | -4.16729345E-6 | 5.73165232E-6 | -4.88171348E-8 | -3.40929394E-6 | -4.20534932E-7 |
| -6.50138113E-8 | -4.72535510E-6 | 6.06781967E-6 | -9.01748207E-8 | 3.68419523E-7 | 1.35801368E-6 | 5.79156989E-8 |
| 2.25082689E-6 | -7.28276329E-8 | -1.11976465E-8 | 2.56568212E-6 | -2.45069480E-8 | -1.24443128E-7 | 3.61611379E-6 |
| 2.36853765E-7 | -3.42418492E-7 | 2.77265640E-6 | -1.38478688E-7 | -1.33580312E-7 | 2.80160904E-6 | -4.20534932E-7 |
| 1.76032970E-7 | 2.82004715E-6 | -2.45233749E-7 | 9.41182921E-8 | 2.58977188E-6 | -1.18613157E-7 | 5.79156989E-8 |
| 2.60934553E-6 | 4.47125718E-8 | -1.66352613E-7 | 1.74917311E-6 | 7.28276329E-8 | 1.11976465E-8 | 5.63660689E-8 |
| -2.55361577E-7 | 2.86332553E-6 | -5.63660689E-8 | -2.04696456E-8 | 7.31158351E-6 | -3.54004288E-6 | -3.42418492E-7 |
| 2.27716830E-5 | -1.37568983E-5 | -2.50587943E-7 | 1.17743625E-5 | -5.93637874E-6 | -2.91676474E-7 | 2.36853765E-7 |
| -5.92281500E-6 | -1.67243986E-7 | 1.32427057E-5 | -7.45674053E-6 | -7.71439188E-8 | 1.06135231E-5 | 1.16664720E-5 |
| -7.64179881E-8 | 1.18004690E-5 | -6.73346209E-6 | 2.55361577E-7 | 1.13667447E-6 | 5.63660689E-8 | -5.66510971E-6 |
| 1.43925117E-7 | -2.73668395E-7 | 2.43885305E-6 | -8.91994411E-8 | -3.56244498E-6 | 5.46755265E-6 | -3.42418492E-7 |
| -1.37568983E-5 | 1.46524132E-5 | -2.67116446E-8 | -5.79391016E-6 | 7.40757833E-6 | 7.26223680E-8 | -5.69634781E-6 |
| 7.61964150E-6 | 1.33345393E-9 | -7.44447357E-6 | 8.73949829E-6 | -1.10330436E-8 | -5.53292822E-6 | 6.49530272E-6 |
| -8.68924037E-8 | -6.76964313E-6 | 7.60877623E-6 | -1.43925117E-7 | 2.73668395E-6 | 1.56114695E-6 | 2.77265640E-6 |
| 2.33330639E-6 | -4.77689970E-8 | -2.38478679E-9 | 2.64303916E-6 | -2.28849922E-7 | 1.31937467E-7 | -6.59831660E-7 |
| -2.50587943E-7 | -2.67116446E-8 | 3.13582297E-6 | -6.86797142E-7 | 3.72612625E-7 | 2.77478150E-6 | -6.59831660E-7 |
| 3.64196473E-7 | 2.60917807E-6 | -3.66078310E-7 | 1.23851360E-7 | 2.44119656E-6 | -2.71438743E-7 | 1.08247296E-7 |

(Continued)

FIGURE 13.3 (*Continued*)  Computer Printout of Least Squares Adjustment

| | | | | | | |
|---|---|---|---|---|---|---|
| 2.51797164E-6 | -1.78246937E-7 | -4.72579674E-8 | 1.66669361E-6 | 4.77689970E-8 | 2.38478679E-9 | -1.38478688E-7 |
| -1.34603208E-7 | 2.99134358E-6 | -1.02299575E-7 | -2.03717204E-7 | 8.13160691E-6 | -4.46461824E-6 | 1.08557413E-5 |
| 1.17743625E-5 | -5.79391016E-6 | -6.86797142E-7 | 1.77255095E-5 | -1.09025694E-5 | -5.98621358E-7 | -4.53441325E-6 |
| -5.18719919E-6 | -3.11091792E-7 | 1.03118881E-5 | -5.26571390E-6 | -2.03210085E-7 | 8.89936669E-6 | -1.33580312E-7 |
| -1.97169402E-7 | 1.05147431E-5 | -5.95447731E-6 | 1.34603208E-7 | 1.00865642E-6 | 1.02299575E-7 | -5.24682076E-6 |
| 6.75031951E-8 | -2.84278309E-7 | 2.60050998E-6 | 1.23491295E-7 | -4.42367301E-6 | 6.74700589E-6 | 5.39555732E-6 |
| -5.93637874E-6 | 7.40757833E-6 | 3.72612625E-7 | -1.09025694E-5 | 1.27948746E-5 | 3.28974846E-7 | 2.80160904E-6 |
| 7.01540449E-6 | 9.22001303E-6 | -5.32454280E-6 | 6.60464706E-6 | 6.12890680E-8 | -4.39383808E-6 | -1.04419553E-6 |
| -3.47265342E-9 | -5.94076007E-6 | 6.84669089E-6 | -6.75031951E-8 | 2.84278309E-7 | 1.39949002E-6 | 5.95566414E-8 |
| 2.18174054E-6 | -6.08022815E-8 | 4.30809387E-9 | 2.61667266E-6 | -1.17836212E-7 | -3.31637603E-8 | -4.20534932E-7 |
| -2.91676474E-7 | 7.26223680E-8 | 2.77478150E-6 | -5.98621358E-7 | 3.28974846E-7 | 3.24255617E-6 | 1.71577785E-5 |
| 6.12846558E-7 | 2.86747764E-6 | -3.65465410E-7 | 1.14607039E-7 | 2.66803872E-6 | -2.03109557E-7 | -7.79688676E-6 |
| 2.73383161E-6 | -1.54272434E-8 | -1.75258746E-7 | 1.81825946E-6 | 6.08022815E-8 | -4.30809387E-9 | 1.76032970E-7 |
| -2.68577345E-7 | 2.62941676E-6 | 2.42114160E-7 | -1.03102662E-7 | 8.82118075E-6 | -4.69116582E-6 | -9.07152371E-6 |
| 1.16664720E-5 | -5.69634781E-6 | -6.59831660E-7 | 1.08557413E-5 | -5.24682076E-6 | -1.04419553E-6 | 8.38162801E-6 |
| -9.07152371E-6 | -4.50123869E-7 | 1.52101605E-5 | -8.51755011E-6 | -2.53252586E-7 | 1.37108135E-5 | 2.82004715E-6 |
| -1.85978341E-7 | 1.70483225E-5 | -1.06929597E-5 | 2.68577345E-7 | 1.37058324E-6 | -2.42114160E-7 | -4.50123869E-7 |
| 1.79038825E-7 | -1.55994478E-7 | 2.08342665E-6 | -3.65912865E-8 | -4.91214086E-6 | 6.51043604E-6 | |
| -5.92281500E-6 | 7.61964150E-6 | 3.64196473E-7 | -5.18719919E-6 | 7.01540449E-6 | 6.12846558E-7 | |
| 1.12133959E-5 | 1.84194449E-7 | -8.73565417E-6 | 9.72124297E-7 | 9.13091035E-8 | -7.88085291E-6 | |
| -4.59600356E-8 | -1.10115377E-5 | 1.10868400E-5 | -1.79038825E-7 | 1.55994478E-7 | 1.91657335E-6 | |
| 2.10628549E-6 | -5.39977630E-8 | -1.05698511E-8 | 2.57596221E-6 | -1.26246235E-7 | -6.71050236E-8 | |
| -1.67243986E-7 | 1.33345393E-9 | 2.60917807E-6 | -3.11091792E-7 | 9.22001303E-8 | 2.86747764E-6 | |

FIGURE 13.3 (**Continued**) Computer Printout of Least Squares Adjustment

*(Continued)*

| | | | | | | |
|---|---|---|---|---|---|---|
| 1.84194449E-7 | 3.22173936E-6 | -2.60527636E-7 | -5.75022070E-8 | 2.87105185E-6 | -2.73085776E-7 | 1.16544288E-7 |
| 2.79306278E-6 | -2.65377254E-7 | 3.63579880E-8 | 1.89371451E-6 | 5.39977630E-8 | 1.05698511E-8 | -2.45233749E-7 |
| -2.06452266E-7 | 2.48846774E-6 | 6.50465644E-8 | -1.65488995E-7 | 8.06089011E-6 | -4.16729345E-6 | 1.52101605E-5 |
| 1.32427057E-5 | -7.44447357E-6 | -3.66078310E-7 | 1.03118881E-5 | -5.32454280E-6 | -3.65465410E-7 | -9.10361549E-6 |
| -8.73565417E-6 | -2.60527636E-7 | 2.26907308E-5 | -1.43634201E-5 | -2.46805699E-7 | 1.59483972E-5 | 9.41182921E-8 |
| -3.57270057E-7 | 1.54451924E-5 | -8.86374861E-6 | 2.06452266E-7 | 1.51153226E-6 | -6.50465644E-8 | -8.51755011E-6 |
| 1.08947787E-7 | -2.47740287E-7 | 1.97607092E-6 | -4.04933363E-8 | -4.24313886E-6 | 5.73165232E-6 | 9.76919379E-6 |
| -7.45674053E-6 | 8.73949829E-6 | 1.23851360E-7 | -5.26571390E-6 | 6.60464706E-6 | 1.14607039E-7 | 2.58977188E-6 |
| 9.72124297E-6 | -5.75022070E-8 | -1.43634201E-5 | 1.53166610E-5 | 6.74253524E-8 | -9.09372997E-6 | -2.53252586E-7 |
| 1.01846121E-7 | -9.00091017E-6 | 9.68554471E-6 | -1.08947787E-7 | 2.47740287E-7 | 2.02392908E-6 | 1.93948744E-7 |
| 1.98521312E-6 | -3.19868348E-8 | -5.06829902E-9 | 2.47910436E-6 | -1.36768220E-7 | -4.88171348E-8 | -1.18613157E-7 |
| -7.71439188E-8 | -1.10330436E-8 | 2.44119656E-6 | -2.03021085E-7 | 6.12890680E-8 | 2.66803872E-6 | 1.37108135E-5 |
| 9.13091035E-8 | 2.87105185E-6 | -2.46805699E-7 | 6.74253524E-8 | 3.38805295E-6 | -3.52057512E-7 | -1.48188594E-5 |
| 2.75024798E-6 | -4.10611277E-7 | 1.57180929E-7 | 2.01478688E-6 | 3.19868348E-8 | 5.06829902E-9 | 5.79156989E-8 |
| -1.49578782E-7 | 2.21835276E-6 | 2.25515225E-7 | -2.31261152E-7 | 7.17426443E-6 | -3.40929394E-6 | -7.79688676E-6 |
| 1.06135231E-5 | -5.53292822E-6 | -2.71438743E-7 | 8.99366669E-6 | -4.39383808E-6 | -2.03109557E-7 | 1.55374217E-5 |
| -7.88085291E-6 | -2.73085776E-7 | 1.59483972E-5 | -9.09372997E-6 | -3.52057512E-7 | 2.40005799E-5 | 2.60934553E-6 |
| -4.99962685E-7 | 1.40174451E-5 | -7.49065788E-6 | 1.49578782E-7 | 1.78164724E-6 | -2.25515225E-7 | -1.85978341E-7 |
| 8.54083646E-8 | -2.54771406E-7 | 1.52692915E-6 | 2.41134976E-8 | -3.61195951E-6 | 4.73253912E-6 | |
| -5.66510971E-6 | 6.49530272E-6 | 1.08247296E-7 | -4.53441325E-6 | 5.39555732E-6 | 5.95566414E-8 | |
| 8.38162801E-6 | 1.16544288E-6 | -9.10361549E-6 | 9.76919379E-6 | 1.93948744E-6 | -1.48188594E-5 | |
| 2.54228851E-7 | -7.57831099E-6 | 8.41016134E-6 | -8.54083646E-8 | 2.54771406E-7 | 2.47307085E-6 | |
| 2.06220315E-6 | -5.06115621E-8 | -4.15590258E-8 | 2.63770488E-6 | -2.15470176E-7 | -6.50138113E-8 | |
| -7.64179881E-8 | -8.68924037E-8 | 2.51797164E-6 | -1.97169402E-7 | -3.47265342E-9 | 2.73383161E-6 | |

(Continued)

FIGURE 13.3 (Continued) Computer Printout of Least Squares Adjustment

| | | | | | | |
|---|---|---|---|---|---|---|
| -4.59600356E-8 | 2.79306278E-6 | -3.57270057E-7 | 1.01846121E-7 | 2.75024798E-6 | -4.99962685E-7 | 2.54228851E-7 |
| 3.00608161E-6 | -6.74699122E-7 | 3.29049227E-7 | 1.93779685E-6 | 5.06115621E-8 | 4.15590258E-8 | 4.47125718E-8 |
| -8.00321968E-8 | 2.43994893E-6 | -5.04741844E-8 | -2.48525168E-7 | 8.72447156E-6 | -4.72535510E-6 | 1.70483225E-5 |
| 1.18004690E-5 | -6.76964313E-6 | -1.78246937E-7 | 1.05147431E-5 | -5.94076007E-6 | -1.54272434E-8 | -7.57831099E-6 |
| -1.10115377E-5 | -2.65377254E-7 | 1.54451924E-5 | -9.00091017E-6 | -4.10611277E-7 | 1.40174451E-5 | -1.66352613E-7 |
| -6.74699122E-7 | 1.75154380E-5 | -9.96307743E-6 | 8.00321968E-8 | 1.56005107E-6 | 5.04741844E-8 | -1.06929597E-5 |
| -4.50111475E-8 | -3.73419489E-7 | 1.85116012E-6 | -7.75339712E-8 | -4.75884420E-6 | 6.06781967E-6 | 8.41016134E-6 |
| -6.73346209E-6 | 7.60877623E-6 | -4.72579674E-8 | -5.95447731E-6 | 6.84669089E-6 | -1.75258746E-7 | 1.74917311E-6 |
| 1.10868400E-5 | 3.63579880E-8 | -8.86374861E-6 | 9.68554471E-6 | 1.57180929E-7 | -7.49065788E-6 | 2.68577345E-7 |
| 3.29049227E-7 | -9.96307743E-6 | 1.11817178E-5 | 4.50111475E-8 | 3.73419489E-7 | 2.14883988E-6 | -8.54083646E-8 |
| 1.36323898E-6 | 5.48656843E-8 | 1.73752545E-8 | 1.72167051E-6 | 1.07866090E-7 | -9.01748207E-8 | 1.74917311E-6 |
| 2.53361577E-7 | -1.43925117E-7 | 1.66669361E-6 | 1.34603208E-7 | -6.75031951E-8 | 1.81825946E-6 | 2.68577345E-7 |
| -1.79038825E-7 | 1.89371451E-6 | 2.06452266E-7 | -1.08947787E-7 | 2.01478688E-6 | 1.49578782E-7 | -8.54083646E-8 |
| 1.93779685E-6 | 8.00321968E-8 | 4.50111475E-8 | 2.63676102E-6 | -5.48656843E-8 | -1.73752545E-8 | 7.28276329E-8 |
| 5.48656843E-8 | 4.76458374E-7 | 4.38046131E-7 | 3.69590972E-8 | 8.50478202E-7 | 3.68419523E-7 | 1.37058324E-6 |
| 1.13667447E-6 | 2.73668395E-7 | 4.77689970E-8 | 1.00865642E-6 | 2.84278309E-7 | 6.08022815E-8 | 2.54771406E-7 |
| 1.55994478E-7 | 5.39977630E-8 | 1.51153226E-6 | 2.47740287E-7 | 3.19868348E-8 | 1.78164724E-6 | 1.11976465E-8 |
| 5.06115621E-8 | 1.56005107E-6 | 3.73419489E-7 | -5.48656843E-8 | 3.52354163E-6 | -4.38046131E-7 | -2.42114160E-7 |
| 1.73752545E-8 | 4.38046131E-7 | 7.59855482E-7 | -2.21002617E-8 | 1.31060303E-7 | 1.35801368E-6 | 2.47307085E-6 |
| 5.63660689E-8 | 1.56114695E-6 | 2.38478679E-9 | 1.02299575E-7 | 1.39949002E-6 | -4.30809387E-9 | |
| 1.91657335E-6 | 1.05698511E-8 | -6.50465644E-8 | 2.02392908E-6 | 5.06829902E-9 | -2.25515225E-7 | |
| 4.15590258E-8 | 5.04741844E-8 | 2.14883988E-8 | -1.73752545E-8 | -4.38046131E-7 | 3.24014452E-6 | |

**FIGURE 13.3 (*Continued*)** Computer Printout of Least Squares Adjustment

*(Continued)*

Delta – the answers—X/Y/Z in meters.

| 124,517.64562 | FOLA | 149,899.212 | WEBE | 115,561.95436 | WATH |
| -4,609,735.81196 | | -4,637,713.16888 | | -4,706,440.30029 | |
| 4,391,808.83078 | | 4,361,716.03796 | | 4,289,090.74715 | |
| 182,409.48216 | SHAN | 153,059.63023 | RASN | 164,796.25211 | SIW2 |
| -4,611,414.93873 | | -4,666,858.81245 | | -4,679,186.50767 | |
| 4,387,983.54545 | | 4,330,615.80371 | | 4,316,889.72405 | |
| 102,706.41607 | WAWN | 114,970.93919 | PAR0 | 170,846.09348 | KEHA |
| -4,657,769.22343 | | -4,680,260.75314 | | -4,702,276.07364 | |
| 4,341,771.60252 | | 4,317,398.78716 | | 4,291,680.31732 | |

Residuals—Also in meters.

| 0.00112 | FOLA | 0.00516 | 6 | 0.00008 | 12 |
| -0.00046 | | -0.02102 | | 0.00153 | |
| 0.00008 | | 0.01019 | | -0.00289 | |
| -0.00146 | 1 | -0.00188 | 7 | 0.00017 | 13 |
| 0.00323 | | -0.00471 | | -0.00115 | |
| -0.00633 | | 0.00765 | | 0.00099 | |
| 0.00245 | 2 | -0.00076 | 8 | 0.00026 | 14 |
| 0.01453 | | 0.00143 | | 0.00137 | |
| 0.00374 | | 0.00075 | | 0.00011 | |
| 0.00137 | 3 | -0.00005 | 9 | 0.00111 | 15 |
| -0.00392 | | 0.00306 | | 0.00265 | |
| 0.00018 | | -0.0064 | | 0.00017 | |
| | | -0.00012 | 10 | | |

**FIGURE 13.3 (Continued)** Computer Printout of Least Squares Adjustment

(Continued)

| -0.00216 | 4 | -0.00022 | | 0.00137 | 16 |
| -0.00615 | | -0.00166 | | -0.01397 | |
| 0.00151 | | | | 0.00727 | |
| -0.00092 | 5 | 0.00096 | 11 | -0.00112 | KEHA |
| -0.00355 | | 0.00031 | | 0.00046 | |
| -0.00344 | | -0.00154 | | -0.00008 | |

Redundancy, R = 27

Reference Variance = $\hat{\sigma}_0^2$ = 2.21472105E0

The covariance matrix of computed parameters is the product of *Ninv* and the reference variance. Note that the covariance matrix of each station has been underlined.

Reference Variance × *Ninv*

FOLA

| 5.83969012E-6 | -1.21512186E-7 | -3.84813418E-8 | 5.04586428E-6 | -2.38893299E-7 | 1.99712073E-7 | 4.98495369E-6 |
|---|---|---|---|---|---|---|
| -5.65554660E-7 | 3.18753985E-7 | 5.16762277E-6 | -2.98108557E-7 | 1.49500747E-7 | 4.83194670E-6 | -5.94823898E-7 |
| 3.96521054E-7 | 4.66483480E-6 | -4.57234178E-7 | 2.41288956E-7 | 4.39669328E-6 | -3.31275277E-7 | 1.89155703E-7 |
| 4.56720471E-6 | -1.77248991E-7 | -9.96871357E-8 | 3.01919406E-6 | 1.21512186E-7 | 3.84813418E-8 | -1.61292891E-7 |
| -1.21512186E-7 | 7.80366180E-6 | -9.70149986E-7 | -8.18540905E-8 | 6.97531221E-6 | -8.15946472E-7 | 5.82342465E-6 |
| 6.34146732E-6 | -6.06099154E-7 | -1.05795003E-7 | 6.62499158E-6 | -6.29597154E-7 | -1.34660092E-7 | -5.64247595E-7 |
| -3.45484253E-7 | -1.19589982E-7 | 5.51126188E-6 | -5.48675628E-7 | -7.08419162E-8 | 4.91303254E-6 | -2.47996633E-8 |
| -1.12090492E-7 | 5.40380625E-6 | -8.27020001E-7 | 1.21512186E-7 | 1.05522239E-6 | 9.70149986E-7 | 5.36215327E-7 |
| -3.84813418E-8 | -9.70149986E-7 | 7.17601626E-6 | 4.89459148E-8 | -2.90262012E-7 | 5.85126270E-6 | 3.38172213E-6 |
| -1.24835119E-7 | 5.40137917E-6 | -5.28163750E-9 | -2.26565021E-7 | 5.75940418E-6 | 9.54122617E-6 | |
| 4.61420885E-6 | -2.34092717E-8 | 1.44059995E-7 | 4.37644586E-6 | -1.12248685E-8 | 4.99453315E-8 | |
| -9.20416491E-8 | -1.11786239E-7 | 4.09980329E-6 | 3.84813418E-8 | 9.70149986E-7 | 1.68286793E-6 | |

(Continued)

**FIGURE 13.3 (*Continued*)** Computer Printout of Least Squares Adjustment

**SHAN**

| | | | | | | |
|---|---|---|---|---|---|---|
| 5.04586428E-6 | -8.18540905E-8 | 4.89459148E-8 | 7.23016526E-6 | -1.76651085E-6 | 1.14309458E-6 | 5.68227019E-6 |
| -4.53345549E-8 | -1.97551880E-7 | 5.85359446E-6 | -4.51176780E-7 | 2.73498771E-7 | 5.79520000E-6 | -2.28343635E-7 |
| -8.10394923E-8 | 5.70503773E-6 | -3.66511959E-7 | -8.96814442E-8 | 5.49052459E-6 | -5.12178939E-7 | 5.34046707E-8 |
| 5.84178050E-6 | -5.50413921E-7 | -1.71716118E-7 | 3.81301991E-6 | 8.18540905E-8 | -4.89459148E-8 | -5.42760535E-8 |
| -2.38893299E-7 | 6.97531221E-6 | -2.90262012E-7 | -1.76651085E-6 | 4.40147083E-5 | -2.76280655E-5 | 1.95364547E-5 |
| 1.61931179E-5 | -7.88982187E-6 | -5.06838739E-7 | 1.80092410E-5 | -9.79720172E-6 | -2.60974339E-7 | -7.99948275E-6 |
| -1.08790217E-5 | -2.79600194E-7 | 1.78526230E-5 | -9.39736895E-6 | -3.02903455E-7 | 1.58889944E-5 | -2.75606814E-7 |
| -4.77206335E-7 | 1.93222708E-7 | -1.05395124E-5 | 2.38893299E-7 | 1.88357197E-6 | 2.90262012E-7 | -1.03896237E-5 |
| 1.99712073E-7 | -8.15946472E-7 | 5.85126270E-6 | 1.14309458E-6 | -2.76280655E-5 | 3.13373319E-5 | 1.04812540E-5 |
| -7.84020747E-6 | 1.21091039E-5 | 2.92204685E-7 | -9.88788399E-6 | 1.49427359E-5 | -7.34484779E-8 | -1.03896237E-5 |
| 1.44187997E-5 | -1.48618908E-7 | -9.22939250E-6 | 1.26940110E-5 | -1.08116336E-7 | -7.55063504E-6 | 1.04812540E-5 |
| -1.43987456E-7 | -1.04653434E-5 | 1.34385279E-5 | -1.99712073E-7 | 8.15946472E-7 | 3.00762148E-6 | |

**WAWN**

| | | | | | | |
|---|---|---|---|---|---|---|
| 4.98495369E-6 | -1.61292891E-7 | -2.47996633E-8 | 5.68227019E-6 | -5.42760535E-8 | -2.75606814E-7 | 8.00868332E-6 |
| 5.24565019E-7 | -7.58361441E-7 | 6.14066049E-6 | -3.06691665E-7 | -2.95843128E-7 | 6.20478250E-6 | -9.31367564E-7 |
| 3.89863923E-7 | 6.24561778E-6 | -5.43124346E-7 | 2.08445762E-7 | 5.73562228E-6 | -2.62695056E-7 | 1.28267117E-7 |
| 5.77897247E-6 | 9.90258738E-8 | -3.68424633E-7 | 3.87393049E-6 | 1.61292891E-7 | 2.47996633E-8 | 5.24565019E-7 |
| -5.65554660E-7 | 6.34146732E-6 | -1.24835119E-7 | -4.53345549E-8 | 1.61931179E-5 | -7.84020747E-6 | 2.58379810E-5 |
| 5.04329255E-5 | -3.04676923E-5 | -5.54982391E-7 | 2.60769285E-5 | -1.31474229E-5 | -6.45982026E-7 | -1.25466377E-5 |
| -1.311173830E-5 | -3.70398775E-7 | 2.93288991E-5 | -1.65146002E-5 | -1.70852261E-7 | 2.35059930E-5 | -7.58361441E-7 |
| -1.69244526E-7 | 2.61347470E-5 | -1.49127402E-5 | 5.65554660E-7 | 2.51741687E-6 | 1.24835119E-7 | -1.26158214E-5 |
| 3.18753985E-7 | -6.06099154E-7 | 5.40137917E-6 | -1.97551880E-7 | -7.88982187E-7 | 1.21091039E-5 | 1.43852836E-5 |
| -3.04676923E-5 | 3.24510079E-5 | -5.91588414E-8 | -1.28318948E-5 | 1.64057196E-5 | 1.60838287E-7 | |
| 1.68753804E-5 | 2.95322848E-9 | -1.64874323E-5 | 1.93555508E-5 | -2.44351139E-8 | -1.22538926E-5 | |
| -1.92442435E-7 | -1.49928711E-5 | 1.68513169E-5 | -3.18753985E-7 | 6.06099154E-7 | 3.45750501E-6 | |

FIGURE 13.3 (*Continued*) Computer Printout of Least Squares Adjustment

*(Continued)*

(Continued)

WEBE

| | | | | | | |
|---|---|---|---|---|---|---|
| 5.16762277E-6 | -1.05795003E-7 | -5.28163750E-9 | 5.85359446E-6 | -5.06838739E-7 | 2.92204685E-7 | 6.14066049E-6 |
| -5.54982391E-7 | -5.91588414E-8 | 6.94497313E-6 | -1.52106409E-6 | 8.25233024E-7 | 6.14536698E-6 | -1.46134306E-6 |
| 8.06593593E-7 | 5.77860159E-6 | -8.10761338E-7 | 2.74296213E-7 | 5.40656940E-6 | -6.01161098E-7 | 2.39737564E-7 |
| 5.57660479E-6 | -3.94767243E-7 | -1.04663215E-7 | 3.69126141E-6 | 1.05795003E-7 | 5.28163750E-9 | |
| -2.98108557E-7 | 6.62499158E-6 | -2.26565021E-7 | -4.51176780E-7 | 1.80092410E-5 | -9.88788399E-6 | -3.06691665E-7 |
| 2.60769285E-5 | -1.28318948E-5 | -1.52106409E-6 | 3.92570590E-5 | -2.41461498E-5 | -1.32577932E-6 | 2.40424386E-5 |
| -1.14881992E-5 | -6.88981538E-7 | 2.28379555E-5 | -1.16620874E-5 | -4.49635069E-7 | 1.97096147E-5 | -1.00424605E-5 |
| -4.36675225E-7 | 2.32872229E-5 | -1.31875062E-5 | 2.98108557E-7 | 2.23389260E-6 | 2.26565021E-7 | -2.95843128E-7 |
| 1.49500747E-7 | -6.29597154E-7 | 5.75940418E-6 | 2.73498771E-7 | -9.79720172E-6 | 1.49427359E-5 | -1.16202444E-5 |
| -1.31474229E-5 | 1.64057196E-5 | 8.25233024E-7 | -2.41461498E-5 | 2.83370780E-5 | 7.28587514E-7 | 1.19496544E-5 |
| 1.55371640E-5 | 2.04197569E-7 | -1.17923770E-5 | 1.46274508E-5 | 1.35738189E-7 | -9.73112566E-6 | |
| -7.69095863E-9 | -1.31571264E-5 | 1.51635104E-5 | -1.49500747E-5 | 6.29597154E-7 | 3.09948001E-6 | |

RASN

| | | | | | | |
|---|---|---|---|---|---|---|
| 4.83194670E-6 | -1.34660092E-6 | 9.54122617E-9 | 5.79520000E-6 | -2.60974339E-7 | -7.34484779E-8 | 6.20478250E-6 |
| -6.45982026E-7 | 1.60838287E-7 | 6.14536698E-6 | -1.32577932E-6 | 7.28587514E-7 | 7.18135740E-6 | -2.31260182E-6 |
| 1.35728417E-6 | 6.35066308E-6 | -8.09403935E-7 | 2.53822621E-7 | 5.90896150E-6 | -4.49831011E-7 | 1.31901347E-7 |
| 6.05467441E-6 | -3.41670407E-8 | -3.88149233E-7 | 4.02693749E-6 | 1.34660092E-7 | -9.54122617E-9 | |
| -5.94823898E-7 | 5.82342465E-6 | 5.36215327E-7 | -2.28343635E-7 | 1.95364547E-5 | -1.03896237E-5 | -9.31367564E-7 |
| 2.58379810E-5 | -1.26158214E-5 | -1.46134306E-6 | 2.40424386E-5 | -1.16202444E-5 | -2.31260182E-6 | 3.79996930E-5 |
| -2.00908945E-5 | -9.96898807E-7 | 3.36862625E-5 | -1.88639975E-5 | -5.60883831E-7 | 3.03656273E-5 | -1.72679292E-5 |
| -4.11890147E-7 | 3.77572787E-5 | -2.36819230E-5 | 5.94823898E-7 | 3.03545954E-6 | -5.36215327E-7 | |
| 3.96521054E-7 | -3.45484253E-7 | 4.61420885E-6 | -8.10394923E-8 | -1.08790217E-5 | 1.44187997E-5 | 3.89863923E-7 |
| -1.31173830E-5 | 1.68753804E-5 | 8.06593593E-7 | -1.14881992E-5 | 1.55371640E-5 | 1.35728417E-6 | -2.00908945E-5 |

**FIGURE 13.3 (Continued)**   Computer Printout of Least Squares Adjustment

(*Continued*)

| | | | | | | |
|---|---|---|---|---|---|---|
| **2.48345440E-5** | 4.07939322E-7 | -1.93470371E-5 | 2.15298414E-5 | 2.02224193E-7 | -1.74538908E-5 | 1.85629679E-5 |
| -1.01788658E-7 | -2.43874844E-5 | 2.45542579E-5 | -3.96521054E-7 | 3.45484253E-7 | 4.24467534E-6 | |
| PAR0 | | | | | | |
| 4.66483480E-6 | -1.19589982E-7 | -2.34092717E-8 | 5.70503773E-6 | -2.79600194E-7 | -1.48618908E-7 | 6.24561778E-6 |
| -3.70398775E-7 | 2.95322848E-9 | 5.77860159E-6 | -6.88981538E-7 | 2.04197569E-7 | 6.35066308E-6 | -9.96898807E-7 |
| 4.07939322E-7 | **7.13525397E-6** | -5.76996038E-7 | **-1.27351348E-7** | 6.35857897E-6 | -6.04808816E-7 | 2.58113088E-7 |
| 6.18585492E-6 | -5.87736590E-7 | 8.05228011E-8 | 4.19404938E-6 | 1.19589982E-7 | 2.34092717E-8 | -5.43124346E-7 |
| -4.57234178E-7 | 5.51126188E-6 | 1.44059995E-7 | -3.66511959E-7 | 1.78526230E-5 | -9.22939250E-6 | 3.36862625E-5 |
| 2.93288991E-5 | **-1.64874323E-5** | 5.02536390E-5 | **2.28379555E-5** | -1.17923770E-5 | -8.09403935E-7 | -2.01619688E-5 |
| -1.93470371E-5 | -5.76996038E-7 | -1.96307306E-5 | **-8.10761338E-7** | -5.46605777E-7 | 3.53212510E-5 | |
| -7.91253516E-7 | 3.42067926E-5 | 4.37644586E-6 | 4.57234178E-7 | 3.34762231E-6 | -1.44059995E-7 | 2.08445762E-7 |
| 2.41288956E-7 | -5.48675628E-7 | **2.74296213E-7** | -8.96814442E-8 | -9.39736895E-6 | 1.26940110E-5 | -1.88639975E-5 |
| -1.65146002E-5 | **1.93555508E-5** | -3.18109688E-5 | **-1.16620874E-5** | 1.46274508E-5 | 2.53822621E-7 | 2.16360391E-5 |
| 2.15298414E-5 | -1.27351348E-7 | 2.14507797E-5 | 3.39221315E-5 | 1.49328347E-7 | -2.01400752E-5 | |
| 2.55560748E-7 | -1.99345052E-5 | | -2.41288956E-7 | 5.48675628E-7 | 4.48243832E-6 | |
| WATH | | | | | | |
| 4.39669328E-6 | -7.08419162E-8 | -1.12248685E-8 | 5.49052459E-6 | -3.02903455E-7 | -1.08116336E-5 | 5.73562228E-6 |
| -1.70852261E-7 | -2.44351139E-8 | 5.40656940E-6 | -4.49635069E-7 | **1.35738189E-7** | **5.90896150E-6** | **-5.60883831E-7** |
| 2.02224193E-7 | 6.35857897E-6 | -5.46605777E-7 | 1.49328347E-7 | 7.50359219E-6 | -7.79709181E-7 | 4.29542366E-7 |
| 6.09103208E-6 | -9.09389437E-7 | 3.48111911E-7 | 4.46219090E-6 | 7.08419162E-8 | 1.12248685E-8 | -2.62695056E-7 |
| -3.31275277E-7 | 4.91303254E-6 | 4.99453315E-7 | -5.12178939E-7 | 1.58889944E-5 | -7.55063504E-6 | **3.03656273E-5** |
| 2.35059930E-5 | -1.22538926E-5 | -6.01161098E-7 | 1.97096147E-5 | **-9.73112566E-6** | **-4.49831011E-7** | -3.28196398E-5 |
| -1.74538908E-5 | -6.04808816E-7 | 3.53212510E-5 | -2.01400752E-5 | -7.79709181E-7 | 5.31545894E-5 | 1.28267117E-7 |
| -1.10727788E-6 | 3.10447307E-5 | -1.65897177E-5 | 3.31275277E-7 | 3.94585164E-6 | -4.99453315E-7 | |
| 1.89155703E-7 | -5.64247595E-7 | 3.38172213E-6 | 5.34046707E-8 | -7.99948275E-6 | 1.04812540E-5 | |

**FIGURE 13.3 (*Continued*)**   Computer Printout of Least Squares Adjustment

(Continued)

| Label | | | | | | | |
|---|---|---|---|---|---|---|---|
| | | 1.31901347E-7 | 1.19496544E-5 | -1.00424605E-5 | 2.39737564E-7 | 1.43852836E-5 | -1.25466377E-5 |
| | -1.72679292E-5 | -3.28196398E-5 | 4.29542366E-7 | 2.16360391E-5 | -2.01619688E-5 | 2.58113088E-7 | 1.85629679E-5 |
| | 3.44110548E-5 | 5.47716206E-6 | 5.64247595E-7 | -1.89155703E-7 | 1.86261613E-5 | -1.67838448E-5 | 5.63045987E-7 |
| SIW2 | 5.77897247E-6 | -1.43987456E-7 | -4.77206335E-7 | 5.84178050E-6 | -9.20416491E-8 | -1.12090492E-7 | 4.56720471E-6 |
| | -4.11890147E-7 | 6.05467441E-6 | -7.69095863E-9 | -4.36675225E-7 | 5.57660479E-6 | -1.92442435E-7 | -1.69244526E-7 |
| | 5.63045987E-7 | -1.10727788E-6 | 6.09103208E-6 | 2.25560748E-8 | -7.91253516E-7 | 6.18585492E-6 | -1.01788658E-7 |
| | 9.90258738E-8 | 9.20416491E-8 | 1.12090492E-7 | 4.29167947E-6 | 7.28752248E-7 | -1.49427035E-6 | 6.65763220E-6 |
| | 3.77572787E-5 | -1.04653434E-5 | 1.93222708E-5 | -5.50413921E-7 | -1.11786239E-7 | 5.40380625E-6 | -1.77248991E-7 |
| | -1.67838448E-5 | -3.41670407E-8 | -1.31571264E-5 | 2.32872229E-5 | -3.94767243E-7 | -1.49928711E-5 | 2.61347470E-5 |
| | -3.68424633E-7 | 3.10447307E-5 | -9.09389437E-7 | -1.99345052E-5 | 3.42067926E-5 | -5.87736590E-7 | -2.43874844E-5 |
| | -2.36819230E-5 | 1.11786239E-7 | 3.45507794E-6 | 1.77248991E-7 | -2.20654373E-5 | 3.87918092E-5 | -1.49427035E-6 |
| | 1.86261613E-5 | 1.34385279E-5 | -1.05395124E-5 | -1.71716118E-7 | 4.09980329E-6 | -8.27020001E-7 | -9.96871357E-8 |
| | 3.87393049E-6 | -3.88149233E-7 | 1.51635104E-5 | -1.31875062E-5 | -1.04663215E-7 | 1.68513169E-5 | -1.49127402E-5 |
| | 5.94823898E-7 | -1.65897177E-5 | 3.48111911E-7 | 2.14507797E-5 | -1.96307306E-5 | 8.05228011E-8 | 2.45542579E-5 |
| | -1.89155703E-7 | 4.75908090E-6 | 8.27020001E-8 | 9.96871357E-8 | 2.47643858E-5 | -2.20654373E-5 | 7.28752248E-7 |
| KEHA | 1.61292891E-7 | -1.99712073E-7 | 2.38893299E-7 | 3.81301991E-6 | 3.84813418E-8 | 1.21512186E-7 | 3.01919406E-6 |
| | 3.03545954E-6 | 4.02693749E-6 | -1.49500747E-7 | 2.98108557E-7 | 3.69126141E-6 | -3.18753985E-7 | 5.65554660E-7 |
| | 5.64247595E-7 | 3.31275277E-7 | 4.46219090E-6 | -2.41288956E-7 | 4.57234178E-7 | 4.19404938E-6 | -3.96521054E-7 |
| | | -3.84813418E-8 | -1.21512186E-7 | 5.83969012E-6 | 9.96871357E-8 | 1.77248991E-7 | 4.29167947E-6 |
| | | 8.15946472E-7 | 1.88357197E-6 | 8.18540905E-8 | 9.70149986E-7 | 1.05522239E-6 | 1.21512186E-7 |
| | | 1.34660092E-7 | 6.29597154E-7 | 2.23389260E-6 | 1.05795003E-7 | 6.06099154E-7 | 2.51741687E-6 |
| | | 3.94585164E-6 | 7.08419162E-8 | 5.48675628E-7 | 3.34762231E-6 | 1.19589982E-7 | 3.45484253E-7 |
| | | -9.70149986E-7 | 7.80366180E-6 | -1.21512186E-7 | 8.27020001E-7 | 3.45507794E-6 | 1.12090492E-7 |

FIGURE 13.3 (Continued) Computer Printout of Least Squares Adjustment

| | | | | | | |
|---|---|---|---|---|---|---|
| 3.84813418E−8 | 9.70149986E−7 | 1.68286793E−6 | −4.89459148E−8 | 2.90262012E−7 | 3.00762148E−6 | 2.47996633E−8 |
| 1.24835119E−7 | 3.45750501E−6 | 5.28163750E−9 | 2.26565021E−7 | 3.09948001E−6 | −9.54122617E−9 | −5.36215327E−7 |
| 4.24467534E−6 | 2.34092717E−8 | −1.44059995E−7 | 4.48243832E−6 | 1.12248685E−8 | −4.99453315E−7 | 5.47716206E−6 |
| 9.20416491E−8 | 1.11786239E−7 | 4.75908090E−6 | −3.84813418E−8 | −9.70149986E−7 | 7.17601626E−6 | |

This is the end of the adjustment. The following section is included in order to clarify interpretation of the covariance matrix of the computed coordinates.

Each station has a symmetric 3 × 3 covariance matrix on the diagonal of the printout above. The correlation between stations is given in the off–diagonal 3 × 3 submatrices. The correlation 3 × 3 submatrices are not themselves symmetrical, but the 3 × 3 correlation submatrices are transpose symmetrical within the overall solution covariance matrix. These properties are important when computing network and local accuracies in Chapter 14.

The generic format of each 3 × 3 submatrix is:

$$\begin{bmatrix} \sigma_1^2 & \sigma_{12} & \sigma_{13} \\ \sigma_{21} & \sigma_2^2 & \sigma_{23} \\ \sigma_{31} & \sigma_{32} & \sigma_3^2 \end{bmatrix}$$

The format of a symmetrical 3 × 3 diagonal submatrix for each station is:

$$\begin{bmatrix} \sigma_X^2 & \sigma_{XY} & \sigma_{XZ} \\ \sigma_{XY} & \sigma_Y^2 & \sigma_{YZ} \\ \sigma_{XZ} & \sigma_{YZ} & \sigma_Z^2 \end{bmatrix}$$

Examples: station covariance matrices for:

FOLA

$$\begin{bmatrix} 5.83969012E-6 & -1.21512186E-7 & -3.84813418E-8 \\ -1.21512186E-7 & 7.8036618E-6 & -9.70149986E-7 \\ -3.84813418E-8 & -9.70149986E-7 & 7.17601626E-6 \end{bmatrix}$$

Example correlation between WEBE and WATH

$$\begin{bmatrix} 5.4065694E-6 & -6.01161098E-7 & 2.39737564E-7 \\ -4.49635069E-7 & 1.97096147E-5 & -1.00424605E-5 \\ 1.35738189E-7 & -9.73112566E-6 & 1.19496544E-5 \end{bmatrix}$$

PAR0

$$\begin{bmatrix} 7.13525397E-6 & -5.76996038E-7 & -1.27351348E-7 \\ -5.76996038E-7 & 5.0253639E-5 & -3.18109688E-5 \\ -1.27351348E-7 & -3.18109688E-5 & 3.39221315E-5 \end{bmatrix}$$

Example correlation between WATH and WEBE

$$\begin{bmatrix} 5.4065694E-6 & -4.49635069E-7 & 1.3573189E-7 \\ -6.01161098E-7 & 1.97096147E-5 & -9.73112566E-6 \\ 2.39737564E-7 & -1.00424605E-5 & 1.19496544E-5 \end{bmatrix}$$

FIGURE 13.3 (*Continued*)    Computer Printout of Least Squares Adjustment

| Geocentric and ECEF Sigma | Geodetic and Local Sigma |
|---|---|
| Station FOLA = 101, NAD 83 (2011) | |
| X = 124,517.646   ± 0.0024 m | ϕ = 43° 47′ 41.″73878 N  ± 0.0026 m (N) |
| Y = −4,609,735.812 ± 0.0028 m | λ = 88° 27′ 09.″75363 W ± 0.0024 m (E) |
| Z = 4,391,808.831  ± 0.0027 m | h = 196.715              ± 0.0029 m (U) |
| Station SHAN = 102, NAD 83 (2011) | |
| X = 182,409.482   ± 0.0027 m | ϕ = 43° 44′ 51.″46198 N ± 0.0031 m (N) |
| Y = −4,611,414.939 ± 0.0066 m | λ = 87° 44′ 05.″22498 W ± 0.0027 m (E) |
| Z = 4,387,983.545  ± 0.0056 m | h = 152.926              ± 0.0081 m (U) |
| Station WAWN = 103, NAD 83 (2011) | |
| X = 102,706.416   ± 0.0028 m | ϕ = 43° 10′ 26.″61357 N ± 0.0032 m (N) |
| Y = −4,657,769.223 ± 0.0071 m | λ = 88° 44′ 12.″48262 W ± 0.0028 m (E) |
| Z = 4,341,771.602 ± 0.0057 m | h = 215.589              ± 0.0085 m (U) |
| Station WEBE = 104, NAD 83 (2011) | |
| X = 149,899.212   ± 0.0026 m | ϕ = 43° 25′ 13.″96916 N ± 0.0031 m (N) |
| Y = −4,637,713.169 ± 0.0063 m | λ = 88° 08′ 55.″47148 W ± 0.0026 m (E) |
| Z = 4,361,716.038 ± 0.0053 m | h = 235.148              ± 0.0076 m (U) |
| Station RASN = 105, NAD 83 (2011) | |
| X = 153,059.630   ± 0.0027 m | **ϕ = 43° 02′ 10.″93118 N ± 0.0033 m (N)** |
| Y = −4,666858.812 ± 0.0062 m | λ = 88° 07′ 17.″52790 W ± 0.0027 m (E) |
| Z = 4,330,615.804 ± 0.0050 m | h = 234.351              ± 0.0072 m (U) |
| Station PAR0 = 106, NAD 83 (2011) | |
| X = 114,970.939   ± 0.0027 m | ϕ = 42° 52′ 25.″77120 N ± 0.0031 m (N) |
| Y = −4,680,260.753 ± 0.0071 m | λ = 88° 35′ 34.″10915 W ± 0.0027 m (E) |
| Z = 4,317,398.787 ± 0.0058 m | h = 233.518              ± 0.0086 m (U) |
| Station WATH = 107, NAD 83 (2011) | |
| X = 115,561.954   ± 0.0027 m | ϕ = 42° 31′ 36.″12588 N ± 0.0032 m (N) |
| Y = −4,706,440.300 ± 0.0073 m | λ = 88° 35′ 36.″39056 W ± 0.0027 m (E) |
| Z = 4,289,090.747 ± 0.0059 m | h = 280.564              ± 0.0088 m (U) |
| Station SIW2 = 108, NAD 83 (2011) | |
| X = 164,796.252   ± 0.0026 m | ϕ = 42° 52′ 04.″53362 N ± 0.0030 m (N) |
| Y = −4,679,186.508 ± 0.0062 m | λ = 87° 58′ 58.″56188 W ± 0.0026 m (E) |
| Z = 4,316,889.714 ± 0.0050 m | h = 191.255              ± 0.0074 m (U) |
| Station KEHA = 109, NAD 83 (2011) | |
| X = 170,846.094   ± 0.0024 m | ϕ = 42° 33′ 32.″29918 N ± 0.0025 m (N) |
| Y = −4,702,276.074 ± 0.0028 m | λ = 87° 55′ 09.″15047 W ± 0.0024 m (E) |
| Z = 4,291,680.317 ± 0.0027 m | h = 204.176              ± 0.0029 m (U) |

**FIGURE 13.4**   Tabulation of Results of Least Squares Adjustment

matrix of the parameters is the product of the reference covariance and the inverse of the normal equations. A small section is added at the end of the computer printout to clarify the significance of the final covariance matrix.

2. Of computed stations—Figure 13.4 shows a summary of the geocentric and geodetic positions of the least squares adjustment in terms of geocentric ECEF reference frame coordinates and their standard deviations along with the local geodetic ϕ/λ/h coordinates and the local perspective standard deviations.

## NOTES PERTAINING TO ADJUSTMENT

1. The purposes of this adjustment are as follows:
   a. Demonstrate formulation of a linear least squares adjustment.
   b. Show how geocentric vectors and station values can be used to populate the $f$, $B$, and $Q$ matrices needed for a least squares adjustment.
   c. Obtain the covariance matrix of the network adjustment.
   d. Provide a basis for additional runs of the same adjustment using different assumptions for the values of datum accuracy for the control points. This adjustment used 0.002 m for all three components on both "control" stations, FOLA and KEHA.
   e. Generate a network covariance matrix that supports computation of both network and local accuracies between all points included in the network (see Chapter 14).
2. Although coming quite close, the purpose was not to duplicate the adjusted values for the WISCORS network as obtained from WisDOT.
3. The information listed in Figure 13.4 is intended to show various derived values that can be obtained directly from the adjusted coordinate values and the adjustment covariance matrix. Additional network and local accuracy information obtained from the adjustment covariance matrix is given in Chapter 14.

## REFERENCES

Ghilani, C.D. 2010. *Adjustments Computations: Spatial Data Analysis*, 5th ed. Hoboken, NJ: John Wiley & Sons.
Mikhail, E. 1976. *Observations and Least Squares*. New York: Donnelly.
Mikhail, E. and G. Gracie. 1981. *Analysis and Adjustment of Survey Measurements*. New York: Van Nostrand Reinhold.

# 14 Computing Network Accuracy and Local Accuracy Using the Global Spatial Data Model

## INTRODUCTION

Neither the ASPRS/ACSM/ASCE *Glossary of the Mapping Sciences* (1994) nor the NGS *Geodetic Glossary* (1986) contains definitions for the terms "network accuracy" or "local accuracy." On the other hand, the terms accuracy and precision are readily understood and defined in numerous sources. Even so, in practice, there appears to be some ambiguity regarding the word "accuracy" and related terms as applied to the use of spatial data. The goal of this chapter is to clarify some of that ambiguity and to provide a comprehensive example of how network accuracy and local accuracy can be computed and used. Additional insights into those ambiguities are left to others.

Part of the ambiguity may be associated with the question, "with respect to what?" At a fundamental level, accuracy is taken to be a measure of "correctness" or nearness to the truth. Precision is more closely related to random error and is taken to be a measure of repeatability within a group of measurements. A smaller standard deviation is associated with greater precision. An intuitive understanding of the difference between accuracy and precision is evidenced by a general nodding of heads when the statement is made that, when using GNSS, it is possible for a user to be precisely wrong.

The mathematical definition of standard deviation as the positive square root of the variance is without question and the definition of variance is unambiguous. Variance is the average squared deviation from the mean. While the preceding is true, more can be said. For some, it may be ambiguous—is it the variance of a population or a sample? For statisticians the difference is important, but spatial data users work almost exclusively with measurements as samples of an infinite population and assume that deviations from the mean are normally distributed. Another source of ambiguity is that, in reality, differences from the mean may contain both random errors and systematic errors. The assumption of a normal distribution is largely valid for random errors but the presence of unidentified systematic error weakens that assumption. In some cases, the effort to identify and remove all systematic errors cannot be justified and the impact of small systematic errors lumped in with random errors is insignificant. In borderline cases, the competence of a spatial data analyst can become critical. Blunders are not part of this discussion.

Strictly speaking, standard deviation of a measurement is a measure of precision but standard deviations are also used to compare accuracies in the context of a standard. If that were not the case, this chapter should address issues of network precision and local precision. As work on standards continues to evolve, the nomenclature may be changed, but, in keeping with conventions adopted by others, the terms network accuracy and local accuracy are used herein. Precision is associated with measurement and is measured in standard deviations. But, the standard deviation of a number in a database is also related to the accuracy of that number.

Equation 1.36 provides concise mathematical definitions for both network accuracy and local accuracy. The example included in this chapter shows how those definitions can be implemented in terms of the global spatial data model (GSDM). In the past, concepts of network accuracy and local accuracy have been embraced and used by various disciplines and organizations. Since spatial data accuracy lies at the heart of many decisions related to the usefulness and applications of spatial data, issues of integrity and overall spatial data quality are becoming more important—especially with the ever-increasing use of drones, intelligent vehicles, and other driverless devices. The functional model component of the GSDM embodies the geometry, reference frames, and coordinate systems critical to knowing where things are. The stochastic model component of the GSDM includes tools for working with both absolute (network) accuracy and relative (local) accuracy.

## BACKGROUND

A summary of spatial data accuracy includes mathematical concepts of calculus, probability and statistics, least squares, confidence intervals, hypothesis testing, error ellipses, and others. These topics have been studied, explained, and used by many over the years. The impact of those efforts should not be discounted, but should be acknowledged, examined, and exploited. When working with spatial data, benefits derived from the use of those tools can be enhanced by implementation of the GSDM as part of the transition from analog to digital. Practices for handling spatial data accuracy continue to evolve and include numerous advances over the years. Other efforts and applications could be cited here to illustrate the importance and potential benefits of using the GSDM as the basis for spatial data accuracy considerations. For brevity in this chapter, numerous references and comments can be found in Appendix E. In all cases, either explicitly or implicitly, "accuracy with respect to what?" is a question that needs to be kept front and center. The GSDM provides tools for explicitly answering this question—based upon user choices with respect to the measurements used and the user's assumptions about "fixed" values.

## SUMMARY OF PERTINENT CONCEPTS

There is a difference between the accuracy of a point and the accuracy of a derived quantity such as a distance or direction. As stated in Chapter 1, datum accuracy is a measure of how well the point is known with respect to the underlying datum.

Network accuracy and local accuracy are specifically applicable to derived quantities on a point-pair basis.

The following concepts are embodied in the example that follows:

1. A covariance matrix is symmetrical.
2. Variances (standard deviation squared) lie on the diagonal of a covariance matrix.
3. Correlation is tracked in the off-diagonal elements of the covariance matrix. If there is no correlation between variables, the off-diagonal elements are zero.
4. When using GSDM there is a covariance matrix for
   a. A point (station $X/Y/Z$ values) referenced to a datum (selected by user or context)
   b. A vector (baseline—$\Delta X/\Delta Y/\Delta Z$ values) either measured or computed from $X/Y/Z$ values
5. A BURKORD™ database stores
   a. Station $X/Y/Z$ geocentric coordinates and covariance values for the station
      i. A local perspective covariance matrix can be derived using Equation 1.32.
      ii. Stations on different datums are not to be stored in the same file.
   b. Correlations between stations in terms of point-pairs
      i. Correlations between stations may exist whether directly connected or not.
      ii. The correlation $3 \times 3$ submatrix "there" to "here" is the transpose of the correlation $3 \times 3$ submatrix "here" to "there.
      iii. It takes (only) nine values to fully describe correlation between two points.
6. The accuracy of a derived quantity is determined by use of information in the covariance matrices.
   a. Local accuracy can be obtained if station covariance values and correlations are used in Equation 1.36. Required data include
      i. Station coordinate values and the covariance matrix for each station; the "p" record in a BURKORD™ database
      ii. Correlation between named endpoints; the "c" record in a BURKORD™ database
   b. Network accuracy is obtained if no correlation between stations exists. Two different accuracy answers are possible:
      i. Full network accuracy is obtained if the full covariance matrix for each station is used and reflects correlation between station coordinate values.
      ii. Nominal network accuracy is obtained if no correlation exists between station coordinate values and only diagonal elements of the station covariance matrix are used.
   c. Derived functional model quantities (directions and distances) are not affected by use (or not) of covariance values.

## DETAILED EXAMPLE BASED ON WISCONSIN NETWORK

Data for the following example were obtained by downloading RINEX files for nine different stations of the WISCORS network. Stations and times were selected to ensure that each of the subsequent 16 baselines was independent of each other. That is, the data at each station was paired with data from another station and used only once to compute a baseline vector. Those vectors became the basis of a least squares adjustment as described in the previous chapter. The ending covariance matrix of adjusted parameters is the basis for computing both network and local accuracies in this example. As a restatement, the purpose of the network adjustment was not to obtain $X/Y/Z$ coordinate values for each station (the results do match WisDOT values rather well) but to obtain the covariance matrix of the results. Correlation values were extracted from this covariance matrix (shown in Figure 13.3), combined with the adjusted $X/Y/Z$ coordinate values for each station, and placed into a BURKORD™ database—see Figure 14.1. Note that the file contains point, "p" records and correlation "c" records. The records can be included in the file in any order. Subsequent computations search the data file for points (by number) and correlations (by number pairs) to compute derived quantities using equations listed in Chapter 1. The format details for a BURKORD™ data file are available at http://www.globalcogo.com/dbformat.html, accessed May 4, 2017.

Equations 1.21, 1.23, and 1.24 are used to compute the distance and direction between a pair of points defined by geocentric $X/Y/Z$ coordinates values. The covariance matrix of the inverse is computed using Equation 1.36. Computational options are discussed in the previous section. Figure 14.2 is an annotated computer printout that shows a number of features as follows:

1. The points included in the project file are listed. This is a summary of the "p" records.
2. The list of stored point-pair correlations is given. This is a summary of the "c" records.
3. All combinations of point-pairs distances are included—both "here" to "there" (forward) and "there" to "here" (reverse).
4. The horizontal distance (HD) is in meters and lies in the tangent plane "here" to "there." This is the same as HD(1) in Burkholder (1991)
5. The forward and reverse HDs are different due to Earth curvature and elevation difference.
6. Other options exist, but a single HD could be computed as the mean of forward and reverse HDs (Burkholder 2016). However, this is not the intent.
7. The message in Figure 14.2 is a comparison of network accuracy and local accuracy.
8. Network accuracies reflect the assumed uncertainties at stations FOLA and KEHA.
9. Local accuracies reflect the quality of each computed point-pair vector—both the observed (directly connected) pairs and those point-pairs having correlations as determined in the network adjustment.

Program name/version: Local Output - version 2016-A, August 30, 2016

The user is: Earl F. Burkholder

Program used on: August 30, 2016

The input file for program is: Ch14Bkfile.002.dat

The name of this output file is: Ch14Local002.txt Renamed: Figure 14.1.docx

| | | | | | | | | | | | |
|---|---|---|---|---|---|---|---|---|---|---|---|
| p | 101 | 124517.6456 | -4609735.812 | 4391808.8308 | 5.83969012E-6 | 7.8036618E-6 | 7.17601626E-6 | -1.21512186E-7 | -3.84813418E-8 | -9.70149986E-7 | FOLA |
| p | 102 | 182409.4822 | -4611414.9387 | 4387983.5454 | 7.23016526E-6 | 4.40147083E-5 | 3.13373319E-5 | -1.76651085E-6 | 1.14309458E-6 | -2.76280655E-5 | SHAN |
| p | 103 | 102706.4161 | -4657769.2234 | 4341771.6025 | 8.00868332E-6 | 5.04329255E-5 | 3.24510079E-5 | 5.24565019E-7 | -7.58361441E-7 | -3.04676923E-5 | WAWN |
| p | 104 | 149899.212 | -4637713.1689 | 4361716.038 | 6.94497313E-6 | 3.92570559E-5 | 2.8337078E-5 | -1.52106409E-6 | 8.25233024E-7 | -2.41461498E-5 | WEBE |
| p | 105 | 153059.6302 | -4666858.8124 | 4330615.8037 | 7.1813574E-6 | 3.7999693E-5 | 2.4834544E-5 | -2.31260182E-6 | 1.35728417E-6 | -2.00089945E-5 | RASN |
| p | 106 | 114970.9392 | -4680260.7531 | 4317398.7872 | 7.13525397E-6 | 5.0253639E-5 | 3.39221315E-5 | -5.76996038E-7 | -1.27351348E-7 | -3.18109688E-5 | PAR0 |
| p | 107 | 115561.9544 | -4706440.3003 | 4289090.7472 | 7.50359219E-6 | 5.31545894E-5 | 3.4410548E-5 | -7.79709181E-7 | 4.29542366E-7 | -3.28196398E-5 | WATH |
| p | 108 | 164796.2521 | -4679186.5077 | 4316889.724 | 6.6576322E-6 | 3.87918092E-5 | 2.47643858E-5 | -1.49427035E-6 | 7.28752248E-7 | -2.20654373E-5 | SIW2 |
| p | 109 | 170846.0935 | -4702276.0736 | 4291680.3173 | 5.83969012E-6 | 7.8036618E-6 | 7.17601626E-6 | -1.21512186E-7 | -3.84813418E-8 | -9.70149986E-7 | KEHA |
| c | 101 | 102 | 5.04586428E-6 | -2.38893299E-7 | 1.99712073E-7 | -8.18540905E-8 | 6.97531221E-6 | -8.15946472E-7 | 4.89459148E-8 | -2.90262012E-7 | 5.8512627E-6 |
| c | 101 | 103 | 4.98495369E-6 | -5.6555466E-7 | 3.18753985E-7 | -1.61292891E-7 | 6.34146732E-6 | -6.06091154E-7 | -2.47996633E-8 | -1.24835119E-7 | 5.40137917E-6 |
| c | 101 | 104 | 5.16762277E-6 | -2.98108557E-7 | 1.49500747E-7 | -1.05795003E-7 | 6.62499158E-6 | -6.29597154E-7 | -5.2816375E-9 | -2.26565021E-7 | 5.75940418E-6 |
| c | 101 | 105 | 4.8319467E-6 | -5.94823898E-7 | 3.96521054E-7 | -1.34660092E-7 | 5.82342465E-6 | -3.45484253E-7 | 9.54122617E-9 | 5.36215327E-7 | 4.61420885E-6 |
| c | 101 | 106 | 4.664834E-6 | -4.57234178E-7 | 2.41288956E-7 | -1.19589982E-7 | 5.51126188E-6 | -5.48675628E-7 | -2.34092717E-8 | 1.44059995E-7 | 4.37644586E-6 |
| c | 101 | 107 | 4.39669328E-6 | -3.31275277E-7 | 1.89155703E-7 | -7.08419162E-8 | 4.91303254E-6 | -5.64247595E-7 | -1.12248685E-8 | 4.99453315E-7 | 3.38172213E-6 |
| c | 101 | 108 | 4.56720471E-6 | -1.77248991E-7 | -9.96871357E-8 | -1.1209049 2E-7 | 5.40380625E-6 | -8.27020001E-7 | -9.20416491E-8 | -1.11786239E-7 | 4.09980329E-6 |
| c | 101 | 109 | 3.01919406E-6 | 1.21512186E-7 | 3.84813418E-8 | 1.21512186E-7 | 1.05522239E-6 | 9.70149986E-7 | 3.84813418E-8 | 9.70149986E-7 | 1.68286793E-6 |
| c | 102 | 101 | 5.04586428E-6 | -8.18540905E-8 | 4.89459148E-8 | -2.38893299E-7 | 6.97531221E-6 | -2.90262012E-7 | 1.99712073E-7 | -8.15946472E-7 | 5.8512627E-6 |
| c | 102 | 103 | 5.68227019E-6 | -4.53344549E-8 | -1.9755188E-7 | -2.38893299E-7 | 1.61931179E-5 | -7.88982187E-6 | -2.75606814E-7 | -7.84020747E-6 | 1.21091039E-5 |
| c | 102 | 104 | 5.85359446E-6 | -5.11176078E-7 | 2.73498771E-7 | -5.06838739E-7 | 1.8009241E-5 | -9.79720172E-6 | 2.922096685E-7 | -9.88788399E-6 | 1.49427359E-5 |
| c | 102 | 105 | 5.7952E-6 | -2.28343635E-7 | -8.10394923E-8 | -2.60974339E-7 | 1.95364547E-5 | -1.08790217E-5 | -7.34484779E-8 | -1.03896237E-5 | 1.44187997E-5 |
| c | 102 | 106 | 5.70503773E-6 | -3.66511959E-7 | -8.96814442E-8 | -2.79600194E-7 | 1.7852623E-5 | -9.39736895E-6 | -1.48618908E-7 | -9.2293925E-6 | 1.269011E-5 |
| c | 102 | 107 | 5.49052459E-6 | -5.12178939E-7 | 5.34046707E-8 | -3.02903455E-7 | 1.58889944E-5 | -7.99948275E-6 | -1.08116336E-7 | -7.55063504E-6 | 1.0481254E-5 |

FIGURE 14.1 BURKORD™ Data File for Network Example

*(Continued)*

| | | | | | | | | | | |
|---|---|---|---|---|---|---|---|---|---|---|
| c | 102 | 108 | 5.8417805E-6 | -5.50413921E-7 | -1.71716118E-7 | -4.7720635E-7 | 1.93222708E-5 | -1.05395124E-5 | -1.43987456E-7 | -1.04653434E-5 | 1.34385279E-5 |
| c | 102 | 109 | 3.81301991E-6 | 8.18540905E-8 | -4.89459148E-8 | 2.38893299E-7 | 1.88357197E-6 | 2.90262012E-7 | -1.99712073E-7 | 8.15946472E-7 | 3.00762148E-6 |
| c | 103 | 101 | 4.98495369E-6 | -1.61292891E-7 | -2.47996633E-8 | -5.6555466E-7 | 6.34146732E-6 | -1.24835119E-7 | 3.18753985E-7 | -6.06099154E-7 | 5.40137917E-6 |
| c | 103 | 102 | 5.68227019E-6 | -5.42760535E-8 | -2.75606814E-7 | -4.53345549E-8 | 1.61931179E-5 | -7.84020747E-6 | -1.9755188E-7 | -7.88982187E-6 | 1.21091039E-5 |
| c | 103 | 104 | 6.14060049E-6 | -3.06691665E-7 | -2.95843128E-7 | -5.54982391E-7 | 2.60769285E-5 | -1.31474269E-5 | -5.31588414E-8 | -1.28318948E-5 | 1.64057196E-5 |
| c | 103 | 105 | 6.2047825E-6 | -9.31367564E-7 | 3.89863923E-7 | -6.45982026E-7 | 2.5837981E-5 | -1.3117383E-5 | 1.60838287E-7 | -1.26158214E-5 | 1.68753804E-5 |
| c | 103 | 106 | 6.24561778E-6 | -5.43124346E-7 | 2.08445762E-7 | -3.70398775E-7 | 2.93288991E-5 | -1.65146002E-5 | 2.95322848E-9 | -1.64874323E-5 | 1.93555508E-5 |
| c | 103 | 107 | 5.73562228E-6 | -2.62695056E-7 | 1.2826717E-7 | -1.70852261E-7 | 2.3505993E-5 | -1.25466377E-5 | -2.44351139E-8 | -1.22538926E-5 | 1.43852836E-5 |
| c | 103 | 108 | 5.77897247E-6 | 9.90258738E-8 | -3.68424633E-7 | -1.69244526E-7 | 2.6134747E-5 | -1.49127402E-5 | -1.92442435E-7 | -1.49928711E-5 | 1.68513169E-5 |
| c | 103 | 109 | 3.87393049E-6 | 1.61292891E-7 | 2.47996633E-8 | 5.6555466E-7 | 2.51741687E-6 | 1.24835119E-7 | -3.18753985E-7 | 6.06099154E-7 | 3.45750501E-6 |
| c | 104 | 101 | 5.16762277E-6 | -1.05795003E-7 | -5.2816375E-9 | -2.98108557E-7 | 6.62499158E-6 | -2.26565021E-7 | 1.49500747E-7 | -6.29597154E-7 | 5.75940418E-6 |
| c | 104 | 102 | 5.85359446E-6 | -5.06838739E-7 | 2.92204685E-7 | -4.5117678E-7 | 1.8009241E-5 | -9.88788399E-6 | 2.73498771E-7 | -9.79720172E-6 | 1.49427359E-5 |
| c | 104 | 103 | 6.14060049E-6 | -5.54982391E-7 | -5.9158841E-8 | -3.06691665E-7 | 2.60769285E-5 | -1.28318948E-5 | -2.95843128E-7 | -1.31474269E-5 | 1.64057196E-5 |
| c | 104 | 105 | 6.14536698E-6 | -1.46134306E-6 | 8.06593593E-7 | -1.32577932E-6 | 2.40424386E-5 | -1.14881992E-5 | 7.28587514E-7 | -1.16202444E-5 | 1.5537164E-5 |
| c | 104 | 106 | 5.77860159E-6 | -8.10761338E-7 | 2.74296213E-7 | -6.88981538E-7 | 2.28379555E-5 | -1.16620874E-5 | 2.04197569E-7 | -1.1792377E-5 | 1.46274508E-5 |
| c | 104 | 107 | 5.4065694E-6 | -6.01161098E-7 | 2.39737564E-7 | -4.49635069E-7 | 1.97096147E-5 | -1.00424605E-5 | 1.35738189E-7 | -9.73112566E-6 | 1.19496544E-5 |
| c | 104 | 108 | 5.57660479E-6 | -3.94767243E-7 | -1.04663215E-7 | -4.36675225E-7 | 2.32872229E-5 | -1.31875062E-5 | -7.69095863E-9 | -1.31571264E-5 | 1.51635104E-5 |
| c | 104 | 109 | 3.69126141E-6 | 1.05795003E-7 | 5.2816375E-9 | 2.98108557E-7 | 2.23389266E-6 | 2.26565021E-7 | -1.49500747E-7 | 6.29597154E-7 | 3.09948001E-6 |
| c | 105 | 101 | 4.8319467E-6 | -1.34660092E-7 | 9.54122617E-9 | -5.94823898E-7 | 5.82348246E-6 | 5.36215327E-7 | 3.96521054E-7 | -3.45484253E-7 | 4.61420885E-6 |
| c | 105 | 102 | 5.7952E-6 | -2.60974339E-7 | -7.34484779E-8 | -2.28343635E-7 | 1.95364547E-5 | -1.03896237E-5 | -8.10394923E-8 | -1.08790217E-5 | 1.44187997E-5 |
| c | 105 | 103 | 6.2047825E-6 | -6.45982026E-7 | 1.60838287E-7 | -9.31367564E-7 | 2.5837981E-5 | -1.26158214E-5 | 3.89863923E-7 | -1.3117383E-5 | 1.68753804E-5 |
| c | 105 | 104 | 6.14536698E-6 | -1.32577932E-6 | 7.28587514E-7 | -1.46134306E-6 | 2.40424386E-5 | -1.16202444E-5 | 8.06593593E-7 | -1.14881992E-5 | 1.5537164E-5 |
| c | 105 | 106 | 6.35066308E-6 | -8.09400935E-7 | 2.53822621E-7 | -9.96898807E-7 | 3.36862625E-5 | -1.88639975E-5 | 4.07939322E-7 | -1.93470371E-5 | 2.15298414E-5 |
| c | 105 | 107 | 5.9089615E-6 | -4.49831011E-7 | 1.31901347E-7 | -5.60883831E-7 | 3.03656273E-5 | -1.72679292E-5 | 2.02224193E-7 | -1.74538908E-5 | 1.85629679E-5 |
| c | 105 | 108 | 6.05467441E-6 | -3.41670407E-8 | -3.88149233E-7 | -4.11890147E-7 | 3.77572787E-5 | -2.3681923E-5 | -1.01788658E-7 | -2.43874844E-5 | 2.45542579E-5 |
| c | 105 | 109 | 4.02693749E-6 | 1.34660092E-7 | -9.54122617E-9 | 5.94823898E-7 | 3.03545954E-6 | -5.36215327E-7 | -3.96521054E-7 | 3.45484253E-7 | 4.24467534E-6 |
| c | 106 | 101 | 4.6648348E-6 | -1.19589982E-7 | -2.34092717E-8 | -4.57234178E-7 | 5.51126188E-6 | 1.44059995E-7 | 2.41288956E-7 | -5.48675628E-7 | 4.37644586E-6 |
| c | 106 | 102 | 5.70503773E-6 | -2.79600194E-7 | -1.48618908E-7 | -3.66511959E-7 | 1.7852623E-5 | -9.2293925E-6 | -8.96814442E-8 | -9.39736895E-6 | 1.2694011E-5 |
| c | 106 | 103 | 6.24561778E-6 | -3.70398775E-7 | 2.95322848E-9 | -5.43124346E-7 | 2.93288991E-5 | -1.64874323E-5 | 2.08445762E-7 | -1.65146002E-5 | 1.93555508E-5 |

FIGURE 14.1 (*Continued*) BURKORD™ Data File for Network Example

*(Continued)*

| | | | | | | | | | | |
|---|---|---|---|---|---|---|---|---|---|---|
| c | 106 | 104 | 5.77860159E-6 | -6.88981538E-7 | 2.04197569E-7 | -8.10761338E-7 | 2.28379555E-5 | -1.1792377E-5 | 2.74296213E-7 | -1.16620874E-5 | 1.46274508E-5 |
| c | 106 | 105 | 6.35063608E-6 | -9.96898807E-7 | 4.07939322E-7 | -8.09403935E-7 | 3.36862625E-5 | -1.93470371E-5 | 2.53822621E-7 | -1.88639975E-5 | 2.15298414E-5 |
| c | 106 | 107 | 6.33857897E-6 | -6.04808816E-7 | 2.58113088E-7 | -5.46605777E-7 | 3.5321251E-5 | -2.01619688E-5 | 1.49328347E-7 | -2.01400752E-5 | 2.16360391E-5 |
| c | 106 | 108 | 6.18585492E-6 | -5.8773659E-7 | 8.05228011E-8 | -7.91253516E-7 | 3.42067926E-5 | -1.96307306E-5 | -2.41288956E-7 | -1.99345052E-5 | 2.14507797E-5 |
| c | 106 | 109 | 4.1904938E-6 | 1.19589982E-7 | 2.34092717E-8 | 4.57234178E-7 | 3.34762231E-6 | -1.44059995E-7 | 5.48675628E-7 | 5.48675628E-7 | 4.48243832E-6 |
| c | 107 | 101 | 4.39669328E-6 | -7.08419162E-8 | -1.12248685E-8 | -3.31275277E-7 | 4.91303254E-6 | 4.99453315E-7 | 1.89155703E-7 | -5.64247595E-7 | 3.38172213E-6 |
| c | 107 | 102 | 5.49052459E-6 | -3.02903455E-7 | -1.08116336E-7 | -5.12178939E-7 | 1.58889944E-5 | -7.55063504E-6 | 5.34046707E-8 | -7.99948275E-6 | 1.0481254E-5 |
| c | 107 | 103 | 5.73562228E-6 | -1.70852261E-7 | -2.44351139E-8 | -2.62695056E-7 | 2.3505993E-5 | -1.22538926E-5 | 1.28267117E-7 | -1.25466377E-5 | 1.43852836E-5 |
| c | 107 | 104 | 5.4056694E-6 | -4.49635069E-7 | 1.35738189E-7 | -6.01161098E-7 | 1.97096147E-5 | -9.73112566E-6 | 2.39737564E-7 | -1.00424605E-5 | 1.19496544E-5 |
| c | 107 | 105 | 5.9089615E-6 | -5.60883831E-7 | 2.02224193E-7 | -4.49831011E-7 | 3.03656273E-5 | -1.74538908E-5 | 1.31901347E-7 | -1.72679292E-5 | 1.85629679E-5 |
| c | 107 | 106 | 6.35857897E-6 | -5.46605777E-7 | 1.49328347E-7 | -6.04808816E-7 | 3.5321251E-5 | -2.01400752E-5 | 2.58113088E-7 | -2.01619688E-5 | 2.16360391E-5 |
| c | 107 | 108 | 6.09103208E-6 | -9.09389437E-7 | 3.48111911E-7 | -1.10727788E-6 | 3.10447307E-5 | -1.65897177E-5 | -1.89155703E-7 | -1.67838448E-5 | 1.86261613E-5 |
| c | 107 | 109 | 4.4621909E-6 | 7.08419162E-8 | 1.12248685E-8 | 3.31275277E-7 | 3.94585164E-6 | -4.99453315E-7 | -9.96871357E-8 | 5.64247595E-7 | 5.47716206E-6 |
| c | 108 | 101 | 4.56720471E-6 | -1.12090492E-7 | -9.20416491E-8 | -1.77248991E-7 | 5.40380625E-6 | -1.11786239E-7 | -1.71716118E-7 | -8.27020001E-7 | 4.09980329E-6 |
| c | 108 | 102 | 5.8417805E-6 | -4.77206335E-7 | -1.43987456E-7 | -5.50413921E-7 | 1.93222708E-5 | -1.04653434E-5 | -3.68424633E-7 | -1.05395124E-5 | 1.34385279E-5 |
| c | 108 | 103 | 5.77897247E-6 | -1.69244526E-7 | -1.92442435E-7 | 9.90258738E-8 | 2.6134747E-5 | -1.49928711E-5 | -1.04663215E-7 | -1.49127402E-5 | 1.68513169E-5 |
| c | 108 | 104 | 5.57660479E-6 | -4.36675225E-7 | -7.69095863E-9 | -3.94767243E-7 | 2.32872229E-5 | -1.31571264E-5 | -3.88149233E-7 | -1.31875062E-5 | 1.51635104E-5 |
| c | 108 | 105 | 6.05467441E-6 | -4.11890147E-7 | -1.01788658E-7 | -3.41670407E-8 | 3.77572787E-5 | -2.43874844E-5 | 8.05228011E-8 | -2.3681923E-5 | 2.45542579E-5 |
| c | 108 | 106 | 6.18585492E-6 | -7.91253516E-7 | 2.25560748E-7 | -5.8773659E-7 | 3.42067926E-5 | -1.99345052E-5 | 3.48111911E-7 | -1.96307306E-5 | 2.14507797E-5 |
| c | 108 | 107 | 6.09103208E-6 | -1.10727788E-6 | 5.63045987E-7 | -9.09389437E-7 | 3.10447307E-5 | -1.67838448E-5 | 9.96871357E-8 | -1.65897177E-5 | 1.86261613E-5 |
| c | 108 | 109 | 4.29167947E-6 | 1.12090492E-7 | 9.20416491E-8 | 1.77248991E-7 | 3.45507794E-6 | 1.11786239E-7 | 8.27020001E-7 | 8.27020001E-7 | 4.7590809E-6 |
| c | 109 | 101 | 3.01919406E-6 | 1.21512186E-7 | 3.84813418E-8 | 1.21512186E-7 | 1.05522239E-6 | 9.70149986E-7 | 9.70149986E-7 | 9.70149986E-7 | 1.68286793E-6 |
| c | 109 | 102 | 3.81301991E-6 | 2.38893299E-7 | -1.99712073E-7 | 8.18540905E-8 | 1.88357197E-6 | 8.15946472E-7 | 3.84813418E-8 | 2.90262012E-7 | 3.00762148E-6 |
| c | 109 | 103 | 3.87393049E-6 | 5.6555466E-7 | -3.18753985E-7 | 1.61292891E-7 | 2.51741687E-6 | 6.06099154E-7 | 2.47996633E-8 | 1.24835119E-7 | 3.45750501E-6 |
| c | 109 | 104 | 3.69126141E-6 | 2.98108557E-7 | -1.49500747E-7 | 1.05795003E-7 | 2.2338926E-6 | 6.29597154E-7 | 5.2816375E-9 | 2.26565021E-7 | 3.09948001E-6 |
| c | 109 | 105 | 4.02693749E-6 | 5.94823898E-7 | -3.96521054E-7 | 1.34660092E-7 | 3.03456954E-6 | 3.45484253E-7 | -9.54122617E-9 | -5.36215327E-7 | 4.24467534E-6 |
| c | 109 | 106 | 4.1904938E-6 | 4.57234178E-7 | -2.41288956E-7 | 1.19589982E-7 | 3.34762231E-6 | 5.48675628E-7 | 2.34092717E-8 | -1.44059995E-7 | 4.48243832E-6 |
| c | 109 | 107 | 4.4621909E-6 | 3.31275277E-7 | -1.89155703E-7 | 7.08419162E-8 | 3.94585164E-6 | 5.64247595E-7 | 1.12248685E-8 | -4.99453315E-7 | 5.47716206E-6 |
| c | 109 | 108 | 4.29167947E-6 | 1.77248991E-7 | 9.96871357E-8 | 1.12090492E-7 | 3.45507794E-6 | 8.27020001E-7 | 9.20416491E-8 | 1.11786239E-7 | 4.7590809E-6 |

**FIGURE 14.1 (Continued)** BURKORD™ Data File for Network Example

Program name/version: Local Output - version 2016-A, August 30, 2016

The user is:                                    Earl F. Burkholder

Program used on:                                August 30, 2016

The input file for program is:                  Ch14BKfile.002.dat

The name of this output file is:                Ch14Local.002.txt                Renamed:   Figure 14.2.docx

Points in the BURKORD™ project file are (see Figure 14.1):

| PT  | X/Y/Z          |                 |                | Station |
|-----|----------------|-----------------|----------------|---------|
| 101 | 124,517.6456   | −4,609,735.8120 | 4,391,808.8308 | FOLA    |
| 102 | 182,409.4822   | −4,611,414.9387 | 4,387,983.5454 | SHAN    |
| 103 | 102,706.4161   | −4,657,769.2234 | 4,341,771.6025 | WAWN    |
| 104 | 149,899.212    | −4,637,713.1689 | 4,361,716.038  | WEBE    |
| 105 | 153,059.6302   | −4,666,858.8124 | 4,330,615.8037 | RASN    |
| 106 | 114,970.9392   | −4,680,260.7531 | 4,317,398.7872 | PAR0    |
| 107 | 115,561.9544   | −4,706,440.3003 | 4,289,090.7472 | WATH    |
| 108 | 164,796.2521   | −4,679,186.5077 | 4,316,889.724  | SIW2    |
| 109 | 170,846.0935   | −4,702,276.0736 | 4,291,680.3173 | KEHA    |

List of stored correlations:

PT to PT

| | | |
|---|---|---|
| 101 102 | 104 101 | 107 101 |
| 101 103 | 104 102 | 107 102 |
| 101 104 | 104 103 | 107 103 |
| 101 105 | 104 105 | 107 104 |
| 101 106 | 104 106 | 107 105 |
| 101 107 | 104 107 | 107 106 |
| 101 108 | 104 108 | 107 108 |
| 101 109 | 104 109 | 107 109 |
| 102 101 | 105 101 | 108 101 |
| 102 103 | 105 102 | 108 102 |
| 102 104 | 105 103 | 108 103 |
| 102 105 | 105 104 | 108 104 |
| 102 106 | 105 106 | 108 105 |
| 102 107 | 105 107 | 108 106 |
| 102 108 | 105 108 | 108 107 |
| 102 109 | 105 109 | 108 109 |
| 103 101 | 106 101 | 109 101 |
| 103 102 | 106 102 | 109 102 |
| 103 104 | 106 103 | 109 103 |
| 103 105 | 106 104 | 109 104 |
| 103 106 | 106 105 | 109 105 |
| 103 107 | 106 107 | 109 106 |
| 103 108 | 106 108 | 109 107 |
| 103 109 | 106 109 | 109 108 |

**FIGURE 14.2** Network and Local Accuracies Based on Standard Deviation of 0.002 m for Each Component at Stations FOLA and KEHA                                    (*Continued*)

*Notes:*

1. The horizontal distance (HD) between points is in meters.
2. The HD lies in the tangent plane from "here" to "there."
3. The HD "there" to "here" is slightly different due to Earth curvature and elevation differences.
4. A single HD can be computed as the mean of forward and back. But, that is not the issue.
5. The issue in this printout is a comparison of network accuracy and local accuracy.
6. The standard deviations of the inverses are given in meters.
7. Network accuracies reflect the assumed uncertainties at the two control stations.
8. Local accuracies reflect the observed quality of each point with respect to its "point-pair."
9. Network accuracy can be one of two values based on use of station covariance information:
   a. Nominal network accuracy is based only on the diagonal values in the station covariance matrix.
   b. Full network accuracy is based on using the full station covariance matrix values.
10. The number of decimal places shown cannot be justified except to show where differences first occur.

| PT to PT | Horizontal Distance | Nominal Network Accuracy | Full network Accuracy | Local Accuracy | Baseline |
|---|---|---|---|---|---|
| Points that are directly connected with baselines | | | | | |
| FOLA-SHAN | 58,041.558 m | 0.003635 m | 0.003617 m | 0.001730 m | BL1 |
| SHAN-FOLA | 58,041.956 m | 0.003635 m | 0.003614 m | 0.001726 m | |
| FOLA-WAWN | 72,708.361 m | 0.006747 m | 0.004094 m | 0.002457 m | BL2 |
| WAWN-FOLA | 72,708.143 m | 0.006732 m | 0.004072 m | 0.002423 m | |
| FOLA-WEBE | 48,296.083 m | 0.005757 m | 0.003936 m | 0.002044 m | BL3 |
| WEBE-FOLA | 48,295.792 m | 0.005750 m | 0.003928 m | 0.002032 m | |
| SHAN-WEBE | 49,381.015 m | 0.006833 m | 0.004052 m | 0.002023 m | BL4 |
| WEBE-SHAN | 49,380.378 m | 0.006822 m | 0.004039 m | 0.002004 m | |
| WAWN-WEBE | 55,019.436 m | 0.005553 m | 0.003980 m | 0.001796 m | BL5 |
| WEBE-WAWN | 55,019.270 m | 0.005567 m | 0.003995 m | 0.001806 m | |
| WAWN-RASN | 52,368.694 m | 0.004397 m | 0.003993 m | 0.001722 m | BL6 |
| RASN-WAWN | 52,368.543 m | 0.004391 m | 0.003987 m | 0.00172 m | |
| WAWN-PAR0 | 35,359.795 m | 0.008612 m | 0.004470 m | 0.002227 m | BL7 |
| PAR0-WAWN | 35,359.697 m | 0.008602 m | 0.004456 m | 0.002217 m | |
| WEBE-RASN | 42,739.449 m | 0.008015 m | 0.004525 m | 0.002088 m | BL8 |
| RASN-WEBE | 42,739.455 m | 0.008005 m | 0.004511 m | 0.002081 m | |
| SHAN-SIW2 | 99,784.717 m | 0.008204 m | 0.004343 m | 0.002758 m | BL9 |
| SIW2-SHAN | 99,784.116 m | 0.008178 m | 0.004306 m | 0.002727 m | |
| RASN-SIW2 | 21,865.935 m | 0.006930 m | 0.004367 m | 0.002367 m | BL10 |
| SIW2-RASN | 21,866.083 m | 0.006924 m | 0.004359 m | 0.002363 m | |
| RASN-PAR0 | 42,485.641 m | 0.005102 m | 0.003856 m | 0.001384 m | BL11 |
| PAR0-RASN | 42,485.647 m | 0.005092 m | 0.003849 m | 0.001382 m | |
| PAR0-SIW2 | 49,838.93 m | 0.003717 m | 0.003708 m | 0.001184 m | BL12 |
| SIW2-PAR0 | 49,839.26 m | 0.003721 m | 0.003709 m | 0.001184 m | |
| PAR0-WATH | 38,562.394 m | 0.009195 m | 0.004487 m | 0.002091 m | BL13 |
| WATH-PAR0 | 38,562.109 m | 0.009183 m | 0.004470 m | 0.002085 m | |
| WATH-SIW2 | 62,764.719 m | 0.006125 m | 0.003963 m | 0.001687 m | BL14 |
| SIW2-WATH | 62,765.598 m | 0.006143 m | 0.003982 m | 0.001698 m | |
| WATH-KEHA | 55,500.287 m | 0.003695 m | 0.003650 m | 0.002088 m | BL15 |
| KEHA-WATH | 55,500.951 m | 0.003699 m | 0.003653 m | 0.002092 m | |

**FIGURE 14.2 (*Continued*)** Network and Local Accuracies Based on Standard Deviation of 0.002 m for Each Component at Stations FOLA and KEHA    (*Continued*)

| PT to PT | Horizontal Distance | Nominal Network Accuracy | Full network Accuracy | Local Accuracy | Baseline |
|---|---|---|---|---|---|
| SIW2-KEHA | 34,716.511 m | 0.006154 m | 0.003976 m | 0.002586 m | BL16 |
| KEHA-SIW2 | 34,716.441 m | 0.006147 m | 0.003968 m | 0.002572 m | |
| Points that are NOT directly connected with baselines | | | | | |
| FOLA-RASN | 88,441.711 m | 0.005989 m | 0.00418 m | 0.002622 m | |
| RASN-FOLA | 88,441.188 m | 0.005974 m | 0.00416 m | 0.002598 m | |
| FOLA-PAR0 | 102,961.780 m | 0.006996 m | 0.004044 m | 0.002638 m | |
| PAR0-FOLA | 102,961.184 m | 0.006976 m | 0.004017 m | 0.002604 m | |
| FOLA-WATH | 141,353.316 m | 0.007123 m | 0.004131 m | 0.002986 m | |
| WATH-FOLA | 141,351.453 m | 0.007093 m | 0.004091 m | 0.002941 m | |
| FOLA-SIW2 | 109,807.701 m | 0.005941 m | 0.003977 m | 0.002647 m | |
| SIW2-FOLA | 109,807.795 m | 0.005922 m | 0.003952 m | 0.002617 m | |
| FOLA-KEHA | 143,989.991 m | 0.003822 m | 0.003606 m | 0.00288 m | |
| KEHA-FOLA | 143,989.821 m | 0.003818 m | 0.003603 m | 0.002872 m | |
| SHAN-WAWN | 103,132.188 m | 0.006415 m | 0.004139 m | 0.002373 m | |
| WAWN-SHAN | 103,131.170 m | 0.006387 m | 0.004104 m | 0.002330 m | |
| SHAN-RASN | 85,007.403 m | 0.007895 m | 0.004429 m | 0.002729 m | |
| RASN-SHAN | 85,006.316 m | 0.007873 m | 0.004402 m | 0.002704 m | |
| SHAN-PAR0 | 119,452.217 m | 0.007674 m | 0.004185 m | 0.002521 m | |
| PAR0-SHAN | 119,450.706 m | 0.007643 m | 0.004143 m | 0.002478 m | |
| SHAN-WATH | 152,562.475 m | 0.008274 m | 0.004370 m | 0.002965 m | |
| WATH-SHAN | 152,559.417 m | 0.008231 m | 0.004311 m | 0.002911 m | |
| SHAN-KEHA | 132,898.399 m | 0.006684 m | 0.004028 m | 0.003188 m | |
| KEHA-SHAN | 132,897.328 m | 0.006663 m | 0.004003 m | 0.003150 m | |
| WAWN-WATH | 72,864.854 m | 0.009045 m | 0.004560 m | 0.002870 m | |
| WATH-WAWN | 72,864.114 m | 0.009023 m | 0.004528 m | 0.002846 m | |
| WAWN-SIW2 | 70,233.801 m | 0.005198 m | 0.004023 m | 0.002027 m | |
| SIW2-WAWN | 70,234.071 m | 0.005184 m | 0.004011 m | 0.002014 m | |
| WAWN-KEHA | 95,564.023 m | 0.005536 m | 0.003948 m | 0.002930 m | |
| KEHA-WAWN | 95,564.198 m | 0.005520 m | 0.003936 m | 0.002907 m | |
| WEBE-PAR0 | 70,669.265 m | 0.007779 m | 0.004172 m | 0.002182 m | |
| PAR0-WEBE | 70,669.283 m | 0.007761 m | 0.004148 m | 0.002165 m | |
| WEBE-WATH | 105,717.578 m | 0.008391 m | 0.004366 m | 0.002798 m | |
| WATH-WEBE | 105,716.823 m | 0.008363 m | 0.004327 m | 0.002773 m | |
| WEBE-SIW2 | 62,858.831 m | 0.007866 m | 0.004334 m | 0.002624 m | |
| SIW2-WEBE | 62,859.264 m | 0.007851 m | 0.004313 m | 0.002611 m | |
| WEBE-KEHA | 97,527.111 m | 0.006294 m | 0.003992 m | 0.003116 m | |
| KEHA-WEBE | 97,527.585 m | 0.006280 m | 0.003977 m | 0.003092 m | |
| RASN-WATH | 68,534.636 m | 0.007529 m | 0.004341 m | 0.002461 m | |
| WATH-RASN | 68,534.139 m | 0.007509 m | 0.004317 m | 0.002455 m | |
| RASN-KEHA | 55,557.524 m | 0.005977 m | 0.004160 m | 0.003185 m | |
| KEHA-RASN | 55,557.787 m | 0.005968 m | 0.004149 m | 0.003166 m | |
| PAR0-KEHA | 65,330.044 m | 0.004743 m | 0.003781 m | 0.002470 m | |
| KEHA-PAR0 | 65,330.344 m | 0.004735 m | 0.003775 m | 0.002457 m | |

**FIGURE 14.2 (*Continued*)**    Network and Local Accuracies Based on Standard Deviation of 0.002 m for Each Component at Stations FOLA and KEHA

10. Network accuracy can be one of two values based upon use of the station covariance matrix:
    a. Nominal network accuracy is based only on the diagonal values in the station covariance matrix.
    b. Full network accuracy is based on using the full covariance matrix at each station.
11. The number of decimal places shown in Figure 14.2 cannot be justified except to show where differences begin to occur.

Data in Figure 14.2 are based upon the assumption of a 0.002 m standard deviation for each component of each control station—FOLA and KEHA. The same network adjustment was run again assuming a standard deviation value of 0.020 m for each component at each control station. The network adjustment results can be duplicated by using control station variances of 0.0004 (instead of 0.000004) for each X/Y/Z coordinate as shown in Chapter 13. The same $f$ vector and $B$ matrix can be used with the revised $Q$ matrix in the same matrix manipulation sequence to obtain the covariance matrix of the parameters. Values of the adjusted parameters and values from the covariance matrix can be used to build a BURKORD™ data file as shown in Figure 14.1. The process for computing network and local accuracies is the same as outlined in this chapter. Figure 14.3 is a partial printout—similar to Figure 14.2—that shows accuracies for two directly connected stations and two station pairs that are not directly connected.

Program name/version: BURKORD™ Local Output - version 2016-A, August 30, 2016
The user is:                          Earl F. Burkholder
Program used on:                      August 31, 2016
The input file for program is:   Chl4bkfile0.020.dat
The name of this output file is:  Chl4local0.020.ans      Annotated and renamed Figure 14.3.docx

| PT to PT | Horizontal Distance | Nominal Network Accuracy | Full network Accuracy | Local Accuracy |
|---|---|---|---|---|
| Directly connected points: | | | | |
| FOLA-SHAN | 58,041.558 m | 0.029326 m | 0.029322 m | 0.001762 m |
| SHAN-FOLA | 58,041.956 m | 0.029327 m | 0.029322 m | 0.001760 m |
| PAR0-SIW2 | 49,838.93 m | 0.029322 m | 0.029320 m | 0.001168 m |
| SIW2-PAR0 | 49,839.26 m | 0.029322 m | 0.029320 m | 0.001168 m |
| Points that were NOT directly connected: | | | | |
| FOLA-RASN | 88,441.710 m | 0.030018 m | 0.029429 m | 0.002707 m |
| RASN-FOLA | 88,441.187 m | 0.030012 m | 0.029423 m | 0.002680 m |
| WATH-WEBE | 105,716.823 m | 0.030336 m | 0.029416 m | 0.002909 m |
| WEBE-WATH | 105,717.578 m | 0.030345 m | 0.029423 m | 0.002934 m |

**FIGURE 14.3**  Network and Local Accuracies Based on Standard Deviation of 0.020 m for Each Component at Stations FOLA and KEHA

## CONCLUSION

The examples shown in Figures 14.2 and 14.3 illustrate the following:

1. The difference between network and local accuracies for points in a network or project.
2. Two separate values of network accuracy are available. Full network accuracy is typically better than nominal network accuracy but requires use of full covariance matrix at each station. If full covariance matrix information is not available, nominal network accuracy is obtained from the standard deviations of the $X/Y/Z$ coordinate values as being independent and having no correlation.
3. Increasing the uncertainty of the control points degrades the network accuracy, but the local accuracy remains "tight" as determined by the quality of the original baseline and network observations.
4. The phrase "with respect to what?" is even more important because the GSDM provides options for the spatial data user that will provide different answers depending on assumptions made by the user.

However, more work remains to be done before promulgating additional accuracy standards. The results reported herein are based on results realized using GNSS baseline processing software from a single vendor. Four different vendor software baseline processing packages were used and compared. There was amazing consistency of results among the four packages for baseline components computed from the RINEX files. It appears that the purpose and philosophy of the RINEX procedures have been met. But, the same cannot be said for baseline covariance matrices vendor to vendor. Several vendor result comparisons were quite similar but, at least in one case, results were not at all acceptable. The goal is that, sometime in the future (within reason), the same covariance matrix for a given baseline processed from the same RINEX files will yield similar results independent of which vendor software is used.

A separate issue concerns the reasonable limit that may exist in downgrading the quality of the control points held. The examples presented herein show results as might be expected. Reasonable agreement has been found using the GSDM procedures for cases in which the standard deviations for the two control points (FOLA and KEHA) were raised to 1.0 and 5.0 m. For some unexplained reason, the algorithm of Soler and Smith (2010) did not deliver consistent results for the 1.0 and 5.0 m cases—see http://www.globalcogo.com/AccuracyComp.pdf, accessed May 4, 2017.

## REFERENCES

American Society of Photogrammetry & Remote Sensing (ASPRS), American Congress on Surveying & Mapping (ACSM), American Society of Civil Engineers (ASCE). 1994. *Glossary of Mapping Sciences*. Bethesda, MD: American Society of Photogrammetry and Remote Sensing.

Burkholder, E.F. 1991. Computation of level/horizontal distances. *Journal of Surveying Engineering* 117 (3): 104–115.

Burkholder, E.F. 2016. Using the global spatial data model (GSDM) to compute combined factors. *Journal of Surveying Engineering* 142 (4): 06016001.

National Geodetic Survey (NGS). 1986. *Geodetic Glossary.* Silver Spring, MD: National Geodetic Survey, National Oceanic and Atmospheric Administration.

Soler, T. and Smith, D. 2010. Rigorous estimation of local accuracies. *Journal of Surveying Engineering* 136 (3): 120–125 (and subsequent Discussion and Closure).

# 15 Using the GSDM—Projects and Applications

## INTRODUCTION

Using the GSDM is primarily a matter of choosing to do so. The technology is already in place and all equations and procedures are in the public domain. Although the GSDM contains no new science, it does represent a different way of organizing and storing spatial data. The intent in defining the GSDM was to begin with the assumption of a single origin for 3-D data and, from that, to build a collection of procedures that can be used to handle 3-D spatial data more efficiently. Due to overlaps with existing practice, the decision to use the GSDM need not be "all or nothing." For many, the transition to using the GSDM will lie between two extremes:

- Start with a small project and build a 3-D database using the GSDM. Anyone can input autonomous latitude, longitude, ellipsoid height, and point name (or station name) and descriptor (metadata) as obtained from a GNSS receiver. Standard deviations are optional—the default value is 0.0. But provable values or reasonable estimates for standard deviation are appropriate. Each point record includes a point number, ECEF coordinates, and a descriptor. The associated covariance matrix information is optional but, especially if working with precise point positioning (PPP) data, that information can be very valuable.

  If using GPS vector data, the beginning control points are defined with ECEF coordinates as obtained from a reliable source and entered into the database. Subsequent field observations are processed to obtain ECEF coordinate differences (a baseline vector) and new points are defined by adding those differences to points already in the 3-D database. The $X/Y/Z$ coordinates of any project (large or small) are compatible with all other ECEF coordinates on the same datum the world over. If included, the positional quality (standard deviation) of each point is described by its covariance matrix.

- Build a 3-D database before using it. Points in existing horizontal and vertical databases are systematically converted to ECEF coordinates and stored in an integrated 3-D database. The challenge will be to make sure that each point has a legitimate ECEF 3-D definition. It may be possible to build a reliable 3-D database with few (or no new) field ties but the mathematical conversion of existing data will need to be done with great care. For example, datum compatibility will be a huge issue and the orthometric height of each vertical point will need to be converted to ellipsoid height using a proven (acceptable) geoid model. But the saving grace is that the GSDM accommodates the standard deviation of each point—whatever it is. So long as a standard deviation reliably reflects the quality of the newly defined point in the 3-D database, there is no need for datum conversions to be "perfect." Well-documented professional judgment and consistent application of adopted policies will serve to protect the reputation of a named 3-D database. Each subsequent person who uses information from a given database must be able to rely on the statistical quality of data obtained therefrom.

Regardless of how the 3-D database is started or established, once information is stored as ECEF coordinates and covariances, then interoperability has been established and any or all users have the option of using that information in a multitude of applications. The user is responsible for subsequent manipulation of the data. Ideally rules of such manipulation will be easy to understand and will be bidirectional so that new data can be added to the 3-D database. (In this context there needs to be a careful discussion of administrative responsibility for databases—master, global, national, regional, agency, discipline, project, task, local and/or temporary. That discussion is beyond the scope of this book.) Of course, any data added to the database will be done under appropriate authority and will meet established quality control criteria. However, it should be understood that a *reliable* standard deviation of any quantity is more important than its magnitude. The GSDM competently handles 3-D data with large standard deviations just as well as it handles data with small standard deviations. The difference lies in the quality of the answer obtained when using those data. Data with large standard deviations will provide answers with large standard deviations.

The actual transition to using the GSDM will vary among persons and organizations. Many will begin using the GSDM only when they see it as being in their self-interest to do so. Arriving at that conclusion or developing such a consensus will take time. At the beginning of a new project or on a given date, it could be decreed that GSDM policies and procedures will henceforth be used. That approach would be the most efficient way to implement the GSDM—given a compatible database has been developed. But, developing a 3-D database will also take time and resources. The presumption is that each user or organization will explore and discover what is best for the circumstance. Many of the individual procedures described in this book are already being used and will continue to be useful. However, as users become more comfortable with how the individual pieces relate to the whole, resistance to using the GSDM will be reduced. The goal is that spatial data practices will evolve in such a manner that the underlying features of the GSDM will be recognized as common ground for practice worldwide. It is conceivable and anticipated that the GSDM will eventually become the global standard for handling geospatial data.

Once we get beyond the "magic" of electronic signal processing and have access to the spatial data components from our GNSS receivers (or other sensors), all we need are the rules of solid geometry to keep track of where we are or where we have been. No, it is not quite that easy as we still need to deal with issues of datums, coordinate systems, units, and whether the data are relative or absolute. But, the GSDM provides a unique bridge between the builders/operators of measurement systems and spatial data users in many disciplines all over the world. Using the same

GSDM and 3-D database, rocket scientists, engineers, photogrammetrists, surveyors, and others can continue working in a rigorous global environment while local users simultaneously enjoy the luxury of working with flat-Earth rectangular components. The GSDM will be useful to novice and expert alike.

## FEATURES

The GSDM has two primary features—the functional model and the stochastic model. The functional model is a collection of equations and geometrical relationships that can be used to describe a unique position anywhere within the birdcage of orbiting GNSS satellites. The optional stochastic model is a set of rules and procedures that can be used to keep track of standard deviations of the observations, the derived measurements, the computed/stored coordinates, and any quantity computed from them.

### The Functional Model

Because geospatial data are connected to the Earth and because spatial data are 3-D, the "big picture" geometrical relationships included in the GSDM involve a lot of geometrical geodesy. Given previous use of 2-D latitude and longitude on the ellipsoid surface, rules of classical geodesy are still very useful and need to be included. However, given that local users view a "flat Earth" from a given point, the underlying model needs to provide simple rectangular plane surveying answers with the same integrity as it provides answers on the ellipsoid. The GSDM does both. Functional model equations are listed in Chapter 1 and derived in Chapter 7—except for the rotation matrix, which is derived in Appendix A.

Following is an oversimplified summary of models. In the first approximation, the world is considered flat. That assumption and model is appropriate for many local applications. A spherical Earth is a better model and useful for many "big picture" applications in geography and navigation. An ellipsoidal Earth model has been used for triangulation computations (and other applications requiring a high level of geometrical integrity) for over 200 years. The GSDM goes beyond those by modeling geospatial data in 3-D space. Given the digital revolution and enormous data storage capability, each user now has the option of viewing geospatial data from any origin or perspective. With simple solid geometry relationships readily available, each user can enjoy the luxury of unique ECEF absolute $X/Y/Z$ coordinates while simultaneously working with local relative coordinate differences—all without sacrificing the geometrical integrity of the observations, the measurements, or the data. And, the case can be made that performing spatial data computations in 3-D space is less onerous or complicated than using the ellipsoidal model or a map projection model. For example, see Burkholder (2016).

### The Stochastic Model

The stochastic model can be used to assign standard deviations to the position of any point in any direction—that is, standard deviation in the north-south direction, in the east-west direction, and in the up-down direction or with respect to the $X/Y/Z$ axes in the ECEF reference frame. With practice and appropriate software, determining those standard deviations can be fairly straightforward. Even so, spatial data users need to be specific when quoting standard deviations by stating the context (i.e., "with respect to what?"). Possible options include

- With respect to the NAD 83 (or subsequent replacement datum)
- With respect to the ITRF
- With respect to the WGS 84
- With respect to the geoid
- With respect to the control held by the user
- With respect to an implied, an unstated, or even an explicitly stated datum

Given the sheer volume of spatial data being generated and given the possible consequences of bad decisions, the issue of spatial data accuracy is becoming increasingly important and statements of spatial data accuracy need to be unambiguous. In the past, the quality of geodetic surveys was often described with adjectives such as first-order, second-order, and third-order. Such categories worked for those familiar with geodetic surveying, but the standards were largely distance dependent and were expressed as a ratio—for example, one part in 100,000 is first-order. While that criterion was appropriate for long triangulation lines, it is very difficult to make short line measurements meeting the first-order specification. One part in 100,000 translates to 0.001 meter in 100 meters.

As shown in Table 15.1, the Federal Geographic Data Committee (FGDC 1998) developed "Geospatial Positioning Accuracy Standards; Part 2—Standards for Geodetic Networks." Several comments are as follows:

- The 1998 standards are part of a larger effort to quantify spatial data accuracy in many categories, not just those used in geodetic surveying.
- The 1998 standards have a better theoretical basis and use positional tolerance criterion rather than ratio of precision to describe spatial data quality.
- The first-order, second-order categories have been replaced with names that are more intuitive.
- The standards are more closely aligned with modern positioning (GNSS) technology and provide a more intuitive "meter-stick" against which to make comparisons.
- The FGDC standards are quoted at the 95% confidence level and are applicable to 3-D data as well as 2-D data.
- The FGDC standards discuss local accuracy and network accuracy without providing a specific mathematical definition. While mostly compatible, the mathematical definition for network accuracy and local accuracy as provided by Equation 1.36 (and repeated in Chapter 12) goes beyond that provided by the FGDC.

Metadata are data about data. Concepts of metadata were developed in parallel with GIS and metadata are a very important part of working with spatial data.

## TABLE 15.1
### Summary of FGDC Accuracy Standards
Horizontal, Ellipsoid Height, and Orthometric Height

| Accuracy Classification | 95 Percent Confidence Less than or equal to: |
| --- | --- |
| 1 millimeter | 0.001 meter |
| 2 millimeter | 0.002 meter |
| 5 millimeter | 0.005 meter |
| 1 centimeter | 0.010 meter |
| 2 centimeter | 0.020 meter |
| 5 centimeter | 0.050 meter |
| 1 decimeter | 0.100 meter |
| 2 decimeter | 0.200 meter |
| 5 decimeter | 0.500 meter |
| 1 meter | 1.000 meter |
| 2 meter | 2.000 meters |
| 5 meter | 5.000 meters |
| 10 meter | 10.00 meters |

One reason for the popularity of metadata is that metadata contain more than just metrical characteristics. For images and photogrammetric data for example, metadata include information such as the name of the organization collecting the data, the equipment used to record the observations, flying height of aircraft, time of day, and other details. Given appropriate metadata, equipment calibration details, and being knowledgeable of the data reduction overall process, reasonable professionals place justifiable reliance on the quality of the resulting information. Typically, metadata apply to a particular data set—often a very large data set. The accuracy of an individual point is stated as being representative of many points in the same data set sharing similar characteristics. Although not exclusive, metadata are often considered to be more appropriate when working with raster data as opposed to vector data.

Certainly metadata will continue to be important but, when working with vector data, the GSDM provides some very powerful advantages. The GSDM contains algorithms for determining the standard deviation of each point in all three directions. Without knowing any of the metadata associated with points in the GSDM database, a numerical filter can be imposed to screen out any data not meeting a positional tolerance criterion selected by the user.

The following statements include some crystal ball speculation. As storage capacity becomes more affordable, it will be feasible to convert more raster data to vector data. At some point in the future, it will be possible and sometimes warranted, for each pixel in a raster image file to have its geospatial location defined by ECEF coordinates and to have the associated point covariance matrix stored for each point. Since raster data are already stored, an alternative would be to develop, store, and use an algorithm for converting raster data to ECEF vector data (ECEF coordinates and covariances) on an "as needed" basis. That would avoid duplicate storage requirements while permitting raster data to be converted and used in a vector environment. With that capability in place, many silos of existing electronic imagery can be readily accessible to spatial data users for a multitude of applications.

## DATABASE ISSUES

A statement of the obvious is—don't mix datums in a 3-D database. One of the challenges of using GNSS data is knowing what datum to choose. In the United States, three obvious choices currently are the NAD 83, the WGS 84, and the ITRF. Of course, the anticipated new datum published by NGS in 2022 is added to the list. At the gross level it does not matter which datum is used. As computed in Chapter 4 and using spherical trigonometry, the distance from New Orleans to Chicago is 1,360 kilometers. Later in Chapter 7, the GSDM was used to find the distance as 1,356 kilometers. That difference is primarily the effect of using an ellipsoidal Earth in place of a spherical Earth. Furthermore the latitude-longitude positions in each city were listed only to the nearest 1 second of arc, giving an implied tolerance of about 30 meters at each end of the line. The GSDM inverse in Chapter 7 assumed the latitude-longitude positions were "exact" and achieved a proven answer within a centimeter on the same line. Now, if the only difference is to recompute the inverse using the WGS 84 as the ellipsoid (on the WGS 84 datum) instead of the GRS 80 ellipsoid (on the NAD 83 datum), the GSDM answer is still within 1 cm (left as an exercise for the reader). So, why does the datum make a difference? The datum makes a difference because the origins of the two datums are not necessarily at the same place. Relative differences on the two datums can be very nearly identical, but the absolute positions are different. In the New Orleans to Chicago example just cited, the latitude-longitude values were taken to be the same but, because of different datums, the same latitude-longitude values represent two different points on the ground in New Orleans and two different points on the ground in Chicago. The egregious mistake would be to inverse from New Orleans on one datum to Chicago on another datum. Of course, even a discrepancy of a meter or two in the distance New Orleans to Chicago would hardly be significant if the comparison were being made to the nearest kilometer. The rule may be conservative but should be honored: don't mix datums in a 3-D database.

Other database issues are related to records, fields, and format. A BURKORD™ database contains two header lines and two kinds of records—a point record and a correlation record. There is no specified order for records in the database but the first field in each record is reserved for flexibility and future use. Any refinements to the generic format listed here will be identified on the Global COGO, Inc. web site http://www.globalcogo.com/dbformat.html, accessed May 4, 2017.

Additional uses may be defined in the future but any such changes are intended to preserve compatibility with the following format. Any record beginning with a "p" is a point record and any record beginning with a "c" is a correlation record. The specified format for each type is as follows:

A point record contains the following fields, space, or comma delimited.

- Attribute field—string characters (no blanks).
  1. First character (required) is reserved.
     a. "p" is a point record.
     b. "c" is a correlation record—see below.
  2. Next three characters (optional) are project identifiers.
  3. Characters 5 to $n$ (also optional) are prerogative of user.
- Point number must be an integer.
- $X/Y/Z$ coordinate values—three double precision fields.
- Variances of $X/Y/Z$—three double precision fields
- Covariances $XY$, $XZ$, $YZ$—three double precision fields.
- Station name—string characters (blanks OK prior to end-of-record).

A point-to-point correlation record contains the following fields, space or comma delimited.

- Attribute field—string characters (no blanks)
  4. First character (required) is reserved.
     a. "p" is a point record—see above.
     b. "c" is a correlation record.
  5. Next three characters (optional) are project identifiers.
  6. Characters 5 to $n$ (also optional) are prerogative of user.
- Two point numbers—two integer fields.
- $X_1X_2$, $X_1Y_2$, $X_1Z_2$ covariances—three double precision fields.
- $Y_1X_2$, $Y_1Y_2$, $Y_1Z_2$ covariances—three double precision fields.
- $Z_1X_2$, $Z_1Y_2$, $Z_1Z_2$ covariances—three double precision fields.

Figure 14.1 is a printout of a program that reads, uses, and outputs the contents of a BURKORD™ file.

## IMPLEMENTATION ISSUES

This section looks specifically at characteristics of the GSDM and discusses some of the implications associated with implementation. Issues include, but are not limited to, the following:

- The GSDM accommodates modern measurement procedures and digital data.
- The GSDM uses proven rules of solid geometry and vector algebra in a global rectangular environment.
- The 3-D database is simple and equally applicable the world over.

- While the underlying standard insures interoperability, geospatial data users in any discipline have complete freedom to be innovative in derivative applications. Traditional 2-D applications are fully supported.
- Orthometric heights are referenced to the elusive geoid while ellipsoid heights are referenced to the Earth's center of mass. Full adoption of the GSDM presumes ellipsoid heights will be used to describe the third dimension. Geoid modeling will still be needed and used, especially by those requiring precise hydraulic grade lines.
- The stochastic feature provides tools by which spatial data accuracy can be established, tracked, and used.
- The GSDM provides a concise mathematical definition of network accuracy and local accuracy.
- Absolute $X/Y/Z$ coordinates for any point in the "birdcage" of satellites are globally unique. As appropriate, any user is free to convert those coordinates to other systems (for internal use) such as latitude/longitude/height, UTM, state plane or other coordinate systems.
- The GSDM does NOT provide for transformations between datums. But, datum-to-datum relationships are best modeled in terms of ECEF coordinates using the standard seven-parameter transformation. In some cases, velocities are also accommodated, in which case the seven-parameter transformation becomes a fourteen-parameter transformation.
- Relative coordinate differences of $\Delta X/\Delta Y/\Delta Z$ are not very intuitive for humans. But, given the ease with which geocentric differences can be rotated to local tangent plane differences, the local flat-Earth user has immediate access to plane surveying rectangular components.
- Distances from standpoint to forepoint are in the local tangent plane and are identical to the HD(1) distance defined in Burkholder (1991). Other distances can be computed, if needed, without disturbing the $X/Y/Z$ coordinates in the database.
- Important concept! The tangent plane from "here" to "there" is not the same as the tangent plane from "there" to "here." Within a very small tolerance, the 3-D azimuth (Burkholder 1997a) is the true geodetic azimuth from standpoint to forepoint. To the geodesist, this is as it should be. However, for plane surveying applications, the P.O.B. datum feature is an attractive alternate because all local $\Delta e/\Delta n/\Delta u$ components are treated as flat-Earth 3-D coordinates with respect to the P.O.B. selected by the user.
- When using the P.O.B. datum all distances between points are in the same tangent plane through the P.O.B. and the azimuths are grid azimuths with respect to the true meridian through the P.O.B. The implication of this feature is that two surveys referenced to separate P.O.B.s and sharing a common line will have two azimuths for the same line. That difference of the two azimuths is the convergence of the meridian between the respective P.O.B.s. That "problem" is resolved by identifying the P.O.B. for the survey on each plat. The underlying ECEF coordinates and their covariances remain unchanged in each case.

- With these features of the GSDM already defined, in place, and universally available, there is no need for a "low-distortion projection" (LDP). The GSDM is a truly three-dimensional model while a LDP is strictly two-dimensional. A LDP is applicable in a carefully defined "local" area while the GSDM is already defined and equally applicable worldwide.
- Ellipsoid heights and their standard deviations are obtained directly from ECEF coordinates in the 3-D database. Geoid modeling procedures can be used to obtain orthometric heights if needed. The broader question is, "Why are orthometric heights needed?" Unless grades are very critical for hydraulic grade lines (in which case dynamic heights and height differences should be used) an ellipsoid height difference readily approximates an ellipsoid height difference. Geoid modeling will still be used by those for whom the difference matters.
- Comments and rhetorical questions:
  1. Satellites orbit the physical center of mass of the Earth.
  2. The center of mass of the Earth is quite stable whereas the geoid moves up and down during the Earth's daily rotation.
  3. CORS fixed to bedrock also go up and down during the day—Earth tides.
  4. Do CORS and the underlying geoid go up and down together? Yes!
  5. GNSS data can be used to monitor the daily motion of a precisely surveyed CORS.
  6. Is the mean ellipsoid height of a CORS preferred to the instantaneous ellipsoid height? How precisely can an instantaneous ellipsoid height be determined?
  7. What difference, if any, does it make that ellipsoid heights are absolute (they move up and down daily relative to the center of mass) while orthometric heights are relative (bench marks move up and down during the day along with the geoid)?

## EXAMPLES AND APPLICATIONS

This section describes various projects in which the GSDM has been used to an advantage. Some projects are more involved than others but all projects illustrate beneficial features of the GSDM.

### EXAMPLE 1—SUPPLEMENTAL NMSU CAMPUS CONTROL NETWORK

The NMSU network described in Chapter 12 is based on single frequency Trimble data collected years ago. That project was first computed on NAD 83 (1992) using several weighting options. Those options include equal weights, weights according to baseline component standard deviations, and finally a comprehensive adjustment using the full covariance matrix of each baseline. The full covariance matrix option solution is reported in Chapter 11 of the First Edition and results of all weighting options are reported at http://www.globalcogo.com/nmsunet1.pdf, accessed May 4, 2017. Since then, the underlying 1992 control network was upgraded by NGS and, using the same vectors, the network was readjusted based upon the NAD 83 (2011). Those results are reported in Chapter 12.

In April 2008 this author was indignant to discover that an important control monument on central campus had been disturbed during routine sidewalk reconstruction. However, NMSU facilities personnel were confident that they could replace the disk "exactly where it had been" because they had made careful reference measurements prior to removal. When challenged, their response was "after the concrete has cured, prove that we did not put it back in the same place."

The brass tablet for station BROMILOW—a key point on campus and part of the reported network—was removed during routine sidewalk reconstruction and replaced in April 2008. Additional GNSS baselines were observed in 2010 to relocate station BROMILOW and to establish two new stations for subsequent use—WAKEMAN and EFB. A portion of the current NMSU GNSS network is shown in Figure 15.1.

Data for the addition to the NMSU network were collected using four identical model Topcon GNSS receivers. It would have been possible, maybe even desirable, to recompute the entire NMSU network holding the two original HARN stations and combining "old" vectors with new observations. However, since the first network was observed using Trimble receivers and the additional vectors were all observed using Topcon receivers, the results of a fully weighted adjustment would have been affected by differences in the manner in which Trimble and Topcon compute and report baseline covariance. Therefore, the enhanced network as reported herein was based upon an adjustment of Topcon vectors while holding stations PSEUDO and REILLY as previously computed/reported. The six baselines used in this adjustment are all independent (no trivial baselines) and the full baseline covariance matrix of each baseline was used in the supplemental adjustment. The published position of station REILLY was held to a very tight positional tolerance (each component ±0.0001 m). The $X/Y/Z$ coordinate values for station PSEUDO as determined in the previous survey were also held.

The enhanced adjustment was performed in the same manner as reported for the original network in Chapter 12. Figure 15.2 is a computer printout that shows the

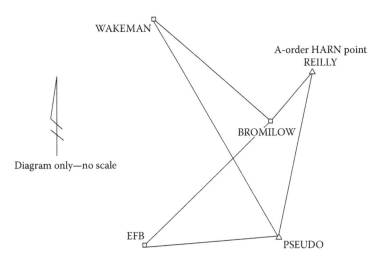

**FIGURE 15.1**   Supplemental GNSS Network at NMSU

Program name/version: LSAdjGPSvectors: Version 2014-A

| The user is: | Earl F. Burkholder |
| --- | --- |
| Program used on: | September 13, 2016 |
| The input file for program is: | NMSU Finial GPS control Sept 2016.dat |
| The name of this output file is: | Final Finial.txt Renamed: Figure 15.2.docx |

| The client is: | Earl F. Burkholder |
| --- | --- |
| The project is: | Finial GPS Control Network - Sept. 13, 2016 |

Three input matrices are F, B, and Q. They are: The name/dimensions of the F vector are: $f_{24,1}$

| 1,556,177.595 | | | | 305.6989 | |
| --- | --- | --- | --- | --- | --- |
| 5,169,235.284 | REILLY | 280.2088 | | -230.3542 | BL 5 |
| -3,387,551.720 | | 72.1905 | BL 3 | -199.1763 | |
| | | 245.7077 | | | |
| -32.1305 | | | | 308.8281 | |
| -51.1740 | BL 1 | -283.3340 | | -344.5944 | BL 6 |
| -94.1862 | | 42.0494 | BL 4 | -370.7117 | |
| | | -74.1750 | | | |
| -29.0029 | | | | 1,556,206.595 | |
| -165.4261 | BL 2 | | | 5,169,400.704 | PSEUDO |
| -265.7117 | | | | -3,387,285.999 | |

**FIGURE 15.2** Computer Printout of Supplemental Network Adjustment

(Continued)

*(Continued)*

The name/dimensions of the B matrix are: $B_{24,15}$

REILLY BROMILOW WAKEMAN EFB PSEUDO REILLY

BL 1

BL 2

BL 3

BL 4

BL 5

BL 6

PSEUDO

**FIGURE 15.2 (*Continued*)** Computer Printout of Supplemental Network Adjustment

The name/dimensions of the Q matrix are: $Q_{24,24}$

REILLY

$$
\begin{matrix}
1E-8 & 0 & 0 \\
0 & 1E-8 & 0 \\
0 & 0 & 1E-8
\end{matrix}
$$

BL 1

$$
\begin{matrix}
1.00128E-7 & 5.97736E-8 & -2.99669E-8 \\
5.97736E-8 & 8.80595E-7 & -2.46688E-7 \\
-2.99669E-8 & -2.46688E-7 & 1.47364E-7
\end{matrix}
$$

BL 2

$$
\begin{matrix}
2.07207E-7 & 2.36705E-7 & -1.86832E-7 \\
2.36705E-7 & 1.0315E-6 & -5.70018E-7 \\
-1.86832E-7 & -5.70018E-7 & 7.46012E-7
\end{matrix}
$$

BL 3

$$
\begin{matrix}
1.24545E-7 & 1.08998E-7 & -5.66011E-8 \\
1.08998E-7 & 9.1604E-7 & -2.72622E-7 \\
-5.66011E-8 & -2.72622E-7 & 1.96728E-7
\end{matrix}
$$

BL 5

$$
\begin{matrix}
1.64641E-7 & 1.73234E-7 & -8.32390E-8 \\
1.73234E-7 & 8.56939E-7 & -2.53085E-7 \\
-8.3239E-8 & -2.53085E-7 & 1.84960E-7
\end{matrix}
$$

BL 6

$$
\begin{matrix}
2.39180E-7 & 2.64158E-7 & -2.33001E-7 \\
2.64158E-7 & 1.04530E-6 & -5.74803E-7 \\
-2.33001E-7 & -5.74803E-7 & 8.03497E-7
\end{matrix}
$$

PSEUDO

$$
\begin{matrix}
1E-8 & 0 & 0 \\
0 & 1E-8 & 0 \\
0 & 0 & 1E-8
\end{matrix}
$$

FIGURE 15.2 (*Continued*)  Computer Printout of Supplemental Network Adjustment

(*Continued*)

*W* is the inverse of the **Q** matrix.

REILLY

| 100,000,000 | 0 | 0 | 0 | 0 | 0 | 0 | 0 |
| 0 | 0 | 0 | 0 | 0 | 0 | 0 | 0 |
| -0 | 0 | 0 | 0 | -0 | 0 | 0 | 0 |
| 0 | 100,000,000 | 0 | 0 | 0 | 0 | 0 | 0 |
| 0 | 0 | 0 | 0 | 0 | 0 | 0 | 0 |
| 0 | -0 | 0 | 0 | -0 | 0 | 0 | 0 |
| 0 | 0 | 100,000,000 | 0 | 0 | 0 | 0 | 0 |
| 0 | 0 | 0 | 0 | 0 | 0 | 0 | 0 |
| 0 | 0 | -0 | 0 | -0 | 0 | 0 | 0 |

BL 1

| 0 | 0 | 0 | 10,656,810.86684 | 0 | −218,972.62028 | 1,800,532.47614 | 0 | 0 |
| 0 | 0 | 0 | 0 | 0 | 0 | 0 | 0 |
| 0 | 0 | 0 | −0 | 0 | 0 | −0 | 0 |
| 0 | 0 | 0 | −218,972.62028 | 0 | 2,142,904.77305 | 3,542,703.52352 | 0 | 0 |
| 0 | 0 | 0 | 0 | 0 | 0 | 0 | 0 |
| 0 | 0 | 0 | 0 | 0 | −0 | 0 | 0 |
| 0 | 0 | 0 | 1,800,532.47614 | 0 | 3,542,703.52352 | 13,082,563.06472 | 0 | 0 |
| 0 | 0 | 0 | 0 | 0 | 0 | 0 | 0 |
| 0 | 0 | 0 | 0 | 0 | −0 | 0 | 0 |

BL 2

| 0 | 0 | 0 | 0 | 0 | 0 | 0 | 6,867,845.44135 | −1,082,675.96014 | 892,731.6371 |
| 0 | 0 | 0 | 0 | 0 | 0 | 0 | 0 | 0 | 0 |
| 0 | 0 | 0 | 0 | 0 | −0 | 0 | 0 | 0 | 0 |
| 0 | 0 | 0 | 0 | 0 | 0 | 0 | −1,082,675.96014 | 1,848,651.80459 | 1,141,384.1726 |
| 0 | 0 | 0 | 0 | 0 | 0 | 0 | 0 | 0 | 0 |

**FIGURE 15.2 (*Continued*)**  Computer Printout of Supplemental Network Adjustment                (*Continued*)

```
                                                                                            892,731.6371    1,141,384.1726  2,436,154.32529
                                                                                                      0               0               0

0   0         0            0               0               -0              0   0    0   0
0   0         0            0               0                0              0   0    0   0
0   0         0            0               0               -0              0   0    0   0

BL 3
0   0         0            0               0                0              0   0    0   0
9,387,475.80265  -533,015.11032  1,962,251.49062   0               0              0   0    0   0
0   0         0            0               0                0              0   0    0   0
0   0        -0            0               0                0              0   0    0   0
-533,015.11032  1,888,152.51252  2,463,211.50373   0               0              0   0    0   0
0   0         0            0               0               -0              0   0    0   0
0   0         0            0               0                0              0   0    0   0
1,962,251.49062  2,463,211.50373  9,061,197.38632   0               0              0   0    0   0
0   0         0            0               0               -0              0   0    0   0

BL 4
0   0         0            0               0                0              0   0    0   0
0   0         0       8,496,693.19557  -1,181,982.34684  1,420,837.36576  0   0    0   0
0   0         0            0               0               -0              0   0    0   0
0   0         0            0               0                0              0   0    0   0
0   0         0       -1,181,982.34684  2,222,280.17005  1,381,202.06312  0   0    0   0
0   0         0            0               0               -0              0   0    0   0
0   0         0            0               0                0              0   0    0   0
0   0         0       1,420,837.36576  1,381,202.06312  3,152,032.66852   0   0    0   0
0   0         0            0               0               -0              0   0    0   0

BL 5
0   0         0            0               0                0              0   0    0   0    0
0   0         0            0               0                0        8,313,559.6061  -966,038.09933  2,419,564.42303
                                                                                                      0               0
```

FIGURE 15.2 (Continued) Computer Printout of Supplemental Network Adjustment

(Continued)

FIGURE 15.2 (*Continued*) Computer Printout of Supplemental Network Adjustment

(*Continued*)

(Continued)

$Ninv_{15 \times 15}$

| | | | | | |
|---|---|---|---|---|---|
| 9.08165998E-9 | 1.10057442E-10 | -1.38577939E-10 | 5.79605273E-9 | 1.65676795E-9 | -6.39840034E-10 |
| 3.65385793E-9 | 8.59178448E-10 | -2.42022952E-10 | 3.59379320E-9 | 1.30143202E-9 | -4.84326019E-10 |
| 9.18340019E-10 | -1.10057442E-10 | 1.38577939E-10 | | | |
| 1.10057442E-10 | 9.74749207E-9 | -1.80310545E-10 | 3.88869432E-11 | 4.45379443E-9 | 6.60999453E-10 |
| -1.57821107E-10 | 2.12077137E-9 | 9.32126270E-10 | -1.04063489E-10 | 1.74664250E-9 | 9.36227804E-10 |
| -1.10057442E-10 | 2.52507929E-10 | 1.80310545E-10 | | | |
| -1.38577939E-10 | -1.80310545E-10 | 9.52383872E-9 | -5.65157663E-10 | -1.25184093E-9 | 7.86747851E-9 |
| -7.85433727E-10 | -1.89998304E-9 | 6.90636154E-9 | -7.04339560E-10 | -1.88930965E-9 | 6.44294702E-9 |
| 1.38577939E-10 | 1.80310545E-10 | 4.76161276E-10 | | | |
| 5.79605273E-9 | 3.88869432E-11 | -5.65157663E-10 | 6.66078885E-8 | 5.04177270E-8 | -2.69919575E-8 |
| 3.99216529E-8 | 2.87172767E-8 | -1.78157795E-8 | 3.92771362E-8 | 3.18754381E-8 | -1.94386005E-8 |
| 4.20394727E-9 | -3.88869432E-11 | 5.65157663E-10 | | | |
| 1.65676795E-9 | 4.45379443E-9 | -1.25184093E-9 | 5.04177270E-8 | 4.32363151E-7 | -1.33742996E-7 |
| 2.80919784E-8 | 2.30427043E-7 | -8.43006654E-8 | 3.08237470E-8 | 1.97887074E-7 | -7.39955345E-8 |
| -1.65676795E-9 | 5.54620557E-9 | 1.25184093E-9 | | | |
| -6.39840034E-10 | 6.60999453E-10 | 7.86747851E-9 | -2.69919575E-8 | -1.33742996E-7 | 1.09123397E-7 |
| -1.93942817E-8 | -8.31651811E-8 | 8.60358766E-8 | -1.98466956E-8 | -7.47462988E-8 | 7.92258128E-8 |
| 6.39840034E-10 | -6.60999453E-10 | 2.13252149E-9 | | | |
| 3.65385793E-9 | -1.57821107E-10 | -7.85433727E-10 | 3.99216529E-8 | 2.80919784E-8 | -1.99342817E-8 |
| 1.20137006E-7 | 1.13717676E-7 | -6.90785520E-8 | 2.55883949E-8 | 1.87773583E-8 | -1.47496499E-8 |
| 6.34614207E-9 | 1.57821107E-10 | 7.85433727E-10 | | | |
| 8.59178448E-10 | 2.12077137E-9 | -1.89998304E-9 | 2.87172767E-8 | 2.30427043E-7 | -8.31651811E-8 |
| 1.13717676E-7 | 5.75957743E-7 | -2.08997777E-7 | 1.83364044E-8 | 1.11340994E-7 | -4.94596318E-8 |
| -8.59178448E-10 | 7.87922863E-9 | 1.89898304E-9 | | | |
| -2.42022952E-10 | 9.32126270E-10 | 6.90636154E-9 | -1.78157795E-8 | -8.43006654E-8 | 8.60358766E-8 |
| -6.90785520E-8 | -2.08997777E-7 | 2.11779464E-7 | -1.40231025E-8 | -5.05528844E-8 | 6.44853094E-8 |
| 2.42022952E-10 | -9.32126270E-10 | 3.09363846E-9 | | | |

**FIGURE 15.2 (Continued)** Computer Printout of Supplemental Network Adjustment

*(Continued)*

| | | | | | |
|---|---|---|---|---|---|
| 3.59379320E−9 | −1.04063489E−10 | −7.04339560E−10 | 3.92771362E−8 | 3.08237470E−8 | −1.98466956E−8 |
| 2.55883949E−8 | 1.83364044E−8 | −1.40231025E−8 | 9.52422489E−8 | 8.67539858E−8 | −5.49488526E−8 |
| 6.40620680E−9 | 1.04063489E−10 | 7.04339560E−10 | | | |
| 1.30143202E−9 | 1.74664250E−9 | −1.88930965E−9 | 3.18754381E−8 | 1.97887074E−7 | −7.47462988E−8 |
| 1.87773583E−8 | 1.11340994E−7 | −5.05528844E−8 | 8.67539858E−8 | 5.07319645E−7 | −1.92876331E−7 |
| −1.30143202E−9 | 8.25335750E−9 | 1.88930965E−9 | | | |
| −4.84326919E−10 | 9.36227804E−10 | 6.44294702E−9 | −1.94386005E−8 | −7.39955345E−8 | 7.92258128E−8 |
| −1.47496499E−8 | −4.94596318E−8 | 6.44853094E−8 | −5.49488526E−8 | −1.92876331E−7 | 1.99412510E−7 |
| 4.84326919E−10 | −9.36227804E−10 | 3.55705298E−9 | | | |
| 9.18340019E−10 | −1.10057442E−10 | 1.38577939E−10 | 4.20394727E−9 | −1.65676795E−9 | 6.39840034E−10 |
| 6.34614207E−9 | −8.59178448E−10 | 2.42022952E−10 | 6.40620680E−9 | −1.30143202E−9 | 4.84326919E−10 |
| 9.08165998E−9 | 1.10057442E−10 | −1.38577939E−10 | | | |
| −1.10057442E−10 | 2.52507929E−10 | 1.80310545E−10 | −3.88869432E−11 | 5.54620557E−9 | −6.60999453E−10 |
| 1.57821107E−10 | 7.87922863E−9 | −9.32126270E−10 | 1.04063489E−10 | 8.25335750E−9 | −9.36227804E−10 |
| 1.10057442E−10 | 9.74749207E−9 | −1.80310545E−10 | | | |
| 1.38577939E−10 | 1.80310545E−10 | 4.76161276E−10 | 5.65157663E−10 | 1.25184093E−9 | 2.13252149E−9 |
| 7.85433727E−10 | 1.89898304E−9 | 3.09363846E−9 | 7.04339560E−10 | 1.88930965E−9 | 3.55705298E−9 |
| −1.38577939E−10 | −1.80310545E−10 | 9.52383872E−9 | | | |

Deltas – answers are X/Y/Z values in meters

| | | | | |
|---|---|---|---|---|
| REILLY | | EFB | | WAKEMAN |
| −1,556,177.59491 | −1,556,489.93138 | | −1,556,515.42315 | |
| −5,169,235.28398 | −5,169,358.65463 | | −5,169,056.10858 | |
| 3,387,551.71983 | 3,387,211.82636 | | 3,387,656.71068 | |
| BROMILOW | | PSEUDO | | |
| −1,556,209.72412 | −1,556,206.59519 | | | |
| −5,169,286.46187 | −5,169,400.70392 | | | |
| 3,387,457.53418 | 3,387,285.99917 | | | |

**FIGURE 15.2 (Continued)** Computer Printout of Supplemental Network Adjustment

Residuals – meters

| | | | | | | | |
|---|---|---|---|---|---|---|---|
| 0.00019 | | | -0.00262 | | 0.00219 | | 0.00014 |
| -0.00008 | REILLY | | -0.00617 | BL 2 | 0.00011 | BL 4 | 0.00093 | BL 6 |
| -0.00017 | | | 0.00897 | | -0.00219 | | -0.00019 |
| | | | | | | | |
| -0.00129 | | | 0.00154 | | -0.00013 | | -0.00019 |
| 0.00389 | BL 1 | | -0.00225 | BL 3 | -0.00091 | BL 5 | 0.00008 | PSEUDO |
| -0.00055 | | | -0.00012 | | 0.00019 | | 0.00017 |

Now, finally redundancy, R = 9
And, SigmaNaughtSguared = 2.74982679E1
The covariance matrix of the computed parameters is
sigmanaughtsguared × *Ninv*

REILLY

| | | | | | |
|---|---|---|---|---|---|
| 2.49729919E-7 | 3.02638902E-9 | -3.81065330E-9 | 1.59381411E-7 | 4.55582489E-8 | -1.75944927E-8 |
| 1.00474764E-7 | 2.36259191E-8 | -6.65521196E-9 | 9.88230882E-8 | 3.57871263E-8 | -1.33181514E-8 |
| 2.52527599E-8 | -3.02638902E-9 | 3.81065330E-9 | | | |
| 3.02638902E-9 | 2.68039148E-7 | -4.95822766E-9 | 1.06932358E-9 | 1.22471632E-7 | 1.81763400E-8 |
| -4.33980707E-9 | 5.83175394E-8 | 2.56318579E-8 | -2.86156569E-9 | 4.80296433E-8 | 2.57446430E-8 |
| -3.02638902E-9 | 6.94353068E-9 | 4.95822766E-9 | | | |
| -3.81065330E-9 | -4.95822766E-9 | 2.61889069E-7 | -1.55408568E-8 | -3.44234573E-8 | 2.16342032E-7 |
| -2.15980670E-8 | -5.22187444E-8 | 1.89912980E-7 | -1.93681179E-8 | -5.19527428E-8 | 1.77169883E-7 |
| 3.81065330E-9 | 4.95822766E-9 | 1.30936103E-8 | | | |

**FIGURE 15.2 (Continued)** Computer Printout of Supplemental Network Adjustment

(Continued)

(Continued)

**BROMILOW**

| | | | | | |
|---|---|---|---|---|---|
| 1.59381411E-7 | 1.06932358E-9 | -1.55408568E-8 | 1.83160156E-6 | 1.38640016E-6 | -7.42232078E-7 |
| 1.09777631E-6 | 7.89675369E-7 | -4.89903078E-7 | 1.08005321E-6 | 8.76519335E-7 | -5.34527843E-7 |
| 1.15601268E-7 | -1.06932358E-9 | 1.55408568E-8 | | | |
| 4.55582489E-8 | 1.22471632E-7 | -3.44234573E-8 | 1.38640016E-6 | 1.18892377E-5 | -3.67770072E-6 |
| 7.72480748E-7 | 6.33634457E-6 | -2.31812228E-6 | 8.47599653E-7 | 5.44155178E-6 | -2.03474903E-6 |
| -4.55582489E-8 | 1.52511046E-7 | 3.44234573E-8 | | | |
| -1.75944927E-8 | 1.81763400E-8 | 2.16342032E-7 | -7.42232078E-7 | -3.67770072E-6 | 3.00070440E-6 |
| -5.48158219E-7 | -2.28689843E-6 | 2.36583758E-6 | -5.45749753E-7 | -2.05539375E-6 | 2.17857262E-6 |
| 1.75944927E-8 | -1.81763400E-8 | 5.86406472E-8 | | | |

**WAKEMAN**

| | | | | | |
|---|---|---|---|---|---|
| 1.00474764E-7 | -4.33980707E-9 | -2.15980670E-8 | 1.09777631E-6 | 7.72480748E-7 | -5.48158219E-7 |
| 3.30355957E-6 | 3.12703912E-6 | -1.89954053E-6 | 7.03636539E-7 | 5.16344829E-7 | -4.05589825E-7 |
| 1.74507915E-7 | 4.33980707E-9 | 2.15980670E-8 | | | |
| 2.36259191E-8 | 5.83175394E-8 | -5.22187444E-8 | 7.89675369E-7 | 6.33634457E-6 | -2.28689843E-6 |
| 3.12703912E-6 | 1.58378403E-5 | -5.74707686E-6 | 5.04219362E-7 | 3.06168448E-6 | -1.36005420E-6 |
| -2.36259191E-8 | 2.16665139E-7 | 5.22187444E-8 | | | |
| -6.65521196E-9 | 2.56318579E-8 | 1.89912980E-7 | -4.89903078E-7 | -2.31812228E-6 | 2.36583758E-6 |
| -1.89954053E-6 | -5.74707686E-6 | 5.82356842E-6 | -3.85611029E-7 | -1.39011676E-6 | 1.77323431E-6 |
| 6.65521196E-9 | -2.56318579E-8 | 8.50696992E-8 | | | |

**EFB**

| | | | | | |
|---|---|---|---|---|---|
| 9.88230882E-8 | -2.86156569E-9 | -1.93681179E-8 | 1.08005321E-6 | 8.47599653E-7 | -5.45749753E-7 |
| 7.03636539E-7 | 5.04219362E-7 | -3.85611029E-7 | 2.61899687E-6 | 2.38558434E-6 | -1.51099827E-6 |
| 1.76159591E-7 | 2.86156569E-9 | 1.93681179E-8 | | | |
| 3.57871263E-8 | 4.80296433E-8 | -5.19527428E-8 | 8.76519335E-7 | 5.44155178E-6 | -2.05539375E-6 |
| 5.16344829E-7 | 3.06168448E-6 | -1.39011676E-6 | 2.38558434E-6 | 1.39504115E-5 | -5.30376501E-6 |

**FIGURE 15.2 (Continued)** Computer Printout of Supplemental Network Adjustment

| | | | | | |
|---|---|---|---|---|---|
| -3.57871263E-8 | 2.26953036E-7 | 5.19527428E-8 | -5.34527843E-7 | -2.03474903E-6 | 2.17857262E-6 |
| -1.33181514E-8 | 2.57446430E-8 | 1.77169883E-7 | -1.51099827E-6 | -5.30376501E-6 | 5.48349863E-6 |
| -4.05589825E-7 | -1.36005420E-6 | 1.77323431E-6 | | | |
| 1.33181514E-8 | -2.57446430E-8 | 9.78127958E-8 | | | |

PSEUDO

| | | | | | |
|---|---|---|---|---|---|
| 2.52527599E-8 | -3.02638902E-9 | 3.81065330E-9 | 1.15601268E-7 | -4.55582489E-8 | 1.75944927E-8 |
| 1.74507915E-7 | -2.36259191E-8 | 6.65521196E-9 | 1.76159591E-7 | -3.57871263E-8 | 1.33181514E-8 |
| 2.49729919E-7 | 3.02638902E-9 | -3.81065330E-9 | | | |
| -3.02638902E-9 | 6.94353068E-9 | 4.95822766E-9 | -1.06932358E-9 | 1.52511046E-7 | -1.81763400E-8 |
| 4.33980707E-9 | 2.16665139E-7 | -2.56318579E-8 | 2.86156569E-9 | 2.26953036E-7 | -2.57446430E-8 |
| 3.02638902E-9 | 2.68039148E-7 | -4.95822766E-9 | | | |
| 3.81065330E-9 | 4.95822766E-9 | 1.30936103E-8 | 1.55408568E-8 | 3.44234573E-8 | 5.86406472E-8 |
| 2.15980670E-8 | 5.22187444E-8 | 8.50696992E-8 | 1.93681179E-8 | 5.19527428E-8 | 9.78127958E-8 |
| -3.81065330E-9 | -4.95822766E-9 | 2.61889069E-7 | | | |

Notes:
1. This output has been annotated to increase readability.
2. The original output was a text file. It was renamed to be consistent with use of Word.
3. Values of the $f$ vector are in meters, $B$ matrix is unitless, and values of $Q$ matrix are in meters squared.
4. Values of coordinates for the stations and residuals of observations are in meters.
5. Values of covariance matrix are in meters squared.
6. The essential input values are the elements of the $f$ vector, the $B$ matrix, and the $Q$ matrix.
7. The observed baselines were chosen to insure use of nontrivial vectors.
8. The order of stations is arbitrary but, once chosen, must be strictly followed.
9. Observation vectors, computed parameters, and residuals were listed in column mode to conserve space.
10. The covariance values for each of the solved stations are underlined to assist locating the same.
11. Correlations between coordinates at each pair of stations can be extracted from covariance matrix.

**FIGURE 15.2 (*Continued*)** Computer Printout of Supplemental Network Adjustment

| Geocentric and ECEF Sigma | Geodetic and Local Sigma |
|---|---|
| **Station REILLY—NAD 83 (2011)** | |
| $X = -1,556,177.595$ m $\pm$ 0.0022 m | $\phi =$  32° 16′ 55.″ 93002 N $\pm$ 0.0024 m (N) |
| $Y = -5,169,235.284$ m $\pm$ 0.0042 m | $\lambda =$ 106° 45′ 15.″16035 W $\pm$ 0.0018 m (E) |
| $Z =$  3,387,551.720 m $\pm$ 0.0038 m | $h =$            1,166.543 m $\pm$ 0.0052 m (U) |
| | |
| **Station WAKEMAN—NAD 83 (2011)** | |
| $X = -1,556,515.423$ m $\pm$ 0.0018 m | $\phi =$  32° 17′ 00.″ 09689 N $\pm$ 0.0016 m (N) |
| $Y = -5,169,056.109$ m $\pm$ 0.0040 m | $\lambda =$ 106° 45′ 29.″49401 W $\pm$ 0.0018 m (E) |
| $Z =$  3,387,656.711 m $\pm$ 0.0024 m | $h =$            1,159.912 m $\pm$ 0.0043 m (U) |
| | |
| **Station EFB -NAD 83 (2011)** | |
| $X = -1,556,489.931$ m $\pm$ 0.0016 m | $\phi =$  32° 16′ 42.″ 99414 N $\pm$ 0.0015 m (N) |
| $Y = -5,169,358.655$ m $\pm$ 0.0037 m | $\lambda =$ 106° 45′ 25.″22846 W $\pm$ 0.0017 m (E) |
| $Z =$  3,387,211.826 m $\pm$ 0.0023 m | $h =$            1,161.020 m $\pm$ 0.0041 m (U) |
| | |
| **Station PSEUDO—NAD 83 (2011)** | |
| $X = -1,556,206.595$ m $\pm$ 0.0017 m | $\phi =$  32° 16′ 45.″ 74754 N $\pm$ 0.0020 m (N) |
| $Y = -5,169,400.704$ m $\pm$ 0.0031 m | $\lambda =$ 106° 45′ 14.″39942 W $\pm$ 0.0014 m (E) |
| $Z =$  3,387,285.999 m $\pm$ 0.0030 m | $h =$            1,165.614 m $\pm$ 0.0039 m (U) |
| | |
| **Station BROMILOW—NAD 83 (2011)** | |
| $X = -1,556,209.724$ m $\pm$ 0.0014 m | $\phi =$  32° 16′ 52.″ 33538 N $\pm$ 0.0015 m (N) |
| $Y = -5,169,286.462$ m $\pm$ 0.0034 m | $\lambda =$ 106° 45′ 15.″77216 W $\pm$ 0.0014 m (E) |
| $Z =$  3,387,457.534 m $\pm$ 0.0017 m | $h =$            1,165.501 m $\pm$ 0.0036 m (U) |

**FIGURE 15.3**   Geocentric/Local Reference Frame Positions and Standard Deviations

values for the $f$ vector, the $B$ coefficient matrix and the $Q$ matrix of baseline covariances as used in the adjustment. Figure 15.2 also includes the results of the least squares adjustment. That printout shows the adjusted coordinate values for the stations, the residuals associated with the observations, and the covariances of the computed positions. Those results—summarized in Figure 15.3—are used in two subsequent projects described later in this chapter—a hypothesis test for determining whether station BROMILOW was replaced "exactly where it was" and a terrestrial survey to determine the location of the FINIAL on Skeen Hall on the NMSU campus.

### Example 2—Hypothesis Testing

The relocation of station BROMILOW provides an excellent opportunity to use concepts of hypothesis testing as described in Chapter 4—see also Chapter 5 of Ghilani (2010). The question to be answered is, "Does the evidence prove that station BROMILOW was not replaced in its original location following reconstruction of the sidewalk in that area?" The $X/Y/Z$ coordinate values and the covariance matrix for

---

**TABLE 15.2**

**Summary of "Before" and "After" Values for Station BROMILOW**

BROMILOW—original position computed on NAD 83 (2011))

$X = -1,556,209.730$ m

$Y = -5,169,286.460$ m

$Z = 3,387,457.523$ m

$$\begin{bmatrix} \sigma_X^2 = 4.5195E - 6 & \text{(triangular and symmetric)} \\ \sigma_{XY} = 4.8554E - 6 & \sigma_Y^2 = 1.1651E - 5 \\ \sigma_{XZ} = -3.6316E - 6 & \sigma_{YZ} = -7.4121E - 6 & \sigma_Z^2 = 9.8355E - 6 \end{bmatrix}$$

BROMILOW—reset, NAD 83 (2011)

$X = -1,556,209.724$ m

$Y = -5,169,286.462$ m

$Z = 3,387,457.534$ m

$$\begin{bmatrix} \sigma_X^2 = 1.83160E - 6 & \text{(triangular and symmetric)} \\ \sigma_{XY} = 1.38640E - 6 & \sigma_Y^2 = 1.18892E - 5 \\ \sigma_{XZ} = -7.42232E - 7 & \sigma_{YZ} = -3.67770E - 6 & \sigma_Z^2 = 3.00070E - 6 \end{bmatrix}$$

---

the undisturbed station are found in Chapter 12 as part of the NAD 83 (2011) adjustment of the NMSU GPS network. The resurveyed location of station BROMILOW is included as part of the NMSU network enhancement survey as described in this chapter. Table 15.2 contains a summary of the computed NAD 83 (2011) position and the associated covariance matrix, both before and after replacement. The monument for station BROMILOW continues to be used with the updated coordinate values.

## EXAMPLE 3—USING TERRESTRIAL OBSERVATIONS IN THE GSDM

This section describes using total station horizontal and vertical angle observations to determine the 3-D location of the top of a FINIAL on Skeen Hall on the NMSU campus—see Figure 15.4. The exercise was a class project completed successfully by two senior Surveying Engineering students—Kyle Spolar and Ford Prather. With regard to Figure 15.5, total station instruments were set up on supplemental network stations WAKEMAN, EFB, and REILLY and angles (both horizontal and vertical) were observed to the top of the FINIAL. The only distances observed were the Height of Instrument (HI) at each occupied station. The GSDM provides a rectangular 3-D environment in which to perform the computations—meaning that the equations for locating station FINIAL can be formulated as a linear problem. Since the objective of this example is to illustrate use of the GSDM as a computational environment and all angles were measured similarly, equal weights were assumed for the observations. A more comprehensive solution could accommodate

**FIGURE 15.4    (See color insert.)** FINIAL on Skeen Hall—NMSU

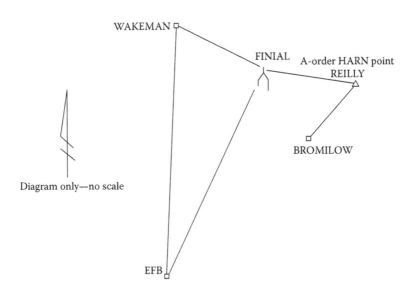

**FIGURE 15.5**    Diagram Showing Relative Location of Station FINIAL

instrument-centering errors and the number of repetitions for each angle measurement to determine appropriate weights.

Admittedly, in this case, formulation of the observation equations and taking the partial derivatives is somewhat onerous. But, it was valuable as a learning exercise. Part of that learning was comparing hand solutions for the partial derivatives with partial derivatives obtained using math software. With such powerful software tools readily available, efficiencies realized by avoiding time-consuming iterations should not be discounted. The fixed data for this problem includes the X/Y/Z geocentric coordinate values at stations REILLY, BROMILOW, WAKEMAN, and EFB. Observations include HIs at each station, vertical angles to the top of the FINIAL, and horizontal angles from respective backsights—BROMILOW, WAKEMAN, and EFB—to the FINIAL.

Three observation equations ($\Delta X$, $\Delta Y$, $\Delta Z$) are written for each vector from each of three known stations—REILLY, WAKEMAN, and EFB. Hence, there are nine equations available to solve for six unknowns—three slope distances to the FINIAL and the X/Y/Z coordinates of station FINIAL. When using GNSS vectors, the $\Delta X$, $\Delta Y$, $\Delta Z$ components are readily available for each vector. In this case, equations for those $\Delta X$, $\Delta Y$, $\Delta Z$ measurements must be derived from the total station observations. Equations 1.25, 1.26, and 1.27 are inserted into Equation 1.23 to obtain the following matrix formulation:

$$\begin{bmatrix} \Delta X \\ \Delta Y \\ \Delta Z \end{bmatrix} = \begin{bmatrix} -\sin\lambda & -\sin\phi\cos\lambda & \cos\phi\cos\lambda \\ \cos\lambda & -\sin\phi\sin\lambda & \cos\phi\sin\lambda \\ 0 & \cos\phi & \sin\phi \end{bmatrix} \begin{bmatrix} S\sin\alpha\sin Z \\ S\cos\alpha\sin Z \\ HI + S\cos Z \end{bmatrix} \quad (15.1)$$

where:

$\Delta X$, $\Delta Y$, and $\Delta Z$ = geocentric components of each vector.

$\phi$ and $\lambda$ = geodetic latitude and longitude of instrument station. (East longitude is used at each instrument station.)

$S$ = the slope distance instrument setup to station FINIAL.

$\alpha$ = geodetic azimuth instrument setup to station FINIAL.

$Z$ = Zenith direction instrument setup to station FINIAL.

$HI$ = Height of instrument at each station.

Equation 15.1 is written for a generic vector. Individual equations use the following subscripts:

$R$ = subscript for station REILLY.
$W$ = subscript for station WAKEMAN.
$E$ = subscript for station EFB.
$F$ = subscript for station FINIAL.
$RF$ = subscript for vector REILLY to FINIAL.
$WF$ = subscript for vector WAKEMAN to FINIAL.
$EF$ = subscript for vector EFB to FINIAL.

The unknown parameters to be found in the solution include three slope distances—one for each vector—and the unknown geocentric X/Y/Z coordinates for station FINIAL. In other words, the geocentric coordinates of station FINIAL are found by adding the geocentric coordinate difference as determined by Equation 15.1 to the beginning geocentric coordinate at each instrument station. Equation 15.2 is written in matrix form for a generic vector. Note that each equation needs a residual at the end to make the pieces fit. Equation 15.5 recasts the generic form to be compatible with the least squares solution by indirect observations ($v + B\Delta = f$) and includes all nine equations with appropriate subscripts.

$$\begin{bmatrix} X_F \\ Y_F \\ Z_F \end{bmatrix} = \begin{bmatrix} X - S\sin\lambda\sin\alpha\sin Z - S\sin\phi\cos\lambda\cos\alpha\sin Z + (HI + S\cos Z)\cos\phi\cos\lambda + v_i \\ Y + S\cos\lambda\sin\alpha\sin Z - S\sin\phi\sin\lambda\cos\alpha\sin Z + (HI + S\cos Z)\cos\phi\sin\lambda + v_i \\ Z + S\cos\phi\cos\alpha\sin Z + (HI + S\cos Z)\sin\phi + v_i \end{bmatrix}$$

$$(15.2)$$

Next, all nine equations are written out in the form $v + B\Delta = f$.
The vector REILLY to FINIAL provides the first three equations:

$$v_1 + X_F + S_{RF}\left(\sin\lambda_R \sin\alpha_{RF} \sin Z_{RF}\right) + S_{RF}\left(\sin\phi_R \cos\lambda_R \cos\alpha_{RF} \sin Z_{RF}\right)$$
$$- S_{RF}\left(\cos\phi_R \cos\lambda_R \cos Z_{RF}\right) = X_R + HI_R \cos\phi_R \cos\lambda_R \qquad (15.3.1)$$

$$v_2 + Y_F - S_{RF}\left(\cos\lambda_R \sin\alpha_{RF} \sin Z_{RF}\right) + S_{RF}\left(\sin\phi_R \sin\lambda_R \cos\alpha_{RF} \sin Z_{RF}\right)$$
$$- S_{RF}\left(\cos\phi_R \sin\lambda_R \cos Z_{RF}\right) = Y_R + HI_R \cos\phi_R \sin\lambda_R \qquad (15.3.2)$$

$$v_3 - S_{RF}\left(\cos\phi_R \cos\alpha_{RF} \sin Z_{EF}\right) - S_{RF}\left(\cos Z_{RF} \sin\phi_R\right) = Z_R + HI_R \sin\phi_R \qquad (15.3.3)$$

The vector WAKEMAN to FINIAL provides three more equations:

$$v_4 + X_W + S_{WF}\left(\sin\lambda_W \sin\alpha_{WF} \sin Z_{WF}\right) + S_{WF}\left(\sin\phi_W \cos\lambda_W \cos\alpha_{WF} \sin Z_{WF}\right)$$
$$- S_{WF}\left(\cos\phi_W \cos\lambda_W \cos Z_{WF}\right) = X_W + HI_W \cos\phi_W \cos\lambda_W \qquad (15.3.4)$$

$$v_5 + Y_W - S_{WF}\left(\cos\lambda_W \sin\alpha_{WF} \sin Z_{WF}\right) + S_{WF}\left(\sin\phi_W \sin\lambda_W \cos\alpha_{WF} \sin Z_{WF}\right)$$
$$- S_{WF}\left(\cos\phi_W \sin\lambda_W \cos Z_{WF}\right) = Y_W + HI_W \cos\phi_W \sin\lambda_W \qquad (15.3.5)$$

$$v_6 - S_{WF}\left(\cos\phi_W \cos\alpha_{WF} \sin Z_{WF}\right) - S_{WF}\left(\cos Z_{WF} \sin\phi_W\right) = Z_W + HI_W \sin\phi_W \qquad (15.3.6)$$

The vector EFB to FINIAL provides the final three equations:

$$v_7 + X_E + S_{EF}\left(\sin\lambda_E \sin\alpha_{EF} \sin Z_{EF}\right) + S_{EF}\left(\sin\phi_E \cos\lambda_E \cos\alpha_{EF} \sin Z_{EF}\right)$$
$$- S_{EF}\left(\cos\phi_E \cos\lambda_E \cos Z_{EF}\right) = X_E + HI_E \cos\phi_E \cos\lambda_E \qquad (15.3.7)$$

$$v_8 + Y_E - S_{EF}\left(\cos\lambda_E\sin\alpha_{EF}\sin Z_{EF}\right) + S_{EF}\left(\sin\phi_E\sin\lambda_E\cos\alpha_{EF}\sin Z_{EF}\right)$$
$$- S_{EF}\left(\cos\phi_E\sin\lambda_E\cos Z_{EF}\right) = Y_E + HI_E\cos\phi_E\sin\lambda_E \tag{15.3.8}$$

$$v_9 - S_{EF}\left(\cos\phi_E\cos\alpha_{EF}\sin Z_{EF}\right) - S_{EF}\left(\cos Z_{EF}\sin\phi_E\right) = Z_E + HI_E\sin\phi_E \tag{15.3.9}$$

With nine equations written in the form $v + B\Delta = f$, the solution proceeds in the same manner as the previous solutions. The $B$ matrix is formed as the partial derivative of each equation with respect to each of the unknowns. The $f$ vector is formed by computing actual numbers for the right side of the nine equations and the $Q$ matrix is identity in this case because all measurements (angles—horizontal and vertical) are considered of equal quality and the $HIs$ are considered to be without error. The pre-processing computations also included provision for refraction affecting the zenith directions. Deflection-of-vertical in this case was evaluated and determined to be of no consequence.

The $B$ matrix is $9 \times 6$ and contains the following elements:

$$B = \begin{matrix} X_F & Y_F & Z_F & S_{RF} & S_{WF} & S_{EF} \end{matrix}$$

$$B = \begin{bmatrix} 1 & 0 & 0 & \dfrac{\partial EQ1}{\partial S_{RF}} & 0 & 0 \\[2mm] 0 & 1 & 0 & \dfrac{\partial EQ2}{\partial S_{RF}} & 0 & 0 \\[2mm] 0 & 0 & 1 & \dfrac{\partial EQ3}{\partial S_{RF}} & 0 & 0 \\[2mm] 1 & 0 & 0 & 0 & \dfrac{\partial EQ4}{\partial S_{WF}} & 0 \\[2mm] 0 & 1 & 0 & 0 & \dfrac{\partial EQ5}{\partial S_{WF}} & 0 \\[2mm] 0 & 0 & 1 & 0 & \dfrac{\partial EQ6}{\partial S_{WF}} & 0 \\[2mm] 1 & 0 & 0 & 0 & 0 & \dfrac{\partial EQ7}{\partial S_{EF}} \\[2mm] 0 & 1 & 0 & 0 & 0 & \dfrac{\partial EQ8}{\partial S_{EF}} \\[2mm] 0 & 0 & 1 & 0 & 0 & \dfrac{\partial EQ9}{\partial S_{EF}} \end{bmatrix} \tag{15.4}$$

The partial derivatives shown in Equation 15.5 are evaluated from Equation 15.3 using the known values listed in Table 15.3 later in this chapter:

$$\frac{\partial EQ1}{\partial S_{RF}} = \sin\lambda_R\sin\alpha_{RF}\sin Z_{RF} + \sin\phi_R\cos\lambda_R\cos\alpha_{RF}\sin Z_{RF} - \cos\phi_R\cos\lambda_R\cos Z_{RF}$$

$$= 0.97164224963 \tag{15.5.1}$$

$$\frac{\partial EQ2}{\partial S_{RF}} = -\cos\lambda_R \sin\alpha_{RF} \sin Z_{RF} + \sin\phi_R \sin\lambda_R \cos\alpha_{RF} \sin Z_{RF} - \cos\phi_R \sin\lambda_R \cos Z_{RF}$$

$$= -0.21897071682 \tag{15.5.2}$$

$$\frac{\partial EQ3}{\partial S_{RF}} = -\cos\phi_R \cos\alpha_{RF} \sin Z_{RF} - \cos Z_{RF} \sin\phi_R$$

$$= -0.08923656150 \tag{15.5.3}$$

$$\frac{\partial EQ4}{\partial S_{WF}} = \sin\lambda_W \sin\alpha_{WF} \sin Z_{WF} + \sin\phi_W \cos\lambda_W \cos\alpha_{WF} \sin Z_{WF} - \cos\phi_W \cos\lambda_W \cos Z_{WF}$$

$$= -0.58087030934 \tag{15.5.4}$$

$$\frac{\partial EQ5}{\partial S_{WF}} = -\cos\lambda_W \sin\alpha_{WF} \sin Z_{WF} + \sin\phi_W \sin\lambda_W \cos\alpha_{WF} \sin Z_{WF} - \cos\phi_W \sin\lambda_W \cos Z_{WF}$$

$$= 0.67997353083 \tag{15.5.5}$$

$$\frac{\partial EQ6}{\partial S_{WF}} = -\cos\phi_W \cos\alpha_{WF} \sin Z_{WF} - \cos Z_{WF} \sin\phi_W$$

$$= 0.44746584350 \tag{15.5.6}$$

$$\frac{\partial EQ7}{\partial S_{EF}} = \sin\lambda_E \sin\alpha_{EF} \sin Z_{EF} + \sin\phi_E \cos\lambda_E \cos\alpha_{EF} \sin Z_{EF} - \cos\phi_E \cos\lambda_E \cos Z_{EF}$$

$$= -0.55937883644 \tag{15.5.7}$$

$$\frac{\partial EQ8}{\partial S_{EF}} = -\cos\lambda_E \sin\alpha_{EF} \sin Z_{EF} + \sin\phi_E \sin\lambda_E \cos\alpha_{EF} \sin Z_{EF} - \cos\phi_E \sin\lambda_E \cos Z_{EF}$$

$$= -0.46949743142 \tag{15.5.8}$$

$$\frac{\partial EQ9}{\partial S_{EF}} = -\cos\phi_E \cos\alpha_{EF} \sin Z_{EF} - \cos Z_{EF} \sin\phi_E$$

$$= -0.88092913221 \tag{15.5.9}$$

The elements of the $f$ vector are computed from the values in Table 15.3 according to the right-hand side of Equations 15.3.1 through 15.3.9 as

$$\begin{bmatrix} f_1 \\ f_2 \\ f_3 \\ f_4 \\ f_5 \\ f_6 \\ f_7 \\ f_8 \\ f_9 \end{bmatrix} = \begin{bmatrix} X_R + HI_R \cos\phi_R \cos\lambda_R \\ Y_R + HI_R \cos\phi_R \sin\lambda_R \\ Z_R + HI_R \sin\phi_R \\ X_W + HI_W \cos\phi_W \cos\lambda_W \\ Y_W + HI_W \cos\phi_W \sin\lambda_W \\ Z_W + HI_W \sin\phi_W \\ X_E + HI_E \cos\phi_E \cos\lambda_E \\ Y_E + HI_E \cos\phi_E \sin\lambda_E \\ Z_E + HI_E \sin\phi_E \end{bmatrix} = \begin{bmatrix} -1,556,178.0081 \\ -5,169,236.6562 \\ 3,387,552.6253 \\ -1,556,515.8421 \\ -5,169,057.5006 \\ 3,387,657.6292 \\ -1,556,490.3463 \\ -5,169,360.0343 \\ 3,387,212.7359 \end{bmatrix} \tag{15.6}$$

**TABLE 15.3**

**Control Values and Observations for Locating FINIAL on Skeen Hall**

| | Known Values at Stations: | | |
|---|---|---|---|
| | **REILLY** | **WAKEMAN** | **EFB** |
| $X =$ | −1,556,177.595 m | −1,556,515.423 m | −1,556,489.931 m |
| $Y =$ | −5,169,235.284 m | −5,169,056.109 m | −5,169,358.655 m |
| $Z =$ | 3,387,551.720 m | 3,387,656.711 m | 3,387,211.826 m |
| $\phi =$ | 32° 16′ 55.″93002 N | 32° 17′ 00.″09689 N | 32° 16′ 42.″99414 N |
| $\lambda =$ | 253° 14′ 44.″83965 E | 253° 14′ 30.″50599 E | 253° 14′ 34.″77154 E |
| $HI =$ | 1.695 m | 1.719 m | 1.704 m |
| **Mean Zenith Directions to FINIAL From** | | | |
| REILLY | = | | 83° 50′ 47.″4 |
| WAKEMAN | = | | 80° 13′ 13.″5 |
| EFB | = | | 85° 41′ 53.″1 |
| **Geodetic Azimuth to FINIAL From** | | | |
| REILLY | = | | 272° 10′ 51.″0 |
| WAKEMAN | = | | 130° 14′ 21.″7 |
| EFB | = | | 4° 08′ 56.″0 |

The $Q$ matrix is the identity matrix because the problem is solved assuming equal weights for the horizontal and vertical angles measured. The solution is shown in Figure 15.6. Figure 15.7 is a summary of the results for Example 3.

## Example 4—Using the GSDM to Develop a 2-D Survey Plat

The 2-D plat example is included in Chapter 12 of the first edition and illustrates the direct connection between 3-D GNSS-derived positions and a 2-D plat of the survey. With reference to Figure 1.4, the direct connection lies in use of the rotation matrix and the P.O.B. datum feature of the GSDM. Very briefly, the project consists of two HARN stations and Section 31, T23S-R1E, New Mexico Principal Meridian (NMPM)—all in the Las Cruces area. Section 31 is BLM property, conveniently lies between the two HARN stations, and, although strictly vacant desert land, enjoys convenient vehicular access. Comprehensive details of the project—summarized here—are posted at http://www.globalcogo.com/3DGPS.pdf, accessed May 04, 2017.

As part of a class project, GNSS data were collected on 3 separate days and trivial vectors were carefully avoided. The NAD 83 (1992) geocentric $X/Y/Z$ coordinates as published by the NGS were held fixed for the two HARN points and a least squares adjustment of the inter-connected baselines provided the geocentric coordinates and standard deviations as shown Table 15.4. The latitude/longitude/ellipsoid heights and standard deviations were computed using the GSDM.

Program name/version: LSAdjGPSvectors: Version 2014-A

| The user is: | Earl F. Burkholder |
|---|---|
| Program used on: | September 18, 2016 |
| The input file for program is: | LSFinial.dat |
| The name of this output file is: | LSFinial.txt |
| Output file annotated/renamed: | Figure 15.6.docx |
| The project is: | Location of Finial at NMSU |

The name/dimensions (in meters) of the $F$ vector are: $f_{9,1}$

```
-1,556,178.0081
-5,169,236.6562
 3,387,552.6253
-1,556,515.8421
-5,169,057.5006
 3,387,657.6292
-1,556,490.3463
-5,169,360.0343
 3,387,212.7359
```

The name/dimensions of the $B$ matrix are: $B_{9,6}$

```
1  0  0        0.97164224963   0  0
0  1  0       -0.21897071682   0  0
0  0  1       -0.08923656150   0  0
1  0  0  0    -0.58087030934      0
0  1  0  0     0.67997353083      0
0  0  1  0     0.44746584350      0
1  0  0  0  0 -0.20393014633
0  1  0  0  0 -0.42705545241
0  0  1  0  0 -0.88092913221
```

The name/dimensions of the $Q$ matrix are: $Q_{9,9}$

```
1 0 0 0 0 0 0 0 0
0 1 0 0 0 0 0 0 0
0 0 1 0 0 0 0 0 0
0 0 0 1 0 0 0 0 0
0 0 0 0 1 0 0 0 0
0 0 0 0 0 1 0 0 0
0 0 0 0 0 0 1 0 0
0 0 0 0 0 0 0 1 0
0 0 0 0 0 0 0 0 1
```

$W_{9,9}$ is the inverse of the $Q$ matrix.

```
1 0 0 0 0 0 0 0 0
0 1 0 0 0 0 0 0 0
0 0 1 0 0 0 0 0 0
0 0 0 1 0 0 0 0 0
0 0 0 0 1 0 0 0 0
0 0 0 0 0 1 0 0 0
0 0 0 0 0 0 1 0 0
0 0 0 0 0 0 0 1 0
0 0 0 0 0 0 0 0 1
```

**FIGURE 15.6** Computer Output of Adjustment for Station FINIAL  (*Continued*)

$Ninv_{6,6}$

| | | | | | |
|---|---|---|---|---|---|
| 6.66275057E−1 | −1.86767426E−1 | −1.20049225E−1 | −6.98990373E−1 | 5.67734232E−1 | −4.96413369E−2 |
| −1.86767426E−1 | 5.36798033E−1 | 2.01882825E−1 | 3.17029501E−1 | −5.63831775E−1 | 3.68999481E−l |
| −1.20049225E−1 | 2.01882825E−1 | 5.76138539E−1 | 2.12263948E−1 | −4.64810325E−1 | 5.69270729E−1 |
| −6.98990373E−1 | 3.17029501E−1 | 2.12263948E−1 | 1.76753046E0 | −7.1657529E−1 | 1.79833463E−1 |
| 5.67734232E−1 | −5.63831775E−1 | −4.64810325E−1 | −7.1657529E−1 | 1.92115739E0 | −5.34474265E−1 |
| −4.96413369E−2 | 3.68999481E−1 | 5.69270729E−1 | 1.79833463E−1 | −5.34474265E−1 | 1.64894704E0 |

Delta - the answers.

| | | |
|---|---|---|
| −1,556,406.814 m | X | |
| −5,169,185.096 m | Y | FINIAL |
| 3,387,573.640 m | Z | |
| 235.4829 m | Slope distance REILLY to FINIAL | |
| 187.6750 m | Slope distance WAKEMAN to FINIAL | |
| 409.6740 m | Slope distance EFB to FINIAL | |

Residuals.

| | |
|---|---|
| 0.0007 m | |
| 0.0036 m | Vector from REILLY |
| −0.0009 m | |
| −0.0133 m | |
| −0.0187 m | Vector from WAKEMAN |
| 0.0112 m | |
| 0.0126 m | |
| 0.0151 m | Vector from EFB |
| −0.0102 m | |

Redundancy, R = 3

SigmaNaughtSquared = 3.85583294E−4

The covariance matrix of the computed parameters is

| | | | | | |
|---|---|---|---|---|---|
| 2.56904531E−4 | −7.20143992E−5 | −4.62889755E−5 | −2.6951901E−4 | 2.18908835E−4 | −1.91408702E−5 |
| −7.20143992E−5 | 2.06980354E−4 | 7.78426447E−5 | 1.22241279E−4 | −2.17404113E−4 | 1.42280035E−4 |
| −4.62889755E−5 | 7.78426447E−5 | 2.22149396E−4 | 8.18454322E−5 | −1.79223096E−4 | 2.19501283E−4 |
| −2.6951901E−4 | 1.22241279E−4 | 8.18454322E−5 | 6.81530217E−4 | −2.7629946E−4 | 6.93407792E−5 |
| 2.18908835E−4 | −2.17404113E−4 | −1.79223096E−4 | −2.7629946E−4 | 7.40766193E−4 | −2.06084348E−4 |
| −1.91408702E−5 | 1.42280035E−4 | 2.19501283E−4 | 6.93407792E−5 | −2.06084348E−4 | 6.35806433E−4 |

Standard deviations computed from the diagonal elements of the covariance matrix above are:

| For Coordinates of FINIAL | For Slope Distance to FINIAL from | |
|---|---|---|
| Sigma X = 0.016 m | REILLY | Sigma = 0.026 m |
| Sigma Y = 0.014 m | WAKEMAN | Sigma = 0.027 m |
| Sigma Z = 0.015 m | EFB | Sigma = 0.025 m |

**FIGURE 15.6 (*Continued*)** Computer Output of Adjustment for Station FINIAL

From the least squares adjustment, the geocentric $X/Y/Z$ coordinates and standard deviations for station FINIAL are:

$X = -1,556,406.814$ m    $\sigma_x = 0.016$ m
$Y = -5,169,185.096$ m    $\sigma_y = 0.014$ m
$Z = 3,387,573.640$ m    $\sigma_z = 0.015$ m

The full covariance matrix of station FINIAL from the adjustment printout is:

$$\Sigma_{Finial} = \begin{bmatrix} \sigma_X^2 = 0.0002569 & (\text{symmetric and triangular}) & \\ \sigma_{XY} = -0.0000720 & \sigma_Y^2 = 0.0002070 & \\ \sigma_{XZ} = -0.0000463 & \sigma_{YZ} = 0.0000778 & \sigma_Z^2 = 0.0002222 \end{bmatrix}$$

Using the data above, the geodetic coordinates and local frame standard deviations are:

$\phi = 32°\ 16'\ 56.''21908$ N $\pm 0.0015$ m (N)
$\lambda = 106°\ 45'\ 24.''09945$ W $\pm 0.0016$ m (E)
$h = 1,193.488$ m $\pm 0.0011$ m (U)

*Note:* The results of this adjustment are based on holding the $X/Y/Z$ coordinate values at stations REILLY, EFB, and WAKEMAN fixed.

**FIGURE 15.7**    Summary of Adjustment Results for Station FINIAL

Using the points in Table 15.4, the SW Corner of Section 31 was chosen as the P.O.B. and local tangent plane coordinates (eastings/northings) were computed for all points in the survey with respect to the SW Corner. Those values are shown in Table 15.5 in meters.

Table 15.6 shows local tangent plane inverses around Section 31 referenced to the true meridian through the SW Corner. Although meter units are the international standard, the distances were converted to feet to illustrate the flexibility of the GSDM with respect to units. When displaying the data, the user is free to use units of choice. Figure 15.8 is a plat of the survey described in Example 4.

### EXAMPLE 5—NEW MEXICO INITIAL POINT AND PRINCIPAL MERIDIAN

The Initial Point for the U.S. Public Land Survey System (USPLSS) in New Mexico was established by John W. Garretson in 1855 on a mesa near the intersection of the Rio Puerco, a small river in central NM, and the Rio Grande Rivers. From that Initial Point, base lines were run east and west, and the meridian—NMPM—was run north and south. It was by design that the NMPM should intersect the United States and Mexico border west of El Paso, Texas. As a side note, an account of the survey of the United States–Mexico border west of El Paso in the mid 1800s is documented by Witcher (2016). Coincidentally, the section of land included in the class project described above lies adjacent to the NMPM.

Not coincidentally, the NGS has routinely collected GNSS data on many points throughout the United States—including the New Mexico Initial Point in 2012 (NGS 2016). The GSDM provides a convenient tool by which the layout of the NMPM may be checked. It is a straightforward procedure to inverse between the NM Initial Point and the SW Corner of Section 31, T23S–R1E, NMPM as determined by the class project survey previously reported. Specifically, the NAD 83 (2011) $X/Y/Z$

**TABLE 15.4**

**3-D GPS Points for 2-D Survey of Section 31**

| | ECEF Frame | | | Local Frame | | |
|---|---|---|---|---|---|---|
| CRUCESAIR: | $X = -1,571,430.6720$ m | Fixed | Lat. | 32° 16′ 54.″63123 N | Fixed |
| (HARN PT) | $Y = -5,164,782.3120$ m | Fixed | Long. | 106° 55′ 22.″24784 W | Fixed |
| | $Z = 3,387,603.1880$ m | Fixed | El Hgt | $h = 1,326.250$ m | Fixed |
| REILLY: | $X = -1,556,177.6150$ m | Fixed | Lat. | 32° 16′ 55.″92906 N | Fixed |
| (HARN PT) | $Y = -5,169,235.3190$ m | Fixed | Long. | 106° 45′ 15.″16070 W | Fixed |
| | $Z = 3,387,551.7090$ m | Fixed | El Hgt | $h = 1,166.570$ m | Fixed |
| NW Cor | $X = -1,568,446.9652$ m | ±0.0033 m | Lat. | 32° 16′ 16.″51587 N | ±0.0054 m |
| | $Y = -5,166,282.9266$ m | ±0.0077 m | Long. | 106° 53′ 16.″50858 W | ±0.0039 m |
| | $Z = 3,386,573.0861$ m | ±0.0044 m | El Hgt | $h = 1,256.511$ m | ±0.0067 m |
| NE Cor | $X = -1,566,906.8273$ m | ±0.0034 m | Lat. | 32° 16′ 16.″50308 N | ±0.0057 m |
| | $Y = -5,166,748.4577$ m | ±0.0080 m | Long. | 106° 52′ 15.″04095 W | ±0.0040 m |
| | $Z = 3,386,571.5363$ m | ±0.0046 m | El Hgt | $h = 1,254.233$ m | ±0.0070 m |
| SW Cor | $X = -1,568,698.0864$ m | ±0.0035 m | Lat. | 32° 15′ 24.″28753 N | ±0.0058 m |
| | $Y = -5,167,107.1198$ m | ±0.0083 m | Long. | 106° 53′ 16.″54155 W | ±0.0041 m |
| | $Z = 3,385,214.0743$ m | ±0.0047 m | El Hgt | $h = 1,259.609$ m | ±0.0072 m |
| SE Cor | $X = -1,567,157.4899$ m | ±0.0035 m | Lat. | 32° 15′ 24.″26696 N | ±0.0058 m |
| | $Y = -5,167,571.1861$ m | ±0.0081 m | Long. | 106° 52′ 15.″08320 W | ±0.0041 m |
| | $Z = 3,385,211.2732$ m | ±0.0047 m | El Hgt | $h = 1,255.365$ m | ±0.0071 m |
| N ¼ Cor | $X = -1,567,682.4363$ m | ±0.0036 m | Lat. | 32° 16′ 16.″72241 N | ±0.0058 m |
| | $Y = -5,166,510.8119$ m | ±0.0080 m | Long. | 106° 52′46.″03144 W | ±0.0042 m |
| | $Z = 3,386,578.0990$ m | ±0.0048 m | El Hgt | $h = 1,255.823$ m | ±0.0070 m |
| W ¼ Cor | $X = -1,568,572.7788$ m | ±0.0035 m | Lat. | 32° 15′ 50.″39908 N | ±0.0058 m |
| | $Y = -5,166,694.9985$ m | ±0.0080 m | Long. | 106° 53′ 16.″53459 W | ±0.0041 m |
| | $Z = 3,385,893.5160$ m | ±0.0048 m | El Hgt | $h = 1,258.017$ m | ±0.0070 m |
| S ¼ Cor | $X = -1,567,928.0513$ m | ±0.0038 m | Lat. | 32° 15′ 24.″28747 N | ±0.0060 m |
| | $Y = -5,167,339.5732$ m | ±0.0083 m | Long. | 106° 52′ 45.″81734 W | ±0.0044 m |
| | $Z = 3,385,213.3104$ m | ±0.0050 m | El Hgt | $h = 1,258.181$ m | ±0.0073 m |
| E ¼ Cor | $X = -1,567,032.6554$ m | ±0.0038 m | Lat. | 32° 15′ 50.″37719 N | ±0.0059 m |
| | $Y = -5,167,160.8706$ m | ±0.0082 m | Long. | 106° 52′ 15.″06861 W | ±0.0043 m |
| | $Z = 3,385,891.8435$ m | ±0.0048 m | El Hgt | $h = 1,255.952$ m | ±0.0072 m |

coordinates of the Initial Point and the NAD 83 (2011) coordinates of said SW Corner are used to compute the direction and distance between them. Several notes related to that computation include the following:

1. The coordinates must be on the same datum and epoch. The class project reported above was completed in 2007 and originally based on the NAD 83 (1992). The $X/Y/Z$ coordinates for the Initial Point as shown on the NGS data sheet are based on the NAD 83 (2011). Therefore, the class project survey was recomputed on the NAD 83 (2011) to obtain compatible coordinate values to be used in the following comparison.

## TABLE 15.5
### Local P.O.B. Coordinate Differences

| Point | Description | P.O.B. East (meters) | P.O.B. North (meters) | P.O.B. Up (meters) |
|-------|-------------|---------------------|----------------------|--------------------|
| 1001 | CRUCESAIR | −3,290.100 | 2,783.992 | 65.183 |
| 1002 | REILLY | 12,598.767 | 2,831.217 | −106.099 |
| 1013 | NW Cor Sec 31 | 0.863 | 1,609.117 | −3.302 |
| 1014 | NE Cor Sec 31 | 1,609.819 | 1,608.850 | −5.783 |
| 1015 | SW Cor Sec 31 | 0.000 | 0.000 | 0.000 |
| 1016 | SE Cor Sec 31 | 1,608.970 | −0.506 | −4.447 |
| 1017 | N Qtr Cor Sec 31 | 798.622 | 1,615.512 | −4.041 |
| 1018 | W Qtr Cor Sec 31 | 0.182 | 804.478 | −1.643 |
| 1019 | S Qtr Cor Sec 31 | 804.355 | 0.030 | −1.479 |
| 1020 | E Qtr Cor Sec 31 | 1,609.224 | 803.931 | −3.910 |

## TABLE 15.6
### Inverse Directions and Distances Based on P.O.B. Values

| | | Azimuth | |
|-------------|-----------|------------|----------|
| From Point | To Point | D M S | Distance |
| SW Cor | Qtr Cor | 0 00 46.7 | 2,639.357 ft |
| W Qtr Cor | NW Cor | 0 02 54.5 | 2,639.889 ft |
| NW Cor | N Qtr Cor | 89 32 26.7 | 2,617.399 ft |
| N Qtr Cor | NE Cor | 90 28 13.7 | 2,661.492 ft |
| NE Cor | E Qtr Cor | 180 02 32.6 | 2,640.808 ft |
| E Qtr Cor | SE Cor | 180 01 05.2 | 2,639.222 ft |
| SE Cor | S Qtr Cor | 270 02 17.4 | 2,639.807 ft |
| S Qtr Cor | SW Cor | 269 59 52.3 | 2,638.955 ft |

2. The definition of horizontal distance may be considered a problem. Possible choices include the following:

   a. Arc distance on the ellipsoid—The USPLSS surveyors measured distances at ground level.

   b. Arc distance between the two points at an average elevation—much closer to what was laid out by the original USPLSS surveyors.

   c. 3-D spatial distance between the two points. But, the two points are at different elevations.

   d. Tangent plane distance between the two points—HD(1) in Burkholder (1991). But, there are two separate tangent planes and using HD(1) assumes that the plumb lines at the two points are parallel—they are not!

   e. Chord distance (see Figure 15.9) between the two points at the elevation of
      i. The Initial Point
      ii. The SW Corner of Section 31

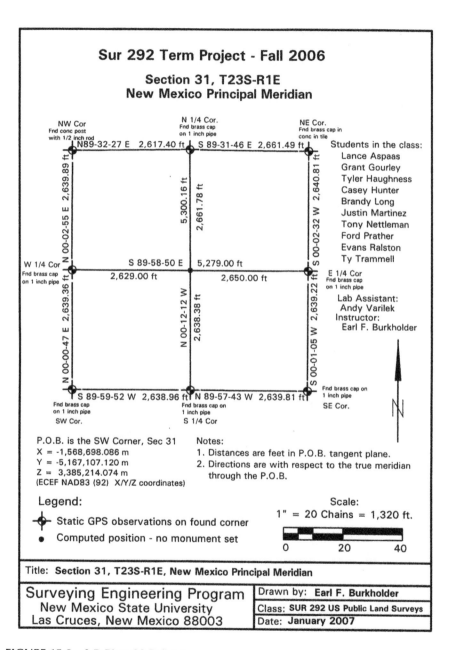

FIGURE 15.8    2-D Plat of 3-D Survey

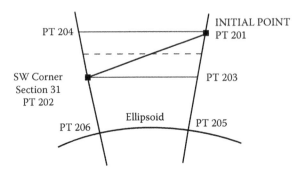

**FIGURE 15.9**  Profile View between New Mexico Initial Point and SW Corner, Section 31, T23S-R1E, NM Principal Meridian

f. A good approximation for computing horizontal distance from ECEF coordinates is given by Armstrong et al. (2017) as

$$D = \sqrt{\Delta X^2 + \Delta Y^2 + \Delta Z^2 - \Delta h^2} \qquad (15.7)$$

3. Determination of the true astronomical meridian in 1855 was with respect to the instantaneous spin axis of the Earth as obtained from astronomical observation. The true geodetic meridian as obtained from GNSS is with respect to the internationally adopted Conventional Terrestrial Pole (CTP). The difference between those two reference meridians—quite small and ignored here—is caused by polar motion and deflection-of-the-vertical (Leick 2015).

Data for computing an inverse between the two points on the NAD 83 (2011) are given in Table 15.7.

With regard to Figure 15.9, GRS 80 geocentric X/Y/Z coordinates for the following points were computed using Equations 1.1, 1.2, 1.3, and 1.4. Note that Points 205 and 206 are on the ellipsoid surface because the ellipsoid height is zero. The X/Y/Z coordinates and ellipsoid heights for those four points are listed in Table 15.8.

---

**TABLE 15.7**
**NAD 83 (2011) Values for Points on the NMPM**

| Initial Point—Point 201 (As published by NGS) | SW Corner Section 31—Point 202 (As surveyed by NMSU class in 2006) |
|---|---|
| $X = -1,533,309.884$ m | $X = -1,568,698.064$ m |
| $Y = -5,050,681.721$ m | $Y = -5,167,107.065$ m |
| $Z = 3,571,149.193$ m | $Z = 3,385,214.088$ m |
| $\phi = 34°\ 15'\ 35.''94618$ N | $\phi = 32°\ 15'\ 24.''28892$ N |
| $\lambda = 106°\ -53'\ 14.''96154$ W | $\lambda = 106°\ 53'\ 16.''54133$ W |
| $h = 1,475.929$ m | $h = 1,259.566$ m |

**TABLE 15.8**

**Geocentric Coordinates for Auxiliary Points**

|      | Point 203 | Point 204 | Point 205 | Point 206 |
|------|-----------|-----------|-----------|-----------|
| X =  | −1,533,257.938 m | −1,568,751.217 m | −1,532,955.528 m | −1,568,388.631 m |
| Y =  | −5,050,510.611 m | −5,167,282.144 m | −5,049,514.482 m | −5,166,087.830 m |
| Z =  | 3,571,027.392 m | 3,385,329.563 m | 3,570,318.320 m | 3,384,541.839 m |
| h =  | 1,259.566 m | 1,475.929 m | 0.000 m | 0.00 m |

- Point 203 uses latitude/longitude of the Initial Point, but the ellipsoid height of the SW Corner.
- Point 204 uses latitude/longitude of the SW Corner, but the ellipsoid height of the Initial Point.
- Point 205 uses latitude/longitude of the Initial Point, but an ellipsoid height of zero.
- Point 206 uses latitude/longitude of the SW Corner, but an ellipsoid height of zero.

In each of the four following computations

$$D = \sqrt{\Delta X^2 + \Delta Y^2 + \Delta Z^2} \qquad (15.8)$$

3-D slope distance, Point 201 to 202, obviously not horizontal $= 222{,}213.968$ m.
Chord distance on ellipsoid, Point 205 to 206 $= 222{,}166.044$ m.
Chord distance at elevation of Initial Point, Point 201 to 204 $= 222{,}217.645$ m.
Chord distance at elevation of SW Corner, Point 202 to 203 $= 222{,}210.080$ m.
Mean chord distance at average ellipsoid height $= 222{,}213.862$ m.
Horizontal distance Point 201 to Point 202 using Equation 15.7 $= 222{,}213.862$ m.

Radius of curvature is needed to compute an arc distance from a chord distance. The mean latitude between the Initial Point and the SW Corner is 33° 15′ 30.″11756 and the mean radius of curvature on the GRS 80 ellipsoid is computed using Equation 7.20 as

$$R = \frac{6{,}378{,}137.0\sqrt{1-0.006694380023}}{\left(1-0.006694380023\sin^2(33°15'30.''11756\right)} = 6{,}369{,}576.807 \text{ m}$$

The radius of curvature at the mean ellipsoid height = 6,370,944.555 m.
The generic arc distance for a curve subtended by a chord is computed as

$$\text{Arc} = R \times 2 \times \theta \quad \text{where } R = \text{radius and } \theta_{rad} = \arcsin\left(\frac{\text{Chord}}{2R}\right) \qquad (15.9)$$

Therefore, the ellipsoid distance between Points 205 and 206 = 222,177.307 m.
And the arc distance at the mean ellipsoid height $= 222{,}225.128$ m.

The NMPM was measured with a Gunter's Chain and a compass. Two chains were specified, one for production and the other for monitoring the length of the production chain on a daily basis when being used. In several areas of impassible terrain, the distance was carried forward using triangulation. Project instructions also stipulated that the crew was to be equipped with a "Burt Improved Solar Compass" and that "you will observe the variation of the needle every township corner." Theoretically, the distance laid out by the USPLSS surveyors is 23 Townships, 6 miles per township, 5,280 feet per mile, and 12.00 meters per 39.37 feet for a total distance of 222,089.92 m. The BLM Manual of Surveying Instructions 1973 ( BLM 1973) stipulates that two independent sets of measurements are to be made when surveying a Principal Meridian and that a line is to be remeasured if the difference between the two measurements exceeds 7 links in 80 chains (that is equivalent to 1.41 m in 1,609.3 m or 1:1,141). Incidentally, according to Roeder (1994) 108 miles of the NMPM north and south of the Initial Point had to be resurveyed because of a misunderstanding in the calibration of the standard length being used. The BLM Manual also stipulates that the alignment of a standard line is not to exceed three minutes of arc from the true cardinal course.

Of the distance choices listed above, the horizontal distance found using Equation 15.6 is probably the easiest to compute and a good approximation of the distance laid out. The real question is how well the theoretical distance matches the GNSS surveyed distance. The ratio of precision is

$$\frac{\text{Theoretical} - \text{Actual}}{\text{Actual}} = \frac{222,089.92 - 222,225.13}{222,225.13} = -0.000557778 \text{ or } 1:1,644$$

That ratio of precision could have been better but, considering the terrain and the specifications used by USPLSS surveyors for distance measurement in the 1850s, the results are quite acceptable.

The direction of the line between the Initial Point and the SW Corner of said Section 31 is computed from the ECEF coordinates by rotating the ECEF coordinate differences to the local perspective before using the standard inverse tangent function for azimuth computation. In this case, the azimuth between the Initial Point and the SW Corner should be due north-south. In general, the direction from "here" to "there" on the Earth is different than the direction "there" to "here" because meridians at the two points are not parallel. That is true unless, as in this case, the two points lie on the same meridian. Even so, the procedures for computing a 3-D azimuth (Burkholder 1997a) are the same regardless of which direction the computation proceeds. First, the geocentric differences are computed. Then those differences are rotated from the geocentric ECEF environment to the local geodetic horizon at the chosen standpoint using Equation 1.21. The resulting local $\Delta e$ and $\Delta n$ components are used to compute the geodetic azimuth as

$$\alpha = \tan^{-1}\left(\frac{\Delta e}{\Delta n}\right) \text{ with due regard to the quadrant} \qquad (15.10)$$

From the Initial Point to the SW Corner, the geocentric differences ("there" minus "here") are

$$\Delta X = -1,568,698.064 - (-1,533,309.884) \quad = -35,388.180 \text{ m}$$
$$\Delta Y = -5,050,681.721 - (-5,167,107.065) \quad = -116,425.344 \text{ m}$$
$$\Delta Z = 3,385,214.088 - 3,571,149.193 \quad = -185,935.105 \text{ m}$$

Using the latitude and longitude at the Initial Point and the geocentric differences in Equation 1.21, the local $\Delta e$, $\Delta n$, $\Delta u$ components are computed as

$$\Delta e = \quad -41.358 \text{ m}$$
$$\Delta n = -222,176.127 \text{ m}$$
$$\Delta u = \quad -4,100.532 \text{ m}$$

And, the 3-D azimuth Initial Point to SW Corner is $180° 00' 38.''4$. These results are impressive when considering some rather rough terrain over a distance of 23 townships. Note that to a very close approximation, the 3-D azimuth is the same as the geodetic azimuth (Burkholder 1997a).

From the SW Corner to the Initial Point, the geocentric differences have the same numerical values but the opposite sign due to going in the opposite direction. But, using Equation 1.21 at the SW Corner, the local components are slightly different because the latitude and longitude of the SW Corner are different than the latitude and longitude of the Initial Point:

$$\Delta e = \quad 40.426 \text{ m}$$
$$\Delta n = \quad 222,183.683 \text{ m}$$
$$\Delta u = \quad -3,668.358 \text{ m}$$

The 3-D azimuth computed from the SW Corner to the Initial Point is $000° 00' 37.''5$. To the nearest second of arc, the two directions are indeed different by $180°$.

## Example 6—State Boundary between Texas and New Mexico along the Rio Grande River

Figure 15.10 shows a portion of the boundary between Texas and New Mexico as originally defined by the meandering course of the Rio Grande River. Not surprisingly, that stretch of boundary was contested when New Mexico became a state in 1912. Resolution of the disagreement consists of a 1929 U.S. Supreme Court decision in response to a lawsuit brought by New Mexico against Texas to settle the boundary. Prior to modern channelization, the Rio Grande River changed courses frequently over the years and the reconstructed boundary courses do not lie within the current Rio Grande River channel. In 1929, brass tablets were embedded in concrete pillars that were established to mark the historical location of the boundary. Many, but not all, of those monuments still exist. Although it is more informative to read the Report of the Boundary Commissioner to the U.S. Supreme Court dated July 17, 1930 (Gannett 1930), only a summary of the case is provided here.

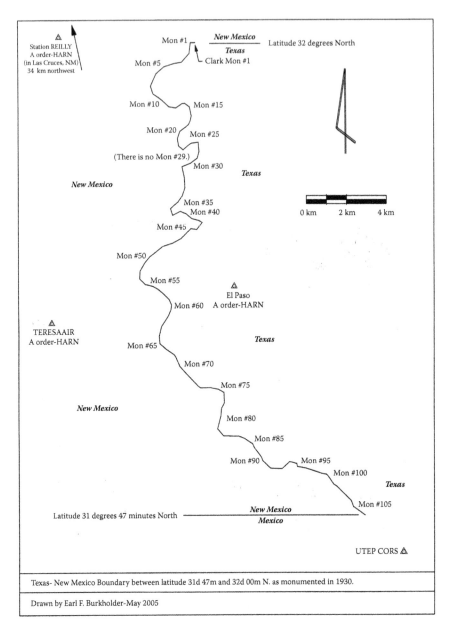

**FIGURE 15.10**   Map Showing Course of Rio Grande River as It Existed in 1850

Many pieces of evidence were collected and evaluated as part of the lawsuit and the decision was made to mark the boundary as being the location of the middle of the Rio Grande River as it existed in 1850. A comprehensive geodetic survey was conducted as part of the project and is the basis of the Boundary Commissioner's Report. The latitude and longitude of each brass tablet was reported (to three decimal places of seconds) on the newly adopted NAD 27 and the elevations were referenced to "mean sea level" (to three decimal places of feet).

In 2005 and 2006, a group of surveyors systematically recovered many of the boundary monuments and collected GNSS data on most of them. It being a volunteer effort, the project was never brought to conclusion in a formal final report. However, some of the collected GNSS data were used in beneficial manner on a vector-by-vector basis. In simplistic terms, comparisons are useful to the extent that apples are compared with apples. The GSDM provides a convenient way to compare historical data as reported in the Supreme Court decision with GNSS observations. One method would be to manipulate the GNSS data to duplicate, to the extent possible, the historical equivalent. The method described herein makes the comparison in terms of ECEF coordinate differences. Understanding that a truly legitimate comparison of the record relative positions with the GNSS observed baselines must be based upon a formally completed geodetic survey, the comparisons described here are acknowledged to be incomplete. But, there is value in looking at the method of comparison as supported by the GSDM. The following comparisons are made on a vector-to-vector basis.

Given the legal stature of information from the Supreme Court document and the geodetic quality of the reported results, those data (both horizontal and vertical) were "brought forward" so the comparisons could be made between directly observed GNSS baselines and vectors computed from the historical data. First, the Supreme Court NAD 27 positions were transformed to current-epoch NAD 83 positions and the mean sea level elevations were converted to NAVD 88 values using CORPSCON 5.11.08—a transformation program implemented by the Corps of Engineers using algorithms developed by the NGS. The NAVD 88 elevations were further converted to ellipsoid heights using GEOID03—the geoid model in use at the time. Table 15.9 contains data from the Supreme Court document. Table 15.10 shows the NAD 83 latitude/longitude positions for the points and the computed ellipsoid height for each point. These geodetic values were input to a BURKORD™ program that computed corresponding ECEF coordinates and stored the results in a 3-D BURKORD™ file.

**TABLE 15.9**
**List of U.S. Supreme Court Report Positions**

| Designation | Latitude D-M-S | Longitude (W) D-M-S | Elevation U.S. Feet |
|---|---|---|---|
| Bdy Mon No 90 | 31 48 49.595 | 106 34 49.949 | 3,738.869 |
| Bdy Mon No 91 | 31 48 37.111 | 106 34 34.807 | 3,738.123 |
| Bdy Mon No 92 | 31 48 36.334 | 106 34 12.833 | 3,736.727 |
| Bdy Mon No 93 | 31 48 45.947 | 106 34 00.044 | 3,735.827 |
| Bdy Mon No 94 | 31 48 47.654 | 106 33 58.085 | 3,736.359 |
| Bdy Mon No 95 | 31 48 44.651 | 106 33 45.110 | 3,736.215 |
| Bdy Mon No 96 | 31 48 39.752 | 106 33 43.873 | 3,735.972 |
| Bdy Mon No 97 | 31 48 37.202 | 106 33 27.567 | 3,734.934 |
| Bdy Mon No 98 | 31 48 28.403 | 106 32 56.747 | 3,730.844 |
| Bdy Mon No 99 | 31 48 27.544 | 106 32 50.008 | 3,732.452 |

**TABLE 15.10**

**NAD 83 Latitude/Longitude, NAVD 88 Elevations, and Ellipsoid Heights of Boundary Monuments**

| PT No. | Point Name | NAD 83 Latitude D-M-S | | | NAD 83 Longitude D-M-S | | | NAVD 88 meters | N | h (meters) |
|---|---|---|---|---|---|---|---|---|---|---|
| 1090 | Bdy Mon No 90 | 31 | 48 | 49.97607 | 106 | 34 | 51.91094 | 1140.110 | −23.985 | 1116.125 |
| 1091 | Bdy Mon No 91 | 31 | 48 | 37.49255 | 106 | 34 | 36.76849 | 1139.880 | −23.981 | 1115.899 |
| 1092 | Bdy Mon No 92 | 31 | 48 | 36.71568 | 106 | 34 | 14.79407 | 1139.453 | −23.972 | 1115.481 |
| 1093 | Bdy Mon No 93 | 31 | 48 | 46.32844 | 106 | 34 | 2.00496 | 1139.179 | −23.966 | 1115.213 |
| 1094 | Bdy Mon No 94 | 31 | 48 | 48.03540 | 106 | 34 | 0.04595 | 1139.341 | −23.965 | 1115.376 |
| 1095 | Bdy Mon No 95 | 31 | 48 | 45.03256 | 106 | 33 | 47.07067 | 1139.296 | −23.961 | 1115.335 |
| 1096 | Bdy Mon No 96 | 31 | 48 | 40.13372 | 106 | 33 | 45.83358 | 1139.221 | −23.961 | 1115.260 |
| 1097 | Bdy Mon No 97 | 31 | 48 | 37.58389 | 106 | 33 | 29.52724 | 1138.903 | −23.956 | 1114.947 |
| 1098 | Bdy Mon No 98 | 31 | 48 | 28.78533 | 106 | 32 | 58.70656 | 1137.652 | −23.949 | 1113.703 |
| 1099 | Bdy Mon No 99 | 31 | 48 | 27.92639 | 106 | 32 | 51.96743 | 1138.143 | −23.947 | 1114.196 |
| 1100 | Bdy Mon No 100 | 31 | 48 | 26.22846 | 106 | 32 | 49.12936 | 1137.977 | −23.947 | 1114.030 |

*These points were determined using the following steps:*

1. The NAD 27 latitude and longitude were obtained from the 1930 Supreme Court document.
2. The elevations are quoted as being "mean sea level." They were used as NGVD 29 U.S. Survey Feet.
3. CORPSCON 5.11.08 was used to obtain NAD 83 (HPGN) latitude/longitude and NAVD 88 elevations in meters. The New Mexico HPGN files were used in CORPSCON.
4. GEOID03 was used to compute the geoid heights (N) at each NAD 83 position.
5. The NAD 83 latitude/longitude and ellipsoid heights as shown here were used as input for BURKORD™. No standard deviations were used but could be assigned later.

Figure 15.11 is a printout of the BURKORD™ program that used the NAD 83 positions and ellipsoid heights from the 3-D file as input to compute geocentric ECEF coordinates for each point. The second part of Figure 15.11 lists a series of 3-D inverses (both geocentric and local) between the "record" points. By way of explanation, no covariance values were used in this case, so the stochastic data are all listed as zeros. Note that the ECEF coordinates and the geodetic coordinates for each point are listed and that the inverse between points is printed between the station data. The third line of each inverse provides the tangent plane ground distance and the 3-D azimuth point to point.

Geocentric coordinate differences computed from the historical data were then compared directly to the observed geocentric $\Delta X/\Delta Y/\Delta Z$ baseline components as shown in Table 15.11. These four comparisons are only a sample of the many comparisons (including stochastic information) that could have been made in a funded (rather than volunteer) effort. Many of the baselines were observed monument to monument, but in other cases the observations were not made directly on sequentially numbered monuments.

GNSS data on this project were collected on several different outings using various brands of equipment. The baselines reported herein were computed from simultaneous observations using same brand equipment. Comments about the comparisons include the following:

1. Figures 15.12 and 15.13 are photos of monuments #93 and #94. Monument #94 has obviously been disturbed as shown by the way it leans.
2. The observed differences are raw GNSS baselines. No adjustment was made on those data.
3. The comparisons include both the geocentric (ECEF) differences and local frame differences.
4. The observed distance between monuments #93 and #94 is short of the record distance by about 0.13 meter. Figure 15.13 shows monument #94 leaning "in the direction of" #93.
5. The distance comparison between monuments #94 and #95 is reasonable but the direction is off by almost 2 minutes of arc—due to monument #94 being disturbed.
6. The 3-D slope distance is quite close to the horizontal distance in each comparison because little elevation difference is involved.
7. Table 15.11 lists comparisons between the record and observed relative positions of the following:
   - Monuments #90 and #93: The agreement of 1:15,144 is impressive.
   - Monument #93 and #94: The agreement is not good. Monument #94 has been disturbed.
   - Monument #94 and #95: The agreement is not really bad, but not very good either.
   - Monument #95 and #96: Knowing that monuments #95 and #96 showed little evidence of being disturbed, the distance and direction comparison between them appears to be indicative of results expected overall.

"BURKORD™" COMPUTES 3-D COORDINATE GEOMETRY POSITIONS FOR SPATIAL DATA UTILIZING GPS VECTORS, LOCAL COORDINATE DIFFERENCES AND 3-D SURVEYING MEASUREMENTS.

COPYRIGHT (C) 1999 AND ALL RIGHTS RESERVED BY:      USE OF BURKORD(TM) LICENSED TO:
EARL F. BURKHOLDER
P.O. BOX 3162
LAS CRUCES, NM 88003

USER:     Earl F. Burkholder
DATE:     September 29, 2016

PROGRAM:     BURKORD(TM) – VERSION 9A. 04, JAN 2004  S/N 9AA03001
DATA FILE:     TXMXRcrd.dat
OUTPUT FILE:     Fig–15.7.BK3     Renamed: Figure 15.11.docx

CLIENT/AGENCY:     Texas - New Mexico Joint GPS Boundary Survey
JOB/PROJECT:     Record Points Compiled by Earl F. Burkholder
*********************************************************

PARTIAL LISTING OF POINTS AS STORED IN THE BURKORD (TM) DATA FILE. NOTE - NO COVARIANCE DATA ARE STORED.

| 1090 | -1548404.1080 | -5200287.8040 | 3343500.3170 | 0.000E+00 | 0.000E+00 | 0.000E+00 | 0.000E+00 | 0.000E+00 | Bdy Mon No 90 |
| 1091 | -1548080.1190 | -5200595.5750 | 3343173.3960 | 0.000E+00 | 0.000E+00 | 0.000E+00 | 0.000E+00 | 0.000E+00 | Bdy Mon No 91 |
| 1092 | -1547529.5610 | -5200772.2210 | 3343152.8380 | 0.000E+00 | 0.000E+00 | 0.000E+00 | 0.000E+00 | 0.000E+00 | Bdy Mon No 92 |
| 1093 | -1547162.5170 | -5200718.3250 | 3343404.3470 | 0.000E+00 | 0.000E+00 | 0.000E+00 | 0.000E+00 | 0.000E+00 | Bdy Mon No 93 |
| 1094 | -1547105.2590 | -5200706.5820 | 3343449.1190 | 0.000E+00 | 0.000E+00 | 0.000E+00 | 0.000E+00 | 0.000E+00 | Bdy Mon No 94 |
| 1095 | -1546791.9920 | -5200850.6010 | 3343370.4880 | 0.000E+00 | 0.000E+00 | 0.000E+00 | 0.000E+00 | 0.000E+00 | Bdy Mon No 95 |
| 1096 | -1546783.4590 | -5200936.0690 | 3343242.2020 | 0.000E+00 | 0.000E+00 | 0.000E+00 | 0.000E+00 | 0.000E+00 | Bdy Mon No 96 |
| 1097 | -1546384.0160 | -5201097.7680 | 3343175.2850 | 0.000E+00 | 0.000E+00 | 0.000E+00 | 0.000E+00 | 0.000E+00 | Bdy Mon No 97 |
| 1098 | -1545647.2310 | -5201464.7120 | 3342944.2880 | 0.000E+00 | 0.000E+00 | 0.000E+0C | 0.000E+00 | 0.000E+00 | Bdy Mon No 98 |
| 1099 | -1545481.3780 | -5201528.9800 | 3342922.0610 | 0.000E+00 | 0.000E+00 | 0.000E+00 | 0.000E+00 | 0.000E+00 | Bdy Mon No 99 |

**FIGURE 15.11**  BURKORD™ Printout for Coordinates and Inverses Point to Point

(Continued)

INVERSE BETWEEN POINTS

1090      Bdy Mon No 90
X =  -1548404.1080    LAT (N+S-)    31 48 49.976071      ±0.000E+00 METERS    N
Y =  -5200287.8040    LON (E+W-)    -106 34 51.910949    ±0.000E+00 METERS    E    STANDARD DEVIATIONS
Z =   3343500.3170    EL HGT        1116.1246 M          ±0.000E+00 METERS    U

DELTA X/Y/Z WITH SIGMAS    1241.5910M ± 0.000E+00M    -430.5210M ± 0.000E+00M    -95.9700M ± 0.000E+00M
DELTA E/N/U WITH SIGMAS    1312.8204M ± 0.000E+00M    -112.2872M ± 0.000E+00M    -1.0476M ± 0.000E+00M
LOCAL PLANE INV: DIST =    1317.6136M ±0.000E+00M                  N AZI. = 94 53 19.26 ± 0.0 SEC

1093      Bdy Mon No 93
X =  -1547162.5170    LAT (N+S-)    31 48 46.328421      ±0.000E+00 METERS    N
Y =  -5200718.3250    LON (E+W-)    -106 34 2.004944     ±0.000E+00 METERS    E    STANDARD DEVIATIONS
Z =   3343404.3470    EL HGT        1115.2129 M          ±0.000E+00 METERS    U

DELTA X/Y/Z WITH SIGMAS    57.2580M ± 0.000E+00M    11.7430M ± 0.000E+00M    44.7720M ± 0.000E+00M
DELTA E/N/U WITH SIGMAS    51.5326M ± 0.000E+00M    52.5859M ± 0.000E+00M    0.1629M ± 0.000E+00M
LOCAL PLANE INV: DIST =    73.6266M ±0.000E+00M                  N AZI. = 44 25 13.42 ± 0.0 SEC

1094      Bdy Mon No 94
X =  -1547105.2590    LAT (N+S-)    31 48 48.035397      ±0.000E+00 METERS    N
Y =  -5200706.5820    LON (E+W-)    -106 34 0.045956     ±0.000E+00 METERS    E    STANDARD DEVIATIONS
Z =   3343449.1190    EL HGT        1115.3762 M          ±0.000E+00 METERS    U

**FIGURE 15.11 (Continued)** BURKORD™ Printout for Coordinates and Inverses Point to Point

```
DELTA X/Y/Z WITH SIGMAS      313.2670M ± 0.000E+00M    -144.0190M ± 0.000E+00M    -78.6310M ± 0.000E+00M
DELTA E/N/U WITH SIGMAS      341.3271M ± 0.000E+00M     -92.5004M ± 0.000E+00M     -0.0512M ± 0.000E+00M
LOCAL PLANE INV: DIST =      353.6390M ± 0.000E+00M          N AZI. = 105 9 47.07 ± 0.0 SEC

    1095      Bdy Mon No 95
X =  -1546791.9920   LAT (N+S-)    31 48 45.032578    ±0.000E+00 METERS   N
Y =  -5200850.6010   LON (E+W-)   -106 33 47.070680   ±0.000E+00 METERS   E    STANDARD DEVIATIONS
Z =   3343370.4880   EL HGT        1115.3349 M        ±0.000E+00 METERS   U


DELTA X/Y/Z WITH SIGMAS        8.5330M ± 0.000E+00M     -85.4680M ± 0.000E+00M   -128.2860M ± 0.000E+00M
DELTA E/N/U WITH SIGMAS       32.5434M ± 0.000E+00M    -150.9166M ± 0.000E+00M     -0.0768M ± 0.000E+00M
LOCAL PLANE INV: DIST =      154.3855M ± 0.000E+00M          N AZI. = 167 49 52.25 ± 0.0 SEC

    1096      Bdy Mon No 96
X =  -1546783.4590   LAT (N+S-)    31 48 40.133703    ±0.000E+00 METERS   N
Y =  -5200936.0690   LON (E+W-)   -106 33 45.833589   ±0.000E+00 METERS   E    STANDARD DEVIATIONS
Z =   3343242.2020   EL HGT        1115.2600 M        ±0.000E+00 METERS   U
```

FIGURE 15.11 (Continued)  BURKORD™ Printout for Coordinates and Inverses Point to Point

## TABLE 15.11
## Comparison of Observed Vectors with Record Inverses

*Mon #90 to Mon #93*

|  | ΔX (m) | ΔY (m) | ΔZ (m) | Mark to Mark Distance |  |
|---|---|---|---|---|---|
| Record | 1,241.591 | −430.521 | −95.970 | 1,317.614 m |  |
| Observed | 1,241.509 | −430.477 | −96.031 | 1,317.527 m |  |
|  |  |  | Differ. | −0.087 m | or **1:15,144** |

|  | Δe (m) | Δn (m) | Δu (m) | Hor. Dist. | 3-D Azimuth |
|---|---|---|---|---|---|
| Record | 1,312.820 | −112.287 | −1.048 | 1,317.614 m | 94° 53′ 19.″26 |
| Observed | 1,312.729 | −112.329 | −1.096 | 1,317.526 m | 94° 53′ 27.″03 |
|  |  | Differ. | 0.048 | −0.088 m | 7.″77 |

*Mon #93 to Mon #94*

|  | ΔX (m) | ΔY (m) | ΔZ (m) | Mark to Mark Distance |  |
|---|---|---|---|---|---|
| Record | 57.258 | 11.743 | 44.772 | 73.627 m |  |
| Observed | 57.096 | 11.740 | 44.772 | 73.501 m |  |
|  |  |  | Differ. | −0.126 m | or **1:584** |

|  | Δe (m) | Δn (m) | Δu (m) | Hor. Dist. | 3-D Azimuth |
|---|---|---|---|---|---|
| Record | 51.533 | 52.586 | 0.163 | 73.627 m | 44° 25′ 13.″42 |
| Observed | 51.379 | 52.560 | 0.204 | 73.501 m | 44° 20′ 55.″54 |
|  |  | Differ. | 0.041 | −0.126 m | −4′ 17.″88 |

*Mon #94 to Mon #95*

|  | ΔX (m) | ΔY (m) | ΔZ (m) | Mark to Mark Distance |  |
|---|---|---|---|---|---|
| Record | 313.267 | −144.019 | −78.631 | 353.639 m |  |
| Observed | 313.292 | −143.907 | −78.485 | 353.583 m |  |
|  |  |  | Differ. | −0.056 m | or **1:6,315** |

|  | Δe (m) | Δn (m) | Δu (m) | Hor. Dist. | 3-D Azimuth |
|---|---|---|---|---|---|
| Record | 341.327 | −92.500 | −0.051 | 353.639 m | 105° 09′ 47.″07 |
| Observed | 341.319 | −92.316 | −0.071 | 353.583 m | 105° 08′ 04.″29 |
|  |  | Differ. | −0.020 | −0.056 m | −1′ 42.″78 |

*Mon #95 to Mon #96*

|  | ΔX (m) | ΔY (m) | ΔZ (m) | Mark to Mark Distance |  |
|---|---|---|---|---|---|
| Record | 8.533 | −85.468 | −128.286 | 154.386 m |  |
| Observed | 8.538 | −85.477 | −128.261 | 154.370 m |  |
|  |  |  | Differ. | −0.016 m | or **1:9,648** |

|  | Δe (m) | Δn (m) | Δu (m) | Hor. Dist. | 3-D Azimuth |
|---|---|---|---|---|---|
| Record | 32.543 | −150.917 | −0.077 | 154.386 m | 167° 49′ 52.″25 |
| Observed | 32.551 | −150.899 | −0.058 | 154.370 m | 167° 49′ 37.″73 |
|  |  | Differ. | 0.019 | −0.016 m | −14.″52 |

**FIGURE 15.12**   **(See color insert.)** Picture of Monument #93 on New Mexico—Texas Boundary

**FIGURE 15.13**   **(See color insert.)** Picture of Monument #94 on New Mexico—Texas Boundary

**TABLE 15.12**

**Comparison of Elevation Differences (from Burkholder 1997b)**

HARN Control Stations Published by NGS

| | (PT. 4410) | WEST BEND GPS | PID OM1264 |
| | (PT. 4412) | MILWAUKEE GPS | PID OM1261 (This station has since been destroyed.) |

USPLSS Corners

| | (PT. 35) | NE Corner, Section 8, T9N-R21E | SEWRPC designation = NE 8-9-21 |
| | (PT. 36) | N ¼ Corner, Section 8, T9N-R21E | SEWRPC designation = N ¼ 8-9-21 |

| Stations From–To | Distance | GNSS $\Delta H$ | Pub. $\Delta H$ | Differ. Meters | Differ. Feet | Coefficient Feet × $\sqrt{\text{Miles}}$ |
|---|---|---|---|---|---|---|
| 35 – 36 | 0.499 mi. | −1.151 m | −1.151 m | 0.000 | 0.000 | 0.000 |
| 35 – 4410 | 12.15 mi. | 13.923 m | 13.929 m | 0.006 | 0.020 | 0.006 |
| 35 – 4412 | 15.43 mi. | −21.555 m | −21.540 m | −0.015 | −0.049 | −0.013 |
| 36 – 4410 | 11.94 mi. | 15.074 m | 15.080 m | 0.006 | 0.020 | 0.006 |
| 36 – 4412 | 15.41 mi. | −20.404 m | −20.389 m | −0.015 | −0.049 | −0.013 |
| 4410 – 4412 | 26.70 mi. | −35.478 m | −35.469 m | −0.009 | −0.030 | −0.006 |

### Example 7 in Wisconsin—Leveling in the Context of the GSDM (Example in Wisconsin)

Appendix E in this book describes the critical role that the Southeastern Wisconsin Regional Planning Commission (SEWRPC) played in development of the GSDM. One of the projects completed for SEWRPC included preparation of a comprehensive report describing how the GSDM could be implemented by SEWRPC (Burkholder 1997b). Pages 57, 58, and 59 of that report describe a control leveling project that included using two HARN stations located about 27 miles apart, two USPLSS corners—separated by half a mile—located about halfway between the HARN stations, GNSS observations, geoid modeling software, and the GSDM to determine the elevation of the USPLSS corners. The combined GNSS and geoid modeling differences were compared to the known second-order elevation differences as shown in Table 15.12. Although not strictly an apples-with-apples comparison, the results listed in the report fall within FGCS first-order leveling criteria.

### Example 8—Determining the NAVD 88 Elevation of HARN Station REILLY

Station REILLY on the NMSU campus was established in 1992 as a HARN station. The position was reobserved in 2000 as part of a network upgrade and NAD 83 (1992) geocentric $X/Y/Z$ coordinates were published by NGS. Although station REILLY is central to the NMSU campus, there is no formal first- or second-order leveling connection to station REILLY from a local line of first-order bench marks; see Figure 15.14. As part of a 2005 class project, student Melvin Pyeatt collected

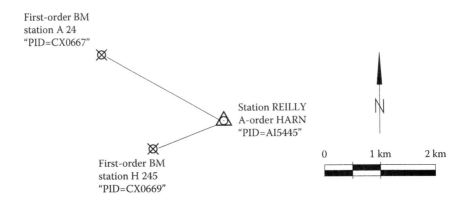

**FIGURE 15.14**   Diagram of GNSS Leveling Network for Station REILLY

GNSS data on two first-order bench marks—"A 245" and "H 245"—and station REILLY. Trivial vectors were carefully avoided and the observed baselines were used to compute a least squares adjustment of the resulting triangle. The ECEF coordinates of the two bench marks were computed holding the ECEF coordinates of station REILLY. GEOID03 was the appropriate geoid model at the time and was used to compute geoid heights at all three stations. The results of that project are reported at http://www.globalcogo.com/ReilElevA.pdf, accessed May 4, 2017. But, that is only part of the story.

Since 2005, the NGS has published improvements to the GEOID03 model known as GEOID09, GEOID12A, and GEOID12B (note—in many, but not all cases, GEOID12A and GEOID12B provide identical results). Furthermore, the underlying NAD 83 (1992) datum was readjusted and published as NAD 83 (2007) and again as NAD 83 (2011). Using the same vectors observed in 2005, the network was recomputed using the NAD 83 (2011) values for station REILLY and using GEOID12B. Note that the NAVD 88 elevations of the two first-order bench marks were not changed during the time interval in question. The "rest of the story" needs to show the impact of those improvements. Following is a comparison taken from Burkholder (2015):

| Station REILLY | Published in 2005 | Published in 2015 | Diff. 2015–2005 |
|---|---|---|---|
| $X =$ | −1,556,177.615 m | −1,556,177.595 m | 0.020 m |
| $Y =$ | −5,169,235.319 m | −5,169,235.284 m | 0.035 m |
| $Z =$ | 3,387,551.709 m | 3,387,551.720 m | −0.011 m |
| **NAVD88 Elevation** | **Published in 2005** | **Published in 2015** | **Diff. 2015–2005** |
| Bench Mark A 245 | 1,186.626 m | 1,186.626 m | 0.000 m |
| Bench Mark H 245 | 1,183.102 m | 1,183.102 m | 0.000 m |
| **Geoid Height at:** | **GEOID03** | **GEOID12B** | **Diff. 2015–2005** |
| Station REILLY = | −23.905 m | −23.943 m | −0.038 m |
| Station A 245 = | −23.957 m | −23.999 m | −0.042 m |
| Station H 245 = | −23.954 m | −23.993 m | −0.039 m |

A least squares network adjustment of the observed GPS vectors gives

| Station A 245 | Computed 2005 | Computed 2015 | Diff. 2015–2005 |
|---|---|---|---|
| X = | −1,558,114.588 m ± 0.0016 m | −1,558,114.568 m ± 0.0016 m | 0.020 m |
| Y = | −5,168,006.589 m ± 0.0042 m | −5,168,006.554 m ± 0.0042 m | 0.035 m |
| Z = | 3,388,522.031 m ± 0.0027 m | 3,388,522.042 m ± 0.0027 m | 0.011 m |
| Station H 245 | | | |
| X = | −1,557,508.610 ± 0.0012 m | −1,557,508.590 ± 0.0012 m | 0.020 m |
| Y = | −5,169,122.541 ± 0.0029 m | −5,169,122.506 ± 0.0029 m | 0.035 m |
| Z = | 3,387,101.071± 0.0020 m | 3,387,101.082 m ± 0.0020 m | 0.011 m |

Note that the ECEF coordinate differences are identical for all three stations, REILLY, A 245, and H 245. The standard deviation of the $X/Y/Z$ values at station REILLY were assumed to be zero, both in 2005 and in 2015. The standard deviations for A 245 and H 245 are the same in the 2015 adjustment and they were in the 2005 adjustment.

The 3-D coordinate geometry and error propagation software, BURKORD™ was used to compute the NAD 83 (2011) latitude/longitude/ellipsoid height at each point. Input includes the geocentric $X/Y/Z$ coordinates and the covariance matrix of each station in the geocentric reference frame. BURKORD™ output includes local e/n/u standard deviations as well as the latitude/longitude/height at each point. The (derived) results are as follows:

**NAD 83 (2011) Station REILLY (Fixed):**

| Latitude | = | 32° 16′ 55.″93001 N | (N) ±0.000 m |
|---|---|---|---|
| Longitude | = | 106° 45′ 15.″16035 W | (E) ±0.000 m |
| Ellipsoid height | = | 1,166.5429 m | (U) ±0.000 m |

**NAD 83 (2011) Station A 245**

| Latitude | = | 32° 17′ 33.″26573 N | (N) ±0.0012 m |
|---|---|---|---|
| Longitude | = | 106° 46′ 39.″57079 W | (E) ±0.0009 m |
| Ellipsoid height | = | 1,162.6198 m | (U) ±0.0050 m |

**NAD 83 (2011) Station H 245**

| Latitude | = | 32° 16′ 38.″78204 N | (N) ±0.0012 m |
|---|---|---|---|
| Longitude | = | 106° 46′ 05.″09654 W | (E) ±0.0011 m |
| Ellipsoid height | = | 1,159.0942 m | (U) ±0.0034 m |

Two methods are available when using ECEF coordinates and ellipsoid heights to compute the orthometric height of a point. The ellipsoid height computed from ECEF coordinates can be combined with the absolute value of geoid height obtained from a geoid model to compute an orthometric height. But, as discussed in Chapter 9 and used in the Wisconsin example, better results can be obtained using relative ellipsoid height differences and relative geoid height differences to find relative orthometric height differences. A relative orthometric height difference is combined with the known elevation at one point to find the unknown orthometric height at another point. With reference to Figure 15.15, that is the process used to compute the NAVD 88 orthometric height of station REILLY from two first-order bench marks in the area.

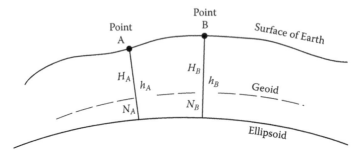

**FIGURE 15.15** Diagram Showing Computation of Height Differences by GNSS

Given:

Known elevation at Point A = $H_A$.
GNSS ellipsoid heights at Points A and B are $h_A$ and $h_B$.
Geoid heights at Points A and B are $N_A$ and $N_B$.

Find: Elevation (orthometric height) at Point B.
Solution:

$\Delta h = h_B - h_A$  $\left(\text{from GNSS results—NAD 83 }(1992)\text{ and NAD 83 }(2011)\right)$

$\Delta N = N_B - N_A$  $\left(\text{from GEOID03 and GEOID12B}\right)$

$\Delta H = \Delta h - \Delta N$ and $H_B = H_A + \Delta H$

The 2015 values and 2005 values are shown for comparison.
Observed Orthometric Height Difference between Published Bench Marks

|  |  | 2005 | 2015 |
|---|---|---|---|
| $\Delta h = h_{StaH} - h_{StaA}$ | = 1,159.1217 m − 1,162.6493 m = | −3.5276 m | |
|  | = 1,159.0942 m − 1,162.6198 m = |  | −3.5256 m |
| $\Delta N = N_{StaH} - N_{StaA}$ | = −23.954 m − (−23.957 m) = | 0.003 m | |
|  | = −23.993 m − (−23.999 m) = |  | 0.006 m |
| $\Delta H = \Delta h - \Delta N$ | = −3.5276 m − 0.003 m = | −3.531 m | |
|  | = −3.5256 m − 0.006 m = |  | −3.532 m |

Elevation at Station REILLY from Station A 245

|  |  | 2005 | 2015 |
|---|---|---|---|
| $\Delta h = h_{REILLY} - h_{StaA}$ | = 1,166.5703 m − 1,162.6493 m = | 3.9210 m | |
|  | = 1,166.5429 m − 1,162.6198 m = |  | 3.9231 m |
| $\Delta N = N_{REILLY} - N_{StaA}$ | = −23.905 m − (−23.957 m) = | 0.052 m | |
|  | = −23.943 m − (−23.999 m) = |  | 0.056 m |
| $\Delta H = \Delta h - \Delta N$ | = 3.9210 m − 0.052 m = | 3.869 m | |
|  | = 3.9231 m − 0.056 m = |  | 3.867 m |
| Elevation at Station REILLY | = 1,186.626 m + 3.869 m | **1,190.495 m** | |
|  | = 1,186.626 m + 3.867 m = |  | **1,190.493 m** |

### Elevation at Station REILLY from Station H 245

|  |  | 2005 | 2015 |
|---|---|---|---|
| $\Delta h = h_{\text{REILLY}} - h_{\text{StaH}}$ | $= 1{,}166.5703 \text{ m} - 1{,}159.1217 \text{ m} =$ | 7.4486 m | |
| | $= 1{,}166.5429 \text{ m} - 1{,}159.0942 \text{ m} =$ | | 7.4487 m |
| $\Delta N = N_{\text{REILLY}} - N_{\text{H}}$ | $= -23.905 \text{ m} - (-23.954 \text{ m}) =$ | 0.049 m | |
| | $= -23.943 \text{ m} - (-23.993 \text{ m}) =$ | | 0.050 m |
| $\Delta H = \Delta h - \Delta N$ | $= 7.449 \text{ m} - 0.049 \text{ m} =$ | 7.400 m | |
| | $= 7.449 \text{ m} - 0.050 \text{ m} =$ | | 7.399 m |
| Elevation at Station REILLY | $= 1{,}183.102 \text{ m} + 7.400 \text{ m} =$ | 1,190.502 m | |
| | $= 1{,}183.102 \text{ m} + 7.399 \text{ m} =$ | | 1,190.501 m |
| Average of the two 2015 values is: $(1{,}190.493 \text{ m} + 1{,}190.501 \text{ m})/2 =$ | | 1,190.498 m | |
| Average of the two 2005 values is: $(1{,}190.495 \text{ m} + 1{,}190.502 \text{ m})/2 =$ | | | 1,190.497 m |

The difference between averages of 2005 and 2015 values is 0.001 meter. Based upon that comparison, it is reasonable to conclude that computing orthometric heights with NAD 83 (XXXX) geocentric coordinates and a consistent geoid model can be successful.

**But the story is not yet complete.**

Additional observations, comments, and questions include the following:

1. Similar computations and comparisons were made for GEOID09 used along with NAD 83 (2007) values for station REILLY. Those results were not consistent with the comparisons presented.
2. A separate comparison could be informative. It has been estimated that absolute geoid modeling in the past (i.e., GEOID03) could not be expected to be better than about 2 cm but that the accuracy of the newer geoid models (GEOID12B) is better—say about 1 cm.
3. The NAVD 88 elevations computed and compared herein are based upon using geoid height *differences*. As described in Chapter 9, using relative differences is better than using the absolute modeled geoid height value at a single station. What would be the NAVD 88 elevation of REILLY using the absolute GEOID03 model and the GEOID12B values?
4. Those values are computed and compared to the average of 1,190.498 m as reported herein. If using the absolute geoid height value from the geoid model, the equation for the NAVD 88 elevation for REILLY is

$$\text{NAVD88 elevation } H_{\text{REILLY}} = h_{\text{REILLY}} - N_{\text{REILLY}}$$

| | Absolute N | Relative N | Comparison |
|---|---|---|---|
| **2005 Data (GEOID03):** | | | |
| $H_{2005} = 1{,}166.570 \text{ m} - (-23.905 \text{ m}) =$ | 1,190.475 m | 1,190.498 m | 0.023 m |
| **2015 Data GEOID12B:** | | | |
| $H_{2015} = 1{,}166.543 \text{ m} - (-23.943 \text{ m}) =$ | 1,190.486 m | 1,190.498 m | 0.012 m |

If this comparison is legitimate, the evidence clearly shows that GEOID12B is significantly better than GEOID03 when using only the absolute GEOID12B values at a station along with the ellipsoid height.

*And the story is still not complete…*

The observed orthometric height difference between the two first-order bench marks was remarkably consistent (within 1 mm) as computed in the two comparisons. But, here is the problem. The observed difference in orthometric height fails to match the published difference by 0.008 m. Coincidentally, the difference in orthometric height computed for station REILLY from those two bench marks differs by a similar amount (0.0075 m). Has either of those bench marks moved since being established? Additional ties to stable bench marks are needed to answer that question. Regretfully, no other recoverable first-order bench marks close enough to be completed as a student project could be found. Budgets are tight and efforts to obtain funding for additional control leveling in the area have not yet been successful.

EXAMPLE 9—DETERMINING THE SHADOW HEIGHT
AT A PROPOSED NEXRAD INSTALLATION

Driving east on Interstate 10 from Tucson, Arizona, and before reaching Benson, the alert motorist will notice a NEXRAD weather radar dome visible above the mountain range southeast of the highway—see Figure 15.16. One criterion when siting the NEXRD facility was to avoid "polluting" the atmosphere with unwanted electronic emissions that might have a detrimental impact on the observations made at any of the various astronomical observatories in the area. In particular, for the site selected, an optical line-of-sight from the radar dome to the observatories located on Mt. Graham had to be avoided. That meant that the NEXRAD dome needed to be built in the shadow of a blocking mountain. A GNSS survey was conducted with data collected on appropriate control points, at the observatory on Mt. Graham, on the peak of the blocking mountain, and at the proposed site. With the network observed and computed, the next phase was to collect additional data to nail down an acceptable site. At each trial location, the current observations were combined with the existing network to answer the question, "How high is the shadow here?" Performing those computations on-site was facilitated by using the ECEF rectangular environment to establish a mathematical plane through the ellipsoid normal on the blocking mountain using the center of the observatory on Mt. Graham as a backsight. That plane was extended to the proposed site and simple computations were performed (with a HP 45 calculator) to determine the shadow height—see Figure 15.17. Once the plane from the top of Mt. Graham through the top of the blocking mountain was established, two separate computations gave essentially the same answer. First, a perpendicular-offset from the line connecting Mt. Graham and the blocking mountain to the proposed site on the ground provided an excellent approximation. Next, a line-line intersection of the baseline from the top of Mt. Graham with the ellipsoid normal at the proposed site provided essentially the same answer. This is because the ellipsoid normal at the proposed site is almost perpendicular to said baseline. An interesting description of the project is available in Schurian, Hodges, and Burkholder (1997).

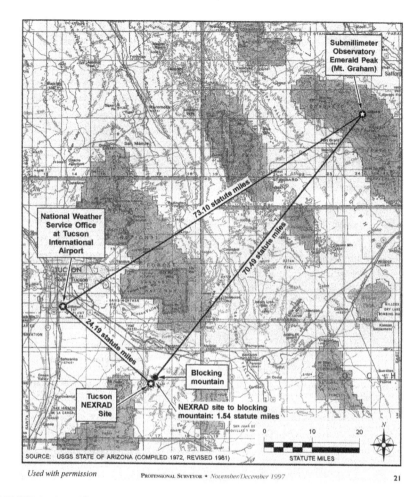

**FIGURE 15.16** **(See color insert.)** Map of NEXRAD Project Near Tucson, AZ

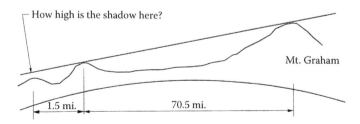

**FIGURE 15.17** Profile View Showing Geometry of Shadow Height Computation

## EXAMPLE 10—COMPARISON OF 3-D COMPUTATIONAL MODELS

The following example consists of a simple observation with a total station instrument set up over station REILLY on the NMSU campus, backsighting station FINIAL, and shooting a single side shot (through a window) to a retro-reflector sitting on the desk of the NMSU Associate Dean of Engineering. Those data were used to compute the latitude, longitude, and orthometric height of the top of the Dean's desk by three different methods—geodetic, state plane, and geocentric (Burkholder 2016). The answer is probably of no consequence to anyone, except maybe the Dean, but much can be learned from a comparison of the three methods.

Briefly, the advantages of the 3-D method as discussed here are as follows:

- Computations are performed in 3-D space, not on the ellipsoid and not on a mapping grid.
- Various geodesy reductions (some involving approximations) are avoided.
- No zone parameters, grid scale factors, or combination factors are needed.
- The 3-D computations use measured quantities directly: HI, HT, horizontal angle, vertical (or zenith) angles, and slope distance.
- The GSDM preserves the 3-D geometrical integrity of spatial data computations without distorting measured angles or distances.

The actual 3-D computations are summarized as

$$\phi_{desk} = \text{Computed from } X/Y/Z \text{ geocentric ECEF coordinate values} \qquad (15.10)$$

$$\lambda_{desk} = \text{Computed from } X/Y/Z \text{ geocentric ECEF coordinate values} \qquad (15.11)$$

$$H_{desk} = h_{Desk} - N_{desk}; h \text{ from } X/Y/Z \text{ values and } N \text{ from GEOID12B} \qquad (15.12)$$

$$X_{desk} = X_{REILLY} + \Delta X \qquad (15.13)$$

$$Y_{desk} = Y_{REILLY} + \Delta Y \qquad (15.14)$$

$$Z_{desk} = Z_{REILLY} + \Delta Z \qquad (15.15)$$

$$\Delta X = -\Delta e \sin \lambda - \Delta n \sin \phi \cos \lambda + \Delta u \cos \phi \cos \lambda \qquad (15.16)$$

$$\Delta Y = \Delta e \cos \lambda - \Delta n \sin \phi \sin \lambda + \Delta u \cos \phi \sin \lambda \qquad (15.17)$$

$$\Delta Z = \Delta n \cos \phi + \Delta u \sin \phi \qquad (15.18)$$

*Note*: Equations 15.16 through 15.18 use north latitude and east longitude at Station REILLY.

$$\Delta e = SD \sin Z \sin \alpha_{Geo} \qquad (15.19)$$

$$\Delta n = SD \sin Z \cos \alpha_{Geo} \tag{15.20}$$

$$\Delta u = SD \cos Z + HI - HT \tag{15.21}$$

where:
SD = slope distance.
Z = zenith direction to reflector.

$\alpha_{Geo}$ = azimuth to reflector on desk = $\alpha_{BS}$ + angle right. $\tag{15.22}$

$\alpha_{BS}$ = azimuth from station REILLY to station FINIAL = 272° 10′ 51″.

NAD 83 (2011) values for station REILLY:

| | |
|---|---|
| $X = -1,556,177.595$ m | $\phi = 32°\ 16'\ 55."93001$ N |
| $Y = -5,169,235.284$ m | $\lambda = 106°\ 45'\ 15."16035$ W |
| $Z = 3,387,551.720$ m | $\lambda = 253°\ 14'\ 44."83965$ E |

Measurements:

| | |
|---|---|
| EDM slope distance to retro-reflector on Dean's desk = | 78.452 m |
| Height of instrument at REILLY = | 1.682 m |
| Height of retro-reflector = | 0.366 m |
| Angle right from station FINIAL = | 269° 23′ 08″ |
| Zenith direction to retro-reflector = | 090° 54′ 08″ |

Computations:

$$\alpha_{Geo} = 272°\ 10'\ 51'' + 269°\ 23'\ 08'' - 360° = 181°\ 33'\ 59''$$

$$\Delta e = 78.452 \sin\left(90°\ 54'\ 08''\right) \times \sin(181°\ 33'\ 59'') = -2.144 \text{ m}$$

$$\Delta n = 78.452 \sin\left(90°\ 54'\ 08''\right) \times \cos(181°\ 33'\ 59'') = -78.413 \text{ m}$$

$$\Delta u = 78.452 \cos\left(90°\ 54'\ 08''\right) + 1.682 - 0.366 = 0.081 \text{ m}$$

$$\begin{aligned}
\Delta X = &-\left(-2.144\right)\sin\left(253°\ 14'\ 44."83965\right) - \left(-78.413\right)\sin\left(32°\ 16'\ 55."93001\right) \\
&\times \cos\left(253°\ 14'\ 44."83965\right) + 0.081\cos\left(32°\ 16'\ 55."93001\right) \\
&\times \cos\left(253°\ 14'\ 44."83965\right) = -14.145 \text{ m}
\end{aligned}$$

$$\Delta Y = (-2.144)\cos(253°\ 14'\ 44."83965) - (-78.413)\sin(32°\ 16'\ 55."93001)$$
$$\times \sin(253°\ 14'\ 44."83965) + 0.081\cos(32°\ 16'\ 55."93001)$$
$$\times \sin(253°\ 14'\ 44."83965) = -39.549\ \text{m}$$

$$\Delta Z = -78.413\cos(32°\ 16'\ 55."93001) + 0.081\sin(32°\ 16'\ 55."93001) = -66.2459\ \text{m}$$

$$X_{desk} = -1,556,177.595 + (-14.145) = -1,556,191.740\ \text{m}$$

$$Y_{desk} = -5,169,235.284 + (-39.549) = -5,169,274.833\ \text{m}$$

$$Z_{desk} = 3,387,551.720 + (-66.249) = 3,387,485.471\ \text{m}$$

These geocentric $X/Y/Z$ values need to be converted to latitude, longitude, and ellipsoid height (and to elevation). The longitude computation is very straightforward but the latitude and ellipsoid height computations are more challenging. Techniques such as one of those described in Meyer (2010) can be used with excellent results. But, the iteration method used here provides fully rigorous results with fewer mathematical gymnastics.

Since the geocentric $X$ and $Y$ values are both negative, the east longitude lies in the third quadrant of the equator and is computed (with due regard to radian units) as

$$\lambda = 180° + a\tan\left(\frac{Y}{X}\right) = \text{East longitude} \tag{15.23}$$

$$\text{Longitude} = 180° + a\tan\left(\frac{-5,169,274.833}{-1,556,191.740}\right) = 253°\ 14'\ 44."75773\ \text{E}$$
$$= 106°45'15."24227\ \text{W}$$

Closed form equations for computing geocentric $X/Y/Z$ coordinates from latitude, longitude, and ellipsoid height are called a BK1 transformation in Burkholder (2008) and given as

$$X = (N+h)\cos\phi\cos\lambda \tag{15.24}$$

$$Y = (N+h)\cos\phi\sin\lambda \tag{15.25}$$

$$Z = \left[N(1-e^2)+h\right]\sin\phi \tag{15.26}$$

where:
$N$ = radius of curvature in Prime Vertical.
$h$ = ellipsoid height.

A mathematical inversion of Equations 15.24 through 15.26 can be used to compute the latitude and ellipsoid height from the $X/Y/Z$ coordinates. That inversion is also closed form but must be solved using iteration. Solving those inverted equations is referred to as a BK2 transformation and given in Chapter 7 as

$$P = \sqrt{X^2 + Y^2}, \quad \text{an intermediate value} \tag{15.27}$$

$$\phi_0 = \arctan\left(\frac{Z}{P\left(1 - e^2\right)}\right), \quad \text{"seed" value for subsequent use} \tag{15.28}$$

$$N_0 = \frac{a}{\sqrt{1 - e^2 \sin^2 \phi_0}}, \quad \text{needed in next step} \tag{15.29}$$

$$\phi_i = \arctan\left[\frac{Z}{P}\left(1 + \frac{e^2 N_0 \sin \phi_0}{Z}\right)\right], \quad \text{second and subsequent values} \tag{15.30}$$

$$N_i = \frac{a}{\sqrt{1 - e^2 \sin^2 \phi_i}}, \quad \text{second and subsequent values} \tag{15.31}$$

Once the solution for latitude has converged sufficiently, the ellipsoid height is computed using the latest values of $\phi$ and $N$:

$$h = \frac{P}{\cos \phi} - N \tag{15.32}$$

Using a spreadsheet, the latitude for the top of the Dean's desk is computed as

| Iteration | Latitude (rad) | Difference | Normal | Difference |
|---|---|---|---|---|
| 0 | 0.563418945242 | | 6,384,570.81481 m | |
| 1 | 0.563418550755 | −0.000000394487 | 6,384,235.29594 m | −335.51888 m |
| 2 | 0.563418390041 | −0.000000160715 | 6,384,235.29283 m | −0.00311 m |
| 3 | 0.563418389270 | −0.000000000770 | 6,384,235.29281 m | −0.00002 m |
| 4 | 0.563418389267 | −0.000000000004 | 6,384,235.29281 m | −0.00000 m |

The latitude is computed by converting radians to degrees-minutes-seconds using the conversion $spr = 206{,}264.806247096$ seconds/radian:

$$\phi = 0.563418389267 \times 206{,}264.806247 = 116{,}213.''384899 = 32° \, 16' \, 53.''38490$$

Twelve decimal places of latitude or longitude in radians translates to 6 decimal places of seconds when expressed as degrees, minutes, and seconds. More decimal places are included in the latitude tabulation than can be justified. This is done to show where differences begin to occur. It is safer to use more iterations than needed than to stop the iteration prematurely. Good judgment is essential in reporting and interpreting results. In this case, 5 decimal places of seconds for latitude and longitude (0."00001 $\cong$ 0.0003 m) is deemed sufficient.

To compute the orthometric height of the top of the Dean's desk, the ellipsoid height must be converted to elevation. In this case, the value of geoid height at a point as obtained from GEOID12B is used. Using NGS interactive software GEOID12B, the value of geoid height on the Dean's desk was found to be −23.944 m.

$$H_{desk} = h_{Desk} - \left(\text{geoid height}_{desk}\right) = 1,166.624 - \left(-23.944\right) \text{ m} = 1,190.578 \text{ m} \quad (15.33)$$

Acknowledged disadvantages of the 3-D method include the following:

- Computing latitude and ellipsoid height from geocentric $X/Y/Z$ coordinates requires iteration—or using any one of several noniterative methods.
- Geoid modeling is needed in order to get orthometric height from ellipsoid height.

Answers for the 3-D position on top of the Dean's desk are

---

Latitude = 32° 16′ 53.″38490 N
Longitude = 106° 45′ 15.″24227 W
Ellipsoid height = 1,166.624 m
GEOID12B Geoid height = −23.944 m
Orthometric height = 1,190.578 m

---

Given that the desk is not bolted to the floor and is subject to movement, that point on the top of the Dean's desk was never marked!

## EXAMPLE 11—UNDERGROUND MAPPING

Many applications involve the use of 3-D spatial data. In some cases, out of sight means out of mind. There are thousands of wells bored into the Earth and knowledge of those well locations is of great importance to many people. In the past, knowing the surface location of the wellhead of a vertical well may have been sufficient but now that modern technology involves directional drilling, the 3-D location of the drill bits and other features within the well are of increasing importance. The age-old question is—with respect to what? Given the huge vertical extent of a well, a flat-Earth local coordinate system centered on the wellhead is not deemed adequate. True 3-D could be tracked (possible, but onerous) using latitude/longitude/ellipsoid height. Pseudo 3-D can be realized using state plane coordinates and orthometric heights but here again, the huge vertical distances involved go way beyond the legitimate limits of a map projection. The problem is especially acute if and when a location in the well must be related to the ground surface—property lines, roads, utilities, and other features. The GSDM is offered as a solution to the underground mapping challenges.

Following is the abstract to a paper presented at *the ASCE Shale Energy Engineering Conference* in Pittsburgh, PA, July 2014 (Burkholder 2014):

Reliable underground mapping is an essential part of efficient energy production. The challenge is to determine reliable 3-D point positions and/or the trajectory of a well-path. Sensor data, e.g., tilt/distance/azimuth, are typically used to compute the location of discrete points in a well and a wellpath is often determined using the method of minimum curvature. Part of the positioning challenge relates to specific identification of the coordinate origin, orientation of the coordinate system, knowing the quality of measurements, and assumptions about map projections. Historically, survey data were horizontal and vertical but modern measurement systems produce digital 3-D spatial data. The global spatial data model (GSDM) will be described which accommodates modern measurements and 3-D digital spatial data. The GSDM uses proven rules of solid geometry to determine each 3-D position and optionally provides the standard deviation of each computed position. The relative position between any two points is readily available and the underlying coordinate positions can be mathematically transformed to any coordinate system of the user's choice. All uses (site drawings, drilling, well development, production, and regulatory activities) of the position data share a common definition and the positional uncertainty of each computed quantity can be determined from point standard deviations.

Figure 15.18 illustrates the problem with using a map projection as a reference coordinate system for underground mapping. In modern drilling practices, the

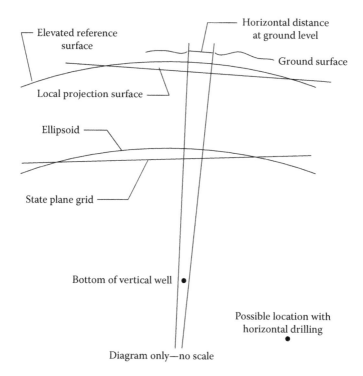

**FIGURE 15.18**  Limited Vertical Extent of Map Projections for Underground Wells

vertical extent of a borehole certainly exceeds the rather narrow vertical range for which a map projection controls the distortion of grid distance compared to actual horizontal distances—at some specified elevation. In directional drilling, the extent of the borehole can easily exceed both the vertical extent and the horizontal extent. The GSDM provides 3-D coordinates of the true location of a mapped point. The user has numerous options available for viewing the data.

### EXAMPLE 12—LAYING OUT A PARALLEL OF LATITUDE USING THE GSDM

A fundamental feature of the USPLSS is that baselines are established as a parallel of latitude. That challenge has been successfully met by various methods in the past and is really no longer an issue—except that surveying on a parallel is still an activity encountered from time to time. Chapter 2 of the 1973 BLM Manual (BLM 1973) contains a description of several methods by which a parallel of latitude may be established. The 2009 Manual (BLM 2009) states in Chapter 3 that "the determination of the alignment of the true latitudinal curve process is described in the record." Although the GSDM greatly facilitates working on a parallel irrespective of BLM considerations, it is convenient to use the NM Initial Point and Baseline in the following example.

The ECEF geocentric $X/Y/Z$ coordinates for the New Mexico Initial Point are published on the NGS Data Sheet—NAD 83 (2011). That "anchor" point is shown in Figure 15.19 as the intersection of the NMPM and the Baseline. Townships of 36 Sections each are laid out east and west of the Principal Meridian and north and south from the Baseline. A geodetic line is the shortest distance between two points on the ellipsoid surface and, at its maximum latitude, crosses the meridian at 90°— see Figures 7.9 and 7.10. With a theodolite (or total station) set up over a point and sighting due east, the prolonged line, to a very close approximation (Burkholder 1997), is a geodetic line. That line is also called a tangent to the parallel of latitude. The distance between a geodetic line and corresponding parallel of latitude is zero at the intersection (Initial Point), is called the tangent-offset, and grows with distance from the instrument station (in this case the Principal Meridian). Since the geometry

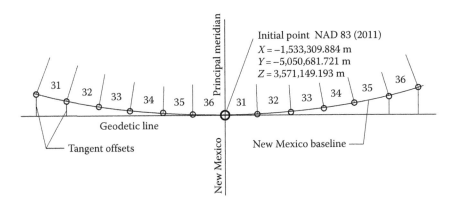

**FIGURE 15.19**   Diagram Showing 3-D Traverse along a Parallel of Latitude

is symmetric with respect to the NMPM, the following example provides numbers only on the east side.

The 1973 Manual (BLM 1973) lists three methods for establishing a true parallel of latitude:

1. The solar method uses an observation at frequent intervals—every 20 or 40 chains—to observe an astronomical azimuth at the current location so that a prolonged line will be with respect to the current meridian. The method is quite good except for logistics of weather, foliage, etc.
2. The tangent method uses precomputing tangent-offset to measure between the geodetic line and the parallel. Estimates for tangent-offsets vary with latitude and distance from the intersection point and are generally obtained from the Standard Field Tables (BLM 1956).
3. Pages 54 and 55 (BLM 1973) states "The designated secant is a great circle which cuts any true parallel of latitude at the first and fifth mile corners ..." "The secant method is recommended for its simplicity of execution and proximity to the true latitude curve ..." A disadvantage of the secant method is heavy reliance on use of the Standard Field Tables.

The GSDM supports computational procedures for all three methods but the following example includes only the first two methods—solar and tangent-offset. Using the GSDM it is easy to precompute local tangent plane coordinates that can be used in modern coordinate computational procedures and applied to either design activities or field layout.

---

Computational Traverse Procedures Included in BURKORD™ Software:

1. Forward by geocentric $\Delta X/\Delta Y/\Delta Z$.
2. Forward by local $\Delta e/\Delta n/\Delta u$ converted to $\Delta X/\Delta Y/\Delta Z$.
   a. Slope distance, zenith angle, and azimuth converted to $\Delta e/\Delta n/\Delta u$.
   b. Slope distance, vertical angle, and azimuth converted to $\Delta e/\Delta n/\Delta u$.
3. Forward by geodetic $\Delta\phi/\Delta\lambda/\Delta h$ converted to $\Delta X/\Delta Y/\Delta Z$.
4. $X_2 = X_1 + \Delta X$; $Y_2 = Y_1 + \Delta Y$; $Z_2 = Z_1 + \Delta Z$.
5. Latitude, longitude, and ellipsoid height at point 2 determined from $X_2/Y_2/Z_2$ using Equations 1.5, 1.6, and 1.7.
6. If standard deviations are used consistently, derived answers can also have standard deviations.

---

### Analogous to Solar Method

With regard to Figure 15.19, a traverse starts at the Initial Point using known ECEF geocentric $X/Y/Z$ coordinates. The first traverse course proceeds 804.674 meters (40 chains) due east from the Initial Point. The second course proceeds 804.674 meters, again on a due east course to the SE Corner of Section 31. The same procedure is repeated for each subsequent section until reaching the SE Corner of Section 36. The ECEF coordinates are stored for each point. Then, using the Initial Point as the P.O.B., local northings and eastings of each point are computed in the tangent plane at the P.O.B. The local northings are an estimate of the tangent-offset to the true parallel.

### Analogous to Tangent-Offset Method

In this case, the longitudinal increment for 80 chains is obtained from the previous method as the difference in longitude between the SE and SW Corners of Section 31. That longitudinal value is used in traversing to each subsequent SE section corner. Latitude difference is zero and ellipsoid height is zero. The points computed in this manner lie on the true parallel. Furthermore, the distance between section corners is ground distance (at elevation of Initial Point) instead of being the distance on the ellipsoid. If ellipsoid distance between section corners is to be held, then the longitudinal difference can be computed on the ellipsoid using Equation 7.87.

Figure 15.20 is a computer printout of the points stored in a BURKORD™ file. The values in Tables 15.11 and 15.12 are obtained from P.O.B. computations of local components with respect to the Initial Point. Those local differences are flat-Earth components (coordinates) in the tangent plane through the P.O.B. Table 15.13 was computed analogous to the Solar Method 40 chains (1/2 mile) at a time. The distance 804.673 m, azimuth 90° (due east) and vertical angle 0° 00' 00." Table 15.14 was computed on the true parallel and provides tangent-offsets from the geodetic line to the true parallel.

| PT No. | X (m) | Y (m) | Z (m) | Covariance Values | PT Designation |
|---|---|---|---|---|---|
| 10 | −1533309.8840 | −5050681.7210 | 3571149.1930 | 0.0 0.0 0.0 0.0 0.0 0.0 | Initial Point |
| 11 | −1532539.9099 | −5050915.4734 | 3571149.1930 | 0.0 0.0 0.0 0.0 0.0 0.0 | S 1/4 Sec 31 |
| 12 | −1531769.9002 | −5051149.1084 | 3571149.1930 | 0.0 0.0 0.0 0.0 0.0 0.0 | SE Section 31 |
| 13 | −1530999.8549 | −5051382.6260 | 3571149.1930 | 0.0 0.0 0.0 0.0 0.0 0.0 | S 1/4 Sec 32 |
| 14 | −1530229.7740 | −5051616.0262 | 3571149.1930 | 0.0 0.0 0.0 0.0 0.0 0.0 | SE Section 32 |
| 15 | −1529459.6576 | −5051849.3090 | 3571149.1930 | 0.0 0.0 0.0 0.0 0.0 0.0 | S 1/4 Sec 33 |
| 16 | −1528689.5055 | −5052082.4744 | 3571149.1930 | 0.0 0.0 0.0 0.0 0.0 0.0 | SE Section 33 |
| 17 | −1527919.3180 | −5052315.5224 | 3571149.1930 | 0.0 0.0 0.0 0.0 0.0 0.0 | S 1/4 Sec 34 |
| 18 | −1527149.0949 | −5052548.4529 | 3571149.1930 | 0.0 0.0 0.0 0.0 0.0 0.0 | SE Section 34 |
| 19 | −1526378.8363 | −5052781.2661 | 3571149.1930 | 0.0 0.0 0.0 0.0 0.0 0.0 | S 1/4 Sec 35 |
| 20 | −1525608.5422 | −5053013.9618 | 3571149.1930 | 0.0 0.0 0.0 0.0 0.0 0.0 | SE Section 35 |
| 21 | −1524838.2127 | −5053246.5401 | 3571149.1930 | 0.0 0.0 0.0 0.0 0.0 0.0 | S 1/4 Sec 36 |
| 22 | −1524067.8477 | −5053479.0009 | 3571149.1930 | 0.0 0.0 0.0 0.0 0.0 0.0 | SE Section 36 |
| 31 | −1531769.8656 | −5051148.9907 | 3571149.1930 | 0.0 0.0 0.0 0.0 0.0 0.0 | True SE Sec. 31 |
| 32 | −1530229.7048 | −5051615.7908 | 3571149.1930 | 0.0 0.0 0.0 0.0 0.0 0.0 | True SE Sec. 32 |
| 33 | −1528689.4018 | −5052082.1213 | 3571149.1930 | 0.0 0.0 0.0 0.0 0.0 0.0 | True SE Sec. 33 |
| 34 | −1527148.9566 | −5052547.9821 | 3571149.1930 | 0.0 0.0 0.0 0.0 0.0 0.0 | True SE Sec. 34 |
| 35 | −1525608.3695 | −5053013.3732 | 3571149.1930 | 0.0 0.0 0.0 0.0 0.0 0.0 | True SE Sec. 35 |
| 36 | −1524067.6406 | −5053478.2946 | 3571149.1930 | 0.0 0.0 0.0 0.0 0.0 0.0 | True SE Sec. 36 |

**FIGURE 15.20**  Computer Printout Showing Computation of Tangent-Offsets

## TABLE 15.13
## Local Plane Coordinates for Tangent-Offsets—Solar Method
P.O.B. Is the Initial Point

| PT No. | Northing | Easting | Up | Description |
|---|---|---|---|---|
| 11 | 0.000 m | 0.000 m | 0.000 m | Initial Point |
| 12 | 0.069 m | 1,609.348 m | −0.101 m | SE Section 31 |
| 14 | 0.414 m | 3,218.696 m | −0.608 m | SE Section 32 |
| 16 | 1.036 m | 4,828.044 m | −1.521 m | SE Section 33 |
| 18 | 1.934 m | 6,437.391 m | −2.840 m | SE Section 34 |
| 20 | 3.108 m | 8,046.737 m | −4.562 m | SE Section 35 |
| 22 | 4.558 m | 9,656.083 m | −6.692 m | SE Section 36 |

## TABLE 15.14
## Local Plane Coordinates for Tangent-Offsets—True Parallel
P.O.B. Is the Initial Point

| PT No. | Northing | Easting | Up | Description |
|---|---|---|---|---|
| 11 | 0.000 m | 0.000 m | 0.000 m | Initial Point |
| 31 | 0.138 m | 1,609.347 m | −0.203 m | SE Section 31 |
| 32 | 0.552 m | 3,218.694 m | −0.822 m | SE Section 32 |
| 33 | 1.243 m | 4,828.040 m | −1.825 m | SE Section 33 |
| 34 | 2.210 m | 6,437.386 m | −3.244 m | SE Section 34 |
| 35 | 3.453 m | 8,046.732 m | −5.069 m | SE Section 35 |
| 36 | 4.972 m | 9,656.076 m | −7.300 m | SE Section 36 |

In this case, the longitudinal difference was 62.88992 seconds of arc per section, the latitude difference was 0° 00′ 00″ and the ellipsoid height difference was 0.000 meter.

## THE FUTURE WILL BE WHAT WE MAKE IT

Chapter 10 contains predictions as found in the first edition. Some predictions have been realized, others have not. Acknowledging the legitimacy of more than one view and recognizing the impossibility of correctly predicting the future, general observations are as follows:

1. The digital revolution is still in its infancy. Additional "futures" are still to be discovered.
2. Learning how to learn (and knowing where to get the information needed) is more important than memorizing details or procedures.
3. Avoiding unintended consequences will continue to be an important criterion in credible decision-making processes.
4. Reduction in privacy is part of the price paid for the luxury of instantaneous convenience.

But spatial data issues coming under the umbrella of more general concepts include the following:

1. The ground-based control network has become obsolete quicker than anticipated. Current practice includes use of a real-time network (RTN) and virtual reference station (VRS). Systems are currently in place and being used in which the end user, armed with a single "connected" instrument, can determine the position of a point within 1–2 cm, or better, based upon satellite observations processed in real time.
2. On the other hand, the ground-based network now consists of a CORS network that provides the integrity needed for successful operation and use of both RTN and VRS operations.
3. Conflicting criteria exist with regard to legal status of "location" of property lines. It is a long-standing principle that the location of the original-undisturbed monument (in the ground) controls the location of property lines and constructed facilities. Conceivably, that concept will never change. On the other hand, it is common knowledge that "everything moves" and the burden of proof for establishing the integrity of the undisturbed monument can be onerous indeed. Issues of absolute position and relative position become integral to successful resolution of such discussions.
4. The National Geodetic Survey (NGS) is responsible for defining, establishing, and providing access to the NSRS. In carrying out that mission, NGS has embarked upon a plan to redefine both the horizontal and vertical datums in the United States by the year 2022. As stated in Chapter 8, Geodetic Datums updates on their progress and status of the same can be reviewed at the NGS web site https://www.ngs.noaa.gov, accessed May 4, 2017.
5. Even with new datums defined and implemented, issues of absolute and relative positioning will need to be addressed. The GSDM provides a consistent environment for addressing those issues.
6. RTN/VRS: An extension of the RTK concept occurs where a CORS network operates in a given area. The network performance is systematically monitored to provide improved positioning capabilities by broadcasting "corrections" to user equipment. This avoids the lack of redundancy encountered in early implementations of RTK surveying. Trimble offers enhanced positioning capability of users of their trademarked Virtual Reference Station (VRS) system in which the system models corrections developed from a CORS network and determines a precise location for a VRS located in proximity to the user position. Two hyperlinks that can be used to obtain additional information are
   a. RTN: https://water.usgs.gov/osw/gps/real-time_network.html, accessed May 4, 2017
   b. VRS: https://www.insidegnss.com/node/2687, accessed May 4, 2017
7. A given consequent of the digital revolution is implementation of a GIS means that land records have been automated and computerized. Society enjoys many benefits of that consequence. But, the future has not yet arrived for those segments of society that cling (however justified) to

outmoded and obsolete processes for handling spatial data. Disruptive innovation forces are inevitable and adjusting to changing conditions is a challenge for many. However, career opportunities for persons contributing to the generation, analysis, and use of spatial data will continue to expand for the foreseeable future. The GSDM provides a unifying standard by which those efforts and data share a common heritage and applications.

http://www.globalcogo.com/DisruptiveInnovation.pdf, accessed May 4, 2017

## REFERENCES

Armstrong, M., R. Singh, and M. L. Dennis. 2014. Oregon Coordinate System—Handbook and user guide. Salem, OR: Oregon Department of Transportation, Highway Division, Geometronics Unit.

Bureau of Land Management (BLM). 1956. Standard field tables and trigonometric formulas. Washington, DC: Bureau of Land Management, U.S. Department of Interior. http://www.ntc.blm.gov/krc/uploads/521/Red_Book.pdf (accessed October 8, 2016).

Bureau of Land Management (BLM). 1973. Manual of surveying instructions 1973. Washington, DC: Bureau of Land Management, U.S. Department of Interior.

Bureau of Land Management (BLM). 2009. Manual of surveying instructions for the survey of the public lands of the United States. Washington, DC: Cadastral Survey, Bureau of Land Management, U.S. Department of Interior.

Burkholder, E.F. 1991. Computation of level/horizontal distances. *Journal of Surveying Engineering* 117 (3): 104–116.

Burkholder, E.F. 1997a. The 3-D azimuth of a GPS vector. *Journal of Surveying Engineering* 123 (4): 139–146.

Burkholder, E.F. 1997b. Definition of a three-dimensional spatial data model for southeastern Wisconsin. Waukesha, WI: Southeastern Wisconsin Regional Planning Commission. http://www.sewrpc.org/SEWRPCFiles/Publications/ppr/definition_three-dimensional_spatial_data_model_for_wi.pdf (accessed May 4, 2017).

Burkholder, E.F. 2014. Underground (well) mapping revisited. *ASCE Shale Energy Engineering Conference Proceedings*, Pittsburgh, PA, July 20–23, 2014. http://globalcogo.com/underground-mapping.pdf (accessed May 4, 2017).

Burkholder, E.F. 2015. The NAVD 88 elevation of station REILLY based upon the HARN values of REILLY in 2005 with Geoid 03 and the HARN adjustment of 2011 with Geoid 12B. *Benchmarks—The Official Publication of the New Mexico Professional Surveyors* 29 (4): 8–14. http://nmps.org/wp-content/uploads/2015/07/BENCHMARKS-july-2015-web.pdf (accessed May 4, 2017).

Burkholder, E.F. 2016. Comparison of geodetic, state plane, and geocentric computational models. *Surveying and Land Information Science* 74 (2): 53–59.

Federal Geographic Data Committee (FGDC). 1998. Geospatial positioning accuracy standards, part 2: Standards for geodetic networks. Reston, VA: Federal Geographic Data Committee, U.S. Geological Survey.

Gannett, S.S., Commissioner. 1930. The State of New Mexico, Complainant vs. The State of Texas, Defendant, Report of the Commissioner. Washington, DC: Supreme Court of the United States, October Term 1930, No. 2, Original, July 17, 1930.

Ghilani, C.D. 2010. *Adjustment Computations Spatial Data Analysis*, 5th ed. Hoboken, NJ: John Wiley.

Leick, A., L. Rapoport, and D. Tatarnikov. 2015. *GPS Satellite Surveying*, 4th ed. Hoboken, NJ: John Wiley.

National Geodetic Survey (NGS). 2016. Data sheet—PID AI5439, "Initial Point". Silver Spring, MD: New Mexico. http://www.ngs.noaa.gov/cgi-bin/ds_pid.prl/1 (accessed May 4, 2017).

Roder, W.E. 1994. *Antepasados—Surveyors in History*. Hobbs, NM: New Mexico Professional Surveyors.

Schurian, D., J. Hodges, and E. Burkholder. November/December 1997. 3D analysis—Siting a NRXRAD weather radar system. *Professional Surveyor* 17 (8): 20–28.

Witcher, T. R. 2016. A fine line: The U.S.–Mexico boundary. *Civil Engineering* 86 (6): 42–45.

**FIGURE 15.4** FINIAL on Skeen Hall—NMSU

**FIGURE 15.12** Picture of Monument #93 on New Mexico—Texas Boundary

**FIGURE 15.13** Picture of Monument #94 on New Mexico—Texas Boundary

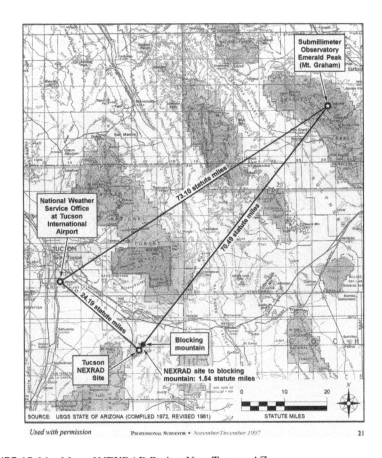

**FIGURE 15.16** Map of NEXRAD Project Near Tucson, AZ

# Appendix A: Rotation Matrix Derivation

A rotation matrix is a collection of equations expressed in matrix form and used to change the perspective associated with spatial data. In the case of the GSDM, the perspective changes from looking at points in the geocentric ECEF reference frame to viewing the same data from the perspective of one occupying a given point (called the point of beginning—P.O.B.) and viewing all other points as if standing at the selected origin.

One way of applying the rotation matrix is to recompute the coordinates of each point with respect to the original origin. Another approach is to apply the rotation matrix to the vector from the P.O.B. to any other point in the database. The GSDM uses the second approach. That means the geocentric coordinate differences are found first. Then the rotation matrix is applied to the ECEF vector components and the result is local "flat-Earth" components from the P.O.B. to the selected point. Although it could be argued that the requirement to compute values of the rotation matrix for each P.O.B. is a disadvantage, several advantages are as follows:

1. The underlying primary data (ECEF coordinates) are not modified and remain available for immediate recomputation—such as selecting a different P.O.B.
2. After using the rotation matrix, the local components of any vector appear to the user as rectangular flat-Earth plane surveying components. These components can be used in a variety of operations as selected by the user.
3. The process is reversible in that local perspective components can also be rotated into the ECEF perspective, making such differences compatible with *X/Y/Z* points already stored in the database. The *X/Y/Z* coordinates of new points are established using simple addition and subtraction operations.
4. If the spatial data user needs more traditional values for a point, such as latitude/longitude or state plane coordinates, those values are also immediately available by rigorous computation (though not involving a rotation matrix) from the geocentric ECEF values for the specified point as stored in the database.

The convention used is that a positive rotation is counterclockwise as viewed looking at the origin from the positive direction along the axis being rotated (Leick 2004). This rule needs to be applied once for each of the three possible axes. This process yields three separate matrices—one for each rotation. This part is generic without regard to being attached to the Earth and each rotation is illustrated in a separate

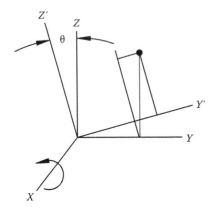

$$X' = X$$

$$Y' = Y \cos \theta + Z \sin \theta$$

$$Z' = -Y \sin \theta + Z \cos \theta$$

In matrix form,

$$\begin{bmatrix} X' \\ Y' \\ Z' \end{bmatrix} = \begin{bmatrix} 1 & 0 & 0 \\ 0 & \cos \theta & \sin \theta \\ 0 & -\sin \theta & \cos \theta \end{bmatrix} \begin{bmatrix} X \\ Y \\ Z \end{bmatrix} \quad (A.1)$$

**FIGURE A.1**   $R_1$ Rotation about the $X$-Axis

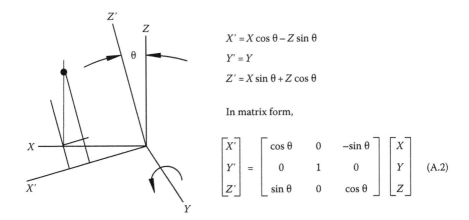

$$X' = X \cos \theta - Z \sin \theta$$

$$Y' = Y$$

$$Z' = X \sin \theta + Z \cos \theta$$

In matrix form,

$$\begin{bmatrix} X' \\ Y' \\ Z' \end{bmatrix} = \begin{bmatrix} \cos \theta & 0 & -\sin \theta \\ 0 & 1 & 0 \\ \sin \theta & 0 & \cos \theta \end{bmatrix} \begin{bmatrix} X \\ Y \\ Z \end{bmatrix} \quad (A.2)$$

**FIGURE A.2**   $R_2$ Rotation about the $Y$-Axis

diagram. Figure A.1 shows a positive rotation ($R_1$) about the $X$-axis. Figure A.2 shows a positive rotation ($R_2$) about the $Y$-axis. And, Figure A.3 shows a positive rotation ($R_3$) about the Z-axis.

With each rotation quantified, the process is applied to vectors (coordinate differences) attached to the Earth. Starting with the $X/Y/Z$ axes of the ECEF reference frame, the first rotation is a positive rotation about the Z-axis ($R_3$) to bring the $Y$-axis into the vertical plane of the local meridian. The angular amount is determined by the longitude of the selected P.O.B. and is computed as east longitude +90°. A full circle can be subtracted if the sum goes over 360°. The second rotation is also a positive rotation about $X$-axis ($R_1$) to bring the $Y$-axis into the local geodetic horizon. The two rotations describe movement of the original $Y$-axis in two rotations—first is a positive rotation about the Z-axis, then a second positive rotation about the $X$-axis. Other rotation sequences could be used to obtain the same result. The original $\Delta X/\Delta Y/\Delta Z$

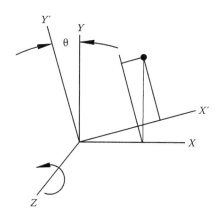

$$X' = X \cos\theta + Y \sin\theta$$

$$Y' = -X \sin\theta + Y \cos\theta$$

$$Z' = Z$$

In matrix form,

$$
\begin{bmatrix} X' \\ Y' \\ Z' \end{bmatrix} =
\begin{bmatrix} \cos\theta & \sin\theta & 0 \\ -\sin\theta & \cos\theta & 0 \\ 0 & 0 & 1 \end{bmatrix}
\begin{bmatrix} X \\ Y \\ Z \end{bmatrix} \quad (A.3)
$$

**FIGURE A.3**   $R_3$ Rotation about the Z-Axis

vector is right handed, and the rotated differences are also right handed if used as $\Delta e/\Delta n/\Delta u$.

Multiplication of the two matrices is shown below. The rules of matrix multiplication require the sequence to be as shown in Equation A.4. In Equation A.5, the actual latitude and longitude of the selected P.O.B. are used, and Equation A.6 is a simplification based upon making substitutions for trigonometric identities. Finally, Equation A.7 is the form of the rotation matrix given in Equation 1.21.

It is stated without proof here that Equation 1.22 uses the transpose of the rotation matrix to convert the local geodetic horizon perspective to the ECEF perspective. Being able to use the transpose for the reverse computation is a consequence of the two right-handed systems, both being orthogonal—see Chapter 3 (Vanicek and Krakiwsky 1986).

Now, using the matrices in Figures A.1 and A.3 above, the process is attached to the Earth in the following matrix statement:

$$
\begin{bmatrix} \Delta e \\ \Delta n \\ \Delta u \end{bmatrix} = R_1\left(90° - \phi\right) R_3\left(\lambda + 90°\right)
\begin{bmatrix} \Delta X \\ \Delta Y \\ \Delta Z \end{bmatrix} \quad (A.4)
$$

$$
\begin{bmatrix} \Delta e \\ \Delta n \\ \Delta u \end{bmatrix} =
\begin{bmatrix} 1 & 0 & 0 \\ 0 & \cos(90°-\phi) & \sin(90°-\phi) \\ 0 & -\sin(90°-\phi) & \cos(90°-\phi) \end{bmatrix}
\begin{bmatrix} \cos(\lambda+90°) & \sin(\lambda+90°) & 0 \\ -\sin(\lambda+90°) & \cos(\lambda+90°) & 0 \\ 0 & 0 & 1 \end{bmatrix}
\begin{bmatrix} \Delta X \\ \Delta Y \\ \Delta Z \end{bmatrix}
$$

$$(A.5)$$

$$
\begin{bmatrix} \Delta e \\ \Delta n \\ \Delta u \end{bmatrix} =
\begin{bmatrix} 1 & 0 & 0 \\ 0 & \sin\phi & \cos\phi \\ 0 & -\cos\phi & \sin\phi \end{bmatrix}
\begin{bmatrix} -\sin\lambda & \cos\lambda & 0 \\ -\cos\lambda & -\sin\lambda & 0 \\ 0 & 0 & 1 \end{bmatrix}
\begin{bmatrix} \Delta X \\ \Delta Y \\ \Delta Z \end{bmatrix} \quad (A.6)
$$

$$\begin{bmatrix} \Delta e \\ \Delta n \\ \Delta u \end{bmatrix} = \begin{bmatrix} -\sin\lambda & \cos\lambda & 0 \\ -\sin\phi\cos\lambda & -\sin\phi\sin\lambda & \cos\phi \\ \cos\phi\cos\lambda & \cos\phi\sin\lambda & \sin\phi \end{bmatrix} \begin{bmatrix} \Delta X \\ \Delta Y \\ \Delta Z \end{bmatrix} \qquad (A.7)$$

## REFERENCES

Leick, A. 2004. *GPS Satellite Surveying*, 3rd ed. New York: John Wiley & Sons.
Vanicek, P. and Krakiwsky, E. 1986. *Geodesy: The Concepts*. New York: North Holland.

# Appendix B: 1983 State Plane Coordinate Zone Constants

**Defining Constants for the 1983 State Plane Coordinate System**

Ref: NOAA Manual NOS NGS 5, State Plane Coordinate System of 1983

Abbreviations for projection types:    (T.M.) = Transverse Mercator

(O.M.) = Oblique Mercator

(L.) = Lambert Conic Conformal

Latitudes and longitudes are given as DD MM and all longitudes are given as west longitudes.

| State and Zone Name | | Zone Code | Projection Type | Central Meridian and Scale Factor (T.M.) or Standard Parallels (L.) | | Grid Origin | | | |
|---|---|---|---|---|---|---|---|---|---|
| | | | | | | Longitude | | Easting | |
| | | | | | | Latitude | | Northing | |
| Alabama: | | | | | | | | | |
| AL | East | 0101 | T.M. | 85 | 50 | 85 | 50 | 200,000 | m |
| | | | | 1:25,000 | | 30 | 30 | 0.0 | m |
| AL | West | 0102 | T.M. | 87 | 30 | 87 | 30 | 600,000 | m |
| | | | | 1:15,000 | | 30 | 30 | 0.0 | m |
| Alaska: | | | | | | | | | |
| AK | Zone 1 | 5001 | O.M. | Axis azimuth = atan (−3/4) | | 133 | 40 | 5,000,000.0 | m |
| | | | | 1:10,000 | | 57 | 00 | −5,000,000.0 | m |
| AK | Zone 2 | 5002 | T.M. | 142 | 00 | 142 | 00 | 500,000.0 | m |
| | | | | 1:10,000 | | 54 | 00 | 0.0 | m |
| AK | Zone 3 | 5003 | T.M. | 146 | 00 | 146 | 00 | 500,000.0 | m |
| | | | | 1:10,000 | | 54 | 00 | 0.0 | m |
| AK | Zone 4 | 5004 | T.M. | 150 | 00 | 150 | 00 | 500,000.0 | m |
| | | | | 1:10,000 | | 54 | 00 | 0.0 | m |
| AK | Zone 5 | 5005 | T.M. | 154 | 00 | 154 | 00 | 500,000.0 | m |
| | | | | 1:10,000 | | 54 | 00 | 0.0 | m |
| AK | Zone 6 | 5006 | T.M. | 158 | 00 | 158 | 00 | 500,000.0 | m |
| | | | | 1:10,000 | | 54 | 00 | 0.0 | m |
| AK | Zone 7 | 5007 | T.M. | 162 | 00 | 162 | 00 | 500,000.0 | m |
| | | | | 1:10,000 | | 54 | 00 | 0.0 | m |
| AK | Zone 8 | 5008 | T.M. | 166 | 00 | 166 | 00 | 500,000.0 | m |
| | | | | 1:10,000 | | 54 | 00 | 0.0 | m |
| AK | Zone 9 | 5009 | T.M. | 170 | 00 | 170 | 00 | 500,000.0 | m |
| | | | | 1:10,000 | | 54 | 00 | 0.0 | m |

*(Continued)*

| State and Zone Name | | Zone Code | Projection Type | Central Meridian and Scale Factor (T.M.) or Standard Parallels (L.) | | Grid Origin | | | |
|---|---|---|---|---|---|---|---|---|---|
| | | | | | | Longitude | | Easting | |
| | | | | | | Latitude | | Northing | |
| AK | Zone 10 | 5010 | L. | 51 | 50 | 176 | 00 | 1,000,000.0 | m |
| | | | | 53 | 50 | 51 | 00 | 0.0 | m |
| Arizona: | | | | | | | | | |
| AZ | East | 0201 | T.M. | 110 | 10 | 110 | 10 | 213,360.0 | m |
| | | | | 1:10,000 | | 31 | 00 | 0.0 | m |
| AZ | Central | 0202 | T.M. | 111 | 55 | 111 | 55 | 213,360.0 | m |
| | | | | 1:10,000 | | 31 | 00 | 0.0 | m |
| AZ | West | 0203 | T.M. | 113 | 45 | 113 | 45 | 213,360.0 | m |
| | | | | 1:15,000 | | 31 | 00 | 0.0 | m |

*Note*: State defines origin in International Feet. 213,360 m = 700,000 International Feet

| State and Zone Name | | Zone Code | Projection Type | Central Meridian and Scale Factor (T.M.) or Standard Parallels (L.) | | Grid Origin | | | |
|---|---|---|---|---|---|---|---|---|---|
| Arkansas: | | | | | | | | | |
| AR | North | 0301 | L. | 34 | 56 | 92 | 00 | 400,000.0 | m |
| | | | | 36 | 14 | 34 | 20 | 0.0 | m |
| AR | South | 0302 | L. | 33 | 18 | 92 | 00 | 400,000.0 | m |
| | | | | 34 | 46 | 32 | 40 | 400,000.0 | m |
| California: | | | | | | | | | |
| CA | Zone 1 | 0401 | L. | 40 | 00 | 122 | 00 | 2,000,000.0 | m |
| | | | | 41 | 40 | 39 | 20 | 500,000.0 | m |
| CA | Zone 2 | 0402 | L. | 38 | 20 | 122 | 00 | 2,000,000.0 | m |
| | | | | 39 | 50 | 37 | 40 | 500,000.0 | m |
| CA | Zone 3 | 0403 | L. | 37 | 04 | 120 | 30 | 2,000,000.0 | m |
| | | | | 38 | 26 | 36 | 30 | 500,000.0 | m |
| CA | Zone 4 | 0404 | L. | 36 | 00 | 119 | 00 | 2,000,000.0 | m |
| | | | | 37 | 15 | 35 | 20 | 500,000.0 | m |
| CA | Zone 5 | 0405 | L. | 34 | 02 | 118 | 00 | 2,000,000.0 | m |
| | | | | 35 | 28 | 33 | 30 | 500,000.0 | m |
| CA | Zone 6 | 0406 | L. | 32 | 47 | 116 | 15 | 2,000,000.0 | m |
| | | | | 33 | 53 | 32 | 10 | 500,000.0 | m |
| Colorado: | | | | | | | | | |
| CO | North | 0501 | L. | 39 | 43 | 105 | 30 | 914,401.8289 | m |
| | | | | 40 | 47 | 39 | 20 | 304,800.6096 | m |
| CO | Central | 0502 | L. | 38 | 27 | 105 | 30 | 914,401.8289 | m |
| | | | | 39 | 45 | 37 | 50 | 304,800.6096 | m |
| CO | South | 0503 | L. | 37 | 14 | 105 | 30 | 914,401.8289 | m |
| | | | | 38 | 26 | 36 | 40 | 304,800.6096 | m |
| Connecticut: | | | | | | | | | |
| CT | | 0600 | L. | 41 | 12 | 72 | 45 | 304,800.6096 | m |
| | | | | 41 | 52 | 40 | 50 | 152,400.3048 | m |
| Delaware: | | | | | | | | | |
| DE | | 0700 | T.M. | 75 | 25 | 75 | 25 | 200,000.0 | m |
| | | | | 1:200,000 | | 38 | 00 | 0.0 | m |

*(Continued)*

| State and Zone Name | | Zone Code | Projection Type | Central Meridian and Scale Factor (T.M.) or Standard Parallels (L.) | | Grid Origin | | |
|---|---|---|---|---|---|---|---|---|
| | | | | | | Longitude | Easting | |
| | | | | | | Latitude | Northing | |
| Florida: | | | | | | | | |
| FL | East | 0901 | T.M. | 81 | 00 | 81    00 | 200,000.0 | m |
| | | | | 1:17,000 | | 24    20 | 0.0 | m |
| FL | West | 0902 | T.M. | 82 | 00 | 82    00 | 200,000.0 | m |
| | | | | 1:17,000 | | 24    20 | 0.0 | m |
| FL | North | 0903 | L. | 29 | 35 | 84    30 | 600,000.0 | m |
| | | | | 30 | 45 | 29    00 | 0.0 | m |
| Georgia: | | | | | | | | |
| GA | East | 1001 | T.M. | 82 | 10 | 82    10 | 200,000.0 | m |
| | | | | 1:10,000 | | 30    00 | 0.0 | m |
| GA | West | 1002 | T.M. | 84 | 10 | 84    10 | 700,000.0 | m |
| | | | | 1:10,000 | | 30    00 | 0.0 | m |
| Hawaii: | | | | | | | | |
| HI | Zone 1 | 5101 | T.M. | 155 | 30 | 155    30 | 500,000.0 | m |
| | | | | 1:30,000 | | 18    50 | 0.0 | m |
| HI | Zone 2 | 5102 | T.M. | 156 | 40 | 156    40 | 500,000.0 | m |
| | | | | 1:30,000 | | 20    20 | 0.0 | m |
| HI | Zone 3 | 5103 | T.M. | 158 | 00 | 158    00 | 500,000.0 | m |
| | | | | 1:100,000 | | 21    10 | 0.0 | m |
| HI | Zone 4 | 5104 | T.M. | 159 | 30 | 159    30 | 500,000.0 | m |
| | | | | 1:100,000 | 21 | 50 | 0.0 | m |
| HI | Zone 5 | 5105 | T.M. | 160 | 10 | 160    10 | 500,000.0 | m |
| | | | | 1:infinity | | 21    40 | 0.0 | m |
| Idaho: | | | | | | | | |
| ID | East | 1101 | T.M. | 112 | 10 | 112    10 | 200,000.0 | m |
| | | | | 1:19,000 | | 41    40 | 0.0 | m |
| ID | Central | 1102 | T.M. | 114 | 00 | 114    00 | 500,000.0 | m |
| | | | | 1:19,000 | | 41    40 | 0.0 | m |
| ID | West | 1103 | T.M. | 115 | 45 | 115    45 | 800,000.0 | m |
| | | | | 1:15,000 | | 41    40 | 0.0 | m |
| Illinois: | | | | | | | | |
| IL | East | 1201 | T.M. | 88 | 20 | 88    20 | 300,000.0 | m |
| | | | | 1:40,000 | | 36    40 | 0.0 | m |
| IL | West | 1202 | T.M. | 90 | 10 | 90    10 | 700,000.0 | m |
| | | | | 1:17,000 | | 36    40 | 0.0 | m |
| Indiana: | | | | | | | | |
| IN | East | 1301 | T.M. | 85 | 40 | 85    40 | 100,000.0 | m |
| | | | | 1:30,000 | | 37    30 | 250,000.0 | m |
| IN | West | 1302 | T.M. | 87 | 05 | 87    05 | 900,000.0 | m |
| | | | | 1:30,000 | | 37    30 | 250,000.0 | m |

*(Continued)*

| State and Zone Name | | Zone Code | Projection Type | Central Meridian and Scale Factor (T.M.) or Standard Parallels (L.) | | Grid Origin | | | |
|---|---|---|---|---|---|---|---|---|---|
| | | | | | | Longitude | | Easting | |
| | | | | | | Latitude | | Northing | |
| Iowa: | | | | | | | | | |
| IA | North | 1401 | L. | 42 | 04 | 93 | 30 | 1,500,000.0 | m |
| | | | | 43 | 16 | 41 | 30 | 1,000,000.0 | m |
| IA | South | 1402 | L. | 40 | 37 | 93 | 30 | 500,000.0 | m |
| | | | | 41 | 47 | 40 | 00 | 0.0 | m |
| Kansas: | | | | | | | | | |
| KS | North | 1501 | L. | 38 | 43 | 98 | 00 | 400,000.0 | m |
| | | | | 39 | 47 | 38 | 20 | 0.0 | m |
| KS | South | 1502 | L. | 37 | 16 | 98 | 30 | 400,000.0 | m |
| | | | | 38 | 34 | 36 | 40 | 400,000.0 | m |
| Kentucky: | | | | | | | | | |
| KY | North | 1601 | L. | 37 | 58 | 84 | 15 | 500,000.0 | m |
| | | | | 38 | 58 | 37 | 30 | 0.0 | m |
| KY | South | 1602 | L. | 36 | 44 | 85 | 45 | 500,000.0 | m |
| | | | | 37 | 56 | 36 | 20 | 500,000.0 | m |
| KY | | 1600 | L. | 37 | 05 | 85 | 45 | 1,000,000.0 | m |
| *Note*: Zone 1600 is a | | | | 38 | 40 | 36 | 20 | 1,500,000.0 | m |
| single new zone. | | | | | | | | | |
| Louisiana: | | | | | | | | | |
| LA | North | 1701 | L. | 31 | 10 | 92 | 30 | 1,000,000.0 | m |
| | | | | 32 | 40 | 30 | 30 | 0.0 | m |
| LA | South | 1702 | L. | 29 | 18 | 91 | 20 | 1,000,000.0 | m |
| | | | | 30 | 42 | 28 | 30 | 0.0 | m |
| LA | Offshore | 1703 | L. | 26 | 10 | 91 | 20 | 1,000,000.0 | m |
| | | | | 27 | 50 | 25 | 30 | 0.0 | m |
| Maine: | | | | | | | | | |
| ME | East | 1801 | T.M. | 68 | 30 | 68 | 30 | 300,000.0 | m |
| | | | | 1:10,000 | | 43 | 40 | 0.0 | m |
| ME | West | 1802 | T.M. | 70 | 10 | 70 | 10 | 900,000.0 | m |
| | | | | 1:30,000 | | 42 | 50 | 0.0 | m |
| Maryland: | | | | | | | | | |
| MD | | 1900 | L. | 38 | 18 | 77 | 00 | 400,000.0 | m |
| | | | | 39 | 27 | 37 | 40 | 0.0 | m |
| Massachusetts: | | | | | | | | | |
| MA | Mainland | 2001 | L. | 41 | 43 | 71 | 30 | 200,000.00 | m |
| | | | | 42 | 41 | 41 | 00 | 750,000.00 | m |
| MA | Island | 2002 | L. | 41 | 17 | 70 | 30 | 500,000.00 | m |
| | | | | 41 | 29 | 41 | 00 | 0.0 | m |
| Michigan: | | | | | | | | | |
| MI | North | 2111 | L. | 45 | 29 | 87 | 00 | 8,000,000.0 | m |
| | | | | 47 | 05 | 44 | 47 | 0.0 | m |

*(Continued)*

| State and Zone Name | | Zone Code | Projection Type | Central Meridian and Scale Factor (T.M.) or Standard Parallels (L.) | | Grid Origin | | | |
|---|---|---|---|---|---|---|---|---|---|
| | | | | | | Longitude | | Easting | |
| | | | | | | Latitude | | Northing | |
| MI | Central | 2112 | L. | 44 | 11 | 84 | 22 | 6,000,000.0 | m |
| | | | | 45 | 42 | 43 | 19 | 0.0 | m |
| MI | South | 2113 | L. | 42 | 06 | 84 | 22 | 4,000,000.0 | m |
| | | | | 43 | 40 | 41 | 30 | 0.0 | m |
| Minnesota: | | | | | | | | | |
| MN | North | 2201 | L. | 47 | 02 | 93 | 06 | 800,000.0 | m |
| | | | | 48 | 38 | 46 | 30 | 100,000.0 | m |
| MN | Central | 2202 | L. | 45 | 37 | 94 | 15 | 800,000.0 | m |
| | | | | 47 | 03 | 45 | 00 | 100,000.0 | m |
| MN | South | 2203 | L. | 43 | 47 | 94 | 00 | 800,000.0 | m |
| | | | | 45 | 13 | 43 | 00 | 100,000.0 | m |
| Mississippi: | | | | | | | | | |
| MS | East | 2301 | T.M. | 88 | 50 | 88 | 50 | 300,000.0 | m |
| | | | | 1:20,000 | | 29 | 30 | 0.0 | m |
| MS | West | 2302 | T. M. | 90 | 20 | 90 | 20 | 700,000.0 | m |
| | | | | 1:20,000 | | 29 | 30 | 0.0 | m |
| Missouri: | | | | | | | | | |
| MO | East | 2401 | T.M. | 90 | 30 | 90 | 30 | 250,000.0 | m |
| | | | | 1:15,000 | | 35 | 50 | 0.0 | m |
| MO | Central | 2402 | T.M. | 92 | 30 | 92 | 30 | 500,000.0 | m |
| | | | | 1:15,000 | | 35 | 50 | 0.0 | m |
| MO | West | 2403 | T.M. | 94 | 30 | 94 | 30 | 850,000.0 | m |
| | | | | 1:17,000 | | 36 | 10 | 0.0 | m |
| Montana: | | | | | | | | | |
| MT | | 2500 | L. | 45 | 00 | 109 | 30 | 600,000.0 | m |
| | | | | 49 | 00 | 44 | 15 | 0.0 | m |
| Nebraska: | | | | | | | | | |
| NE | | 2600 | L. | 40 | 00 | 100 | 00 | 500,000.0 | m |
| | | | | 43 | 00 | 39 | 50 | 0.0 | m |
| Nevada: | | | | | | | | | |
| NV | East | 2701 | T.M. | 115 | 35 | 115 | 35 | 200,000.0 | m |
| | | | | 1:10,000 | | 34 | 45 | 8,000,000.0 | m |
| NV | Central | 2702 | T.M. | 116 | 40 | 116 | 40 | 500,000.0 | m |
| | | | | 1:10,000 | | 34 | 45 | 6,000,000.0 | m |
| NV | West | 2703 | T.M. | 118 | 35 | 118 | 35 | 800,000.0 | m |
| | | | | 1:10,000 | | 34 | 45 | 4,000,000.0 | m |
| New Hampshire: | | | | | | | | | |
| NH | | 2800 | T.M. | 71 | 40 | 71 | 40 | 300,000.0 | m |
| | | | | 1:30,000 | | 42 | 30 | 0.0 | m |
| New Jersey: | | | | | | | | | |
| NJ | (NY East) | 2900 | T.M. | 74 | 30 | 74 | 30 | 150,000.0 | m |
| | | | | 1:10,000 | | 38 | 50 | 0.0 | m |

*(Continued)*

| State and Zone Name | | Zone Code | Projection Type | Central Meridian and Scale Factor (T.M.) or Standard Parallels (L.) | | Grid Origin | | | |
|---|---|---|---|---|---|---|---|---|---|
| | | | | | | Longitude | | Easting | |
| | | | | | | Latitude | | Northing | |
| New Mexico: | | | | | | | | | |
| NM | East | 3001 | T.M. | 104 | 20 | 104 | 20 | 165,000.0 | m |
| | | | | 1:11,000 | | 31 | 00 | 0.0 | m |
| NM | Central | 3002 | T.M. | 106 | 15 | 106 | 15 | 500,000.0 | m |
| | | | | 1:10,000 | | 31 | 00 | 0.0 | m |
| NM | West | 3003 | T.M. | 107 | 50 | 107 | 50 | 830,000.0 | m |
| | | | | 1:12,000 | | 31 | 00 | 0.0 | m |
| New York: | | | | | | | | | |
| NY | East | 3101 | T.M. | 74 | 30 | 74 | 30 | 150,000.0 | m |
| | (New Jersey) | | | 1:10,000 | | 38 | 50 | 0.0 | m |
| NY | Central | 3102 | T.M. | 76 | 35 | 76 | 35 | 250,000.0 | m |
| | | | | 1:16,000 | | 40 | 00 | 0.0 | m |
| NY | West | 3103 | T.M. | 78 | 35 | 78 | 35 | 350,000.0 | m |
| | | | | 1:16,000 | | 40 | 00 | 0.0 | m |
| NY | Long Island | 3104 | L. | 40 | 40 | 74 | 00 | 300,000.0 | m |
| | | | | 41 | 02 | 40 | 10 | 0.0 | m |
| North Carolina: | | | | | | | | | |
| NC | | 3200 | L. | 34 | 20 | 79 | 00 | 609,601.22 | m |
| | | | | 36 | 10 | 33 | 45 | 0.00 | m |
| North Dakota: | | | | | | | | | |
| ND | North | 3301 | L. | 47 | 26 | 100 | 30 | 600,000.0 | m |
| | | | | 48 | 44 | 47 | 00 | 0.0 | m |
| ND | South | 3302 | L. | 46 | 11 | 100 | 30 | 600,000.0 | m |
| | | | | 47 | 29 | 45 | 40 | 0.0 | m |
| Ohio: | | | | | | | | | |
| OH | North | 3401 | L. | 40 | 26 | 82 | 30 | 600,000.0 | m |
| | | | | 41 | 42 | 39 | 40 | 0.0 | m |
| OH | South | 3402 | L. | 38 | 44 | 82 | 30 | 600,000.0 | m |
| | | | | 40 | 02 | 38 | 00 | 0.0 | m |
| Oklahoma: | | | | | | | | | |
| OK | North | 3501 | L. | 35 | 34 | 98 | 00 | 600,000.0 | m |
| | | | | 36 | 46 | 35 | 00 | 0.0 | m |
| OK | South | 3502 | L. | 33 | 56 | 98 | 00 | 600,000.0 | m |
| | | | | 35 | 14 | 33 | 20 | 0.0 | m |
| Oregon: | | | | | | | | | |
| OR | North | 3601 | L. | 44 | 20 | 120 | 30 | 2,500,000.0 | m |
| | | | | 46 | 00 | 43 | 40 | 0.0 | m |
| OR | South | 3602 | L. | 42 | 20 | 120 | 30 | 1,500,000.0 | m |
| | | | | 44 | 00 | 41 | 40 | 0.0 | m |

*(Continued)*

| State and Zone Name | | Zone Code | Projection Type | Central Meridian and Scale Factor (T.M.) or Standard Parallels (L.) | | Grid Origin | | | |
|---|---|---|---|---|---|---|---|---|---|
| | | | | | | Longitude | | Easting | |
| | | | | | | Latitude | | Northing | |
| Pennsylvania: | | | | | | | | | |
| PA | North | 3701 | L. | 40 | 53 | 77 | 45 | 600,000.0 | m |
| | | | | 41 | 57 | 40 | 10 | 0.0 | m |
| PA | South | 3702 | L. | 39 | 56 | 77 | 45 | 600,000.0 | m |
| | | | | 40 | 58 | 39 | 20 | 0.0 | m |
| Rhode Island: | | | | | | | | | |
| RI | | 3800 | T.M. | 71 | 30 | 71 | 30 | 100,000.0 | m |
| | | | | 1:160,000 | | 41 | 05 | 0.0 | m |
| South Carolina: | | | | | | | | | |
| SC | | 3900 | L. | 32 | 30 | 81 | 00 | 609,600.0 | m |
| | | | | 34 | 50 | 31 | 50 | 0.0 | m |
| South Dakota: | | | | | | | | | |
| SD | North | 4001 | L. | 44 | 25 | 100 | 00 | 600,000.0 | m |
| | | | | 45 | 41 | 43 | 50 | 0.0 | m |
| SD | South | 4002 | L. | 42 | 50 | 100 | 20 | 600,000.0 | m |
| | | | | 44 | 24 | 42 | 20 | 0.0 | m |
| Tennessee: | | | | | | | | | |
| TN | | 4100 | L. | 35 | 15 | 86 | 00 | 600,000.0 | m |
| | | | | 36 | 25 | 34 | 20 | 0.0 | m |
| Texas: | | | | | | | | | |
| TX | North | 4201 | L. | 34 | 39 | 101 | 30 | 200,000.0 | m |
| | | | | 36 | 11 | 34 | 00 | 1,000,000.0 | m |
| TX | North Central | 4202 | L. | 32 | 08 | 98 | 30 | 600,000.0 | m |
| | | | | 33 | 58 | 31 | 40 | 2,000,000.0 | m |
| TX | Central | 4203 | L. | 30 | 07 | 100 | 20 | 700,000.0 | m |
| | | | | 31 | 53 | 29 | 40 | 3,000,000.0 | m |
| TX | South Central | 4204 | L. | 28 | 23 | 99 | 00 | 600,000.0 | m |
| | | | | 30 | 17 | 27 | 50 | 4,000,000.0 | m |
| TX | South | 4205 | L. | 26 | 10 | 98 | 30 | 300,000.0 | m |
| | | | | 27 | 50 | 25 | 40 | 5,000,000.0 | m |
| Utah: | | | | | | | | | |
| UT | North | 4301 | L. | 40 | 43 | 111 | 30 | 500,000.0 | m |
| | | | | 41 | 47 | 40 | 20 | 1,000,000.0 | m |
| UT | Central | 4302 | L. | 39 | 01 | 111 | 30 | 500,000.0 | m |
| | | | | 40 | 39 | 38 | 20 | 2,000,000.0 | m |
| UT | South | 4303 | L. | 37 | 13 | 111 | 30 | 500,000.0 | m |
| | | | | 38 | 21 | 36 | 40 | 3,000,000.0 | m |
| Vermont: | | | | | | | | | |
| VT | | 4400 | T.M. | 72 | 30 | 72 | 30 | 500,000.0 | m |
| | | | | 1:28,000 | | 42 | 30 | 0.0 | m |

(*Continued*)

| State and Zone Name | | Zone Code | Projection Type | Central Meridian and Scale Factor (T.M.) or Standard Parallels (L.) | | Grid Origin | | | |
|---|---|---|---|---|---|---|---|---|---|
| | | | | | | Longitude | | Easting | |
| | | | | | | Latitude | | Northing | |
| Virginia: | | | | | | | | | |
| VA | North | 4501 | L. | 38 | 02 | 78 | 30 | 3,500,000.0 | m |
| | | | | 39 | 12 | 37 | 40 | 2,000,000.0 | m |
| VA | South | 4502 | L. | 36 | 46 | 78 | 30 | 3,500,000.0 | m |
| | | | | 37 | 58 | 36 | 20 | 1,000,000.0 | m |
| Washington: | | | | | | | | | |
| WA | North | 4601 | L. | 47 | 30 | 120 | 50 | 500,000.0 | m |
| | | | | 48 | 44 | 47 | 00 | 0.0 | m |
| WA | South | 4602 | L. | 45 | 50 | 120 | 30 | 500,000.0 | m |
| | | | | 47 | 20 | 45 | 20 | 0.0 | m |
| West Virginia: | | | | | | | | | |
| WV | North | 4701 | L. | 39 | 00 | 79 | 30 | 600,000.0 | m |
| | | | | 40 | 15 | 38 | 30 | 0.0 | m |
| WV | South | 4702 | L. | 37 | 29 | 81 | 00 | 600,000.0 | m |
| | | | | 38 | 53 | 37 | 00 | 0.0 | m |
| Wisconsin: | | | | | | | | | |
| WI | North | 4801 | L. | 45 | 34 | 90 | 00 | 600,000.0 | m |
| | | | | 46 | 46 | 45 | 10 | 0.0 | m |
| WI | Central | 4802 | L. | 44 | 15 | 90 | 00 | 600,000.0 | m |
| | | | | 45 | 30 | 43 | 50 | 0.0 | m |
| WI | South | 4803 | L. | 42 | 44 | 90 | 00 | 600,000.0 | m |
| | | | | 44 | 04 | 42 | 00 | 0.0 | m |
| Wyoming: | | | | | | | | | |
| WY | East | 4901 | T.M. | 105 | 10 | 105 | 10 | 200,000.0 | m |
| | | | | 1:16,000 | | 40 | 30 | 0.0 | m |
| WY | East Central | 4902 | T.M. | 107 | 20 | 107 | 20 | 400,000.0 | m |
| | | | | 1:16,000 | | 40 | 30 | 100,000.0 | m |
| WY | West Central | 4903 | T.M. | 108 | 45 | 108 | 45 | 600,000.0 | m |
| | | | | 1:16,000 | | 40 | 30 | 0.0 | m |
| WY | West | 4904 | T.M. | 110 | 05 | 110 | 05 | 800,000.0 | m |
| | | | | 1:16,000 | | 40 | 30 | 100,000.0 | m |
| Puerto Rico and Virgin Islands: | | | | | | | | | |
| PR | | 5200 | L. | 18 | 02 | 66 | 26 | 200,000.0 | m |
| | | | | 18 | 26 | 17 | 50 | 200,000.0 | m |

# Appendix C: 3-D Inverse with Statistics

Used to compute values in Table 12.1.

| GRS 80 ellipsoid: | | | | Eccentricity squared= | 0.006694380022903 |
|---|---|---|---|---|---|
| | $a =$ | **6,378,137.000** m | | | |
| | $1/f =$ | **298.257222100883** | | Seconds per radian= | 206,264.806247096 |

| Standpoint: | USPA | | | ECEF Covariance Matrix for Standpoint | | |
|---|---|---|---|---|---|---|
| $X =$ | **−1,555,678.5590** | m ± | 0.0022 m | 0.00000465 | 0.00000483 | −0.00000308 |
| $Y =$ | **−5,169,961.3603** | m ± | 0.0042 m | 0.00000430 | 0.00001763 | −0.00001031 |
| $Z =$ | **3,386,700.1017** | m ± | 0.0038 m | −0.00000308 | −0.00001031 | 0.00001419 |
| | (User inputs these values.) | | | (Default values = zero implies point is held fixed.) | | |

| Forepoint: | PSEUDO | | | ECEF Covariance Matrix for Forepoint | | |
|---|---|---|---|---|---|---|
| $X =$ | **−1,556,206.5951** | m ± | 0.0017 m | 0.00000279 | 0.00000276 | −0.00000189 |
| $Y =$ | **−5,169,400.7041** | m ± | 0.0031 m | 0.00000276 | 0.00000950 | −0.00000571 |
| $Z =$ | **3,387,285.9993** | m ± | 0.0030 m | −0.00000180 | −0.00000571 | 0.00000882 |
| | (User inputs these values.) | | | (Default values = zero implies point is held fixed.) | | |

For network accuracy, the correlation submatrix is zero and implies the point positions are independent. Local accuracy can be computed if correlations are present. Correlation of the Standpoint with respect to the Forepoint is the transpose of the correlation of the Forepoint with respect to the Standpoint.

| Correlation of Forepoint wrt Standpoint | | | Correlation of Standpoint wrt Forepoint | | |
|---|---|---|---|---|---|
| 2.5692E−06 | 2.6423E−06 | −1.8362E−06 | 2.5692E−06 | 2.6621E−06 | −1.8196E−06 |
| 2.6621E−06 | 8.9506E−06 | −5.2988E−06 | 2.6423E−06 | 8.9506E−06 | −5.2854E−06 |
| −1.8196E−06 | −5.2854E−06 | 7.6153E−06 | −1.8362E−06 | −5.2988E−06 | 7.6153E−06 |

| Standpoint: | | | USPA | | | From local co- | |
|---|---|---|---|---|---|---|---|
| | D | M | Sec. | | | variance matrix | |
| $\phi =$ | 32 | 16 | 23.001232 | N | ± | 0.0024 m | |
| $\lambda =$ | 253 | 15 | 11.092169 | E | ± | 0.0018 m | |
| | 106 | 44 | 48.907831 | W | ± | 0.0018 m | |
| $h =$ | | | 1,177.9884 | m | ± | 0.0052 m | |

$$\lambda = \tan^{-1}\left(\frac{Y}{X}\right); \quad 0°\text{–}360° \text{ East}$$

$$\tan\phi = \frac{Z}{P}\left(1 + \frac{e^2 \, N \sin\phi}{Z}\right); \quad P = \sqrt{X^2 + Y^2}$$

| Forepoint: | | | Pseudo | | | | |
|---|---|---|---|---|---|---|---|
| | D | M | Sec. | | | | |
| $\phi =$ | 32 | 16 | 45.747539 | N | ± | 0.0020 m | |
| $\lambda =$ | 253 | 14 | 45.600582 | E | ± | 0.0016 m | |
| | 106 | 45 | 14.399418 | W | ± | 0.0016 m | |
| $h =$ | | | 1,165.6137 | m | ± | 0.0039 m | |

$$h = \frac{P}{\cos\phi} - N \text{ Must iterate to find latitude and height.}$$

$$\phi_0 = \tan^{-1}\left(\frac{Z}{P\,(1 - e^2)}\right) \quad N_0 = \frac{a}{\sqrt{1 - e^2 \sin^2\phi_0}}$$

$$R = \begin{bmatrix} -\sin\lambda & \cos\lambda & 0 \\ -\sin\phi\,\cos\lambda & -\sin\phi\,\sin\lambda & \cos\phi \\ \cos\phi\,\cos\lambda & \cos\phi\,\sin\lambda & \sin\phi \end{bmatrix}$$

$$\phi_i = \tan^{-1}\left[\frac{Z}{P}\left(1 + \frac{e^2 \, N_{i-1} \sin\phi_{i-1}}{Z}\right)\right] \text{ and } N_i = \frac{a}{\sqrt{1 - e^2 \sin^2\phi_i}}$$

Local covariance = $\Sigma_{e/n/u} = R\,\Sigma_{X/Y/Z}\,R^t$

| $R =$ | Standpoint | | $R =$ | Forepoint | |
|---|---|---|---|---|---|
| 0.957586858 | −0.288144773 | 0.000000000 | −0.534048032 | −0.288263116 | 0.000000000 |
| 0.153856283 | 0.511308094 | 0.845513026 | 0.15394635 | 0.511378355 | 0.845454138 |
| −0.243630159 | −0.809652162 | 0.533954795 | −0.243713244 | −0.809565658 | 0.534048032 |

| $R(t) =$ | | | $R(rt) =$ | | |
|---|---|---|---|---|---|
| 0.957586858 | 0.153856283 | −0.243630159 | −0.534048032 | 0.15394635 | −0.243713244 |
| −0.288144773 | 0.511308094 | −0.809652162 | −0.288263116 | 0.511378355 | −0.809565658 |
| 0.000000000 | 0.845513026 | 0.533954795 | 0.000000000 | 0.845454138 | 0.534048032 |

**Local Reference Frame Covariance:**

| | Standpoint | | | Forepoint | |
|---|---|---|---|---|---|
| 0.0000032085 | 0.0000002800 | −0.0000004033 | 0.000002435 | −0.000000261 | 0.000005385 |
| −0.0000000030 | 0.0000058660 | 0.0000024207 | −0.000000302 | 0.000003872 | 0.000001821 |
| 0.0000000449 | 0.0000024207 | 0.0000273955 | 0.000005360 | 0.000001847 | 0.000015414 |

**Local Tangent Plane Inverse from Standpoint to Forepoint:**                        USPA            to  PSEUDO

$$\begin{bmatrix} \Delta e \\ \Delta n \\ \Delta u \end{bmatrix} = R_{stdpt} \begin{bmatrix} \Delta X \\ \Delta Y \\ \Delta Z \end{bmatrix}$$

$$R = \begin{matrix} 0.957586858 & -0.288144773 & 0.000000000 \\ 0.153856283 & 0.511308094 & 0.845513026 \\ -0.243630159 & -0.809652162 & 0.533954795 \end{matrix}$$

$\Delta X = 528.036$
$\Delta Y = 560.656$
$\Delta Z = 585.898$

$$\text{Dist} = \sqrt{\Delta e^2 + \Delta n^2}$$
$$\text{Azi} = \tan^{-1}\left(\frac{\Delta e}{\Delta n}\right)$$

$\Delta e =$    −667.191 m
$\Delta n =$    700.810 m
$\Delta u =$    −12.448 m

| Distance = 967.6149 m |
| Azimuth = 316 24 28.12 |

Use ATAN2 function:   Azi = 0.760827213    5.522358094   radians

Equation 1.36 repeated as Equation 12.9 is used as the basis for computing both network accuracy and local accuracy standard deviations. If the submatrices on the upper right and lower-left are zero, the computed answer will be network accuracy. Local accuracy is obtained if the full covariance matrix is used.

$$\Sigma_\Delta = \begin{bmatrix} -1 & 0 & 0 & 1 & 0 & 0 \\ 0 & -1 & 0 & 0 & 1 & 0 \\ 0 & 0 & -1 & 0 & 0 & 1 \end{bmatrix} \begin{bmatrix} \sigma^2_{X_1} & \sigma_{X_1Y_1} & \sigma_{X_1Z_1} & \sigma_{X_1X_2} & \sigma_{X_1Y_2} & \sigma_{X_1Z_2} \\ \sigma_{X_1Y_1} & \sigma^2_{Y_1} & \sigma_{Y_1Z_1} & \sigma_{Y_1X_2} & \sigma_{Y_1Y_2} & \sigma_{Y_1Z_2} \\ \sigma_{X_1Z_1} & \sigma_{Y_1Z_1} & \sigma^2_{Z_1} & \sigma_{Z_1X_2} & \sigma_{Z_1Y_2} & \sigma_{Z_1Z_2} \\ \sigma_{X_1X_2} & \sigma_{Y_1X_2} & \sigma_{Z_1X_2} & \sigma^2_{X_2} & \sigma_{X_2Y_2} & \sigma_{X_2Z_2} \\ \sigma_{X_1Y_2} & \sigma_{Y_1Y_2} & \sigma_{Z_1Y_2} & \sigma_{Y_2X_2} & \sigma^2_{Y_2} & \sigma_{Y_2Z_2} \\ \sigma_{X_1Z_2} & \sigma_{Y_1Z_2} & \sigma_{Z_1Z_2} & \sigma_{Z_2X_2} & \sigma_{Z_2Y_2} & \sigma^2_{Z_2} \end{bmatrix} \begin{bmatrix} -1 & 0 & 0 \\ 0 & -1 & 0 \\ 0 & 0 & -1 \\ 1 & 0 & 0 \\ 0 & 1 & 0 \\ 0 & 0 & 1 \end{bmatrix} = \begin{bmatrix} \sigma^2_{\Delta X} & \sigma_{\Delta X \Delta Y} & \sigma_{\Delta X \Delta Z} \\ \sigma_{\Delta X \Delta Y} & \sigma^2_{\Delta Y} & \sigma_{\Delta Y \Delta Z} \\ \sigma_{\Delta X \Delta Z} & \sigma_{\Delta Y \Delta Z} & \sigma^2_{\Delta Z} \end{bmatrix}$$

$$\Sigma_{3D-INV} = \begin{bmatrix} \sigma^2_{\Delta e} & \sigma_{\Delta e \Delta n} & \sigma_{\Delta e \Delta u} \\ \sigma_{\Delta e \Delta n} & \sigma^2_{\Delta n} & \sigma_{\Delta n \Delta u} \\ \sigma_{\Delta e \Delta u} & \sigma_{\Delta n \Delta u} & \sigma^2_{\Delta u} \end{bmatrix} = R \begin{bmatrix} \sigma^2_{\Delta X} & \sigma_{\Delta X \Delta Y} & \sigma_{\Delta X \Delta Z} \\ \sigma_{\Delta X \Delta Y} & \sigma^2_{\Delta Y} & \sigma_{\Delta Y \Delta Z} \\ \sigma_{\Delta X \Delta Z} & \sigma_{\Delta Y \Delta Z} & \sigma^2_{\Delta Z} \end{bmatrix} R^t \quad \text{where } R = \begin{bmatrix} -\sin\lambda & \cos\lambda & 0 \\ -\sin\phi\cos\lambda & -\sin\phi-\sin\lambda & \cos\phi \\ \cos\phi\cos\lambda & \cos\phi\sin\lambda & \sin\phi \end{bmatrix}$$

$$S = \sqrt{\Delta e^2 + \Delta n^2}$$
$$\alpha = \tan^{-1}\left(\frac{\Delta e}{\Delta n}\right)$$

$$J_3 = \begin{bmatrix} \frac{\partial S}{\partial \Delta e} & \frac{\partial S}{\partial \Delta n} & \frac{\partial S}{\partial \Delta u} \\ \frac{\partial \alpha}{\partial \Delta e} & \frac{\partial \alpha}{\partial \Delta n} & \frac{\partial \alpha}{\partial \Delta u} \end{bmatrix} = \begin{bmatrix} \frac{\Delta e}{S} & \frac{\Delta n}{S} & 0 \\ \frac{\Delta n}{S^2} & \frac{-\Delta e}{S^2} & 0 \end{bmatrix}$$

$$\begin{bmatrix} \sigma^2_S & \sigma_{S\alpha} \\ \sigma_{S\alpha} & \sigma^2_\alpha \end{bmatrix} = J_3 \Sigma_{3D-INV} J^t$$

Network accuracy standpoint to forepoint is computed assuming no correlation between points. Local accuracy is computed using the full covariance matrix (including correlation) input by user.

| USPA To | PSEUDO | | Network Accuracy | | Local Accuracy |
|---|---|---|---|---|---|
| Tangent Plane Distance = | 967.6149 m | ± | 0.00269 m | ± | 0.0015 m |
| 3-D Azimuth = | 316 24 28.12 | ± | 0.59 sec | ± | 0.35 sec |

# Appendix D: Development of the Global Spatial Data Model (GSDM)

The concept of an integrated 3-D database for spatial data grew out of discussions with Dr. Alfred Leick during the author's 1990/1991 sabbatical at the University of Maine. Leick's description of the 3-D Geodetic Model (Leick 1990) is all-inclusive and presented at a high level. The challenge, against a backdrop of surveying engineering applications and anticipating increasing use of GNSS for positioning, appeared to be capturing the benefits of modern technology for 2-D and flat-Earth users without destroying the inherent rigor of 3-D measurements. As it turns out, GNSS technology has been embraced by the geomatics community since the 1980s, and many spatial data applications are competently accomplished using flat-Earth concepts. But, in part because map projections are strictly two-dimensional and because using state plane coordinates is an obvious application for GNSS, the third dimension did not get the attention it deserved in those early years. Instead, the concern of many focused on the difference between grid distance as defined by state plane coordinates and the ground distance as used in construction activities and on survey plats. In an attempt to understand and to address the grid/ground distance issues, a questionnaire was sent to all 50 state DOTs in 1990 requesting feedback. Those thoughtful responses are reported in Appendix C of a paper presented at the 1991 *ASCE Transportation Specialty Conference* (Burkholder 1993). That paper also lays the groundwork for what became known as the 3-D Global Spatial Data Model (GSDM).

A model is successful to the extent that it is both appropriate and simple. The 3-D Geodetic Model proposed by Alfred Leick (1990) is very comprehensive and includes concepts of both geometrical geodesy and physical geodesy. Without faulting the 3-D Geodetic Model, the question must be asked whether or not that model is appropriate for addressing the grid/ground distance issue. Concepts of physical geodesy and gravity are important as scientific inquiries and are readily handled by a comparatively small number of professionals. At the high end, the demand for appropriateness overrides the need for simplicity and the formal 3-D model should be used. But, that model is viewed as overkill for many surveying and engineering applications. The GSDM incorporates the geometrical concepts of the comprehensive 3-D Geodetic Model but the GSDM presumes that issues related to gravity are accommodated before spatial data are brought into the GSDM.

It has been said that GNSS (and other technologies) has solved the location problem. Many in the spatial data community need and use location data productively. The GSDM is both appropriate and simple for those applications. It is readily acknowledged that understanding why a point is where it is or understanding how or why a point moves as it does involves more complicated concepts of gravity and physical geodesy. In those cases, the appropriateness of the 3-D Geodetic Model becomes the overriding criterion.

Although the conceptual foundation for the GSDM was developed during the 1990/1991 sabbatical experience, recommendations for practical application emerged in work performed for the Southeastern Wisconsin Regional Planning Commission (SEWRPC) in the 1990s. Beginning in 1964, the SEWRPC embarked on an ambitious project to establish reliable state plane coordinates and elevations on all section and quarter section corners within the seven-county region. The networks were built on the NAD 27 horizontal datum and the NGVD 29 vertical datum and have provided reliable survey control to support land and engineering surveys within the seven-county region for over 50 years. Additionally, the SEWRPC control system has provided an invaluable basis for development of computerized land information and public works management systems. The new horizontal and vertical datums—NAD 83 and NAVD 88—developed by NGS in the 1980s offered little benefit to SEWRPC constituents because reliable networks were already serving the user community.

Nonetheless, some state and federal agencies, utilities, and private sector users did make the transition to using the new datums and brought varying forms of pressure on SEWRPC to make the transition as well. SEWRPC was not convinced of the benefits of making the transition to the NAD 83 and NAVD 88 and was resolute in continuing use of the two proven networks. But, recognizing their obligation to their constituents, SEWRPC commissioned preparation of local transformations that could be used by others to make reliable datum transformations. Two separate technical reports were prepared for and published by the Commission (SEWRPC 1994, 1995)—one for horizontal and one for vertical.

While discussing the scope of those reports, the suggestion was made that an integrated 3-D database populated with transformed data from the existing control networks would be an efficient way to capture the value of the existing networks while supporting the user community in uses of 3-D digital spatial data. That suggestion was deemed impractical and two separate reports were prepared and implemented (Bauer 2005). However, following completion of those two reports, the author was asked to prepare a follow-up report outlining the 3-D concepts and procedures by which the value of 3-D digital spatial data could be captured and preserved. That report, "Definition of a three-dimensional spatial data model for Southeastern Wisconsin" (Burkholder 1997b), builds on the concepts described in Burkholder (1993) and references a formal definition and description of the 3-D Global Spatial Data Model (GSDM) filed with the U.S. Copyright Office (Burkholder 1997a). A book devoted to the GSDM was published by CRC Press in 2008 (Burkholder 2008).

## REFERENCES

Bauer, K.W. 2005. A control survey and mapping project for an urbanizing region (A study in persistence). *Surveying and Land Information Science* 65 (2): 75–83.

Burkholder, E.F. 1993. Using GPS results in true 3-D coordinate system. *Journal of Surveying Engineering* 119 (1): 1–21.

Burkholder, E.F. 1997a. Definition and description of a global spatial data model (GSDM). Washington, DC: U.S. Copyright Office. http://www.globalcogo.com/gsdmdefn.pdf (accessed May 4, 2017).

Burkholder, E.F. 1997b. Definition of a three-dimensional spatial data model for southeastern Wisconsin. Waukesha, WI: Southeastern Wisconsin Regional Planning Commission. http://www.sewrpc.org/SEWRPCFiles/Publications/ppr/definition_three-dimensional_spatial_data_model_for_wi.pdf.

Burkholder, E.F. 2008. *The 3-D Global Spatial Data Model: Foundation of the Spatial Data Infrastructure.* Boca Raton, FL: CRC Press.

Leick, A. 1990. *GPS Satellite Surveying.* New York: John Wiley.

Southeastern Wisconsin Regional Planning Commission (SEWRPC). 1994. A mathematical relationship between NAD 27 and NAD 83 (91) state plane coordinates in southeastern Wisconsin. Technical Report No. 34. Waukesha, WI: Southeastern Wisconsin Regional Planning Commission.

Southeastern Wisconsin Regional Planning Commission (SEWRPC). 1995. Vertical datum differences in Southeastern Wisconsin. Technical Report No. 35. Waukesha, WI: Southeastern Wisconsin Regional Planning Commission.

# Appendix E: Evolution of Meaning for Terms

## *Network Accuracy and Local Accuracy**

The concept of spatial data accuracy is not new but, given the digital revolution of the past 50-plus years, meanings of the terms "network accuracy" and "local accuracy" as related to spatial data have evolved due to the transition in practice from analog to digital applications. Normal procedure is to reference a source within a document and to list the sources alphabetically at the end under "references." This document is a list of references (with comments) in chronological order to show sequence of development. Others are invited to share insights and to offer clarifications.

E.1. U.S. Bureau of Budget. 1947. United States National Map Accuracy Standards. http://nationalmap.gov/standards/pdf/NMAS647.PDF (accessed May 4, 2017)

The National Map Accuracy Standards (NMAS) have been beneficial for more than 50 years as related to the quality of information the public could expect on a map showing both planimetric features for location and contours for elevation. The map provided analog storage for spatial data and human consumption of those data was likewise analog. The digital revolution has changed all that and issues of "Disruptive Innovation" (http://www.globalcogo.com/DisruptiveInnovation.pdf, accessed May 4, 2017) have become a challenge for many. When using the NMAS, accuracy is tested by comparing the location of points in a data set with "positions as determined by surveys of a higher accuracy."

E.2. Mikhail, E.M. 1976. *Observations and Least Squares*. New York: Harper & Row.

Mikhail, E.M. and G. Gracie. 1981. *Analysis and Adjustment of Survey Measurements*. New York: Van Nostrand Reinhold.

The term "precision estimation" is used in these two books to describe the *a posteriori* covariance matrix of the estimates of the parameters (computed coordinates). Subsequent practice uses the term "accuracy" when discussing standard deviations (square root of the variance from the *a posteriori* covariance matrix) in the context of a standard.

E.3.  Federal Geodetic Control Committee (FGCC). John Bossler—Chairman. 1984. *Standards and Specifications for Geodetic Control Networks.* Rockville, MD: Federal Geodetic Control Committee. http://maps. gis.co.brown.wi.us/web_documents/LIO/PDF/LION/Geographic Frameworks.pdf (accessed May 4, 2017).

The FGCC standards and specifications are written for horizontal control, vertical control, and gravity networks. Designations assigned to discriminate levels of quality include Orders (first, second, third) and Classes (I, II)—the lower numbers being more precise. In each case the designators were applicable to the network in question—horizontal, vertical, gravity—but terms like global, relative, provisional, intended, absolute, and local were used as *adjectives* in the document to add clarity to concepts being discussed. Although these standards were written to be used with respect "to previously established control," Appendix B of this reference includes sections on "Global Variance Factor Estimation" and "Local Variance Factor Estimation."

E.4.  Federal Geodetic Control Committee (FGCC). Wesley V. Hull—Chairman. 1988/9. *Geometric geodetic accuracy standards and specifications for using GPS relative positioning techniques*, Version 5.0. Rockville, MD: Federal Geodetic Control Committee. https://www.ngs. noaa.gov/PUBS_LIB/GeomGeod.pdf (accessed May 4, 2017).

The FGCC 1988/9 standards and specifications were written specifically for 3-D relative GPS positioning and supplement the 1984 FGCC document by adding three orders (AA, A, and B) more rigorous than first-order. The 1984 standards were also improved by replacing the "distance accuracy standard" with relative positional tolerance. This permitted inclusion of a base error component to the standard in addition to the line-length dependent error component. Vertical is also discussed in terms of relative accuracies. Two possible "final" classifications are described as (1) a "geometric" classification for a relative positioning network and (2) a "NGRS" classification for surveys tied into the local network survey control system. The concepts are discussed but the terms "network accuracy" and "local accuracy" were not found in the document.

E.5.  American Society of Photogrammetry & Remote Sensing (ASPRS). 1990. *ASPRS accuracy standards for large-scale maps.* Bethesda, MD: American Society of Photogrammetry & Remote Sensing. http://www. asprs.org/a/society/committees/standards/1990_jul_1068-1070.pdf (accessed May 4, 2017).

One emphasis of these standards is that accuracy is to be applicable at ground scale. Horizontal accuracy and vertical accuracy are addressed separately with frequent reference to Item E.4. The ASPRS Standard contains the following quotes:

*Horizontal*: "When a horizontal control is classified with a particular order and class, NGS certifies that the geodetic latitude and longitude of that control point bear a relation of specific accuracy to the coordinates of all other points in the horizontal network."

*Vertical*: "When a vertical control point is classified with a particular order and class, NGS certifies that the orthometric elevation at that point bears a specific relation of specific accuracy to the elevations of all other points in the vertical control network."

Although not addressing relative accuracy in contrast to network or local accuracy, the ASPRS Standard states "map features are intended to possess accuracies relative to all other points appearing on the map."

E.6. Leick, A. 1993. Accuracy standards for modern three-dimensional geodetic networks. *Surveying and Land Information Systems* 53 (2): 111–116.

Alfred Leick was the Chair of the ACSM Ad Hoc Committee on Geodetic Accuracy Standards and prepared the report at the request of the Federal Geodetic Control Subcommittee (FGCS) to review possible revisions to existing standards and specifications. The introduction of the paper states that *Standards* specify the absolute or relative accuracy of a survey and *Specifications* contain the rules as to how the standards can be met. The committee focused on the *Standards* portion, leaving discussion of the *Specifications* to another time. The committee met at the ACSM convention in San Jose, California on November 8, 1992. Burkholder was a member of the committee and participated in that meeting.

Two quotes from that 1993 report are as follows:

The length-dependent principle has developed into a cornerstone in the philosophy and perception of geodetic networks. It is well suited for describing in simple terms the principle of neighborhood in geodetic networks (p. 112).

The length-dependent characterization of geodetic networks, surprisingly, tolerates translation and even systematic errors. For example, a translation of the whole network does not affect the quality and shape of the internal geometry. If systematic errors build up slowly as the size of the network increases, large absolute position errors eventually may occur, even though in small regions the line-dependent accuracy might still be acceptable for many users. For many surveying projects, the network accuracy in the immediate vicinity of the project area is important, in order to assure relative position accuracy (p. 112).

The 1993 report includes other issues such as use of Active Control Points (now known as CORS) and proposed position accuracy standards (millimeter, centimeter, decimeter, submeter, meter, and multimeter) as given in Table 4.

Report Conclusion: "The primary recommendation of ACSM's Ad Hoc Committee on Geodetic Accuracy Standards is to supplement current relative positioning standards, which are distance-dependent, with point-position accuracy standards."

E.7. Burkholder, E.F. 1997. *Definition and description of a global spatial data model (GSDM)*. Washington, DC: Filed with U.S. Copyright Office. http://www.globalcogo.com/gsdmdefn.pdf (accessed May 4, 2017).

This paper identifies concise mathematical definitions for both network accuracy and local accuracy. Those definitions were intended to be compatible with Leick (E.5). Those same definitions have been used by the author in subsequent items.

E.8. Southeastern Wisconsin Regional Planning Commission (SEWRPC). 1997. *Definition of a three-dimensional spatial data model for southeastern Wisconsin.* Waukesha, WI: Southeastern Wisconsin Regional Planning Commission. http://www.sewrpc.org/SEWRPCFiles/Publications/ppr/definition_three-dimensional_spatial_data_model_for_wi.pdf (accessed May 4, 2017).

Burkholder prepared this report for the Southeastern Wisconsin Regional Planning Commission (SEWRPC) published in 1997. An important part of the document is a paper (E.7) defining and describing the 3-D global spatial data model (GSDM).

E.9. Federal Geographic Data Committee (FGDC). 1998. *Geospatial positioning accuracy standards part 2: Standards for geodetic networks.* Reston, VA: Federal Geographic Data Committee, U.S. Geological Survey. http://www.fgdc.gov/standards/standards_publications (accessed May 4, 2017).

Part 2 of the FGDC geospatial positioning accuracy standards "provide a common methodology for determining and reporting the accuracy of horizontal coordinate values and vertical coordinate values for geodetic control points." The standards apply to horizontal positions, ellipsoid heights, and orthometric heights. The document emphasizes the importance of high-quality surveys of points to be included in the National Spatial Reference System (NSRS). But, more importantly, the document adopts accuracy classifications (with slight modification) as proposed by Leick (E.5) and states the following:

• Local accuracy is best adapted to check relations between nearby points.
• Network accuracy measures how well coordinates approach an ideal error-free datum.

E.10. Federal Geographic Data Committee (FGDC). 1998. *Geospatial positioning accuracy standards part 3: National standard for spatial data accuracy.* Reston, VA: Federal Geographic Data Committee, U.S. Geological Survey. http://www.fgdc.gov/standards/standards_publications (accessed May 4, 2017).

Part 3 of the FGDC geospatial positioning accuracy standards supersedes, but does not necessarily replace, the 1947 National Map Accuracy Standards (E.1). The root-mean-square-error (RSME) is used to estimate positional accuracy, which is tested by comparison of an observed position to that as determined by an independent source of higher accuracy. Accuracy is reported at the 95% (two sigma) confidence level and, typically, one estimate applies to all points within a given data set.

E.11. Burkholder, E.F. 1999. Spatial data accuracy as defined by the GSDM. *Surveying & Land Information Systems* 59 (1): 26–30. http://www.globalcogo.com/accuracy.pdf (accessed May 4, 2017).

This peer-reviewed paper summarizes the definition and characteristics of spatial data accuracy as contained in the 1997 defining document (E.7). One goal for this paper was to consider the question, "accuracy with respect to what?" It appears that consensus in using the stochastic model portion of the GSDM is still evolving.

E.12. US Forest Service and Bureau of Land Management USFS/BLM. 2001. *Standards and guidelines for cadastral surveys using global positioning system methods.* Washington, DC. https://www.blm.gov/cadastral/ Manual/pdffiles/CadGPSstd.pdf (accessed May 4, 2017).

This document was prepared cooperatively by the U.S. Forest Service and the Bureau of Land Management to provide guidance to Government cadastral surveyors (and others) in a time when GNSS surveying capability was evolving from static to fast static to kinematic to real-time kinematic. It provides an excellent overview of GNSS capability at the time within the context of the accuracy needed for cadastral surveying. Appendix A of this document (E.12) references the FGDC 1998 accuracy standards (E.9) and includes detailed descriptions of both network accuracy and local accuracy.

E.13. Craig, B.A. and J.L. Wahl. 2003. Cadastral survey accuracy standards. *Surveying and Land Information Science* 63 (2): 87–106.

Craig and Wahl note that "current cadastral survey accuracy standards are inadequate and need to be changed to reflect the way modern land surveys are conducted ..." The paper contains a review and commentary on the FGDC use of Network Accuracy and Local Accuracy. Although Appendix A of the paper (E.13) contains an excellent summary of concepts related to Positional Accuracy, the authors conclude that implementation of "local accuracy" needs additional study with consideration given to options that resemble the ALTA/NSPS standards (E.23).

E.14. American Society of Photogrammetry & Remote Sensing (ASPRS). 2004. *ASPRS guidelines, vertical accuracy reporting for Lidar data.* Bethesda, MD: American Society of Photogrammetry & Remote Sensing. https://www.asprs.org/a/society/committees/standards/Vertical_ Accuracy_Reporting_for_Lidar_Data.pdf (accessed May 4, 2017).

These guidelines were written specifically for use with Lidar technology. Nonetheless the concepts are similar to conventional accuracy considerations. The guidelines discuss both relative and absolute vertical accuracies and the importance of recognizing characteristics of each. The importance of horizontal accuracy is discussed as related to the repeatable location of test points and as related to side slope conditions. The glossary section includes definitions for horizontal accuracy as being with respect to a horizontal datum and for vertical accuracy as being with respect to a vertical datum.

E.15. Burkholder, E.F. 2004. *Fundamentals of Spatial Data Accuracy and the Global Spatial Data Model (GSDM).* Washington, DC: U.S. Copyright Office. http://www.globalcogo.com/fsdagsdm.pdf (accessed May 4, 2017).

This paper was written in response to a call for papers from ASPRS for papers to be included in a special issue of PE&RS journal. This paper (E.15) was reviewed but not published by ASPRS. Much of the material included in this paper can be found in Chapter 12 of the second edition of "The 3-D Global Spatial Data Model: Principles and Applications."

E.16.  Burkholder, E.F. 2008. *The 3-D Global Spatial Data Model: Foundation of the Spatial Data Infrastructure.* Boca Raton, FL: CRC Press. https://www.crcpress.com/The-3-D-Global-Spatial-Data-Model-Foundation-of-the-Spatial-Data-Infrastructure/Burkholder-Burkholder/p/book/9781420063011 (accessed May 4, 2017).

The first edition, (E.16), contains the mathematical definitions of network and local accuracies as given by Burkholder (E.7). Chapter 12 of this edition contains an example that is updated from the first edition and Chapter 14 contains a new comprehensive example of computing and using the mathematical definition of network accuracy and local accuracy.

E.17.  Soler, T. and D. Smith. 2010. Rigorous estimation of local accuracies. *Journal of Surveying Engineering* 136 (3): 120–125.

Soler and Smith note that "the concept of local accuracy, although intuitive, has not received the mathematical attention it deserves." The paper (E.17) generalizes the treatment of network and local accuracy as given by Burkholder (E.16) by deriving "exact" equations for computing local accuracy based upon the errors of each *X/Y/Z* component at each of two points defining a vector.

E.18.  Burkholder, E.F. 2012. Discussion of "Rigorous estimation of local accuracies" by T. Soler and D. Smith. *Journal of Surveying Engineering* 138 (1): 46–48.

The material on local accuracies as published in Burkholder (E.16) is defended as being complete, correct, and rigorous.

E.19.  Soler, T. and D. Smith. 2012. Closure to discussion of "Rigorous estimation of local accuracies" by T. Soler and D. Smith. *Journal of Surveying Engineering* 138 (1): 48–50.

In a closure to the discussion of rigorous estimation, Soler and Smith mount forceful arguments in favor of the "exact" method for computing local accuracy. It seems that their computation of "exact" local accuracy uses a different functional model equation than that used by Burkholder in (E.16).

E.20.  Soler, T., J. Han, and D. Smith. 2012. Local accuracies. *Journal of Surveying Engineering* 138 (2): 77–84.

According to the authors, "The objective of this study is to evaluate the different approximations in the technical literature that are used to compute the variance-covariance matrix of local accuracy." The discussion is extensive.

E.21.  Burkholder, E.F. 2013. *Standard deviation and network/local accuracy of geospatial data.* Washington, DC: U.S. Copyright Office. http://www.globalcogo.com/StdDevLocalNetwork.pdf (accessed May 4, 2017).

This paper suggests that the authors of the Discussion and the Closure (E.18 and E.19) talked past each other in promoting competing arguments related to computing local accuracy. The mathematical definition of local accuracy in Burkholder (E.16) is validated by using an independent method not requiring use of rotation matrices. (E.21) also provides specific examples for short, medium, and long lines (1, 20, and 100 km) that show very nearly identical results for both the methods espoused by Burkholder and by Soler and Smith. As an aside, the paper also shows conclusively that, using either method, local accuracies can be significantly better than network accuracy.

E.22. American Society of Photogrammetry & Remote Sensing (ASPRS). 2014. *Positional accuracy standards for digital geospatial data*, Version 1.0. Bethesda, MD: American Society for Photogrammetry and Remote Sensing. http://www.asprs.org/wp-content/uploads/2015/01/ASPRS_Positional_ Accuracy_Standards_Edition1_Version100_November2014.pdf (accessed May 4, 2017).

The stated objective of this document "is to replace the existing ASPRS Accuracy Standards for Large-Scale Maps (E.14) and the ASPRS Guidelines, Vertical Accuracy Reporting for Lidar Data (E.15) to better address current technologies. This standard includes positional accuracy standards for digital orthoimagery, digital planimetric data and digital elevation data." It is a comprehensive document and includes definitions for accuracy, horizontal accuracy, local accuracy, network accuracy, positional accuracy, and others. It also states, "all accuracies are assumed to be relative to a published datum and ground control network used for the data set as specified in the metadata." These standards represent significant advances in use of modern/digital technologies but stop short of recognizing evolving practice with respect to use of an integrated 3-D spatial data model. With only minor modifications, these standards appear to be compatible with definitions of network and local accuracies as defined by the GSDM. Specifically, the GSDM contains additional detail showing how network and local accuracies are determined.

E.23. American Land Title Association and National Society Professional Surveyors (ALTA/NSPS). 2016. *Minimum standard detail requirements for ALTA/NSPS land title surveys*. Washington, DC/Frederick, MD: American Land Title Association (ALTA)/National Society of Professional Surveyors. http://c.ymcdn.com/sites/www.nsps.us.com/resource/resmgr/ ALTA_Standards/2016_Standards.pdf (accessed May 4, 2017).

Members of the American Land Title Association® (ALTA®) need reliable documents as part of the basis for insuring title to land purchased by a client. Among others, the measurement standard is to meet the Relative Positional Precision estimated by the results of a correctly weighted least squares adjustment of the survey. Because of ambiguity "with respect to what" some confusion exists as to whether that standard can be met using network accuracy as obtained from the least squares adjustment. It appears (see Chapter 14) that the intent of the ALTA/NSPS Relative Positional Precision standard can be met more appropriately using local accuracy as defined and computed using the GSDM.

# Index

## A

AASHTO, *see* American Association of State
    Highway Transportation Officials
Absolute quantities, 311–312
Accuracy
    absolute, 310
    datum, 366
    definition of, 365
    of derived quantity, 366
    errors and, 75–76
    network and local, *see* Network and
        local accuracy
    of point, 366
    and precision, 365
    precision and, 76
    relative, 310
    spatial data, 366
ACSM, *see* American Congress on Surveying
    and Mapping
Active Control Points, 467
Adams, Oscar S., 275
Addition, 52
Advanced processing, 259–262
Aircraft landings, automated, 253
Alaska, 267, 301
    Juneau, 302
    Lambert conic conformal projection, 277
    oblique Mercator projection, 277, 302
    transverse Mercator projection, 277
    zones and, 277
Alexandria, 139
    distance to Syene, 139
Algebra, 55–56
    boolean, 56
Algorithm for functional model, 10–14
Algorithms for traditional map projections,
    285–302
    Lambert conic conformal projection, 286–289
    low-distortion projections, 302
    oblique Mercator projection, 297–302
    transverse Mercator projection, 289–296
Al Mamun, Caliph Abdullah, 140
ALTA/NSPS, *see* American Land Title
    Association and National Society
    Professional Surveyors
Altitude, 205
American Association of State Highway
    Transportation Officials (AASHTO), 110
American Congress on Surveying and Mapping
    (ACSM), 197

American Land Title Association and National
    Society Professional Surveyors
    (ALTA/NSPS), 471
American Society of Civil Engineers (ASCE),
    24, 197, 216
American Society of Photogrammetry and
    Remote Sensing (ASPRS), 38, 216,
    466–467, 469, 471
American Standard Code for Information
    Interchange (ASCII), 50
    characters, 253
Angle-right on radial stakeout, 120
Angles, 36, 58
    Euclidean geometry and, 58
    geometry of, 58
Anti-spoofing (A-S), 248
Approximations for $\pi$ based on area of unit
    circle, 113
Arabs
    decimal system, 48–49
Architects, 128
Arctic Circle, 142
Area
    adjacent to a spiral, 117–118
    of a circle, 111
    by coordinates, 104–106
    formed by curves, 112
    of a sector and a segment, 111
Aristotle, 139
Arithmetic, mathematical concepts of, 55
A-S, *see* Anti-spoofing
ASCE, *see* American Society of
    Civil Engineers
Ascension Island, South Atlantic, 242
*ASCE Shale Energy Engineering
    Conference*, 439
ASCII, *see* American Standard Code for
    Information Interchange
ASPRS, *see* American Society of
    Photogrammetry and Remote Sensing
ASPRS Positional Accuracy Standards for
    Digital Geospatial Data (ASPRS
    2014), 24
Astronomy, geodetic, 132
Aswan Dam, 139
Atomic clocks, 247
Augmentation, 260
Australia, 199
Autonomous processing, 255–256
Axioms of addition (for real numbers *A, B,*
    and *C*), 56

Axioms of equality (for real numbers *A, B,* and *C*), 56
Axioms of multiplication (for real numbers *A, B,* and *C*), 56
Azimuth, 94–95

**B**

Babylonians, 50
Baseline, 330
Bestor, O. B., 275
Binary system
   mathematical concepts of, 50
BIPM, *see* Bureau International des Poids et Mesures
Birdseye, Colonel C. H., 275
BK1 transformation, 157
BK2 transformation, 437
   example of, 151–153
   iteration, 158
   Vincenty's method
      ellipsoid surface area, 167–169
      example of, 162–163
      length of a parallel, 166
      meridian arc length, 163–165
      once-through, 159
      surface area of sphere, 166–167
BK10 and BK11 transformations
   Lambert conic conformal projection for, 288–289
   for Lambert projections, example of, 289–290
   for oblique Mercator projection, 299–302
   for oblique Mercator projection, example of, 301
   for transverse Mercator projection, 292–295
   for transverse Mercator projections, example of, 296
BK18 by integration, 178
BK18 numerical integration printout, 179–180
BLM, *see* Bureau of Land Management
BLM Manual of Surveying Instructions, 416
Blunder checks, 338, 341–342
Blunders, 74
*B* matrix, 343, 345–348
Boolean algebra, 56
Brazil, 267
BROMILOW, 388, 400–401
Bureau International des Poids et Mesures (BIPM), 136
Bureau of Land Management (BLM), 134
Burkholder, Earl F., 144
BURKORD™ database, 21, 233, 384, 429
   BURKORDT data file, for network example, 368–371
   correlations between stations, 367
   GSDM computational environment and, 259
   images and, 270

software, 18
   station covariance values, 367
   station *X/Y/Z* geocentric coordinates, 367
   types of records and, 231–232
BURKORD™ program, 419, 421
Burt Improved Solar Compass, 416

**C**

Calculus
   differential, 70
   example, 68–70
   integral, 70–71
   mathematical concepts of, 68–70
California, 277, 279
California Science Center in Los Angeles, 133
Canada, 197, 206, 223
Canberra, Australia, 199
Canopus (constellation), 140
Cape Canaveral, Florida, 242
Carrier phase, 250–251
Cartesian coordinate system, *see* Coordinate systems
Cartesian models, two-dimensional, 97–99
   engineering/surveying reference system, 98–99
   math/science reference system, 98
Cartography, 267
Cassini, Jacques, 142
Cassini, Jean Dominique, 142
Cellular telephones, GNSS, 145
Chicago, 65, 67, 188–191, 384
China, 50
Circle, 58
Circular curves
   area formed by curves, 111–112
   area of unit circle, 112–113
   degree of curve, 106–107
   elements and, 107–109
   equations and, 107–109
   metric considerations, 110
   stationing, 109–110
Circular trigonometric functions, 64
Clairaut's constant, 178–179, 182–183, 188
Claire, Charles N., 275
Clarke Spheroid of 1866, 151, 196–197, 228
   used in North America, 150
Clinton, Bill, 134
Coalition of Geospatial Organizations (COGO), 24–25, 134
Coarse acquisition (C/A) code, 242, 248, 260
   elapsed time and, 244
   receivers, 244
Coarse acquisition (C/A) code handheld receivers
   datum for, 256
   display, 256
   time and, 256
   units and, 256

Coast Survey, 278
COGO, *see* Coordinate geometry
Cold War, satellite-positioning technology and, 237
Collioure, 142
Colorado, 242
Colorado Springs, Colorado, 242
Columbus, Christopher, 137
Combined factor for a line, 284
Committee on the North American Datum 1971, 198
Compass Rule, 285
Computational designations, 5–10
Computation network accuracy and local accuracy, example, 323–327
Computations
  three-dimensional, 124
  two-dimensional, 124
Computer printout, least squares adjustment, 345, 351–362, 364
Computer science, 50
Computers, development of, 132
Conditional adjustment, 332
Confidence intervals, 77–78
Conformal map projection, 269
Conformal projections, state plane coordinate system (SPCS), 281
Conic sections, 61–62
Constants (1983), State plane coordinate system (SPCS), 451–458
Continuously operating reference station (CORS), 204–205, 261, 313, 387
  data transformations, 204
  position, 224
Conventional Terrestrial Pole (CTP), 4, 31, 95, 199, 210, 226, 414
Conventions, 48, 68
  decimal, 48–49
  fractions, 48
  numbers, 48
Convergence, 90
  of meridians, 281–282
Conversions
  mathematical concepts of, 51
Coordinated Universal Time (UTC), 247, 253, 256
  coarse acquisition (C/A) code handheld receivers and, 256
Coordinate geometry (COGO)
  area by coordinates, 104–106
  forward, 99–100
  forward computation equations, 115
  intersections, 100–102
  inverse, 100
  operations, 99
  perpendicular offset, 104
  procedures, 99
  three-dimensional programs, 34

Coordinates, arbitrary local, 255
Coordinates, geodetic
  absolute, 255
Coordinate systems, 51–52
  Cartesian, 30, 51, 93, 195, 305
  diagram showing relationship of, 4
  geodetic, 31
  *latitude/longitude/heigh*, 386
  relationship of, 6
  spatial data and, 30
  three-dimensional rectangular, 51
  two-dimensional, three-dimensional, and geodetic, 94
Coordinating Committee on Great Lakes Basic Hydraulic and Hydrologic Data, 207
Corpscon Windows-based program, 208
Corrections to previous direction and distance, 184
CORS, *see* Continuously operating reference station
Council of State Governments, 277
Covariance matrix, 331–332, 334–335, 367
CRUCESAIR, 319–320
CTP, *see* Conventional Terrestrial Pole
Cube, 60
Current status: NAD83 state plane coordinate systems, 279

**D**

Database records, formats of
  point record, 385
  point-to-point correlation record, 385
Data base, three-dimensional, 305
Data, geospatial
  three-dimensional (3-D), 268
Data transformations
  horizontal time-dependent positioning (HTDP), 208
  NAD83(86) to high-precision geodetic network (HPGN), 208
  NAD27 to NAD83 (86), 208
  NGVD29 to NAVD88, 208
  seven-or (fourteen-) parameter transformation, 207–208
  software sources, 208–209
Datum
  absolute, 310
  accuracy, 366
  autonomous processing, 255
Datum accuracy, 319
Datums, geodetic, 195–210
  data transformations, 207–210
  horizontal, 196–205
  three-dimensional, 209–210
  vertical, 205–207

Death Valley, California, 227
Decimal, 48–49
Decimal degrees, coarse acquisition (C/A) code
    handheld receivers and, 256
Decimal meter, 143
Deflection-of-the-vertical, 147, 155, 219, 222, 405
    direction of gravity and, 218
    Laplace correction and, 219
    Laplace equations and, 220
Degree measurement, 139
Degree of curve, 106–107
Degree of curve radii, comparison of, 106–107
Degrees, minutes, and seconds (decimal)
    coarse acquisition (C/A) code handheld
        receivers and, 256
Delambre, Chevalier, 143
Department of Transportation (DOT), 110
Derived physical quantities, 34
Descartes, René
    Cartesian coordinate system, 305
    rules of logic and, 55
Designations for spatial data computations and
        transformations, 10
DGPS, see Differential code phase system,
        see Differential GPS positioning
Diego Garcia, Indian Ocean, 242
Differencing
    double, 252
    option, 252
    single, 252
    triple, 252
Differential code phase system (DGPS), 37
Differential GPS positioning (DGPS), 260
    augmentation and, 261
Digital maps, 269, 307
    transition to, 307
Digital revolution, consequences of, 306
    increased knowledge of location, 306
    more access to information, 306
    more efficient movement of people, products,
        and resources, 306
    nanotechnology, 306
    reduction in privacy, 306
Digital revolution, forces driving
    development of information technology,
        science, and management, 306
    electronic signal processing, 306
    enhanced spatial literacy, 306
    miniaturization of circuits and physical
        devices, 306
    satellite positioning, 306
    transistors, 306
Digits, Arabic, 48
Dimension, geometry, 57
Display, autonomous processing, 256
Distance distortion, grid scale factor, 273
Distance, geometry and, 57

Distortion, zones and, 277
Division, 53–54
DMD, see Double-meridian-distance
DOD, see U.S. Department of Defense
Doppler, Christian, 244
Doppler data, 199
    use of with elapsed time, 245
Doppler effect, 240, 252
Doppler measurements, 241
Doppler orbitography and radio positioning
        integrated by satellite (DORIS), 200
Doppler positioning, 241
Doppler shift, 245
    GNSS positioning and, 244
    for signal received (example), 245
DORIS, see Doppler orbitography and radio
        positioning integrated by satellite
DOT, see Department of Transportation
Double differencing, 252
Double-meridian-distance (DMD), 104
Dowd, Charles F.
    Saratoga Springs, N.Y., 136
    standard time zones and, 136
Dracup, Joseph
    U.S. Coast and Geodetic Survey, 275
    "U.S. Horizontal Datums,", 197
Dunkirk, 142
Dynamic height, 206, 217

E

Earth, 213
    center of mass, 3, 198–201, 226, 311, 386
    curvature, 8, 93
    elevation, 136
    ellipsoidal model, 60, 128, 148, 384
    equator, 31, 49, 148
    flat, 139
    as flat, 218, 268
    flattening of at the poles, 147, 215
    gravity and, 131–132, 222
    latitude, 135
    longitude, 135
    magnetic field of, 95
    measuring, 131, 163
    Meridian Section Quadrant, 156
    North Pole and, 49
    projection figures and, 270–271
    radius of, 67, 155
    rotation of, 214
    shape of, 137, 147, 151, 195
    size of, 139, 147, 151
    spherical model of, 67, 128, 139,
        267–268, 381
    spin axis of, 61, 148, 151, 172, 311, 414
    surface and, 31
    systems, 132

tides of, 2, 387
view of satellites from, 240
view of the ellipsoid, 156
Earth-centered Earth-fixed (ECEF) coordinates, 4,
        133, 200–201, 219, 310–311, 386, 414
    absolute values and, 429
    baseline processing and, 264
    coarse acquisition (C/A) code handheld
        receivers and, 256
    coordinates, 386, 416
    datum to datum relationships and, 386
    digital technology and, 156
    of DOD Global Control Stations, 243
    geocentric system, 31, 186
    GSDM and, 30, 264, 268, 379
    image file and, 383
    interferometry and, 246
    metric in GSDM, 279
    rectangular, 155, 200, 224, 280
    reference frame, 382
    rotation matrix derivation, 447–449
    spatial data and, 5, 94
    system, 199, 244
    three-dimensional datum and, 209
    vector data, 379
    $X/Y/Z$ coordinates, 381
Earthquakes, geodesy, 133
Earthquake zones, 201
East/northings, 99
ECEF, see Earth-centered Earth-fixed
ECEF covariances, vector data and, 384
ECHO I satellite, 241
ECHO II satellite, 241
EDMI, see Electronic distance measuring
        instrument
EFB network station, 388, 401, 403–404
Egypt, 139
Elapsed time, GNSS positioning and, 244
Electromagnetic spectrum, 247–248
Electronic distance measuring instrument
        (EDMI), 34–36, 73
Elements and equations, 107–109
Elevation, 255
    generic, 216
    as relative value, 312
Elevation reduction factor, 229
    uncertainty and, 229
Eleventh General Conference on Weights and
        Measures (1960)
    (SI) International System of Units and,
        49, 279
    redefinition of meter and, 278
Ellipse
    eccentricity, 149
    flattening of, 149
    geometry of, 58
    semimajor and semiminor axes, 150

Ellipsoid, 60, 147–148
    equation of, 155–156
    geoid and, 148
    geometrical mean radius, 155
    model of Earth and, 128
    normal section radius of curvature, 155
    radii of curvature, 154–155
    surface area, 167–169
Ellipsoid heights, 217, 317
    as derived quantity, 312
    GPS-derived, 224
    as measurable physical quantity, 224–225
    measurements and computations, 224–225
    orthometric heights and, 227
    referenced to Earth's center of mass, 386
Ellipsoid radii of curvature, 154–155
    equator and, 154–155
    poles and, 154–155
Ellipsoid, rotational, 155–157
    equation of ellipsoid, 155–156
    geocentric and geodetic coordinates,
        156–157
Ellipsoid surface area, 167–169
El Paso, Texas, 410
Engineering/surveying reference system, 98–99
Engineers, 128
England, 196
Equation of an ellipsoid centered on the origin, 61
Equation of a plane in space, 60–61
Equation of a sphere in space, 61
Equation-of-time (motion of earth vs. atomic
        time), 136, 226
Equation-of-Time model, 136
Equator
    determining distance from pole, 143
    maps and, 267
Equidistant map projection, 269
Equipotential surface, 216
Equivalent map projection, 269
Eratosthenes, 139–140
    measurement of the Earth, 140
    size of the Earth and, 195
Error ellipses
    2-D plane coordinates and, 87
    mathematical concepts of, 87
Error propagation
    equation, 86
    mathematical concepts of, 81–87
Errors, 73–75
    bias and, 74
    blunders, 74
    environmental sources of, 75
    instrumental sources of, 75
    personal sources of, 74–75
    precision and, 75–76
    random, 74
    systematic, 74

Euclidean geometry, 280
Euclidean space, 57
Euler's formula, 155, 182
European merchants, decimal system, 49
European Space Agency, 244
Everest, George, 144
Excel spreadsheet, 323
Executive Summary, 24

## F

FAA, *see* Federal Aviation Administration
Father's Point/Rimouski
    benchmark for IGLD85, 207
    Quebec, Canada, 207, 223
Federal Aviation Administration (FAA), 261
    wide-area augmentation system
        (WAAS), 261
Federal Geodetic Control Committee
        (FGCC), 466
Federal Geodetic Control Subcommittee (FGCS),
        427, 467
Federal Geographic Data Committee (FGDC), 24,
        382–383, 468
    NILS, 134
    standard, 311
Federal Register Notice (FRN), 278–279
FGCC, *see* Federal Geodetic Control
        Committee
FGCS, *see* Federal Geodetic Control
        Subcommittee
FGDC, *see* Federal Geographic Data
        Committee
FINIAL on Skeen Hall, 401–402, 407
First General Adjustment of geodetic leveling in
        the US, 223
Flamsteed, John, 226
"Flat-Earth" components, 447
Flat-Earth model, 195, 265, 310
    applications of, 308
    computations, 307
    GSDM and, 237–238
    relationships, 93
    uses of, 128, 147, 231
Florida, 242, 277
Foot, 278–279
FORTRAN, 285
Forward, 99–100
Forward (BK3), 186–187
Fractions, 48
France, 143, 278
French Academy of Science, 49
    arc of triangulation from Paris to
        Amiens, 141
    geodetic surveying expeditions in Peru and
        Lapland, 142
    standardized measurement and, 143

French Revolution, 143
FRN, *see* Federal Register Notice
Full network accuracy, 367
Functional model component of GSDM, 3–6, 381
    algorithm for functional model, 10–14
    computational designations, 5–10
Fundamental wave measurement, 251
*f* vector, 341–345

## G

Galilei, Galileo, 141
GALILEO (European satellite-position system),
        238, 244, 249, 254
GALILEO satellite-position system, 244
Gamma rays, 248
Gauss, Carl Friedrich, 143
Gauss mean radius, 155
General error ellip, 87
General least squares method, 332
Generic triangle, 64–65
Geocentric and geodetic coordinates, 4, 156
Geocentric gravitational constant, 199
Geocentric/local reference frame positions, 400
Geodesic correction, 175
Geodesy
    earthquakes and, 133
    fields of, 131–132
    forecast for the twenty-first century, 145
    goals of, 132–137
    historical perspective, 137
    nineteenth and twentieth century
        developments, 143–144
    religion, 138
    science, 131, 138–139
Geodesy, geometrical, 241
    BK1 transformation, 156–157
    BK2 transformation, 156, 158–169
    ellipse, 149–154
    ellipsoid, 154–155
    geodetic line, 169–175
    geodetic position computation: forward and
        inverse, 175–193
    GSDM, 381
    rotational ellipsoid, 155–157
Geodesy, historical perspective of, 137–143
    degree measurement, 139
    religion, science and geodesy, 138–139
*Geodesy in the Year 2000*, 133
Geodesy, physical, 213–234
    definitions, 215–218
    geoid modeling and, 227–234
    global spatial data model (GSDM) and, 231–232
    interpolation and extrapolation, 221–222
    measurements and computations, 221–225
    processing GPS data, 253–262
    time, 225

Geodesy, satellite, 131, 239–240
    differencing, 251–252
    the future of survey control networks,
        262–265
    global navigation satellite systems (GNSS)
        and, 237–265
    history of satellite positioning, 240–244
    modes of positioning, 244–246
    Receiver Independent Exchange (RINEX)
        format, 252–253
    satellite signals, 247–251
Geodetic astronomy, 132
Geodetic azimuths, 172–174
    geodesic correction, 172–174
    target height correction, 174
Geodetic forward (BK18), 176
*Geodetic Glossary*, 197
Geodetic inverse (BK19) computation, 66
Geodetic line
    around the Earth, 169
    Clairaut's constant, 170–172
    geodetic azimuths, 172–174
Geodetic line azimuth using Clairaut's
        constant, 171
Geodetic line numerical integration, 179
Geodetic position computation: forward
        and inverse
    geodetic position computations using state
        plane coordinates, 185–186
    GSDM 3-D geodetic position computations,
        186–193
    numerical integration, 178–184
    Puissant Forward (BK18), 176–177
    Puissant Inverse (BK19), 177–178
Geodetic reference system 1980 (GRS80), 151,
        228–229, 285–286, 289, 297
    ellipsoid, 200
    North American Datum of 1983 (NAD83)
        and, 197–199
    World Geodetic System of 1984 (WGS84)
        and, 200
"Geographic Information for the 21st Century"
        (NAPA 1998), 23, 238
Geographic information system (GIS), 21,
        132, 238, 255
    applications, 238
    basic geodetic control and, 313
    community, 280
    databases and, 313
    metadata, 382
Geoid, 147–148, 216–217, 312
    ellipsoid and, 148
GEOID09, 428, 431
GEOID12A, 428
GEOID12B, 428, 431–432, 438
"Geoid Contours in North America," U.S. Army
        Map Service, 223

Geoid heights, 217–218, 222
    difference in, 232
GEOID03 model, 428, 432
Geoid modeling, 228
    global spatial data model (GSDM) and,
        231–232
    using, 232–234
Geomatics, 131
Geometrical consistency of computational
        environment, 136
Geometrical geodesy, 131
    BK1 transformation, 157, 160
    BK2 transformation, 158–169
    ellipse, 149–154
    geodetic line, 169–175
    geodetic position computation: forward and
        inverse, 175–193
Geometrical mean radius, 155
Geometrical models for spatial data computations
    circular curves, 106–113
    conventions, 94–97
    coordinate geometry, 99–106
    radial surveying, 118–121
    spiral curves, 113–117
    three-dimensional models for spatial data,
        124–128
    two-dimensional cartesian models, 97–99
    vertical curves, 121–124
Geometry, 56–59
    coordinate, *see* Coordinate geometry (COGO)
    of intersections, 101
    solid, 60–62
Geopotential number, 217
Geospatial Positioning Accuracy Standards;
        Part 2-Standards for Geodetic
        Networks, 382
*Geospatial Solutions*, 238
Germany, 247
GIS, *see* Geographic information system
Global COGO, Inc., 384
    web site, 384
Global navigation satellite systems (GNSS), 4–5,
        21, 23, 159, 237–239, 254, 306, 313,
        319, 379–380, 419, 421, 461
    angles and, 36
    cellular telephones, 145
    computations, 264
    data and, 75
    equipment, 37, 118
    geodetic meridian, 414
    height differences by, 430
    measurement of distance and, 88
    positioning, 73, 75, 382
    precise $\Delta X/\Delta Y/\Delta Z$ components, 5
    procedures, 118
    statewide WISCORS Network, 336
    surveying, 239

surveys, 75, 132
technology, 261, 264
Wisconsin network, 336–338
Global navigation satellite systems (GNSS) data
    choices in the United States, 384
    ITRF, 384
    North American Datum of 1983
        (NAD83), 384
    WGS84, 384
Global navigation satellite systems (GNSS)
        positioning, 239, 244, 247
    Doppler shift, 244–245
    elapsed time, 244
    interferometry, 246
Global positioning system (GPS), 22–23, 227,
        233, 254, 320; see also Navigation
        satellite timing and ranging
        (NAVSTAR) system
    accuracy, 253–254, 263
    aircraft landings and, 253, 255, 261
    angles and, 36
    antenna, 73
    baselines, 239, 258
    base-station networks, 239
    carrier phase data, 252
    cellular telephones, 144
    coarse acquisition (C/A) code, 242
    concept, 238
    control segment, 241
    CORS network, 242
    earth-moving equipment and, 253
    equipment, 37, 118
    International Terrestrial Reference Frame
        (ITRF) and, 200–201
    as interpolation system for spatial
        data, 224
    mapping, 240
    measurement of distance and, 88
    military purposes of, 241
    navigation, 253
    network, 224
    observations, 220
    operation of, 242
    as passive system, 242
    positioning, 247
    RTK surveying practices, 253
    satellite geodesy and, 237–238
    space segment, 241
    spatial data and, 238, 253–254
    stochastic model and, 240
    surveys, 202, 243
    uncertainty and, 254
    unmanned aerial vehicle (UAV), 263
    user segment, 241
    U.S. government and, 244
    vector processing, 253
    vectors, 429

very long baseline interferometry
        (VLBI) and, 247
    wide area differential GPS (WADGPS)
        procedures, 201
Global positioning system (GPS) data monitor
        daily motion of CORS station, 387
Global positioning system (GPS) data processing,
        217, 253–262
    autonomous processing, 255–256
    collection of, 379
    spatial data types and, 253
    vector processing, 256–257
Global positioning system (GPS) traditional
        networks, 259
Global positioning system (GPS) processing, 260
    absolute and relative considerations, 260
    applications to, 255
    Galileo and, 261
    global navigation satellite systems (GNSS), 261
    global spatial data model (GSDM), 262
    GLONASS, 261
    International Terrestrial Reference Frame
        (ITRF), 262
    North American Datum of 1983
        (NAD83), 262
Global positioning system (GPS) receivers, 204,
        244, 253, 259
    for civilian use, 243
    handheld, 250
    observing sessions for nontrivial vectors and, 258
Global positioning system (GPS) satellites, 200,
        224, 232, 244–246, 248
    Earth's center of mass, 201
    orbit parameters and, 313
    orbits express in ITRF reference frame, 314
    spatial data computations, 224
    stored $X/Y/Z$ locations and, 317
Global Positioning System (GPS)
        signal, 243–244
    L1 frequency, 248
    L2 frequency, 248
    radio waves and, 248
    structure of, 248
Global positioning system (GPS)
        technology, 237
    precise $\Delta X/\Delta Y/\Delta Z$ components, 5
Global positioning system (GPS) time, 247, 253
    Coordinated Universal Time (UTC), 247
Global positioning system (GPS) tracking
        stations, 242
    Ascension Island, South Atlantic, 242
    Cape Canaveral, Florida, 242
    Colorado Springs, Colorado, 242
    Diego Garcia, Indian Ocean, 242
    Hawaii, Pacific Ocean, 242
    Kwajalein, North Pacific Ocean, 242
    Master Control Station of GPS system, 242

Global positioning system (GPS) vector, 253
Global spatial data infrastructure (GSDI), 3, 29, 134
Global spatial data model (GSDM), 2–3, 29, 144,
          174, 237–238, 264–265, 419, 467, 471
  accuracy, 29
  algorithms for determining standard deviation
          of each point in three directions, 383
  applications and, 90–91
  assumptions and, 231
  BURKORD™ software and database, 18
  COGO, 24–25
  as a computational environment, 259
  consequences, 308–309
  conventions used by, 97
  coordinate systems and, 52
  covariance matrices, 14–16
  covariance matrix, 367
  database issues, 384–385
  data measurements must be three
          dimensional, 307
  development of, 461–462
  digital geospatial data and, 268
  digital revolution and, 305
  digital technology, 21
  distance distortion and, 277
  dynamic environments, 25
  elevation as derived quantity, 309
  engineering, 309
  errorless spatial data and, 40
  error propagation concepts, 311
  example: New Orleans to Chicago, 188
  features
          functional model, 381
          stochastic model, 381–384
  flat-Earth differences and, 148
  functional model component, 3–14, 366
  geocentric *X/Y/Z* coordinates and, 230
  geodesy and, 132
  geodetic inverse, 193
  geospatial data and, 253
  GNSS-derived data, 253
  GPS and, 260, 280
  hypothesis testing, 400–401
  implementation and, 281
  implementation issues, 385–387
  inventory, 309
  laying out a parallel of latitude, 440–443
  linear least squares GNSS network, *see* Least
          squares adjustment
  low-distortion projections and, 302
  mathematical concepts and, 90–91
  metric units and, 279
  NAPA, 22–23
  NAVD 88 Elevation of HARN Station
          REILLY, 427–432
  navigation, 309
  networks and, 202

New Mexico Initial Point, 410–417
NEXRAD weather radar, 432–433
NOAA, 23
  photogrammetrists, 381
  planning, 309
  point-of-beginning (P.O.B.) datum
          coordinates, 255
  policies and procedures, 380
  Principal Meridian, 410–417
  rotation matrix derivation, 447
  single origin assumed for geospatial
          data, 309
  spatial data accuracy, 366
  spatial data and, 34, 43, 72, 200, 238,
          240, 444
  spatial relationships and, 39
  standard deviation, 310
  static environments, 25–26
  static or dynamic, 26
  stochastic model and, 6, 310
  stochastic model component of, 366
  supplemental NMSU Campus Control
          Network, 387–400
  surveying, 309
  Texas and New Mexico along the Rio Grande
          River, 417–427
  3-D computational models, 434–438
  3-D database, 379
  three-dimensional datums and not supported,
          201, 209–210
  three-dimensional geometrical relationships
          and, 133
  three-dimensional model and, 280
  three-dimensional spatial data and, 305
  2-D survey plat, 407–413
  uncertainty and, 254, 310
  underground mapping, 438–440
  using points stored in the *X/Y/Z* database,
          317–319
  using terrestrial observations,
          401–407
  vector data and, 379
  in Wisconsin, 427
  World Vertical Datum yyyy, 309
Globe and mercator grid, 267–268
*Glossary of the Mapping Sciences*, 197
Gnomonic projection, 269–270
GNSS, *see* Global navigation satellite systems
GNSS Survey Network on NMSU Campus, 319
Google, 253, 255
Gore, Al
  speech at California Science Center in Los
          Angeles, 133
  1998 speech "The Digital Earth Understanding
          Our Planet in the 21st Century,", 134
GPS, *see* Global positioning system
*GPS World*, 238

Grade, 121
Gravity
    absolute, 223
    computations and, 223
    measurement of, 222–223
    physical geodesy and, 214
    shape of the geoid and, 218–219
Great Lakes, 206
Great Lakes System, 137
Great Trigonometric Survey (GTS), 144
Greeks, ancient, 136, 226
Greenwich, England, 160
Greenwich Mean Time, 136
Greenwich Meridian, 97, 151, 154, 311
    ECEF, 31
    functional model component, 3  4
    geometry of rotational ellipsoid and, 148
    section, 154
Greenwich meridian, 196, 199
Greenwich Observatory, 136, 226
Grid azimuth, 281
Grid distance, 281–284
    ellipsoid distance and, 281
Grid-ground distance difference, 277, 309
Grid scale factors, 282
    comparison of, 273
    distortion and, 273, 277
GRS80, see Geodetic reference system 1980
GRS80 ellipsoid, 415
GSDI, see Global spatial data infrastructure
GSDM, see Global spatial data model
GSDM datum, three-dimensional
    defined, 311
    Earth-centered Earth-fixed (ECEF)
        coordinates, 311
GSDM 3-D geodetic position computations,
        186–193
    forward (BK3), 186–187
    GSDM inverse example: New Orleans to
        Chicago, 188–193
    inverse (BK4), 187–188
GSDM 3-D inverse, 16
The GSDM Facilitates Existing Initiatives, 25
GTS, see Great Trigonometric Survey
"Guidelines for Establishing GPS-Derived
        Orthometric Heights Version 1.4"
        (NGS 2006b), 232
Gunter's Chain, 141, 416

**H**

HARN, see High-accuracy reference network
Hassler, Ferdinand, 143
Hawaii, 242
Hayford, John, 197
Height, 205
"Height modernization,", 192

Helmert orthometric heights, 207, 217
High-accuracy reference network (HARN),
        202–203, 232, 313, 407
    CORS stations and, 204–205
High-precision geodetic networks (HPGN)
    high-accuracy reference network (HARN)
        and, 203
Hindu-Arabic number system, 48
History
    maps and, 275
Horizontal datums, 196–205
    brief history of, 196–197
    Continuously Operating Reference Station
        (CORS) stations, 204–205
    high-accuracy reference network (HARN),
        202–203
    International Terrestrial Reference Frame
        (ITRF), 200–202
    NAD27, 144
    North American Datum of 1983 (NAD83),
        198–199
    World Geodetic System of 1984 (WGS84),
        199–200
Horizontal distance
    ellipsoid and, 282–283
    reliable, 282
    sea level distance and, 282–283
Horizontal distance (HD), 368
Horizontal time-dependent positioning (HTDP),
        204, 314
    converting X/Y/Z coordinates from one 3-D
        datum to another, 314
    program, 208–209
HTDP, see Horizontal time-dependent
        positioning
Hudson Bay Region, 206
Huygens, Christian, 142
Hydraulic head computations, 206
Hypothesis testing, 78–79

**I**

IAG, see International Association of Geodesy
IERS, see International Earth Rotation Service
IGLD, see International Great Lakes
        Datum 1955
IGLD85, see International Great Lakes Datum
        of 1985
Image files, 383
    geospatial location and, 383
India, 48
Indirect observations, 332–335
Infrared rays, 247–248
Initial operational capability (IOC), 241
Initial point, 223
Inside GNSS, 238
Interference of light waves, 246

Interferometry, 246
  global positioning system (GPS) carrier
      frequency, 246
  GNSS positioning and, 247
International Association of Geodesy (IAG), 135, 197
International Atomic Time (TAI), 247
International Earth Rotation Service (IERS),
      199–200, 210
International Ellipsoid, 197
International Foot, 278–279
  coarse acquisition (C/A) code handheld
      receivers and, 256
  NATO and, 278
International Great Lakes Datum (IGLD) 1955, 206
International Great Lakes Datum of 1985
      (IGLD85), 206
International Meridian Conference, 97, 136, 226
International system for units of measure (SI),
      49–50
  fundamental physical quantities, 34
International Terrestrial Reference Frame (ITRF),
      157, 200–202, 224, 231, 311, 315
  coarse acquisition (C/A) code handheld
      receivers and, 256
  CORS stations and, 204–205
  datum, 285
  GSDM and, 136, 254, 264
  horizontal time-dependent positioning
      (HTDP) software and, 208
  monuments, 314
  option for handheld C/A receivers, 256
  stochastic model of GSDM, 382
  updates for, 201
International Terrestrial Reference Service
      (ITRS), 200
International Union of Geodesy and Geophysics,
      meeting in Canberra, Australia, 199
Interpolation and extrapolation, 221–222
  measurements and computations, 221–222
Intersections
  circle-circle, 101, 103–104
  line-circle, 101–103
  line-line, 100–101
Inverse, 100
Inverse (BK4), 187–188
Inverse (BK19) computations, 176
Iterations, 89, 158, 330

**J**

Jank-Kivioja method, 178, 184, 186
Jefferson, Thomas, 143
Johns Hopkins University, tracking of
      Sputnik, 240
Joseph, Jean Baptist, 143
*Journal of Surveying Engineering*, 470
Juneau, Alaska, 196–198, 302

**K**

Kansas, 223
Kentucky, 274
King Louis XIV of France, 142
Klamath Falls, Oregon, 161
Krypton 86 gas, 278
Kwajalein, North Pacific Ocean, 242

**L**

Lambert conic conformal projection, 272, 277,
      286–289
  algorithms for, 286
  algorithms for traditional map projections
      and, 285
  BK10 transformation for, 288
  BK11 transformation for, 288–289
  SPCS, 274
  zone length and, 271
Lambert coordinates, 275
Lambert, Johann Heinrich, 270–271
Lambton, Captain William, 144
Laplace correction, 219–220
  Laplace station and, 219
Laplace equations, 220
Laplace stations, 219, 222–223
Lapland (modern-day Finland), 143, 153
  geodetic surveying expeditions in, 142
Latitude, geodetic, 31
Latitude/longitude/height coordinate systems, 386
Law of cosines, 65
Law of sines, 64–65
L-band radio frequencies, 248
Least squares, 88–90
  adjustment, 316
  linearization, 89
  procedure for nonlinear solution, 90
Least squares adjustment
  baseline, 330
  blunder checks, 338, 341–342
  *B* matrix, 343, 345–348
  computer printout, 345, 351–362, 364
  covariance matrix, 331–332, 334–335
  *f* vector, 341–345
  GNSS Network in Wisconsin, 336–338
  indirect observations, 332–335
  linearization, 329–330
  matrix manipulation, 332
  NAD83 (2011) values for control stations,
      336–338
  objective of, 329
  observations and measurements, 330
  observations-only method, 332
  parameters, 329
  purposes of, 364
  *Q* matrix, 343, 345, 349–350

RINEX files, 338–341
statewide WISCORS network, 336
tabulation of results, 363–364
unified approach to least squares
    adjustment, 332
vector, 330
weight matrix, 331
Legendre, Adrien-Marie, 143
Length of a parallel, 166
Leveling, 35
Level surfaces, 216
    as not parallel, 216
L1 frequency, 241, 253
L2 frequency, 242, 253
L5 frequency (civilian frequency), 249
L'Hopital's Rule, 168
Light Detection and Ranging (LiDAR), 38
Linearization, 89, 329–330
Linear least squares GNSS network, GSDM, *see*
    Least squares adjustment
Local accuracy, 16, 319, 323–327, 382,
    465–471
Local coordinate systems, 31–32
Local P.O.B. coordinate differences, 412
Local (east/north/up) reference frame, 310
Location of light sources for
    projection, 270
Logic, 54–55
    René Descartes and, 55
    rules of, 55
Longitude, geodetic, 30
Loop traverse, 285
Louisiana, 277
Louisiana Territory, 143
Louis XIV, King of France, 142
Low-distortion projection (LDP), 14, 94,
    302, 387
Lunar laser ranging (LLR), 199

**M**

Magnavox MX 1502 Doppler receiver, 241
Maine, 197
Map projections, 231, 269, 280
    equidistant, 269
    equivalent, 269
    graphical construction of, 269
    imaginary light ray and, 269
    mathematical equations and, 269
    projection surface and, 270
    range and domain, 269
    stereographic, 269–270
    as two-dimensional models, 280
Map projections and state plane coordinates, 255,
    267–302
    algorithms for traditional map projections,
    285–302

permissible distortion and area covered,
    273–274
procedures, 281–285
projection criteria, 268–270
projection figures, 270–272
U.S. state plan coordinate system (SPCS),
    274–280
Map projections, conformal, 269
    mathematical projections, 270
Map projection (state plane) spatial data
    model, 128
Map projection surfaces, three apex locations,
    270–271
Maps, 128
    concepts of construction, 268
    digitizing, 307
    flat portrayal of spherical Earth, 267
    NEXRAD Project Near Tucson, AZ, 433
    photogrammetric, 38
    storage in digital format, 307
Maryland, Station Principio, 197
Master Control Station of GPS system in
    Colorado Springs CO, 242
Matching signals from receiver and
    satellite, 249
Mathematical concepts, 269
    applications to the global spatial data model
    (GSDM), 90–91
    conventions, 48–51
    of geometry, 56–59
    set theory, 269
Math/science and engineering/surveying
    coordinate systems, 98
Math/science reference system, 98
Matrix, 329–330
    *B* matrix, 343, 345–348
    covariance, 331–332, 334–335
    *f* vector, 341–345
    *Q* matrix, 343, 345, 349–350
    weight, 331
Matrix algebra, 79–80
Meades Ranch, Kansas, 196–198, 223
Mean sea level (MSL), 216, 312
    as reference for elevation, 205
Mean sea level (MSL) datum of 1929
    (now NGVD29), 205–206,
    215, 223
Measurements, 330
Measurements and computations,
    221–225
    ellipsoid heights, 224–225
    gravity, 222–223
    interpolation and extrapolation,
    221–222
    physical geodesy and, 221
    tide readings, 223
    time, 225

Méchain, Pierre, 143
Mercator BK10 and BK11 transformations, 295
Mercator, Gerardus, 140–141, 270, 305
Mercator projection, 267, 274
  cylindrical, 270
  geodetic line and, 169
Mercator projection, oblique, 274, 277, 285,
    297–302
  BK10 transformation for, 299
  BK11 transformation for, 300–302
Mercator projection, transverse, 272, 274, 277,
    285, 289, 291–296
  algorithms for, 289
  BK10 transformation for, 292–294
  BK11 transformation for, 294–295
  first tables for in New Jersey, 275
  Lambert, Johann Heinrich and, 270–271
  zone length and, 271
Mercator world map, 267
Meridian
  arc length, 163–165
  maps and, 267
Meridian coefficients and quadrant arc length for
    selected ellipsoids, 165
Meter, 143, 278–279
  Canada, 278
  coarse acquisition (C/A) code handheld
      receivers and, 256
  England, 278
  geodetic surveys in U.S. and, 278
  Krypton 86 gas, 278
  length of determined, 278
  redefinition of, 278
  speed of light and, 278
  United States, 278
Michigan, 279
Michigan Legislature, 277
  state plane coordinate law, 277
Microwaves, 247
Military users, access to precision (P)
    code, 248
Minnesota, 9
Misclosures in the Trail BK18 Computation, 183
Missiles, satellite geodesy, 131
Models
  functional, 80
  mathematical concepts of, 80–81
  stochastic, 80–81
Modes of positioning, 244–246
  doppler shift, 244–245
  elapsed time, 244
  interferometry, 246
Montana, 274
  zones and, 279
Mt. Graham, 432
Multiple vectors, 257–258
Multiplication, 53–54

**N**

NAD, *see* North American Datum
NAD27, *see* North American Datum of 1927
NAD83, *see* North American Datum of 1983
NAD83(86), 199
  high-precision geodetic networks (HPGN)
      and, 208
NAD83(2007) (North American Datum of 1983
      based on 2007 adjustment of the
      NSRS), 205
NADCON, *see* North American Datum
      conversion software
NAD83(CORS) coordinates, 204
NAPA, *see* U.S. National Academy of Public
      Administration
National Geodetic Reference System (NGRS)
  National Spatial Reference System (NSRS)
      and, 203
National Geodetic Survey (NGS) 1986, 197
National Geodetic Survey (NGS), 216, 222, 260,
    280, 444
  continuously operating reference station
      (CORS) and, 204
  Doppler data and, 241
  height modernization and, 204
  high-precision geodetic networks (HPGN)
      and, 203
  NAD83 (North American Datum of 1983), 311
  observations and adjustments by, 202
  USC&GS (U.S. Coast and Geodetic Survey),
      275, 278
  web site, 234
National Geodetic Vertical Datum of 1929
    (NGVD29), 137, 208, 462
National Imagery and Mapping Agency (NIMA),
    200, 207
"National Integrated Land System (NILS)"
  BLM, 134
  U.S. Forest Service, 134
National Map Accuracy Standards (NMAS), 465
National Oceanic and Atmospheric Administration
    (NOAA), 23, 134
National Ocean Service, 238
National Research Council (NCR), 133
  *Geodesy in the Year 2000*, 133
National Spatial Data Infrastructure (NSDI),
    23, 133
  Bill Clinton and, 134
  Executive Order establishing (1994), 134
National Spatial Reference System (NSRS), 93,
    97, 203, 224, 243, 262, 468
  height component of, 204
NATO, 278
Nautical miles, 256
NAVD88, *see* North American Vertical Datum
    of 1988

Navigation satellite timing and ranging
(NAVSTAR) system, 144, 202, 241
global positioning system (GPS) and, 237, 241
Nebraska, 274
zones and, 279
Network accuracy, 17, 313, 319, 323–327, 382,
465–471
Network and local accuracy
BURKORDT data file, 368–371
derived quantities, point-pair basis, 367
stations FOLA and KEHA, standard deviation
0.002 m for, 368, 372–376
vendor software baseline processing
packages, 376
WISCONSIN network, 368–375
Networks, traditional, 259
New England Datum, 197
New Jersey, 275
New Mexico, 295
Principal Meridian, 319
transverse Mercator projection, 295
New Mexico Principal Meridian (NMPM), 407,
410, 414, 416, 440–441
New Mexico State University (NMSU), 295, 319
New Orleans, 65, 67, 188–191, 384
Newton, Isaac, 141, 147, 153, 215
flattening of Earth at the poles, 218
Galileo and, 141
gravitational attraction and, 214
Picard, Jean, 141
*Principia Mathematica*, 141
shape of the Earth, 143
theory of universal gravitation, 141
Newton's logic for a flattened Earth, 142
New York, 277
NEXRAD weather radar, 432–433
NGRS, *see* National Geodetic Reference System
NGVD29 to NAVD88, 208
Nile River, 139
NIMA, *see* National Imagery and Mapping Agency
N Inverse Matrix from Least Squares
Adjustment, 324
NMPM, *see* New Mexico Principal Meridian
NMSU Associate Dean of Engineering, 434
NMSU network, 387–400
"NOAA Manual NOS NGS 5, State Plane Coordinate
System of 1983" (Stem 1989), 285
Nominal network accuracy, 367
Nonlinear problem, process for solving, 89
Normal distribution curve, 72
Normal section radius of curvature, 155
Normal sections and the geodetic line, 155
North America, 150, 206, 223
coast of, 207
Great Lakes region, 206
techtonic plate, 201
tectonic plate, 200

North American Datum (NAD), 197–198
North American Datum conversion software
(NADCON), 208
North American Datum of 1927 (NAD27), 198,
419, 462
Clarke Spheroid of 1866 and, 150, 198, 228
datum transformations and, 208
geoid and, 228
"Geoid Contours in North America" (map)
and, 223
high-accuracy reference network (HARN)
coordinates and, 204
NAD83 and, 277–279
network adjustments, 144
North American Datum of 1983 (NAD83)
and, 205
readjustment of, 279
SPCS zones and, 274, 279
traditional three-D spatial data models
and, 128
WGS 84, 151
North American Datum of 1983 (NAD83),
197–200, 208, 314, 387, 410–411,
414, 462
adjustment of NAD27 datum and, 279
coarse acquisition (C/A) code handheld
receivers and, 256
control stations, data for, 336–338
datum transformations and, 208
development of, 197
ellipsoid and, 228
ellipsoid heights and, 224
error propagation concepts, 311
as fixed to North American plate, 201
as a global geodetic datum, 199
global navigation satellite systems (GNSS)
computations and, 264
global spatial data model (GSDM) and,
231, 254
GRS 80, 286, 289, 297
Michigan state plane coordinate law, 277
NAD27 and, 277–279
network adjustments, 144
readjustment of, 203, 259
reference position of CORS station
and, 204
spatial data, 135
SPCS zones and, 274
station "Reilly" and station
"Crucesair,", 320
stochastic model of GSDM, 382
traditional three-D spatial data models
and, 128
user/selected option for handheld C/A
receivers, 256
World Geodetic System of 1984 (WGS84)
datum coordinates and, 151

North American Datum of 1983 based on 2007
    adjustment of the NSRS (NAD
    83(2007)), 204
North American Vertical Datum of 1988 (NAVD88),
    137, 210, 223, 419, 431, 462
    elevation for REILLY, 431
    Helmert orthometric heights, 207
North Carolina, 275
    Lambert coordinates and, 275
    zones and, 277
North Carolina Geodetic Survey, 275
North pole, 49, 172, 226, 267, 278
    CTP, 95
    Earth's spin axis and, 61
    geodetic latitude and, 152
    surface area of sphere, 167
Numbers
    mathematical concepts of, 48
    prefixes for, 49

**O**

Observations, 330
Observations-only method, 332
Once-through Vincenty method, 159
On-line User Positioning Service (OPUS), 262
    NGS and, 261–262
Open Geospatial Consortium (OGC), 134
OPUS-RS (rapid static), 262
Oregon, 289
Oregon Institute of Technology, 289
Orthographic projection, 269–270
Orthometric height, 217
    determining using GPS and geoid modeling, 234

**P**

Pacific Ocean, 242
Parametric adjustment, 332
Paris, 142
Paris lines, 143
Paris Observatory, 142
Passive Geodetic Satellite (PAGEOS), 241
Permissible distortion and area covered, 273–274
Perpendicular offset
    to a line, 102
Peru, 153
    geodetic surveying expeditions in, 142
Photogrammetry, 38
Physical geodesy, 131
Physical quantities, fundamental, 34
Physical quantities, measurable, 221
PI, *see* Points of intersection
Picard, Jean, 141–142
Plane, 57
    Earth and, 57
    GSDM and, 57

Plot of sine and cosine functions, 64
Plot of volume, 69
Point-of-beginning (P.O.B.), 8, 318, 386
    absolute local coordinates and, 255
    accuracy and, 17
    coordinate differences, 412
    flat-Earth plane surveying components and, 309
    inverse directions and distances based on, 412
    rotation matrix derivation, 447–448
Point-of-beginning (P.O.B.) datum coordinates,
    195, 209
    absolute local coordinates and, 255
    GSDM and, 270, 279
Point-pair application, 318
Points of intersection (PI), 106
Points stored in the *X/Y/Z* database, using, 317–319
Point-to-point traverse, 285
Polar motion, corrections for, 172
Polar radius of curvature, 150, 200
Polar wandering, 148, 226
Pole-zenith-star (PZS) triangle, 66
Polygon, 59
Polyhedron, 60
Poseidonius, 140
Position, GPS-derived, 224
Posteriori reference variance, 334–335
Prather, Ford, 401
Precise point positioning (PPP), 37, 379
Precision
    and accuracy, 365
    definition of, 365
    measurement, 366
Precision code (P-code), 248
Principal Meridian, 410–417
*Principia Mathematica* in 1687, 141
Priori reference variance, 334
Prismoidal formula, 126–127
Probability and statistics, 71–80
    accuracy and precision, 75–76
    computing standard deviations, 76
    confidence intervals, 77–78
    errors, 73–74
    error sources, 74–75
    hypothesis testing, 78–79
    matrix algebra, 79–80
    measurement, 73
    standard deviation, 72–73, 76–77
Procedures
    grid azimuth, 281
    grid distance, 281–284
    traverses, 284–285
Projection criteria, 268–270
Projection figures, 270–272
Pseudo 3-D, 438
Ptolemy, Claudius
    degrees and, 50
    *Geographica*, 196

Public Land Survey System, *see* U.S. Public Land
        Survey System
Puissant Forward (BK18), 176–177
Puissant Inverse (BK19), 177–178
Pyramid, 60
Pythagoras, 139
Pythagorean Theorem, 59, 63, 249

**Q**

*Q* matrix, 343, 345, 349–350
Quadrilateral, 58–59
Quantities, absolute, 311–312

**R**

Radian, 49
Radii of curvature: M and N, comparisons of, 155
Radio waves, 247
Random errors, 365
Raster data, conversion to vector data, 383
Real number line, 48
Real-time GPS CORS network (RTN), 204
Real-time kinematic (RTK)
    GPS surveying procedures, 37
    surveying, 25, 444
Real-time-kinematic (RTK) GPS surveying, 239
    procedures, 202, 204, 242, 261
Real-time network (RTN), 444
Receiver Independent Exchange (RINEX) format,
        252–253
    meteorological file, 253
    navigation file, 253
    observation file, 253
    WISCORS RINEX files, 338–341
Rectangle
    for area computation, 53
Redundancy, 333
REILLY, 403–404
Relative quantity, 311–312
Religion, geodesy, 138
Remote sensing, 38
Request for information (RFI), 23, 134
Rhodes, 140
Right triangle relationships, 63
Root-mean-square-error (RSME), 468
Rotational ellipsoid
    equation of ellipsoid, 155–156
    geocentric and geodetic coordinates, 156–157
Rotation matrix derivation, 447–450
    earth-centered Earth-fixed (ECEF) coordinates
        and, 447–449
    "flat-Earth" components and, 447
    global spatial data model (GSDM) and, 447
    point-of-beginning (P.O.B.) datum coordinates
        and, 447–449
    *X/Y/Z* coordinates and, 447–448

Round earth, flat map, 267–268
$R_1$ rotation about the X axis, 448
$R_2$ rotation about the Y axis, 448
$R_3$ rotation about the Z axis, 449
RTN/VRS, 444
Russia (Soviet Union), 237, 244
    launch of first artificial satellite to orbit
        earth, 144
Russian Global Navigation Satellite System
        (GLONASS), 238, 249, 254
    compared to GPS, 244
    global navigation satellite systems (GNSS)
        and, 237–238

**S**

Satellite geodesy, 131
Satellite laser ranging (SLR), 199
Satellite positioning, brief history of, 240–244
Satellite-positioning technology, Cold War and, 237
Satellite programs, geodetic research and, 241
Satellites
    Earth-orbiting, 237
    Earth's center of mass and, 386
Satellite signals, 247–251
    C/A code, 249–250
    carrier phase, 250–251
Satellite systems
    NAVSTAR satellite system, 237
    TRANSIT satellite System, 237
Science, geodesy, 138
Sea level, 135
Secant projection, 271–273
Sector, curves and, 111
Segment, curves and, 112
Selected geometrical geodesy ellipsoids, 151
Selective availability (SA), 248
    enabled, 256
Seven-or (fourteen-) parameter
        transformation, 209
Sexagesimal system
    mathematical concepts of, 50
SI (international system for units of measure), 270
Significant figures, 52
    examples using addition and subtraction, 52
    mathematical concepts of, 52
Single differencing, 252
    Doppler effect, 252
Snellis, Willebrord, 141
"SOCS-UTM and Oscar S. Adams"
    Joseph Dracup and, 275
Software sources, 208–209
Solar method, 441–442
Solid geometry
    mathematical concepts of, 60–62
South Carolina, 274
    zones and, 277, 279

Southeastern Wisconsin Regional Planning
Commission (SEWRPC), 144, 427,
462, 468
South Pole, 152, 167, 267
Earth's spin axis and, 61
Spatial data, 29, 43–44
absolute, 30
accuracy, 366
absolute and relative quantities,
311–312
blunder checks, 321–322
confidence intervals, 310
control values and observed vectors,
320–321
definitions, 311
error ellipses, 310
error propagation, 315
least squares solution, 322–323
measurements, 315
observations, 315
positional tolerance, 310
statements, 313
stochastic model, 310
uncertainty, 310, 315–317
using points stored in the *X/Y/Z* database,
317–319
consequences, 308–309
coordinate systems and, 30–32, 237
derived, by computation from primary spatial
data, 42–43
digital revolution and, 305–306
errorless, 39–40
forces driving change, 305–306
measurement and, 29–44
measurements
angles, 36
EDMI, 35–36
errorless spatial data and, 39–40
leveling, 35
LiDAR, 38
logistics, 38–39
photogrammetry, 38
remote sensing, 38
taping, 35
three-dimensional, 307
primary
computation of derived spatial data
and, 42–43
errorless qualities, 42
observations and measurements,
41–42
quality of, 240
relative, 30
three-dimensional, 238, 280
transition, 306–309
types of, 32–33
visualization as well-defined, 34

Spatial data types, 254–255
absolute geocentric *X/Y/Z* coordinates,
254–255
absolute geodetic coordinates of *latitude/
longitude/height*, 255
absolute local coordinates, 255
arbitrary local coordinates, 255
relative geocentric coordinate differences, 254
relative geodetic coordinate and ellipsoid
height differences, 255
relative local coordinate differences, 255
Spatial data, using, 261
Sphere, 60
Spherical law of sines, 66
Spherical trigonometry, 65–67, 384
Spheroid, ellipsoid, 151
Spiral curves, 113–117
computing area adjacent to a spiral, 117–118
intersecting a line with a spiral, 116–117
spiral geometry, 113–116
Spiral elements and geometry, 114
Spiral geometry, 113–114
symbols, 114
Spiral intersection elements and geometry, 116
Spolar, Kyle, 401
Sputnik I, 237, 240
Earth orbit of, 240
Russia and, 240
Square, 59
Standard deviation, 72–73
computing, 76
definition of, 365
of measurement, 366
Standard positioning service (SPS), 248
Standard time zones, in United States, 136, 226
State plane coordinate system (SPCS), 271,
274–281
advantages of, 280
algorithms, 285
constants (1983), 451–458
disadvantage of, 280
distortion and, 277
features, 275–277
grid distance and ellipsoid distance, 281
law, 277
NAD27, 277
NGS, 277
procedures used with, 281
projections, 280
"Special Publication 235" (Mitchell and
Simmons), 275
zone parameters, 279
State plane coordinate traverse, 284
Station covariance values, 367
Station equation
example, 110
policies, 109

Stationing, 109–110
Station Principio Maryland, 197
Station "Reilly,", 319–320
    New Mexico State University campus, 295
Statute miles, coarse acquisition (C/A) code
        handheld receivers and, 256
Stereographic projection, 269–270
Stochastic model, 381–384
Stochastic model component
    GSDM covariance matrices, 14–16
    GSDM 3-D inverse, 16–18
Stochastic model of GSDM, 310
    geoid and, 382
    International Terrestrial Reference Frame
        (ITRF) and, 382
    North American Datum of 1983 (NAD83)
        and, 382
    standard deviations and, 382
    World Geodetic System of 1984
        (WGS84), 382
Stored X/Y/Z locations
    application modes for using, 317
    "cloud" (mapping), 317
    point-pair, 317
    single point, 317
"String theories,", 57
The Structure of Scientific Revolutions, 308
Subtraction, 52
Supplemental NMSU campus control network,
        387–400
Supreme Court NAD 27 positions, 419
Survey control networks, 262–265
Surveying & Land Information Systems, 468
Surveying, radial, 118–121
Survey measurements of a tank, 83
Survey of the Coast, 143, 205
Surveyors, 128
SW Corner, 410–411, 414–415, 417
Syene, 140
Syme, George F., 275
Systematic errors, 365

T

TAI, see International Atomic Time
Tangent-offset method, 442–443
Tangent projection, 271–273
Taping, 35
Target height correction, 173–174
Taylor series approximation, 89
Tennessee, 271
Tetrahedron, 60
Texas, 277
"The Global Positioning System: Charting the
        Future" (NAPA), 238
Theory of least squares, 143
Three-dimensional azimuth, 174

Three-dimensional datums, 209–210
Three-dimensional datums with origin at Earth's
        center of mass
    International Terrestrial Reference Frame
        (ITRF), 201
    North American Datum of 1983
        (NAD83), 201
    World Geodetic System of 1984
        (WGS84), 201
Three-dimensional ellipsoid
    ellipsoid radii of curvature, 154–155
    geometrical mean radius, 155
    normal section radius of curvature, 155
Three-dimensional geodetic model, 461
Three-dimensional global spatial data model
        (3-D GSDM), 379, 470
    COGO, 24–25
    digital technology, 21
    dynamic environments, 25
    NAPA, 22–23
    NOAA, 23
    static environments, 25–26
    static or dynamic, 26
Three-dimensional GPS Points for 2-D
        survey, 411
Three-dimensional inverse with statistics,
        459–460
Three-dimensional models for spatial data,
        124–128, 307
    prismoidal formula, 126–127
    the three-D GSDM, 128–129
    traditional three-D spatial data models, 128
    volume of a rectangular solid, 124
    volume of a sphere, 124–125
    volume of cone, 125–126
Tide readings, 207, 223
    coast of North America, 207
    measurements and computations, 223
Time
    absolute location of the geoid and, 225
    autonomous processing, 256
    definition of, 225
    as measurable physical quantity, 225
    measurements and computations, 225
    standard time zones, 226
Time elapsed, distance and, 244
Time zone offset, 256
Traditional state plane grid distances, computation
        of, 283–284
Transcontinental traverse (TCT), 199
TRANSIT Doppler satellite positioning system,
        237, 240–241, 245
    Doppler data, 241, 245
    U.S. Polaris submarine fleet, 241
Transverse Mercator projection, 271–272, 289,
        291–296
Trapezoid, 59

Traverses, 284–285
 loop, 285
 point-to-point, 285
Trial Geodetic Line New Orleans to Chicago
  (BK18), 191
Triangles
 spherical, 66
Trigonometry, 62–65
 definitions, 63
 identities, 63–64
 law of cosines, 65
 law of sines, 64–65
Triple differencing, 252
 Doppler effect and, 252
Trivial vectors, 330
Tropic of Cancer, 139
T23S-R1E, 407, 410
Turkey, 140
Turkey point, 197
2-D plat of 3-D survey, 410, 413

U

Ultraviolet light, 247
Uncertainty
 direct and indirect measurements
  and, 34–35
 Earth's radius, 229
 ellipsoid height and, 229
 errors in measurement, 35
 fundamental physical constants held
  exact, 34
 of a single point, 317
 standard deviations, 317
 variances, 317
United States, 21, 48, 136, 242, 248, 261
 arcs of high-order triangulation and, 143
 baseline to Germany, 247
 datums used in, 197
 eighteenth century, 143
 elevation and, 136
 ellipsoid height and orthometric height, 217
 first-order triangulation of, 197
 geoid and ellipsoid, 227–228
 geoid height and, 227
 mean sea level datum of 1928, 205
 standard time, 136, 226
 tide gauge readings, 223
 TRANSIT Doppler satellite positioning
  system, 245
Units, autonomous processing, 256
Universal Time Coordinated (UTC), 136
Universal Transverse Mercator (UTM)
  coordinates, 8, 271
 system, 279
U.S. Army Corps of Engineers, 208
U.S. Army Map Service, 223

U.S. Bureau of Land Management (BLM), 238, 241
 NILS, 134
U.S. Coast and Geodetic Survey (USC&GS), 144,
  197, 275
 Michigan Legislature and, 277
 "model law,", 277
 National Geodetic Survey and, 144
 Survey of the Coast and, 144
U.S. Congress, 238
U.S. Department of Defense (DOD), 22, 132,
  199, 241
 Global Positioning System (GPS), 242
 GPS tracking network, 201
 navigation satellite timing and ranging
  (NAVSTAR) satellite system, 241
 updates for GPS tracking network, 201
U.S. Department of Transportation (DOT), 224
U.S. Forest Service, 238
 NILS, 134
US Forest Service and Bureau of Land
  Management (USFS/BLM), 469
U.S. Geological Survey, 207, 238
U.S. National Academy of Public Administration
  (NAPA), 238
U.S. National Academy of Public Administration
  (NAPA) reports
 GSDM, 22–23
U.S. National Imagery and Mapping Agency, 4
U.S. Polaris submarine fleet, 241
 TRANSIT satellite system, 241
U.S. Public Land Survey System (USPLSS), 48,
  143, 313, 410, 412, 416
U.S. Standard Datum, 197
U.S. State plan coordinate system (SPCS),
  274–280
 current status, NAD83 SPCS, 279
 features, 275–277
 history, 275
 NAD27 and NAD83, 277–279
 zones in U.S., 274
U.S. Supreme Court, 417
U.S. Supreme Court Report positions, 419
U.S. Survey Foot, 278–279
 coarse acquisition (C/A) code handheld
  receivers and, 256
U.S. TRANSIT satellite system, 241
Utah, 277
UTC, see Coordinated Universal Time

V

Variance
 covariance matrix, 367
 definition, 365
Vector, 330
Vector components of gravity, 214
Vector processing, 256–257

advanced processing, 261
  multiple vectors, 257–258
  traditional networks, 259
Vectors, 62
Vendor-specific baseline processing software,
    331, 376
VERTCON, *see* Vertical datum conversion
    software
Vertical Accuracy Reporting for Lidar Data, 471
Vertical curves, 121–124
  geometry, 121–122
  notes about derivation and solution of
      problems, 122
Vertical datum conversion software
    (VERTCON), 208
Vertical datums, 205 207
  international great lakes datum, 206
  mean sea level datum of 1929 (now
      NGVD29), 205
  North American Vertical Datum of 1988
      (NAVD88), 206
Very long baseline interferometry (VLBI),
    199–200, 247
Vincenty, 159
Vincenty's method (same point), 162–163
Virtual reference station (VRS), 444
Virtual reference system (VRS), 37
Visible light, 247
VLBI, *see* Very long baseline interferometry
Voltaire, 142
Volume
  of a cone, 125–126
  of a rectangular solid, 124
  of a sphere, 124–125
  of tank, 69

**W**

WAAS, *see* Wide area augmentation system
WAKEMAN, 401, 403–404
Walbeck 1819 spheroid, 197
Waldo, 197
WBK software, *see* BURKORD™ database

Weight matrix, 331
Wide area augmentation system (WAAS), 261
  civil aviation and, 261
Wisconsin, 9
Wisconsin network, 336–338
  BURKORDT data file, 368–371
  stations FOLA and KEHA, standard deviation
      0.002 m for, 368, 372–376
WISCORS network, 336
World Geodetic System 1984 (WGS 84), 151,
    157, 315
  stochastic model of GSDM, 382
World Geodetic System of 1960 (WGS60), 200
World Geodetic System of 1972 (WGS72), 200
World Geodetic System of 1984 (WGS84),
    200–201, 264
  datum, 285
  Earth-centered Earth-fixed (ECEF)
      coordinates, 242
  Earth's center of mass and, 201
  ellipsoid heights and, 224
  global spatial data model (GSDM) and, 231
  GNSS computations expressing location and,
      264
  GSDM (global spatial data model) and, 254
  ITRF CORS positions and, 204
  NAVSTAR satellite system and, 202
World Vertical Datum (WVD), 136–137, 145, 309
  error propagation concepts and, 311
World War II, 278
  radar band portion of spectrum, 247

**X**

X-rays, 247
X/Y coordinates, 99
*X/Y/Z* coordinates, 239
*X/Y/Z* geocentric coordinates, 316

**Z**

Zero, invention of, 48
Zone, 271–272

Printed and bound by CPI Group (UK) Ltd, Croydon, CR0 4YY

01/11/2024

01782617-0018